Advances in Agricultural Systems Modeling
Transdisciplinary Research, Synthesis, and Applications
Volume 5

Lajpat R. Ahuja, Series Editor

Advances in Agricultural Systems Modeling
Transdisciplinary Research, Synthesis, and Applications
Volume 5

Lajpat R. Ahuja, Series Editor

Practical Applications of Agricultural System Models to Optimize the Use of Limited Water

Lajpat R. Ahuja, Liwang Ma, and Robert J. Lascano, Editors

American Society of Agronomy, Inc.
Crop Science Society of America, Inc.
Soil Science Society of America, Inc.
5585 Guilford Road, Madison, WI 53711-5801 USA

agronomy.org | soils.org | crops.org
dl.sciencesocieties.org
SocietyStore.org

ISBN: 978-0-89118-343-3 (print)
ISBN: 978-0-89118-344-0 (digital)
doi:10.2134/advagricsystmodel5
Library of Congress Control Number: 2014956736

Advances in Agricultural Systems Modeling
ISSN: 2163-2790 (online)
ISSN: 2163-2774 (print)

Cover design: Patricia Scullion
Cover photo: "Knowledge rise." Thomas Trout, USDA-ARS.

Many of the photos in this volume were taken by Dr. Thomas J. Trout of USDA-ARS, Water Management Research Unit, on the Limited Irrigation Research Farm, near Greeley, Colorado from 2008 to 2014.

Printed in the United States of America.

Contents

Foreword

Throughout the drier regions of the world, the competition for water between urban, industrial, and agricultural uses is becoming intense. Increasing population and changing weather and climate will continue to increase the pressure on water resources, demanding that all water use become more efficient. Agriculture will always remain a target as water resources are prioritized, since agriculture uses the largest fraction of available water in water limited regions. In addition to being more efficient in using water supplies developed for irrigated agriculture, there is a great need for research seeking ways for crops to be more efficient in the use of soil-stored water derived from precipitation. We remind the audience that globally about 80% of agriculture is rainfed agriculture; therefore, meeting future agricultural production water needs will be largely dependent on using existing water supplies more efficiently.

This book examines a number of different approaches to enhancing water use for a number of important crops, including cultural practices such as circular planting to enhance rainfall capture, tillage, and cover crops, in a range of different cropping systems, climates, and production systems. The title *Practical Applications of Agricultural System Models to Optimize the Use of Limited Water* is very descriptive of the range of subjects covered. This book connects models and modeling to the real world, addressing many of the tools available to conserve and enhance water use, while utilizing experimental data where available.

This volume represents a significant effort in the development of tested, process-based models to enhance our understanding and ability to manage this key, limited resource.

David B. Mengel, 2014 ASA President
David D. Baltensperger, 2014 CSSA President
Jan W. Hopmans, 2014 SSSA President

Preface

Semiarid to arid regions of the world are major producers of food and fiber. Water is becoming a major limiting factor for agriculture in these regions because of increased urban and industrial use, more frequent droughts, climate warming, and limits to this precious natural resource. At the same time, there is a need to produce even more food for the growing population, which requires more water. To obtain the most production from available water, while maintaining natural resources, we need whole system–based quantitative knowledge and tools to help select appropriate crops and manage water and associated inputs at different locations on a site-specific basis under variable and changing climate. These tools should help guide decisions on appropriate crop sequences and optimal use of limited rainfall and irrigation water with respect to the amounts and timings of rainfall, critical crop growth stages, and weather conditions, as well as optimum use of nitrogen. Information is also needed to help determine potential alternative crops that would be suitable during periods of droughts and help producers determine when to plant them. There is currently great excitement about growing bioenergy crops, including oil seed crops and irrigated corn or other biomass crops. Effective decision support tools should also be able to evaluate the long-term economics of bioenergy crops and provide estimates and guidance to ensure that enough crop residues remain on the soil surface to minimize erosion and maintain soil organic matter and productivity.

Site-specific experimental information is available for limited locations, limited periods of time, and limited management options. Well-tested process models of cropping systems can extend field research results to long-term weather conditions, other climates and soils, and potential new management options and thus provide the whole system–based knowledge and management guides for different locations over variable weather and climate. The contributions to this volume present the applications of crop system models to help optimize the use of limited water and associated nutrients and explore adaptations and mitigations for climate change under a variety of conditions.

Previous volumes of the ASA, SSSA, CSSA Advances in Agricultural Systems Modeling have been devoted to the transdisciplinary knowledge-gap areas of science, conceptual synthesis, and parameterization for agricultural system models. We will further address these areas in future volumes. Volume 5 is completely devoted to application of models to help optimize the use of limited water and associated management practices under various climates. Case studies of applications presented in this volume will inspire further useful applications of models in the above areas, as well as in numerous other areas.

Beginning with this volume, we have also placed greater emphasis on originality and quality of contributions by requiring that each contribution should include new useful applications of models, a description of applications thus far with new synthesis of knowledge and experience, new or improved concepts emerging from this synthesis, and/or new directions for model applications and research. Each submission has been reviewed anonymously by two to three reviewers and an editor and accepted on the basis of originality and the final contribution addressing any comments of the reviewers. This rigorous peer review is done to ensure contributors will get credit for their articles like a journal article. An attractive feature of this series compared with journals is that all articles in a volume are focused on one subject area. We openly solicited original contributions for this volume with the above-noted criteria via a call for contributions in the *CSA News* and via emails to several ASA Communities.

Lajpat R. Ahuja, Liwang Ma, and Robert J. Lascano, Editors

Contributors

Aggarwal, Pramod K. Climate Change Agriculture and Food Security - South Asia, International Water Management Institute, NASC Complex, Pusa, New Delhi 110 012, India (P.K.Aggarwal@cgiar.org)

Ahuja, Lajpat R. USDA-ARS, Agricultural Systems Research Unit, 2150 Centre Ave., Bldg. D, Ste. 200 Fort Collins, CO 80526 (laj.ahuja@ars.usda.gov)

Andales, A.A. Dep. of Soil and Crop Sciences, Colorado State University, 1170 Campus Delivery, Fort Collins, CO 80523-1170 (Allan.Andales@colostate.edu).

Brun, François ACTA, INRA, UMR 1248 AGIR, 24 Chemin de Borde Rouge - Auzeville, CS 52627, 31326 Castanet Tolosan, Cedex, France (francois.brun@acta.asso.fr)

Colaizzi, Paul D. USDA-ARS, Conservation and Production Research Lab., P.O. Drawer 10, 2300 Experiment Station Rd., Bushland, TX 79012-0010 (Paul.Colaizzi@ars.usda.gov).

Daudin, Gabriel Vita Consult, ZA du Pré Neuf, 19 rue du Pré Neuf, 44190 Gorges, France (gdaudin@vitaconsult.fr)

Delpuech, Xavier Inst. Français de la Vigne et du Vin, Pôle Rhône-Méditerranée, Domaine de Donadille, 30230 Rodilhan, France (xavier.delpuech@vignevin.com)

Derner, J.D. USDA-ARS, High Plains Grasslands Research Station, 8408 Hildreth Rd., Cheyenne, WY 82009-8899 (Justin.Derner@ars.usda.gov)

El-Mokh, F. Institut des Régions Arides, 4119 Médenine, Tunisia (F.mokh@yahoo.fr)

Fang, Quan X. Agronomy College, Qingdao Agricultural University, Changcheng Rd. 700, Chengyang District, Qingdao, Shandong, China, 266108 (fangqx@igsnrr.ac.cn)

Fritschi, F.B. Division of Plant Sciences, University of Missouri, 1-33 Agriculture Building, Columbia, MO 65211 (fritschif@missouri.edu)

Fujimaki, H. Arid Land Research Ctr., Tottori University, 1390 Hamasaka, 680-0001 Tottori, Japan (fujimaki@alrc.tottori-u.ac.jp)

Grassini, P. Dep. of Agronomy and Horticulture, Univ. of Nebraska, 377M, PLSH, 202 Keim Hall, Lincoln, NE 68583 (pgrassini2@unl.edu)

Inoue, M. Arid Land Research Ctr., Tottori University, 1390 Hamasaka, 680-0001 Tottori, Japan (m11minoue@cpost.plala.or.jp)

Islam, Adlul NRM Division, Indian Council of Agricultural Research, KAB-II, IARI Campus, Pusa, New Delhi 110 012, India (adlulislam@yahoo.com)

Kakani, V.G. Dep. of Plant and Soil Sciences, 368 Agricultural Hall, Oklahoma State University, Stillwater, OK 74078-6028 (v.g.kakani@okstate.edu)

Kiniry, J.R. USDA-ARS, Grassland, Soil and Water Research Lab., 808 E. Blackland Rd., Temple, TX 76502 (jim.kiniry@ars.usda.gov)

Kumar, Soora Naresh Centre For Environment Science and Climate Resilient Agriculture, Indian Agricultural Research Institute, Pusa, New Delhi 100 012, India (nareshkumar.soora@gmail.com)

Lascano, R.J. USDA-ARS, Wind Erosion and Water Conservation Research Unit, Cropping Systems Research Lab., 3810 4th St., Lubbock, TX 79415 (robert.lascano@ars.usda.gov)

Lecler, N.L. N.L. Lecler, Zimbabwe Sugar Association Experiment Station, P/Bag 7006, Trinagle Ngundu Rd., Chiredzi, Zimbabwe; also, Bioresources Engineering, University of KwaZulu-Natal, P/Bag X01, Scottsville 3209, South Africa (nlecler@zsaes.org.zw)

Ma, L. USDA-ARS, Agricultural Systems Research Unit, 2150 Centre Ave., Bldg. D, Ste. 200 Fort Collins, CO 80526 (Liwang.Ma@ars.usda.gov)

Meki, M.N. Blackland Research Lab., Texas Agrilife., 720 E. Blackland Rd., Temple, TX 76502 (nmeki@brc.tamus.edu)

Mercau, J.L. Grupo de Estudios Ambientales del Instituto de Matemática Aplicada, Universidad Nacional de San Luis (GEA-IMASL) and Consejo Nacional de Investigaciones Científicas y Tecnológicas (CONICET), San Luis, Argentina (jorgemercau@gmail.com)

Nagaz, K. Inst. des Régions Arides, 4119 Médenine, Tunisia (nagaz.kameleddine@ira.rnrt.tn)

Nelson, J.R. North Central Kansas Experiment Field, Dep. of Agronomy, Kansas State University, 1300 60 Rd., Courtland, KS 66939

Nielsen, David C. USDA-ARS, Central Plains Resources Management Research Unit, 40335 County Rd. GG, Akron, CO, 80720-0400 (david.nielsen@ars.usda.gov)

Ollenburger, M. Plant Production Systems Group, P.O. Box 430, 6700 AK Wageningen,Wageningen University, Wageningen, the Netherlands (mollenburger@gmail.com)

Otegui, M.E. Inst. de Investigaciones Fisiológicas y Ecológicas Vinculadas a la Agricultura (IFEVA-CONICET) and Departamento de Producción Vegetal, Facultad de Agronomía, Universidad de Buenos Aires, Av. San Martin 4453 (C1417DSE), Buenos Aires, Argentina (otegui@agro.uba.ar)

Okazaki, M. Arid Land Research Ctr., Tottori University, 1390 Hamasaka, 680-0001 Tottori, Japan (alian_johns_wpc@yahoo.co.jp)

Roux, Sébastien INRA, UMR System, 2 Place Pierre Viala, 34060 Montpellier Cedex 2, France (sebastien.roux@supagro.inra.fr)

Saito, T. Faculty of Agriculture, Tottori University, 4-101 Koyamacho-Minami, Tottori 680-8550, Japan (tadaomi@muses.tottori-u.ac.jp)

Saseendran, S.A. USDA-ARS, Agricultural Systems Research Unit, 2150 Centre Ave., Bldg. D, Ste. 200 Fort Collins, CO 80526 (Saseendran.anapalli@ars.usda.gov)

Schumacher, T.E. Plant Science, South Dakota State University, Box 2207A, Agricultural Hall–SAG 004, Brookings, SD 57007 (thomas.schumacher@sdstate.edu)

Shibata, M. Arid Land Research Ctr., Tottori University, 1390 Hamasaka, 680-0001 Tottori, Japan (viva-shiba@h2.dion.ne.jp)

Shirsath, Paresh Bhaskar Climate Change Agriculture and Food Security - South Asia, International Water Management Institute, NASC Complex, Pusa, New Delhi 110 012, India (p.bhaskar@cgiar.org)

Sikka, Alok K. NRM Division, Indian Council of Agricultural Research, KAB-II, IARI Campus, Pusa, New Delhi 110 012, India (aloksikka@yahoo.co.in)

Snapp, S. Dep. of Plant Soil and Microbial Sciences, Michigan State University, 1066 Bogue St., East Lansing, MI 48824 (snapp@msu.edu)

Subash, Nataraja Project Directorate for Farming Systems Research, Modipuram, Meerut 250 110, Meerut, Uttar Pradesh, India (nsubashpdfsr@gmail.com)

Tokumoto, I. Faculty of Agriculture, Saga University, 1 Honjo-machi, Saga 840-8502, Japan (yasu@cc.saga-u.ac.jp)

Trout, Thomas J. USDA-ARS Water Management Research Unit, 2150 Centre Ave., Bldg. D, Ste. 320 Fort Collins, CO 80526 (Thomas.Trout@ars.usda.gov)

Wallach, Daniel INRA, UMR 1248 AGIR, 24 Chemin de Borde Rouge–Auzeville, CS 52627, 31326 Castanet Tolosan, Cedex, France (wallach@toulouse.inra.fr)

Wang, Enli CSIRO Land and Water, Christian Lab., Clunies Ross St., Black Mountain ACT 2601, Australia (Enli.Wang@csiro.au)

Wery, Jacques Montpellier SupAgro, UMR System, 2 Place Pierre Viala, 34060 Montpellier Cedex 2, France (jacques.wery@supagro.inra.fr)

Yang, H. Dep. of Agronomy and Horticulture, Univ. of Nebraska, 377M, PLSH,, 202 Keim Hall, Lincoln, NE 68583 (hyang2@unl.edu)

Zilverberg, C.J. Dep. of Natural Resource Management, South Dakota State University, Northern Plains Biostress Lab. (SNP), Box 2140B, Brookings, SD 57007 (czilverberg@brc.tamus.edu)

Quantifying Corn Yield and Water Use Efficiency under Growth Stage–Based Deficit Irrigation Conditions

Quan X. Fang,* L. Ma, David C. Nielsen, Thomas J. Trout, and L.R. Ahuja

Abstract

An agricultural system model can help optimize limited irrigation for higher grain yield and water use efficiency (WUE) across varied climatic conditions. In this study, the Root Zone Water Quality Model (RZWQM2) was first calibrated to simulate soil water, ET, and corn (*Zea mays* L.) yield under a range, 40 to 100% of crop ET under well-watered conditions (i.e., ET_c, calculated by the reference ET and crop coefficient), of irrigation treatments from 2008 to 2011 in Colorado. The model was then used to explore grain yield responses to irrigation levels of 40, 60, 80, and 100% ET_c (the Shuttleworth–Wallace ET, ET_{sw}, was used as a surrogate for ET_c in the model) under 300, 400, and 500 mm of total irrigation water and to provide guidelines to manage limited irrigation using weather data from 1992 to 2013. With 500 mm of irrigation water, high grain yield and WUE were obtained from the 100% ET_{sw} for the vegetative stage and 80 to 100% ET_{sw} for the reproductive stage. With 400 mm of irrigation water, high grain yield and WUE were simulated at 80 to 100% ET_{sw} irrigation targets between the vegetative and reproductive stages. With 300 mm of irrigation water, however, meeting 100% ET_{sw} at the reproductive and 60% ET_{sw} at the vegetative stage achieved the highest grain yield and WUE. Simulations showed that meeting the crop water requirement during the reproductive stage is more important than during the vegetative stage to achieve high grain yield and WUE under water-limited conditions.

Abbreviations: CoAgMet, Colorado Agricultural Meteorological Network; CV, coefficient of variance; LAI, leaf area index; MD, mean difference; ME, model efficiency; RZWQM2, Root Zone Water Quality Model (version 2.6); UAN, urea–ammonium nitrate; WUE, water use efficiency.

Q.X. Fang, Agronomy College, Qingdao Agricultural University, Changcheng Road 700, Chengyang District, Qingdao, Shandong, China, 266108. *Corresponding author (fangqx@igsnrr.ac.cn).

L. Ma (Liwang.Ma@ARS.USDA.GOV) and L.R. Ahuja (laj.ahuja@ars.usda.gov), USDA-ARS-NPA-ASRU 2150 Centre Ave., Bldg. D, STE. 200 Fort Collins, CO 80526.

D.C. Nielsen, USDA-ARS, Central Plains Resources Management Research, 40335 County Rd GG, Akron, CO 80720-0400 (david.nielsen@ars.usda.gov).

T.J. Trout, USDA-ARS Water Management Research Unit, 2150 Centre Avenue, Bldg D, Suite 320 Fort Collins, CO 80526 (Thomas.Trout@ars.usda.gov).

doi:10.2134/advagricsystmodel5.c1

Practical Applications of Agricultural System Models to Optimize the Use of Limited Water
Lajpat R. Ahuja, Liwang Ma, and Robert J. Lascano, Editors
Advances in Agricultural Systems Modeling, Volume 5. Lajpat R. Ahuja, Series Editor

USDA-ARS

Water is often the limiting factor for crop yield in semiarid areas or water-limited regions. Increasing WUE (crop yield/evapotranspiration, kg m^{-3}) is the key to mitigating water crises in these regions. Previous field results have shown high variability in WUE due to varied seasonal weather variability, especially precipitation, for regions such as the U.S. Great Plains (Angus and van Herwaarden, 2001; Howell, 2001) and the North China Plain (Fang et al., 2010). Yearly variations in WUE have limited the extension of field experimental results to guide irrigation management. Recently, Saseendran et al. (2014) generalized average corn crop water production functions across years and locations and provided useful guidelines for managing limited water for higher corn production, but the yearly variation in WUE needs to be further investigated.

Besides weather variability, the intensity of water stress (the degree of meeting crop ET demand) and time (phenological stage) can influence WUE under deficit irrigation conditions (e.g., Lorite et al., 2007; Pereira et al., 2009). Field studies have explored the response of WUE to various irrigation levels or water stress levels and provide useful options to improve WUE in different regions (e.g., Fereres and Soriano, 2007; Howell, 2001), but these results were limited because of weather variations. Improper irrigation amounts or times may produce severe water stress due to the uneven rainfall distribution and different sensitivities of corn to water stress at different growth stages (Shaw, 1977; Nielsen et al., 2010). Variable irrigation based on soil water deficiency or crop water requirement (such as crop ET) can be a more precise and suitable method for deficit irrigation. Geerts and Raes (2009) reviewed deficit irrigation methods (application of water below full crop water requirements) to maximize WUE in dry areas and found deficit irrigation can increase WUE for various crops without substantially reducing crop yield, but effective use of deficit irrigation required precise knowledge of crop response to water stress at different phenological stages. Molden et al. (2010) also discussed the need for a better understanding of the biophysical and socioeconomic impacts for developing better deficit irrigation strategies. Using field experiments to explore optimal irrigation schedules is rather difficult, costly, and time-consuming because the response of WUE to water stress (or irrigation) becomes complex due to the effect of weather variations (such as rainfall) on both soil water deficits (timing and intensity) and crop yield.

Integrating soil, crop, and climate factors is required to quantify the response of WUE to water stress. System modeling can aid in the design of field experiments in quantifying crop responses under various climate and irrigation conditions. Stockle and James (1989) used the Cropping Systems Simulation Model (CROPSYST) to explore the effect of different irrigation strategies on corn yield and found that a slight water deficit produced greater grain yields than full irrigation, but severe deficits were detrimental to plant growth and grain yield. Various crop models or agricultural system models have been used to explore deficit irrigation strategies for improving WUE across various climate conditions in different regions, such as the Agricultural Production Systems Simulator (APSIM) (Heng et al., 2007; Kloss et al., 2012), Crop Environment Resource Synthesis (CERES)-Wheat model housed in the Decision Support System for Agrotechnology Transfer (DSSAT) (Behera and Panda, 2009; Singh et al., 2008), CROPSYST (Singh et al., 2008), AquaCrop (Steduto et al., 2009), and RZWQM (Fang et al., 2010; Saseendran et al., 2008). These simulation results provided useful guidelines for deficit irrigation strategies under various weather conditions. However, most of these simulation studies used fixed irrigation amounts and timing and did not consider (i) the water availability for the crop season in a region, which may influence how to optimize the limited water for higher WUE, and (ii) the variable irrigation amounts or timing based on phenological growth stages, such as vegetative and reproductive stages. The objective of this study was to use the RZWQM2, calibrated for northeastern Colorado conditions, to simulate corn ET and grain yield under various growth stage–based irrigation schedules to meet a proportion of ET as calculated from the Shuttleworth–Wallace equation (ET_{sw}) and under various seasonal water limitation.

Materials and Methods

Field Experiment and Data Measurements

The field experiment was initiated in 2008 near Greeley, CO (40.45° N, 104.64° W). The soil is an alluvial sandy loam with coarser and finer textured layers in the 2-m soil profile. Weather data were recorded on site (station designation GLY04) with a standard Colorado Agricultural Meteorological Network (CoAgMet) (http://ccc.atmos.colostate.edu/~coagmet/) weather station. From 2008 to 2011, the growing season precipitation was 245, 237, 211, and 237 mm, respectively. A detailed description of the experiment can be found in Trout et al. (2010), Ma et al. (2012), and Saseendran et al. (2014).

Six irrigation treatments (area in 9 by 44 m, and microirrigation with surface drip tubing adjacent to each row) replicated four times (randomized complete block) were designed to meet a certain percentage of ET_c of a nonstressed, well-watered crop during the growing seasons calculated on the basis of the reference

ET (ET_r) for a tall crop [alfalfa (*Medicago sativa* L.)] and crop coefficient (Allen et al., 1998, 2005, 2007). The treatments were 100% (T1), 85% (T2), 70% (T3), 70% (T4), 55% (T5), and 40% (T6) of ET_c. For all treatments except for T1 and T3, about 20% of the estimated target irrigation amounts during the vegetative growth period were saved and added during the reproductive stage. Corn ('Dekalb 52–59') was planted at an average rate of 81,000 seed ha^{-1} with 0.8-m row spacing on 12 May 2008, 11 May 2009, 11 May 2010, and 2 May 2011 and harvested on 6 Nov. 2008, 12 Nov. 2009, 19 Oct. 2010, and 22 Oct. 2011. Fertilizer as urea–ammonium nitrate (UAN) was applied uniformly to all treatments at planting and then with irrigation water during the growing seasons as needed on the basis of estimated plant requirements for the 100% ET_c treatment. Total N applied was 134 kg N ha^{-1} in 2008, 160 kg N ha^{-1} in 2009, 146 kg N ha^{-1} in 2010, and 146 kg N ha^{-1} in 2011.

Leaf area index (LAI) was estimated from canopy ground cover (F_c) measured with a nadir view digital camera according to Farahani and DeCoursey (2000). Soil water content was measured twice a week during the growing season with a portable time domain reflectometry water meter for the surface 0- to 0.15-m soil layer and with a neutron attenuation water meter between 0.15 and 2 m below the soil surface at 0.3-m intervals. Actual ET was estimated from the soil water balance (ET = irrigation + rainfall + profile soil water difference between measurement times). Soil water drainage below the soil profile and water runoff were assumed to be zero in the water balance calculation of actual ET, but some water drainage below the soil profile or runoff may have occurred because of high rain events such as during an 86-mm rainfall occurring on 15 and 16 August 2008. Additional measurement information can be found in Ma et al. (2012).

Description and Parameterization of the Root Zone Water Quality Model

The RZWQM2 (version 2.6) with the DSSAT 4.0 crop modules were used in this study (Ma et al., 2006). On the basis of Ma et al. (2012), we used soil water content at 333 kPa ($\theta_{1/3}$) and 15,000 kPa (θ_{15}) to estimate Brooks–Corey parameters (Brooks and Corey, 1964), where θ_{15} was assumed equal to one-half of $\theta_{1/3}$ (Ma et al., 2012).

The equations relating volumetric soil water content (θ, m^3 m^{-3}) and matric suction head (ψ, where $\psi > 0$ for negative soil water pressures) in Brooks and Corey (1964) are:

$$\theta = \theta_s \qquad \text{for } \psi < \psi_b$$

$$\frac{\theta - \theta_r}{\theta_s - \theta_r} = \left(\frac{\psi}{\psi_b}\right)^{-|\lambda|} \qquad \text{for } \psi \geq \psi_b, \qquad [1]$$

where θ_s and θ_r are saturated and residual soil water contents, ψ_b is the air entry water suction (negative "bubbling pressure"), and λ is the slope of the log(θ)-log(ψ)

curve or the "pore-size distribution index." Similarly, assuming the log-log slope of the water retention curve is linearly related to the log-log slope of the unsaturated conductivity curve, the hydraulic conductivity K versus suction head is

$$K(\psi)=K_s \quad \text{for} \quad \psi<\psi_b$$
$$K(\psi)=K_s\left(\frac{\psi}{\psi_b}\right)^{-(2+3\lambda)} \quad \text{for} \quad \psi\geq\psi_b \qquad [2]$$

where K_s is the saturated hydraulic conductivity in cm h^{-1}.

The Brooks–Corey parameters can be calculated from $\theta_{1/3}$ and θ_{15} (Ma et al., 2012):

$$\lambda=\frac{\ln\left[\left(\theta_{1/3}-\theta_r\right)/\left(\theta_{15}-\theta_r\right)\right]}{\ln\left(15{,}000/333\right)}. \qquad [3]$$

$$\psi_b=\exp\left[\frac{\ln\left(\theta_{1/3}-\theta_r\right)-\ln\left(\theta_s-\theta_r\right)+\lambda\ln\left(333\right)}{\lambda}\right]. \qquad [4]$$

The K_s was estimated using an empirical relationship with effective porosity based on the formula by Ahuja et al. (1989):

$$K_s=764.5\left(\theta_s-\theta_{1/3}\right)^{3.288}. \qquad [6]$$

Data from 2008 to 2010 were previously used to calibrate the model via trial and error method (Ma et al., 2012; Saseendran et al., 2014), where the highest irrigation treatment (100% ET_c) was used for calibration and the remaining treatments were used for validation. In this study, we used the measured data from all six treatments from 2008 to 2011 to calibrate the model using PEST, an automatic optimization parameter estimation software (Doherty, 2010). The model was run from 1 January to 31 December every year for all the treatments with the actual irrigation amount implemented. The RMSE from each treatment and year were combined to define an objective function for parameter optimization (Nolan et al., 2011).

To cope with the sensitivity of PEST to initial values and ranges of the model parameters (Doherty, 2010), we used initial parameter values from a previous study at this site (Ma et al., 2012). The ranges for these parameters were obtained from either field experiment measurements (Table 1–1) or from the calibrated database released in the DSSAT model (Table 1–2) (Ma et al., 2012). The measured data of soil water content, LAI, and final aboveground biomass and grain yield from 2008 to 2011 were used to parameterize the RZWQM2 with PEST. The calibrated soil and crop genetic parameters along with their ranges are presented in Table 1–1 and Table 1–2. Model performance was quantified with statistical measures of RMSE, mean difference (MD), coefficient of determination (r^2), and the Nash Sutcliffe model efficiency (ME; Nash and Sutcliffe, 1970).

Long-Term Simulations for Deficit Irrigation Management

After calibrating the model for all the treatments and years with actual irrigation amounts as scheduled in the field, the RZWQM2 was used for long-term simulations (1992–2013) of corn under various irrigation management scenarios. To simulate crop ET (ET_c) in the model to schedule irrigation, we would need a crop

Table 1–1. Initial ranges and final soil water content at 333 kPa ($\theta_{1/3}$) and the final bulk density (ρ_b), saturated water content (θ_s), air entry water suction (ψ_b), and pore size distribution index (λ) used for calibrating the Root Zone Water Quality Model (version 2.6) with the parameter estimation software PEST (Doherty, 2010).†

Soil depth	ρ_b	Initial $\theta_{1/3}$	Min. $\theta_{1/3}$	Max. $\theta_{1/3}$	Final $\theta_{1/3}$	Final θ_s	Final λ	Final ψ_b
m	10^3 kg m^{-3}	m^3 m^{-3}						cm
0.00–0.15	1.492	0.262	0.18	0.32	0.231	0.44	0.23	15.34
0.15–0.30	1.492	0.249	0.14	0.32	0.242	0.44	0.23	18.73
0.30–0.60	1.492	0.220	0.13	0.38	0.230	0.44	0.23	15.06
0.60–0.90	1.568	0.187	0.13	0.37	0.206	0.41	0.24	13.19
0.90–1.20	1.568	0.173	0.13	0.32	0.205	0.41	0.24	12.97
1.20–1.50	1.617	1.62	0.13	0.31	0.263	0.39	0.23	47.36
1.50–2.00	1.617	0.198	0.13	0.31	0.310	0.39	0.22	103.35

† The initial values were based on a previous study at the experimental site (Ma et al., 2012), and the ranges were based on the field measured minimum and maximum soil water content from 2008 to 2011.

Table 1–2. Initial ranges and final values for crop cultivar genetic parameters for calibrating the Root Zone Water Quality Model (version 2.6) with the parameter estimation software PEST (Doherty, 2010).

Crop cultivar genetic parameter	Parameter description	Initial	Ranges	Final
P1	degree days (base temperature of 8°C) from seedling emergence to end of juvenile phase (thermal degree days)	250	200–290	245.6
P2	day length sensitivity coefficient [the extent (days) that development is delayed for each hour increase in photoperiod above the longest photoperiod (12.5 h) at which development proceeds at maximum rate]	0.2	0.05–0.6	0.1562
P5	degree days (base temperature of 8°C) from silking to physiological maturity (thermal degree days)	600	550–670	704.4
G2	potential kernel number	900	800–1000	994.1
G3	potential kernel growth rate (mg $kernel^{-1}$ d^{-1})	6	3–10	6.239
PHINT	degree days required for a leaf tip to emerge (phyllochron interval) (thermal degree days)	50	45–55	52.89

coefficient that is not estimated in the RZWQM2. However, the Shuttleworth–Wallace ET (ET_{sw}) (Shuttleworth and Wallace, 1985) simulated in the RZWQM2 was intended to take care of crop effects (i.e., LAI) on ET and can be used as a surrogate for ET_c in the model. Ma et al. (2012) showed that when the ET_{sw} was used as the basis for irrigation to replace ET_c, model simulated irrigation amounts matched actual irrigation amounts for all the irrigation levels. Therefore, we used ET_{sw} to replace ET_c in the long-term simulation study.

Irrigation began 3 d after planting at a 20 mm h^{-1} rate and ended on or before 16 Sept. each year. Irrigation was scheduled every 3 d to target a certain percentage of ET_{sw} (100, 80, 60, and 40%) calculated for the past 3 d, minus any precipitation, which was similar to the irrigation experiment. These irrigation targets, defined as 100%ET_{sw}, 80%ET_{sw}, 60%ET_{sw}, and 40%ET_{sw}, were implemented separately during the vegetative (before tasseling) and reproductive (after tasseling) stages. On the basis of the experimental data, we assumed the reproductive stage started 77 d after planting (28 July). All combinations of these ET_{sw}–based irrigation targets resulted in 16 irrigation scenarios (e.g., 100%–40%ET_{sw} denotes an irrigation level of 100%ET_{sw} at the vegetative stage and 40%ET_{sw} at the reproductive stage). N was applied at 150 kg N ha^{-1} as UAN at planting. Corn was planted on 12 May and harvested at harvest maturity as determined by the model, which was generally between late September and late October.

Each of these 16 scenarios was simulated with an unlimited seasonal irrigation condition and with a 300-, 400-, and 500-mm irrigation limit, for a total of 64 scenarios (i.e., 16 × 4). Seasonal water limitations such as these are imposed on farmers by river flows that decline as the summer progresses until flows drop below the level that allows diversions for irrigation. Model simulations for each of the 64 scenarios were run continuously for 22 yr from 1 Jan. 1992 to 17 Dec. 2013. Weather data for the simulations, including precipitation, were taken from nearby CoAgMet weather stations.

Results and Discussion

Model Calibration

Soil Water and ET Simulations

As shown in Table 1–3, a better soil water content simulation was obtained for the deeper soil layers with lower RMSE values for the 0.9- to 1.2-m soil layers than for the 0.3- to 0.6-m soil layers. For the deepest layer of 1.2 to 2.0 m, overestimations of soil water content occurred, with a high MD value of 0.016 m^3 m^{-3}. The overall MD and RMSE values for total soil water storage in the profile were 20 and 35 mm across the six treatments from 2008 to 2011 (Table 1–3). The overestimated soil

Table 1–3. Statistical results for simulated soil water content (θ) at the different depths and soil water storage in a 0- to 2-m layer across the six deficit irrigation treatments from 2008 to 2011.†

Soil depth	Measured θ average	Simulated θ average	MD	RMSE	ME	r^2
m	$m^3\ m^{-3}$					
0.00–0.15	0.169	0.181	0.013	0.053	0.04	0.46
0.15–0.30	0.177	0.181	0.005	0.040	−0.05	0.58
0.30–0.60	0.163	0.160	−0.003	0.037	−0.44	0.55
0.60–0.90	0.147	0.154	0.007	0.023	0.05	0.62
0.90–1.20	0.150	0.153	0.003	0.020	0.25	0.53
1.20–1.50	0.156	0.180	0.024	0.035	−0.19	0.66
1.50–2.00	0.184	0.201	0.016	0.032	−0.40	0.30
0.00–2.00 (SWS)‡	321.6	342.0	20	35	0.41	0.81

† MD, mean deviation; ME, model efficiency; r^2, coefficient of determination.
‡ SWS, soil water storage (mm).

water in the profile was mainly due to overestimating soil water content below 1.2-m depth, and the model only slightly overestimated soil water in the 0.0- to 1.2-m soil profiles with an MD of 5 mm and an RMSE of 20 mm. These results were comparable with the previous simulation study using data from 2008 to 2010 with an RMSE value of 32 mm for profile soil water (Ma et al., 2012).

Both RZWQM2-simulated and measured ET_a showed similar decreases with deficit irrigation from T1 to T6 across the 4 yr (Fig. 1–1). Underestimation of ET_a occurred in 2010 and 2011 and a closer agreement to measured ET_a was obtained in 2008 and 2009. Across the 4 yr, the MD, RMSE, ME, and r^2 values for simulated ET_a were −41 mm, 57 mm, 0.61, and 0.81, respectively (Table 1–4). Both simulated and measured ET_a decreased slightly from T3 to T4 in 2009 and 2011 (Fig. 1–1).

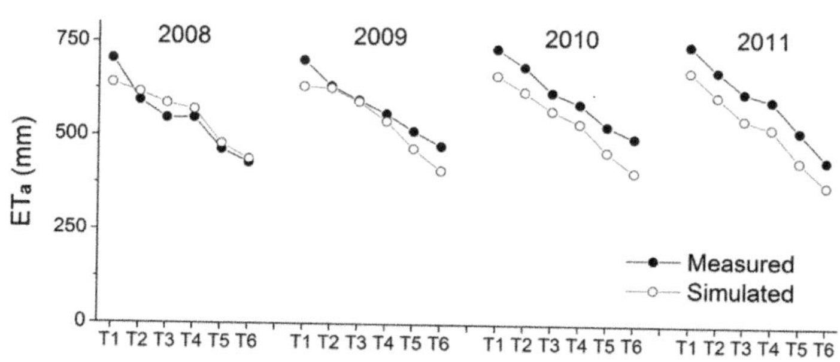

Fig. 1–1. Comparisons between measured and simulated seasonal crop actual evapotranspiration (ET_a) across six irrigation treatments [T1 (100%ET_c), T2 (85%ET_c), T3 (70%ET_c without moving 20% of vegetative stage irrigation to the reproductive stage), T4 (70%ET_c), T5 (55%ET_c), and T6 (40% ET_c), where ET_c is crop ET for a well-watered crop calculated from reference ET and crop coefficient] from 2008 to 2011.

Table 1–4. Statistical results for simulated seasonal ET, leaf area index, final grain yield, aboveground biomass, and water use efficiency for the six irrigation treatments from 2008 to 2011.†

Variables‡	Measured average	Simulated average	MD	RMSE	ME	r^2
ET (mm)	585.4	544.3	−41.1	56.5	0.61	0.81
GY (kg ha^{-1})	8491	8577	86	354	0.97	0.98
LAI (m^2 m^{-2})	2.09	2.15	0.06	0.77	0.69	0.70
AGB (kg ha^{-1})	17102	16941	−160	1203	0.90	0.90
WUE (kg m^{-3})	1.44	1.55	0.11	0.16	0.67	0.89

† MD, mean deviation; ME, model efficiency; r^2, coefficient of determination.

‡ GY, final grain yield; LAI, leaf area index; AGB, aboveground biomass; WUE, water use efficiency.

Leaf Area Index, Crop Yield, and Water Use Efficiency Simulations

The model-calculated LAI agreed well with the measured data with mean values of 2.2 m^2 m^{-2} and 2.1 m^2 m^{-2} across the 4 yr. The corresponding RMSE, ME, and r^2 values are 0.9 m^2 m^{-2}, 0.7, and 0.70, respectively. Both measured and simulated grain yield decreased with the deficit irrigation treatments from T1 to T6 across the 4 yr (Fig. 1–2a), with good agreement between measured and simulated (MD = 86 kg ha^{-1}, RMSE = 354 kg ha^{-1}, and ME = 0.97, Table 1–4). In 2008, grain yield was overestimated for treatment T4. From 2009 to 2011, both measured and calculated grain yields showed similar trends between T3 and T4, which suggested that the model was able to simulate the differences in grain yield due to differential irrigation scheduling between vegetative and reproductive growth stages. Similar simulation results were obtained for biomass across the 4 yr with MD, RMSE, and ME values of −160 kg ha^{-1}, 1203 kg ha^{-1}, and 0.90, respectively (Table 1–4). These simulation results for LAI, grain yield, and biomass were generally better than

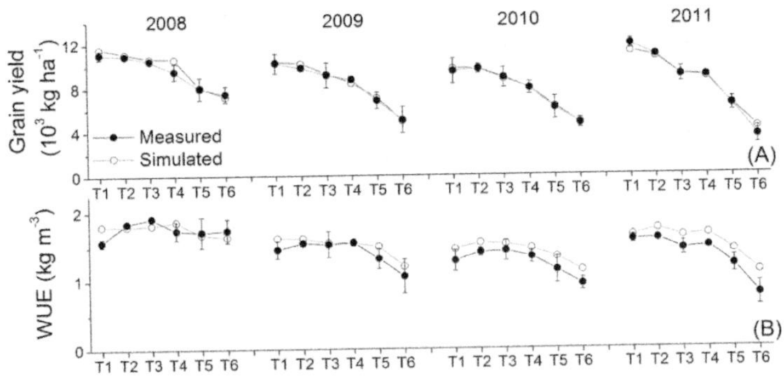

Fig. 1–2. Comparisons between measured and simulated grain yield and water use efficiency (WUE) for six irrigation treatments [T1 (100%ET$_c$), T2 (85%ET$_c$), T3 (70%ET$_c$ without moving 20% of vegetative stage irrigation to the reproductive stage), T4 (70%ET$_c$), T5 (55%ET$_c$), and T6 (40% ET$_c$), where ET$_c$ is crop ET for a well-watered crop calculated from reference ET and crop coefficient] from 2008 to 2011.

those obtained by Ma et al. (2012), who calibrated the RZWQM2 via a trial and error method using data from 2008 to 2010.

The simulated WUE also showed similar trends with measured data across these treatments from 2008 to 2011 (Fig. 1–2b), but overcalculation of WUE was obtained mainly because of the under-simulated ET (Fig. 1–1). In 2008, similar to grain yield, the model oversimulated WUE for treatment T4. The MD, RMSE, ME, and r^2 values for the simulated WUE were 0.11 kg m^{-3}, 0.16 kg m^{-3}, 0.67, and 0.89 across all the 4 yr, which were also better than a previous simulation study in North China Plain using the model (Fang et al., 2010).

Long-Term Simulation of Crop Water Use and Production

Unlimited Irrigation Water Available

With the unlimited irrigation water available condition, the average irrigation amounts increased from 73 to 285 mm during the vegetative stage and from 50 to 224 mm during the reproductive stage (Fig. 1–3a), with the irrigation targets varying from 40%ET_{sw} to 100%ET_{sw}. Correspondingly, the average ET increased from 195 to 362 mm during the vegetative stage and from 122 to 316 mm during the reproductive stage (Fig. 1–3b). Greater irrigation amounts were required during the vegetative stage than during the reproductive stage in part because of the slightly longer vegetative growing season (Fig. 1–3a). The simulated average seasonal irrigation amount increased from 123 to 509 mm (Table 1–5) as irrigation targets varying from 40%ET_{sw} to 100%ET_{sw}. The yearly variations in the irrigation amount due to weather variations (SD across seasons) ranged from 61 to 97 mm, which were much lower than the range of irrigation amounts for these irrigation scenarios (Table 1–5).

Average crop water stress, which is defined as [1 – (potential water uptake/ potential transpiration)] (Ma et al., 2006), decreased with increased irrigation targets at both the vegetative and reproductive stages and was generally lower for the vegetative stage than for the reproductive stage at the same irrigation targets (Fig. 1–4a vs. Fig. 1–4b). The simulated lower water stress at the vegetative stage than at the reproductive stage under the lowest irrigation target (40%ET_{sw}) indicated severe water deficits during the reproductive stage. Conversely, at the lower irrigation targets for the reproductive stage (e.g., 40%ET_{sw} or 60%ET_{sw}), lower water stress occurred at the highest irrigation target (100%ET_{sw}) at the vegetative stage (e.g., 80%ET_{sw}) (Fig. 1–4b), which indicated that maintaining higher irrigation targets during the vegetative stage can increase soil water supply and alleviate plant water stress at the reproductive stage.

Grain yield increased with irrigation targets at both vegetative and reproductive growth stages, but the response was more pronounced by increasing irrigation at the vegetative stage than at the reproductive stage (Table 1–6). This result was different from previous results where corn yield was greatly influenced by water availability during the tasseling stage (Nielsen et al., 2009, 2010). Such a discrepancy may be explained by the simulated soil water contents among the irrigation targets shown in Fig. 1–5. Although the soil profile available water increased after 28 July (end of vegetative stage) for the irrigation targets of 40%–100%ET_{sw}, it was still lower than the irrigation targets of 100%–40%ET_{sw} before 10 August during the tasseling stage. High irrigation targets (100%ET_{sw})

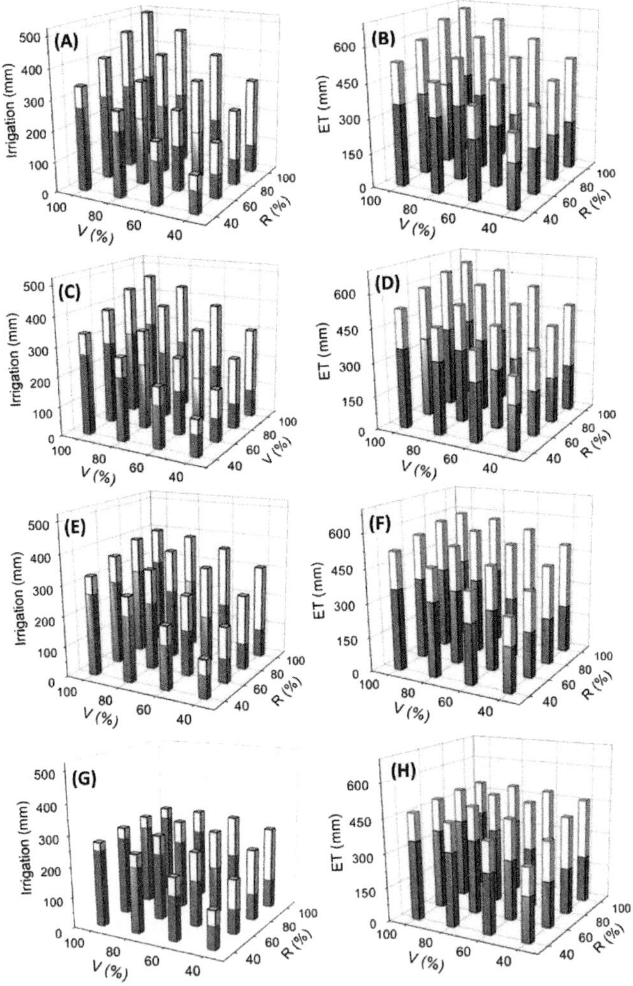

Fig. 1–3. Simulated long-term (1992–2013) averaged irrigation amounts, ET, between the vegetative (V) and the reproductive (R) stages (black bar, irrigation or ET for V; white bar, irrigation or ET for R) as influenced by irrigation targets (40%ET_{sw} to 100%ET_{sw}) between V and R stages under unlimited (a and b), 500-mm (c and d), 400-mm (e and f), and 300-mm (g and h) seasonal irrigation water limit conditions.

during the vegetative stage resulted in high soil profile available water during the early reproductive stage (tasseling stage) and benefited the final grain yield (Table 1–6).

Grain yield generally maximized at irrigation targets >80%ET_{sw} for both vegetative and reproductive stages, but showed a reduced response to irrigation targets from 80%ET_{sw} to 100%ET_{sw} (Table 1–6). However, simulated average grain yield with the highest irrigation target of 100%–100%ET_{sw} (Table 1–6) was slightly lower than measured grain yield from T1 (Fig. 1–2), which was partly caused by the simulated crop N stress under the highest irrigation target (100%–100%ET_{sw}) where 150 kg N ha^{-1} was applied at planting as compared with fertigation in the T1 treatment. The weather variations from 1992 to 2013 also contributed to the differences between measured and long-term simulated corn yield. The WUE also increased with the increase in irrigation targets, but showed less response to irrigation than grain yield (Table 1–7). No significant ($P > 0.01$) difference in simulated WUE was found for irrigation levels of 80%–100%ET_{sw}, 100%–80%ET_{sw}, and 100%–100%ET_{sw} (Table 1–7), but significantly ($P < 0.01$) higher grain yield was obtained for the 100%–100%ET_{sw} (Table 1–6).

Table 1–5. Long-term (1992–2013) simulated average irrigation amounts and standard deviations across seasons for different irrigation targets (40% to 100% of ET_{sw}), between vegetative and reproductive growth stages under various irrigation water availability conditions.†

Water limit	Targets for reproductive stage	Targets for vegetative stage			
		40%ET_{sw}	60%ET_{sw}	80%ET_{sw}	100%ET_{sw}
		mm			
Unlimited water	40%ET_{sw}	123 ± 61	204 ± 71	277 ± 80	338 ± 93
	60%ET_{sw}	180 ± 65	263 ± 75	337 ± 83	399 ± 92
	80%ET_{sw}	243 ± 72	323 ± 80	395 ± 88	463 ± 96
	100% ET_{sw}	307 ± 76	380 ± 89	452 ± 92	**509 ± 97**
500-mm seasonal water limit	40%ET_{sw}	123 ± 61	204 ± 71	277 ± 80	338 ± 93
	60%ET_{sw}	180 ± 65	263 ± 75	337 ± 83	394 ± 86
	80%ET_{sw}	243 ± 72	323 ± 80	392 ± 84	438 ± 66
	100%ET_{sw}	307 ± 76	379 ± 86	435 ± 71	**466 ± 49**
400-mm seasonal water limit	40%ET_{sw}	123 ± 61	204 ± 71	276 ± 79	324 ± 78
	60% ET_{sw}	180 ± 65	263 ± 75	328 ± 7.1	35.8 ± 5.1
	80%ET_{sw}	243 ± 72	318 ± 73	359 ± 50	387 ± 29
	100%ET_{sw}	304 ± 73	353 ± 60	**383 ± 35**	395 ± 19
300-mm seasonal water limit	40%ET_{sw}	123 ± 61	203 ± 69	253 ± 54	274 ± 37
	60%ET_{sw}	180 ± 65	247 ± 58	282 ± 34	294 ± 16
	80%ET_{sw}	236 ± 63	277 ± 36	295 ± 16	300 ± 01
	100%ET_{sw}	271 ± 42	**292 ± 21**	299 ± 06	300 ± 00

† Numbers in bold are the most reasonable irrigation targets with the highest grain yield and water use efficiency and the lowest yearly variations for each water limit condition. ET_{sw}, Shuttleworth–Wallace calculated ET.

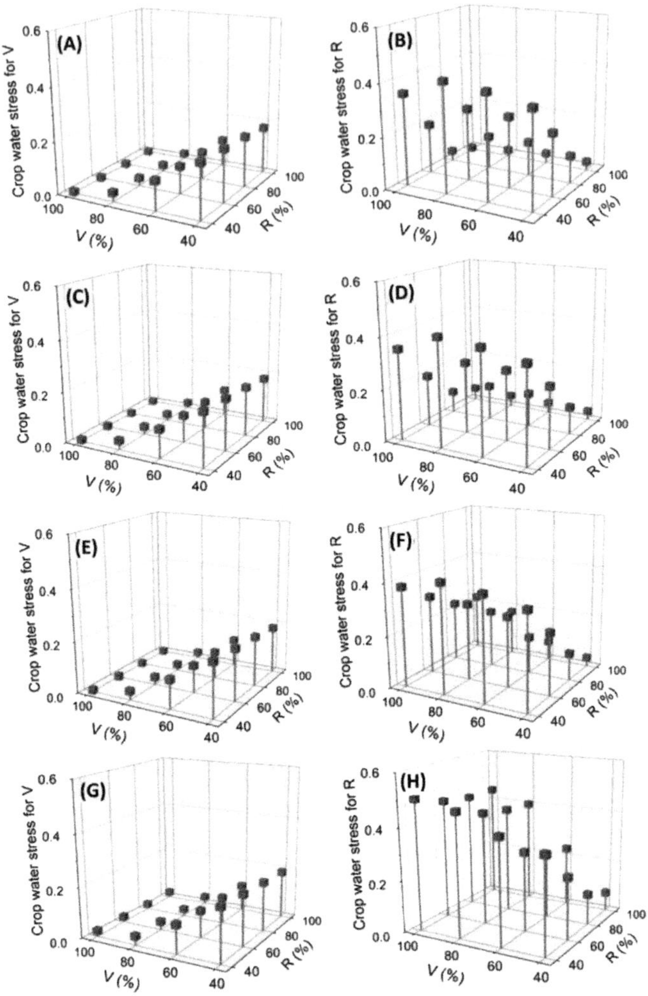

Fig. 1–4. Simulated long-term (1992–2013) averaged crop water stress (1 — potential water uptake/ potential transpiration) between the vegetative (V) and the reproductive (R) stages as influenced by irrigation targets (40%ET_{sw} to 100%ET_{sw}) between V and R stages under unlimited (a and b), 500-mm (c and d), 400-mm (e and f), and 300-mm (g and h) seasonal irrigation water limit conditions.

The coefficient of variance (CV, SD/mean) for grain yield across seasons due to the weather variations ranged from 0.26 (SD = 857 kg ha^{-1}) for 40%–40%ET_{sw} to 0.10 (SD = 900 kg ha^{-1}) for 100%–100%ET_{sw} (Table 1–6). The yearly grain yield variations were generally higher at the lowest irrigation target (40%ET_{sw}) at vegetative (CV = 0.25) or reproductive growth stages (CV = 0.23) than at other irrigation targets (CV = 0.14) (Table 1–6). To achieve both high grain yield and WUE with low yearly varia-

Table 1–6. Long-term (1992–2013) simulated average corn yield and standard deviations across seasons for different irrigation targets (40% to 100% of ET_{sw}), between vegetative and reproductive growth stages under various water availability conditions.†

Water limit	Targets for reproductive stage	Targets for vegetative stage‡			
		40%ET_{sw}	60%ET_{sw}	80%ET_{sw}	100%ET_{sw}
		kg ha^{-1}			
Unlimited water	40%ET_{sw}	3340 ± 857j	4447 ± 1004i	5674 ± 1653gh	7245 ± 1689def
	60%ET_{sw}	4363 ± 965i	5854 ± 1023H	7261 ± 1476ef	8941 ± 1408c
	80%ET_{sw}	5582 ± 1163h	7249 ± 1058E	8834 ± 1263cd	10,051 ± 1009b
	100%ET_{sw}	6584 ± 1505fg	8260 ± 1210cd	9846 ± 992b	**10,256 ± 900a**
500-mm seasonal water limit	40%ET_{sw}	3340 ± 857j	4447 ± 1004i	5674 ± 1653gh	7245 ± 1689cde
	60%ET_{sw}	4363 ± 965i	5854 ± 1023g	7261 ± 1476de	8854 ± 1367c
	80%ET_{sw}	5582 ± 1163h	7249 ± 1058d	8782 ± 1280c	9791 ± 1086b
	100%ET_{sw}	6584 ± 1505ef	8251 ± 1219c	9749 ± 995b	**10,016 ± 994a**
400-mm seasonal water limit	40% ET_{sw}	3340 ± 857j	4447 ± 1004i	5666 ± 1652gh	6903 ± 1593de
	60%ET_{sw}	4363 ± 965i	5854 ± 1023g	7124 ± 1527de	7962 ± 1592bc
	80%ET_{sw}	5582 ± 1163h	7211 ± 1104d	8241 ± 1599b	8665 ± 1789ab
	100%ET_{sw}	6554 ± 1445ef	8048 ± 1302b	**8834 ± 1621a**	8800 ± 1809a
300-mm seasonal water limit	40%ET_{sw}	3340 ± 857j	4427 ± 1014hi	5292 ± 1839g	5088 ± 2088ghi
	60%ET_{sw}	4363 ± 965hi	5654 ± 1137efg	5963 ± 2259c	5730 ± 2654cde
	80%ET_{sw}	5519 ± 1137efg	6707 ± 1540a	6456 ± 2499ab	5930 ± 2809bcde
	100%ET_{sw}	6298 ± 1351abc	**7113 ± 1800a**	6620 ± 2649a	5940 ± 2803bcde

† Numbers in bold are the most reasonable irrigation targets with the highest grain yield and water use efficiency and the lowest yearly variations for each water limit condition. ET_{sw}, Shuttleworth–Wallace calculated ET.

‡ The different letters behind these numbers for each water limit conditions mean significant difference at $P <$ 0.01 level based on a paired *t* test.

tions, the most reasonable irrigation strategy should be 100%–100%ET_{sw} between the vegetative and reproductive stage (Table 1–6 and Table 1–7).

High Seasonal Irrigation Water Available

The grain yield results obtained with the 500-mm seasonal irrigation water availability condition were similar to the grain yields obtained from the unlimited irrigation water condition (Fig. 1–3c, 1–3d, 1–4c, and 1–4d). The irrigation amounts and grain yields were also similar to the unlimited irrigation water simulations except for 80%–100%ET_{sw}, 100%–80%ET_{sw}, and 100%–100%ET_{sw}, when irrigation water became limited at the reproductive stage (Table 1–6). Similar to the unlimited water simulations, the 100%–100%ET_{sw} produced the highest grain yield and WUE with relatively low yearly variations and can be considered as the most reasonable irrigation strategy under the 500-mm seasonal irrigation condition (Table 1–6 and Table 1–7).

Table 1–7. Long-term (1992–2013) simulated average water use efficiency and standard deviations across seasons for different irrigation targets (40% to 100% of ET_{sw}), between vegetative and reproductive stages under various water availability conditions. †

Water limit	Targets for reproductive stage	Targets for vegetative stage‡ 40%ET_{sw}	60%ET_{sw}	80%ET_{sw}	100%ET_{sw}
		kg m^{-3}			
Unlimited water	40%ET_{sw}	1.06 ± 0.21f	1.12 ± 0.19ef	1.20 ± 0.29def	1.35 ± 0.22cd
	60%ET_{sw}	1.18 ± 0.23e	1.29 ± 0.20d	1.37 ± 0.23cd	1.51 ± 0.18ab
	80%ET_{sw}	1.29 ± 0.23d	1.42 ± 0.22bc	1.51 ± 0.19ab	1.54 ± 0.13a
	100%ET_{sw}	1.35 ± 0.25cd	1.48 ± 0.22abc	1.55 ± 0.14a	**1.51 ± 0.12ab**
500-mm seasonal water limit	40%ET_{sw}	1.06 ± 0.21h	1.12 ± 0.19gh	1.20 ± 0.29efgh	1.35 ± 0.22bc
	60%ET_{sw}	1.18 ± 0.23g	1.29 ± 0.20ef	1.37 ± 0.23bcd	1.51 ± 0.18ab
	80%ET_{sw}	1.29 ± 0.23f	1.42 ± 0.22b	1.51 ± 0.19ab	1.55 ± 0.15a
	100%ET_{sw}	1.35 ± 0.25bcd	1.48 ± 0.22a	1.56 ± 0.13a	**1.54 ± 0.13a**
400-mm seasonal water limit	40%ET_{sw}	1.06 ± 0.21f	1.12 ± 0.19ef	1.20 ± 0.29def	1.32 ± 0.24bc
	60%ET_{sw}	1.18 ± 0.23de	1.29 ± 0.20cd	1.36 ± 0.24bc	1.44 ± 0.23b
	80%ET_{sw}	1.29 ± 0.23cd	1.42 ± 0.22b	1.49 ± 0.22b	1.48 ± 0.22b
	100%ET_{sw}	1.35 ± 0.25b	1.49 ± 0.22ab	**1.53 ± 0.19a**	1.49 ± 0.21b
300-mm seasonal water limit	40%ET_{sw}	1.06 ± 0.21g	1.11 ± 0.19efg	1.16 ± 0.32defg	1.06 ± 0.36fg
	60%ET_{sw}	1.18 ± 0.23def	1.28 ± 0.21cde	1.23 ± 0.35cde	1.13 ± 0.42efg
	80%ET_{sw}	1.30 ± 0.23c	1.42 ± 0.25ab	1.29 ± 0.37bcd	1.15 ± 0.43defg
	100%ET_{sw}	1.36 ± 0.24bcd	**1.46 ± 0.26a**	1.30 ± 0.39cd	1.16 ± 0.43defg

† Numbers in bold are the most reasonable irrigation targets with the highest grain yield and water use efficiency and the lowest yearly variations for each water limit condition. ET_{sw}, Shuttleworth–Wallace calculated ET.

‡ The different letters behind these numbers for each water limit conditions mean significant difference at $P <$ 0.01 level based on a paired *t* test.

Medium Seasonal Irrigation Water Available

The simulated average seasonal irrigation amounts and ET ranged from 123 to 395 mm and from 195 to 359 mm with the irrigation varying from 40%ET_{sw} to 100%ET_{sw}, respectively (Table 1–5 and Fig. 1–3e and 1–3f). Compared with the 500-mm seasonal irrigation available conditions, the irrigation amount during the reproductive stage was reduced noticeably at the high irrigation targets (80%ET_{sw} or 100%ET_{sw}) when the irrigation targets increased from 60%ET_{sw} to 100%ET_{sw} during the vegetative stages (Fig. 1–3c vs. Fig. 1–3e), mainly caused by the limited seasonal available water (400-mm total application). Similar trends were also obtained for the simulated ET (Fig. 1–3d vs. Fig. 1–3f).

Similar responses of crop water stress during the vegetative stage were found among the unlimited, 500-, and 400-mm irrigation water available conditions (Fig. 1–4a to 1–4f). During the reproductive stage, however, greater crop water stress was simulated with the high irrigation targets of 80%ET_{sw} and 100%ET_{sw} during both vegetative and reproductive stages compared with the unlimited or 500-mm seasonal irrigation water available conditions (Fig. 1–4c and 1–4d vs. Fig. 1–4e and 1–4f). This result was mainly due to less irrigation water available during the reproductive stage under high irrigation targets at the vegetative stage (Fig. 1–3c and 1–3d vs. Fig. 1–3e and 1–3f).

The grain yield and WUE responses to irrigation targets were similar to the 500-mm seasonal irrigation water availability conditions (Table 1–6 and Table 1–7). The yearly grain yield variations across seasons were higher than the 500-mm seasonal water availability condition (Table 1–6). No significant ($P > 0.01$) difference in simulated grain yield was found among the irrigation levels of 80%–100%ET_{sw}, 100%–80%ET_{sw}, and 100%–100%ET_{sw} (Table 1–6), but significantly ($P < 0.01$) higher WUE was obtained for the irrigation level of 80%–100%ET_{sw} (Table 1–7), which also produced low yearly variations (SD) in both grain yield and WUE and can be considered as the most reasonable irrigation strategy when only 400 mm of water is available.

Low Seasonal Irrigation Water Available

Under the 300-mm seasonal irrigation water available condition, the simulated average seasonal irrigation amounts increased from 123 to 300 mm and simulated average seasonal ET increased from 317 to 495 mm with the irrigation varying from 40%ET_{sw} to 100%ET_{sw} (Table 1–5 and Fig. 1–3g and 1–3h). An obvious decline in simulated irrigation amount and ET during the reproductive stage was simulated with the increase of irrigation targets during the vegetative stage (Fig. 1–3g and 1–3h), as compared with the 500- and 400-mm seasonal irrigation water available conditions (Fig. 1–3c to 1–3f). Simulated water stress during the vegetative stage showed similar responses to irrigation targets as that under the 400-mm seasonal irrigation water available conditions (Fig. 1–4e vs. Fig. 1–4g), but greater crop water stress was simulated during the reproductive stage due to limited water availability, especially when large amounts of irrigation water were applied during the vegetative stage (Fig. 1–4f vs. Fig. 1–4h).

Different from the unlimited and 500-mm seasonal irrigation water available simulations, higher grain yield and WUE with lower yearly variations were generally obtained for lower irrigation targets during the vegetative stage and higher irrigation targets during the reproductive stage (Table 1–6). The response of grain yield or WUE to irrigation was more pronounced by increasing irrigation at the reproductive stage than at the vegetative stage, which was also different from the unlimited and 500-mm seasonal irrigation water available conditions (Table 1–6 and Table 1–7). This result was mainly associated with the simulated higher soil profile water content after 4 August (tasseling, the most water-sensitive growth stage) for 40%–100%ET_{sw} than for 100%–40%ET_{sw} irrigation targets (Fig. 1–5). This result was consistent with previous results where corn yield was greatly influenced by water availability during the tasseling stage (Nielsen et al., 2009, 2010). The 60%–100%ET_{sw} irrigation level produced the highest grain yield and WUE with low yearly variations (SDs) across the seasons (Table 1–6 and Table 1–7). Compared with the unlimited, 500-, 400-, and 300-mm seasonal irrigation water available conditions, the higher grain yield and WUE can be obtained with more irrigation applied at the reproductive stage than at the vegetative stage when seasonal irrigation amounts decreased from unlimited and 500- to 300-mm water available conditions.

Crop Water Production Response to Irrigation Water Availability

Since there was an obvious linear relationship between the long-term (1992–2013) simulated average irrigation and ET_a ($y = 0.98x + 19.7$, $r^2 = 1$, without forcing to through the origin), we only present the relationships between grain yield and ET_a for the various irrigation conditions (Fig. 1–6). The simulated average grain yield increased linearly with increasing ET_a under the various seasonal irrigation water availability conditions ($r^2 > 0.76$, Fig. 1–6a to 1–6d). Similar relationships were also found between measured experimental grain yield and ET_a from 2008 to 2011 ($r^2 = 0.60$, Fig. 1–6a). The slope (kg ha^{-1} mm^{-1}) of the relationship between corn yield and ET_a declined as seasonal irrigation water available varied from 500 to 300 mm. Similarly, simulated WUE also increased linearly with ET ($r^2 > 0.60$, Fig. 1–6e to 1–6g), except for the 300-mm seasonal irrigation water condi-

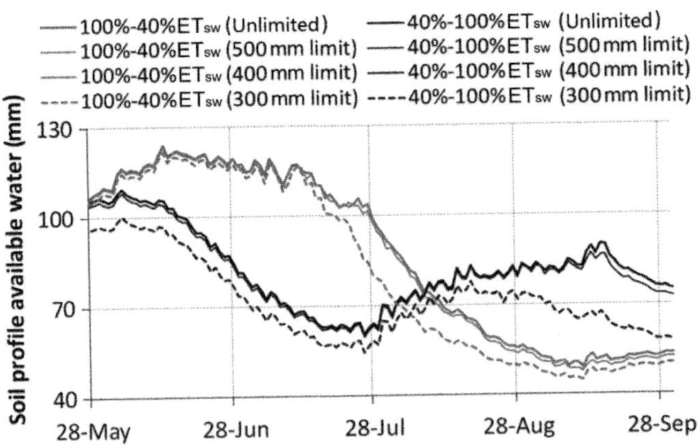

Fig. 1–5. Simulated long-term (1992–2013) averaged soil profile available water [soil water storage minus soil water at wilting point (15,000 kPa) in 0- to 2-m depth] during the vegetative stage (before 28 July) and the reproductive stage (after 28 July) of corn at irrigation targets of 100%–40%ET_{sw} and 40%–100%ET_{sw} under unlimited, 500-, 400-, and 300-mm irrigation water available conditions.

tion ($r^2 = 0.15$, Fig. 1–6h). The above results suggested that the long-term simulated grain yield and WUE were mainly related to ET_a (or irrigation), and the different irrigation targets at vegetative or reproductive stages had less effect on average grain yield and WUE across these seasons under higher irrigation available conditions, especially considering the high yearly variations of grain yield and WUE. Under low (300-mm) irrigation water conditions, however, great differences in crop yield and WUE may occur among the irrigation targets (Fig. 1–6d and 1–6h),

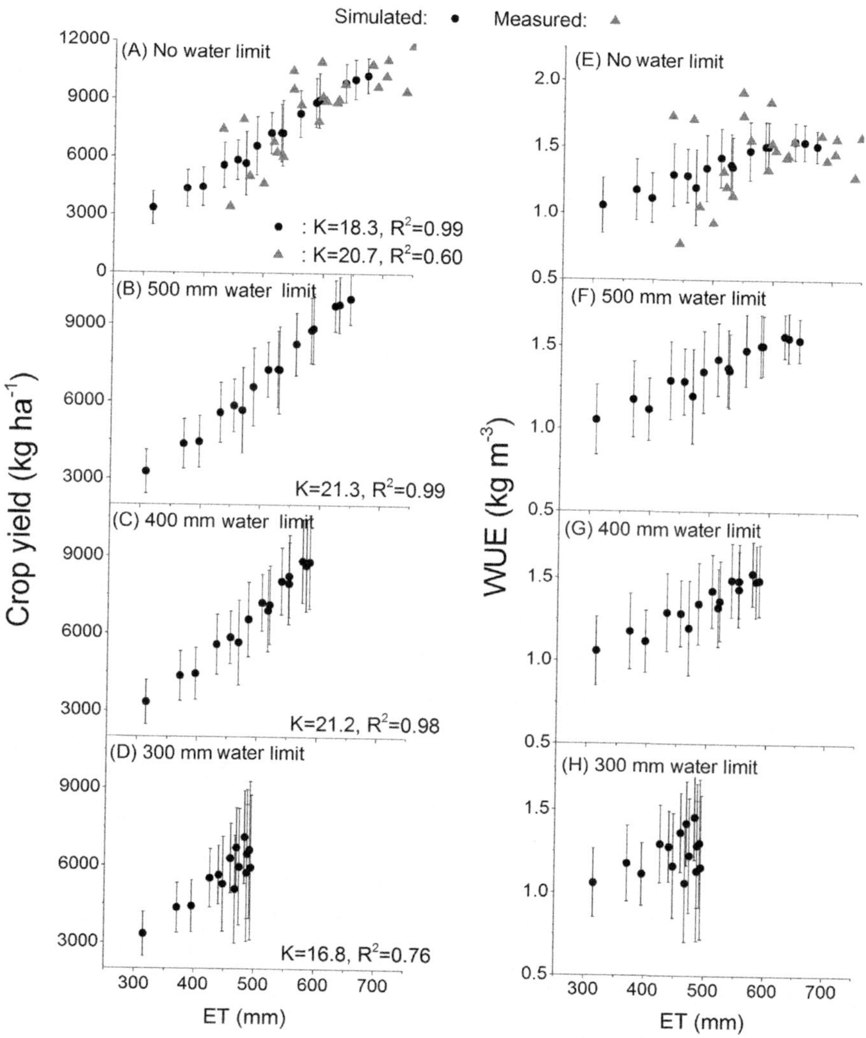

Fig. 1–6. Responses of long-term (1992–2013) simulated corn yield or water use efficiency (WUE) to actual evapotranspiration (ET_a) as influenced by different irrigation targets between the vegetative and reproductive stages under various seasonal irrigation water availabilities. (K = slope of the relationship between ET_a and corn yield, kg ha^{-1} mm^{-1}).

suggesting that higher crop yield and WUE can be achieved with a better irrigation ratio between the vegetative and reproductive stages (more irrigation water at reproductive stage than at vegetative stage as discussed above, see also Table 1–6 and Table 1–7). Therefore, under limited irrigation water conditions, higher corn yield and WUE may be obtained by partitioning irrigation water more to the reproductive stage.

Best Irrigation Strategies for Corn in the Region

On the basis of the results of the long-term simulations from 1992 to 2013, the most reasonable irrigation strategies based on meeting a certain targeted percent of ET_{sw} at vegetative and reproductive stages were selected for these different seasonal irrigation water limitations (i.e., 300, 400, and 500 mm) and are shown in Table 1–5 and Table 1–6. For the most reasonable irrigation strategies, meeting crop water requirement during the reproductive stage is most important for higher corn yield and WUE under water-limited conditions as confirmed by previous corn production studies (Çakir, 2004; Domínguez et al., 2012, Saseendran et al., 2008). The long-term simulation results also showed that allocating more of the limited irrigation water to the reproductive growth stage produced lower yearly variability and thus lower risk due to the weather variations. These irrigation strategies were based on the long-term average weather conditions. However, specific irrigation amount for a given year depended greatly on the distributions of rain during the growing seasons.

As shown in Fig. 1–7, both 100%–80%ET_{sw} under unlimited water conditions and 100%–100%ET_c under 500-mm irrigation water conditions produced higher grain yield than the 80%–100%ET_{sw} irrigation under 400-mm irrigation limit conditions, but all produced similar WUE across the 1992 to 2013 seasons. These irrigation strategies produced higher corn yield than the average regional corn yield (7842 ± 737 kg ha^{-1}) in northeastern Colorado for the same period (Fig. 1–7a, http://quickstats.nass.usda.gov/). Under 300-mm seasonal irrigation water available conditions, the 60%–100%ET_{sw} produced higher corn yield than the regional survey data in 9 of the 22 yr. The 60%–100%ET_{sw} under the 400-mm seasonal irrigation water available limitation (350-mm actual irrigation amount) achieved similar corn yields across these seasons (8047 ± 1302 kg ha^{-1}) to the average corn yield in northeastern Colorado. On the basis of the irrigation water levels at a local area, the farmers can partition irrigation water between the vegetative and reproductive growth stages for higher grain yield and WUE according to the long-term simulation results. Conversely, the farmers should be aware of the high yearly variability in grain yield and WUE associated with the weather variations as shown in Fig. 1–7.

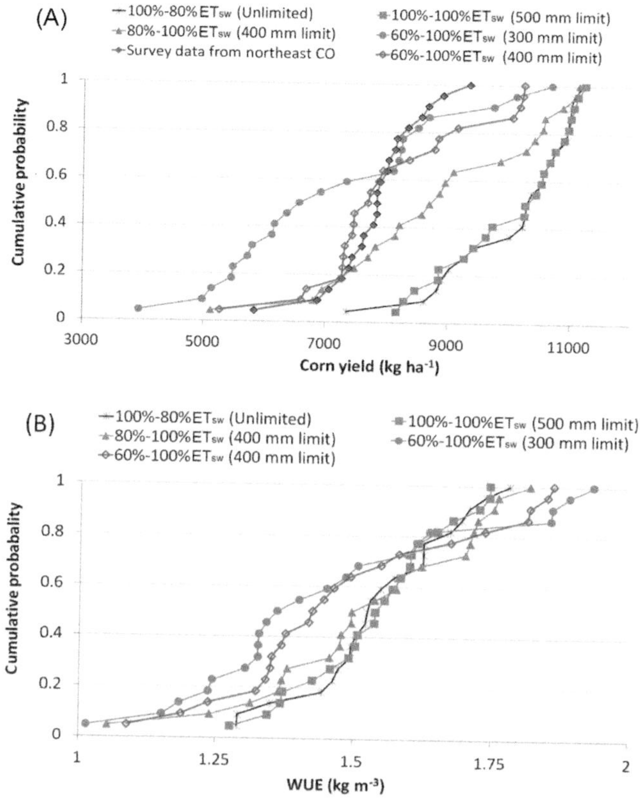

Fig. 1–7. Accumulative probabilities for grain yield and water use efficiency (WUE) from 1992 to 2013 simulated by the Root Zone Water Quality Model (version 2.6) for the selected most reasonable irrigation targets at vegetative and productive stages (Table 1–6) under these different irrigation water available levels (unlimited, 500-, 400-, and 300-mm irrigation water available conditions). The survey corn yield data for the northeast Colorado (including Boulder, Jefferson, Larimer, Logan, and Morgan counties) were obtained from the National Agricultural Statistics Service (http://quickstats.nass.usda.gov/).

Conclusions

Simulations obtained with the calibrated RZWQM2 and thereafter used to evaluate different irrigations strategies to maximize corn grain yield and WUE for northeastern Colorado showed acceptable model performance in simulating ET, grain yield, LAI, and aboveground biomass. A summary of the MD, RMSE, ME, and r^2 between measured and simulated values is given in Table 1–4. The long-term simulations with various irrigation targets showed a higher water requirement at the vegetative stage than at the reproductive stage. Under the unlimited and the 500-mm irrigation water available conditions, the higher irrigation targets at both vegetative and reproductive stages produced the highest corn yield and WUE. Under 400- and 300-mm irrigation water available conditions, maintaining higher irrigation targets during the reproductive stage and lower irrigation targets at the vegetative stage can produce high grain yield and WUE. These results were consistent with the experimental results from Saseendran et al. (2008) and Domínguez et al. (2012) in the same region and from Fang et al. (2010) in the North China Plain.

Irrigation strategies for different irrigation water available conditions provide useful guidelines to optimize the selection of an irrigation strategy to achieve high WUE and grain yield. When there is an unlimited irrigation water condition or a 500-mm irrigation water condition for a season, the best grain yield and WUE can be obtained by an irrigation level of 100%–100%ET_{sw} at both vegetative and reproductive growth stages. Under a 400-mm seasonal irrigation water available condition, the highest grain yield and WUE were simulated at 80%–100%ET_{sw} irrigation levels. With a 300-mm seasonal irrigation water available condition, the highest corn yield and WUE were obtained at irrigation level of 60%–100%ET_{sw}. Meeting 60%ET_{sw} at the vegetative stage and 100%ET_{sw} at the reproductive stage under a 400-mm irrigation water available condition can achieve similar average corn yield as reported in annual surveys in northeastern Colorado (http://quickstats.nass.usda.gov/). The simulated irrigation water was about 350 mm, averaged from 1992 to 2013, which was lower than the average available irrigation water for corn in Colorado (about 430 mm for a normal year) (Frank and Carlson, 1999; http://www.ext.colostate.edu/pubs/crops/04718.html).

References

Ahuja, L.R., D.K. Cassel, R.R. Bruce, and B.B. Barnes. 1989. Evaluation of spatial distribution of hydraulic conductivity using effective porosity data. Soil Sci. 148:404–411. doi:10.1097/00010694-198912000-00002

Allen, R.G., L.S. Pereira, D. Raes, and M. Smith. 1998. Crop evapotranspiration: Guidelines for computing crop water requirements. Irrig. Drain. Pap. 56. FAO, Rome.

Allen, R.G., L.S. Pereira, M. Smith, D. Raes, and J.L. Wright. 2005. FAO-56 dual crop coefficient method for estimating evaporation from soil and application extensions. J. Irrig. Drain. Eng. 131:2–13. doi:10.1061/(ASCE)0733-9437(2005)131:1(2)

Allen, R.G., J.L. Wright, W.O. Pruitt, L.S. Pereira, and M.E. Jensen. 2007. Water requirements. In: G.J. Hoffman et al., editors, Design and operation of farm irrigation systems. 2nd ed. Chap. 8. ASAE, St. Joseph, MI. p. 208–288.

Angus, J.F., and A.F. van Herwaarden. 2001. Increasing water use and water use efficiency in dryland wheat. Agron. J. 93:290–298. doi:10.2134/agronj2001.932290x

Behera, S.K., and R.K. Panda. 2009. Integrated management of irrigation water and fertilizers for wheat crop using field experiments and simulation modeling. Agric. Water Manage. 96:1532–1540. doi:10.1016/j.agwat.2009.06.016

Brooks, R.H., and A.T. Corey. 1964. Hydraulic properties of porous media. Hydrol. Pap. 3, Colorado State Univ., Fort Collins.

Çakir, R. 2004. Effect of water stress at different development stages on vegetative and reproductive growth of corn. Field Crops Res. 89:1–16. doi:10.1016/j.fcr.2004.01.005

Doherty, J. 2010. PEST: Model-independent parameter estimation. Watermark Numerical Computing, Brisbane, Australia. http://www.pesthomepage.org (accessed 22 May 2014).

Domínguez, A., J.A. de Juan, J.M. Tarjuelo, R.S. Martínez, and A. Martínez-Romero. 2012. Determination of optimal regulated deficit irrigation strategies for maize in a semi-arid environment. Agric. Water Manage. 110:67–77. doi:10.1016/j.agwat.2012.04.002

Fang, Q., L. Ma, Q. Yu, L.R. Ahuja, R.W. Malone, and G. Hoogenboom. 2010. Irrigation strategies to improve the water use efficiency of wheat–maize double cropping systems in North China Plain. Agric. Water Manage. 97:1165–1174. doi:10.1016/j.agwat.2009.02.012

Farahani, H.J., and D.G. DeCoursey. 2000. Evaporation and transpiration processes in the soil-residue-canopy system. In: L.R. Ahuja et al., editors, The root zone water quality model. Water Resour. Publ., Highlands Ranch, CO. p. 51–80.

Fereres, E., and M.A. Soriano. 2007. Deficit irrigation for reducing agricultural water use. J. Exp. Bot. 58:147–159. doi:10.1093/jxb/erl165

Frank, A., and D. Carlson. 1999. Colorado's net irrigation requirements for agriculture, 1995. Colorado Department of Agriculture, Denver.

Geerts, S., and D. Raes. 2009. Deficit irrigation as an on-farm strategy to maximize crop water productivity in dry areas. Agric. Water Manage. 96:1275–1284. doi:10.1016/j.agwat.2009.04.009

Heng, L.K., S. Asseng, K. Mejahed, and M. Rusan. 2007. Optimizing wheat productivity in two rain-fed environments of the West Asia–North Africa region using a simulation model. Eur. J. Agron. 26:121–129. doi:10.1016/j.eja.2006.09.001

Howell, T.A. 2001. Enhancing water use efficiency in irrigated agriculture. Agron. J. 93:281–289. doi:10.2134/agronj2001.932281x

Kloss, S., R. Pushpalatha, K.J. Kamoyo, and N. Schuetze. 2012. Evaluation of crop models for simulating and optimizing deficit irrigation systems in arid and semi-arid countries under climate variability. Water Resour. Manage. 26:997–1014.

Lorite, I.J., L. Mateos, F. Orgaz, and E. Fereres. 2007. Assessing deficit irrigation strategies at the level of an irrigation district. Agric. Water Manage. 91:51–60. doi:10.1016/j.agwat.2007.04.005

Ma, L., G. Hoogenboom, L.R. Ahuja, J.C. Ascough, II, and S.A. Saseendran. 2006. Evaluation of the RZWQM-Ceres-Maize hybrid model for maize production. Agric. Syst. 87:274–295. doi:10.1016/j.agsy.2005.02.001

Ma, L., T.J. Trout, L.R. Ahuja, W.C. Bausch, S.A. Saseendran, R.W. Malone, and D.C. Nielsen. 2012. Calibrating RZWQM2 model for maize responses to deficit irrigation. Agric. Water Manage. 103:140–149. doi:10.1016/j.agwat.2011.11.005

Molden, D., T. Oweis, P. Steduto, P. Bindraban, M.A. Hanjra, and J. Kijne. 2010. Improving agricultural water productivity: Between optimism and caution. Agric. Water Manage. 97:528–535. doi:10.1016/j.agwat.2009.03.023

Nash, J.E., and J.V. Sutcliffe. 1970. River flow forecasting through conceptual models. Part I—A discussion of principles. J. Hydrol. 10:282–290. doi:10.1016/0022-1694(70)90255-6

Nielsen, D.C., A.D. Halvorson, and M.F. Vigil. 2010. Critical precipitation period for dryland maize production. Field Crops Res. 118:259–263. doi:10.1016/j.fcr.2010.06.004

Nielsen, D.C., M.F. Vigil, and J.G. Benjamin. 2009. The variable response of dryland corn grain yield to soil water content at planting. Agric. Water Manage. 96:330–336. doi:10.1016/j.agwat.2008.08.011

Nolan, B.T., R.W. Malone, L. Ma, C.T. Green, M.N. Fienen, and D.B. Jaynes. 2011. Inverse modeling with RZWQM2 to predict water quality. In: L.R. Ahuja and L. Ma, editors, Methods of introducing system models into agricultural research. Adv. Agric. Syst. Model. 2. ASA, CSSA, and SSSA, Madison, WI. p. 327–363.

Pereira, L.S., P. Paredes, E.D. Sholpankulov, O.P. Inchenkova, P.R. Teodoro, and M.G. Horst. 2009. Irrigation scheduling strategies for cotton to cope with water scarcity in the Fergana Valley, Central Asia. Agric. Water Manage. 96:723–735. doi:10.1016/j.agwat.2008.10.013

Saseendran, S.A., L.R. Ahuja, L. Ma, D.C. Nielsen, T.J. Trout, A.A. Andeles, J.L. Chavez, and J. Ham. 2014. Enhancing the water stress factors for simulation of corn in RZWQM2. Agron. J. 106:81–94. doi:10.2134/agronj2013.0300

Saseendran, S.A., L.R. Ahuja, D.C. Nielsen, T.J. Trout, and L. Ma. 2008. Use of crop simulation models to evaluate limited irrigation management options for corn in a semiarid environment. Water Resour. Res. 44:W00E02. doi:10.1029/2007WR006181

Shaw, R.H. 1977. Climatic requirement. In: G.F. Sprague, editor, Corn and corn improvement. ASA, CSSA, and SSSA, Madison, WI. p. 609–638.

Shuttleworth, W.J., and J.S. Wallace. 1985. Evaporation from sparse crops—An energy combination theory. Q. J. R. Meteorol. Soc. 111:839–855. doi:10.1002/qj.49711146910

Singh, A.K., R. Tripathy, and U.K. Chopra. 2008. Evaluation of CERES-Wheat and CropSyst models for water–nitrogen interactions in wheat crop. Agric. Water Manage. 95:776–786. doi:10.1016/j.agwat.2008.02.006

Steduto, P., T.C. Hsiao, D. Raes, and E. Fereres. 2009. AquaCrop—The FAO crop model to simulate yield response to water: I. Concepts and underlying principles. Agron. J. 101:426–437. doi:10.2134/agronj2008.0139s

Stockle, C.O., and L.G. James. 1989. Analysis of deficit irrigation strategies for corn using crop growth simulation. Irrig. Sci. 10:85–98. doi:10.1007/BF00265686

Trout, T.J., W.C. Bausch, and G.W. Buchleiter. 2010. Water production functions for Central Plains crops. In: Proceedings of the 5th National Decennial Irrigation Conference, Phoenix, AZ. 5–8 Dec. 2010. ASABE, St. Joseph, MI.

Modeling for Best Management of the Effects of Irrigation Frequencies, Initial Water, and Nitrogen on Corn

S.A. Saseendran,* L.R. Ahuja, L. Ma, and T.J. Trout

Abstract

Crop water and nitrogen production functions (CW-NPF) used in irrigation (I) management need to integrate the effects of climate, initial soil water levels at planting (I_w), amounts and frequencies of irrigations, and N levels. Field experiments integrating these factors and their interactions on crop responses are lacking as they are expensive to conduct. The objective of this modeling study was to study the effects of these factors and explore ways of enhancing the corn (*Zea mays* L.) irrigation water use efficiency (IWUE) and nitrogen use efficiency (NUE) and minimizing residual soil and leached N. The study was an extension of the corn experiments conducted in a sandy loam soil near Greeley, CO from 2008 to 2011, using the Root Zone Water Quality Model (RZWQM2). We simulated corn response to varying I_w, I amounts and frequencies, and N levels. For each of four levels of I_w (low, medium, high, and average), irrigation was applied at frequencies of 3, 7, 10, and 14 d to supplement rainfall to meet the potential crop ET demand from 0 to 100% at 10% intervals. N rates were 0 to 300 kg ha^{-1} at 50 kg ha^{-1} intervals. The simulated grain yield was higher at higher I frequencies and I_w. NUE increased with I up to 60% ET replacement but plateaued thereafter. For higher IWUE and NUE with minimum soil residual N, optimum N rates for grain yield for I levels of 40, 50, 60, 70, 80, 90, and 100% ET replacement were 50, 100, 100, 150, 200, 200, and 200 kg N ha^{-1}, respectively. For optimum grain yield, 3- to 7-d irrigations to replace 90% ET losses and N at 200 kg ha^{-1} were the best management choices simulated. The results demonstrate the value of validated process-based crop system models in developing site-specific irrigation and N recommendations for enhanced IWUE and NUE in corn production.

Abbreviations: CW-NPF, crop water and nitrogen production functions; CWPF, crop water production functions; IWUE, irrigation water use efficiency; LAI, leaf area index; IRF, Limited Irrigation Research Farm; NUE, nitrogen use efficiency; NRE, nitrogen recovery efficiencies; PWS, plant water supply; RZWQM2, Root Zone Water Quality Model.

S.A. Saseendran, L.R. Ahuja, and L. Ma, Agricultural Systems Research Unit, USDA-ARS, Fort Collins, CO 80526; T.J. Trout, Water Management Research Unit, USDA-ARS, Fort Collins, CO 80526. *Corresponding author (Saseendran.anapalli@ars.usda.gov).

doi:10.2134/advagricsystmodel5.c2

Practical Applications of Agricultural System Models to Optimize the Use of Limited Water
Lajpat R. Ahuja, Liwang Ma, and Robert J. Lascano, Editors
Advances in Agricultural Systems Modeling, Volume 5. Lajpat R. Ahuja, Series Editor

USDA-ARS

When water supply is limited, increasing demands for water resources from various human enterprises (urbanization, agriculture, and energy production) challenge agriculturists to develop technologies for producing more with less water. Notwithstanding, the freshwater resources available to us today are not being used sustainably according to needs and demands across the user sectors (United Nations Educational, Scientific and Cultural Organization, 2012). Irrigated agriculture, by far, is the largest consumer of all the fresh water we have in the world, comprising about 66% (Falkenmark and Lannerstad, 2005; Scanlon et al., 2007). Improvement of water management practices in agriculture was identified as one of the main strategies to deal with the competing sectoral demands for water (Gordon et al., 2010; Gleick, 2003).

In the semiarid climate of Colorado's Great Plains where precipitation received during the growing season is inadequate to support crop growth for economic production and sustainability, the practice of "limited irrigation" is gaining importance in agriculture (Halvorson et al., 2004; Klocke et al., 2004; Trout et al., 2010). Water is the most important natural resource limiting corn production in these regions (Halvorson et al., 2004), and corn occupies the second position in consuming all the irrigation water supplies in Colorado (Frank and Carlson, 1999). Adaptation to limited irrigation practices in the region can contribute substantially to the long-term sustainability of corn production in the region. The limited irrigation practice of imposing predetermined drought for saving water without compromising economic yield outcome requires research data quantifying accurate responses of the crop to various levels of inputs (crop water production functions [CWPF]). Such CWPF should be derived from long-term field experiments to account for the variations in precipitation and other climatic variables, amounts and frequencies of irrigations, initial soil water at planting at a location, and N inputs and availabilities for plant uptake. However, such long-term average functions for irrigation, based on empirical data at a specific location are very expensive and time consuming to obtain, and hence not easily available in the region but location specific short term field irrigation trials are conducted from time to time. For example, an irrigation trial was conducted at the Limited Irrigation Research Farm (LIRF) in Greeley, CO and quantified response of corn to limited irrigation (Bausch et al., 2011; Trout et al., 2010). Irrigation levels from about fully irrigated (100% ET replacement) to 40% of full irrigation were tested. Simulation models can synthesize and integrate data collected from such available short- or long-term field studies and present a way to extrapolate results to other soils and climates (Knisel and Turtola, 2000; Matthews et al., 2002).

Saseendran et al. (2014) simulated the LIRF corn experiments successfully (across 4 yr, grain yield was simulated with an average RMSE of 623 kg ha^{-1}), using RZWQM2, and integrated it with long-term climate data and developed long-term averaged CWPF for corn in various soils and locations in Colorado. This study demonstrated how a comprehensive agricultural system simulation model can be integrated with location-specific field experiments and long-term climate data to identify limited water irrigation management practices for various locations, soils, and climates across the world. However, this study did not include the effects of some other factors that could influence the CWFPs, such as the frequency of irrigations at any of the ET-based level of irrigation, initial soil water content at planting, and the interacting N levels. In this context, the harvested crop yield is the outcome of dynamic growth and development processes, which depends not only on temporal allocation of water but also on N inputs throughout the growing period. Such responses of crop yield to inputs of both water and N are termed crop water and nitrogen production functions (CW-NPF). The Saseendran et al. (unpublished data, 2014) study was confined only to developing long-term averaged CWPF for corn at unlimited N levels (no nitrogen stress). In addition to water, the management of the associated and interacting nutrients, especially N, can also be addressed with the models.

Data reported in the literature strongly established that the differences in water availability and uptake by the crop due to the imposed drought can lead to substantial differences in soil N supply and demand by the crop, e.g., water stress modified carbon and N metabolism in plants resulting in a decline in N uptake, photosynthesis, and growth (Cornic and Massacci, 1996; Mwanamwenge et al., 1999). Residual N in the soil finds its way into the atmosphere and surface and groundwater bodies through various chemical and physical processes, leading to environmental pollution and degradation and increased N_2O emission potential (Ferguson et al., 1991; Grassini and Cassman, 2012). High NO_3–N not only affects natural systems, but its concentrations in drinking water are a human health hazard, causing methemoglobinemia (Frink et al., 1999). Hence, it is of paramount important to understand water–N interactions and apply the right amount of N commensurate with the amount of water available for uptake by the crop under different soil and climatic regimes. The objectives of this endeavor were to extend the study of Saseendran et al. (unpublished data, 2014) to simulate the effects of irrigation frequency, initial soil water at planting, and various levels of N applications on the long-term averaged corn grain yield, N uptake, residual soil N, and N leaching responses at different levels of ET-based irrigations at LIRF. The RZWQM2 calibrated and validated by Saseendran et al. (2014) along with 21 yr (1992–2012) of weather data continuously recorded at the location was used for the objectives.

Materials and Methods

Experiments

Field trials to quantify corn responses to limited irrigations conducted from 2008 to 2011 at LIRF near Greeley in the Central Great Plains of Colorado were used in the study of calibration and evaluation of the model (Bausch et al., 2011; Trout et al., 2010). For simulating the various processes, the agricultural system models use parameters that remain constant for a given cropping system. Some of these parameters may not be measured in experiments that are being simulated because they are difficult to measure. In this context, calibration is the procedure by which unknown parameters are estimated by matching measured crop performance (e.g., grain yield, biomass, leaf area index [LAI], phenology) variables with their simulated values. Usually this is done with a portion of the crop performance data collected in the experiments. The data not used in the calibration are used for comparing with the model simulations for assessing the accuracies of simulations of the calibrated model. Details of the LIRF experiments were previously described (Ma et al., 2012; Trout et al., 2010). In brief, in the experiments, six irrigation treatments to replace crop ET losses at 100%F, 85%V, 70%F, 70%V, 55%V, and 40%V levels in 4- to 7-d intervals were conducted (V denotes that 20% of the estimated weekly amounts of irrigation required for that treatment during the vegetative growth period were withheld and added to the weekly amounts during the reproductive growth time; F denotes that full estimated weekly amounts were applied without splitting between stages). Soil at the site was predominantly sandy loam under three types, Nunn (fine, smectitic, mesic Aridic Argiustolls), Olney (fine-loamy, mixed, superactive, mesic Ustic Haplargids), and Otero (coarse-loamy, mixed, superactive, calcareous, mesic Aridic Ustorthents). Corn ('Dekalb 52-59') with 102-d relative maturity was planted in a randomized block design with four replicates at an average rate of 81,000 seeds ha^{-1} with 0.76 m row spacing on 12 May 2008, 11 May 2009, 11 May 2010, and 3 May 2011 and harvested on 6 Nov. 2008, 12 Nov. 2009, 19 Oct. 2010, and 25 Oct. 2011, respectively. Nitrogen applications were based on soil samples taken for soil fertility status. Irrigation water was delivered to the corner of each plot and applied through polyethylene header pipes to drip irrigation tubing (16-mm diameter, thick walled tubing with 1.1 L h^{-1} conventional inline emitters spaced 30 cm apart) laid on the surface near each plant row. Urea–ammonium nitrate fertilizer was applied at planting and then added to irrigation water and applied during the growing seasons as needed to eliminate N stress in the experiments. Daily weather data from 1992 to 2012 recorded at the Colorado Agricultural Meteorological Network station (http://ccc.atmos.colostate.edu/~coagmet/) at the site was used. The mean annual precipitation (1992–2010) at the location was 27 cm, of which 20 cm was received during the corn growing season from May to October.

RZWQM2 Model

The Root Zone Water Quality Model (RZWQM2) was used in the study. It is process oriented and simulates the effects of water (irrigation amount and interval), soil (tillage and crop residue and other C), fertilizers (N), pesticides (herbicides and insecticides), and crop (species and cultivars) management practices on crop production and water quality on a daily time step (Ahuja et al., 2000; Ma et al., 2009). The CSM-CERES-Maize v4.0 model available for simulation within RZWQM2 was used for simulation of corn (Hoogenboom et al., 1991; http://ars.usda.gov/services/software/download.htm?softwareid=412; Jones et al., 2003; Ma et al., 2005, 2006, 2009). The model was extensively tested using data from corn field experiments conducted in the Great Plains and extended the results for managing dryland and irrigated cropping systems (Ma et al., 2003; Saseendran et al., 2005, 2008, 2009). Less than normal crop photosynthetic and transpiration (gas exchange) responses to water input levels were reported in some of these studies. Subsequently, Saseendran et al. (2014) obtained more accurate simulations in corn responses to limited water irrigations by modifying the water stress factor for processes related to photosynthesis (SURFAC) in RZWQM2-CSM-CERES using the daily potential root water uptake (TRWUP) calculated by the approach of Nimah and Hanks (1973) and accounted for water stress due to additional heating of canopy from unused energy of potential soil evaporation. We used the modified model for simulating yield responses to water and N inputs in this study.

The RZWQM2 model needs inputs of soil water, N, and carbon content by soil layer at the start of the simulation. Other input requirements of RZWQM2 include weather (driving variables), soil physical and hydraulic parameters, soil, crop, water, and N management information, and soil initial conditions. The minimum weather needed for the simulations are daily solar irradiance, maximum and minimum temperature, wind speed, relative humidity, and precipitation representing the experimental area; the soil physical properties required are soil profile depth and horizons (layers), soil texture, bulk density, and organic matter content; and the soil hydraulic properties required are water retention curves and saturated hydraulic conductivity of each soil horizon represented in the form of the Brooks and Corey equations (Brooks and Corey, 1964). The management data required are tillage dates and methods for soil; planting date, density, depth, and row spacing for the crop; dates and amounts of irrigation for water; and amount and type of fertilizer for N applications.

The minimum experimental data needed for calibrating the model for simulating a crop cultivar are grain yield and biomass at maturity; crop biomass and LAI at different growth stages; phenology dates, rooting depth, and distribution in the profile; and soil water content measurements. Simulating cropping systems using RZWQM requires careful iterative calibration of the soil water, followed by the nitrogen (N), and finally the plant growth component (CSM-CERES-Maize 4.0 model). To simulate a specific corn hybrid, the CERES-Maize 4.0 model requires six cultivar parameters (Jones et al., 2003). If the simulation of crop growth at a calibration step is not satisfactory, the whole sequence of calibration is repeated to obtain more accurate simulations (Ma et al., 2011). In this study, we used the CERES-Maize 4.0 model calibrated and tested within RZWQM2 for simulations of the LIRF experiments by Saseendran et al. (2014). Therefore, no further calibration or validation of the model was attempted in this study.

Development of CWPF and CW-NPF for Corn Using Long-Term Simulations

The corn crop was simulated for 21 yr using the RZWQM2 calibrated and validated model for the location and soil as described above. Weather data recorded from 1992 to 2012 in Greeley by the Colorado Agricultural Meteorological Network (http://ccc.atmos.colostate.edu/~coagmet/) were used in the simulations. Initial conditions, irrigation frequencies and amounts, and N rates used in the simulations are described below.

Irrigation Frequency, Initial Soil Water, and N Rates

In the simulations, corn was planted every year on 2 May. In each crop season, I_w and N initial conditions were reset to the same values at planting. The I_w for the long-term simulations were termed average low, medium, and high on the basis of the average measured plant available water content (soil water above plant wilting point limited to field capacity) at planting in the soil layers in six irrigation treatments from 2008 to 2011 in the LIRF experiments. When low I_w and medium I_w were defined as 50% and 75% of average I_w, high I_w was assumed to be 150% of average I_w limited to the field capacity water content of the soil layer.

Irrigations and N applications were also started on 2 May (day of planting) and continued until 10 September every year as followed in the LIRF experiments. Simulated irrigations were based on the amount of water needed for meeting the modeled ET demands of 0 (dryland or no irrigation), 10, 20, 30, 40, 50, 60, 70, 80, 90, and 100% accounting for precipitation in the previous 3, 7, 10, and 14 d. In RZWQM2, ET on a daily basis is calculated by adding together the actual crop water uptake and soil evaporation (Ahuja et al., 2000). Actual crop water uptake (transpiration) is calculated as the minimum of potential crop water uptake and potential transpiration on a daily basis. Potential crop water uptake is simulated using the Nimah and Hanks (1973) approach (Saseendran et al., 2014). Potential transpiration and soil evaporation are modeled using the "extended Shuttleworth–Wallace ET model," which is the Pennman–Monteith reference ET model modified for incomplete canopy cover and soil residue cover (Farahani and Ahuja, 1996: Farahani and DeCoursey, 2000). Actual soil evaporation is calculated as the minimum of potential soil evaporation and actual soil evaporation potential. Actual soil evaporation potential is determined by modeling the soil water redistribution using the Richards equation (Ahuja et al., 2000). Irrigations in the model simulations were applied with 100% efficiency by treating them as low-intensity rain events [i.e., effective irrigation (I_{eff}) = applied irrigation (I)]. Simulated N treatments were at rates of 0, 50, 100, 150, 200, 250, and 300 kg ha^{-1} and were applied in allotments of 50 kg ha^{-1} every 20 d starting at planting. Irrigation intervals (frequencies) were simulated at 3, 7, 10, and 14 d.

Development of Average CW-NPF

For the development of CW-NPF, we computed average and standard deviations of the corn growing season precipitation, grain yield, biomass, ET, amount of irrigation, plant available water in the soil at planting, runoff, plant N uptake (NUP), residual N in the soil profile at harvest and N leached beyond the root zone (2 m) from the 21 yearly simulations at different levels and frequencies of irrigation, I_w, and N application rates. Effective precipitation and plant water supply to the crop (PWS) were calculated as

$$P_{eff} = P - R - D \quad [1]$$

$$\text{PWS} = I_{eff} + \text{PAW} + P_{eff} \quad [2]$$

where P_{eff} is effective precipitation, P is precipitation, R is runoff, D is deep percolation, I_{eff} is effective irrigation taken as equal to the simulated irrigation assuming negligible amounts of surface runoff and deep percolation losses in the semiarid climate of the location, and PAW is plant available water in the soil.

The IWUE, NUE, and nitrogen recovery efficiencies (NRE) were calculated as

$$\text{IWUE} = \frac{\left(Y_{\text{i}} - Y_{\text{Noi}}\right)}{I} \quad [3]$$

$$\mathrm{NUE} = \frac{\left(Y_{\mathrm{Ni}} - Y_{\mathrm{No}}\right)}{N_{\mathrm{i}}} \quad [4]$$

$$\mathrm{NRE} = \frac{\left(\mathrm{NUP}_{\mathrm{Ni}} - \mathrm{NUP}_{\mathrm{No}}\right)}{N_{\mathrm{i}}} \quad [5]$$

where Y_i is arithmetic average corn yield in response to irrigation (I) from 1992 to 2012; Y_{Noi} is average yield in response to zero irrigation; N_i is N application rate; Y_{Ni} is average yield in response to a given N rate of N_i; Y_{No} is average yield in response to zero N application; $\mathrm{NUP}_{\mathrm{Ni}}$ is average NUP at a given N rate of N_i; and $\mathrm{NUP}_{\mathrm{No}}$ is NUP at zero N application.

Results and Discussion

Model Calibration and Validation for Simulation of Corn in the LIRF Experiments

Precise quantification and simulation of corn growth responses to water stress in system models is key for its applications in limited irrigation management decision support. In this context, Saseendran et al. (2014) modified water stress simulations in RZWQM2 for more accurate corn crop responses to water input variations as encountered in the LIRF experiments. They further tested the model and achieved reasonable accuracy, taking into account the uncertainties in measurements, in its detailed simulations of the LIRF experiments from 2008 to 2011 (e.g., grain yield in Fig. 2–1). A detailed comparison of the simulated and measured crop growth and yield results is not repeated in this paper. In brief, in the 4-yr study with six irrigation treatments each year, grain yields were simulated with an average RMSE of 623 kg ha^{-1} (Fig. 2–1). The 4-yr simulations of grain yield responses to ET (CWPF in terms of yield vs. ET) as well as the field estimated ET–grain yield responses were close to each other (Fig. 2–2). The simulated 21-yr average CWPF in response to different levels of ET-based weekly irrigation applications, in addition to precipitation received during the crop period, followed the measured CWPF (yield vs. irrigations plus precipitation) from 2008 to 2011 in the LIRF experiments accurately. Results of the simulations rendered confidence in the use of the model for its applications in predictions of crop response to soil–crop–water–nutrient management outside those conducted in the LIRF experiments. Saseendran et al. (unpublished data, 2014) developed long-term average CWPF for corn at the location for irrigation management. As noted earlier, the results presented below are an extension of that simulation study of corn CW-NPF, N uptake, residual soil N at harvest and N leaching responses to varying initial planting soil water levels, N rates, and irrigation frequencies in the

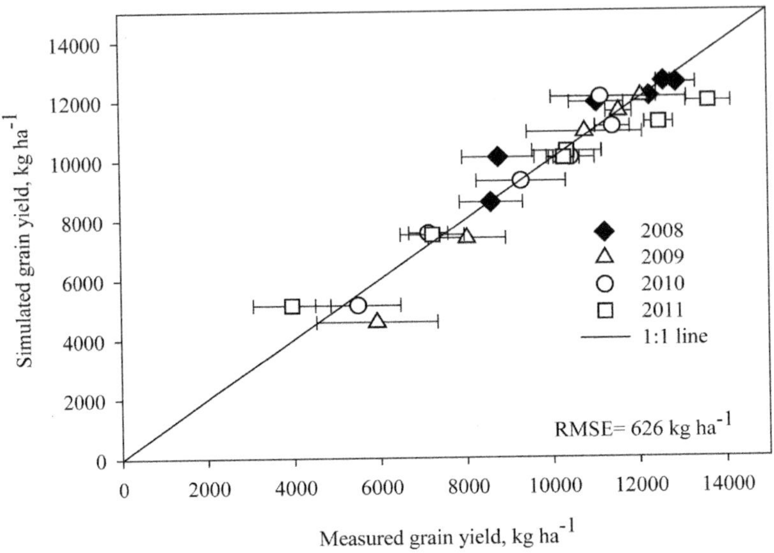

Fig. 2–1. Measured and simulated corn yield from 2008 to 2011 in the Limited Irrigation Research Farm experiments in Greeley, CO. Error bars indicate one standard deviation in the measured data. Adapted with permission from Saseendran et al. (2014).

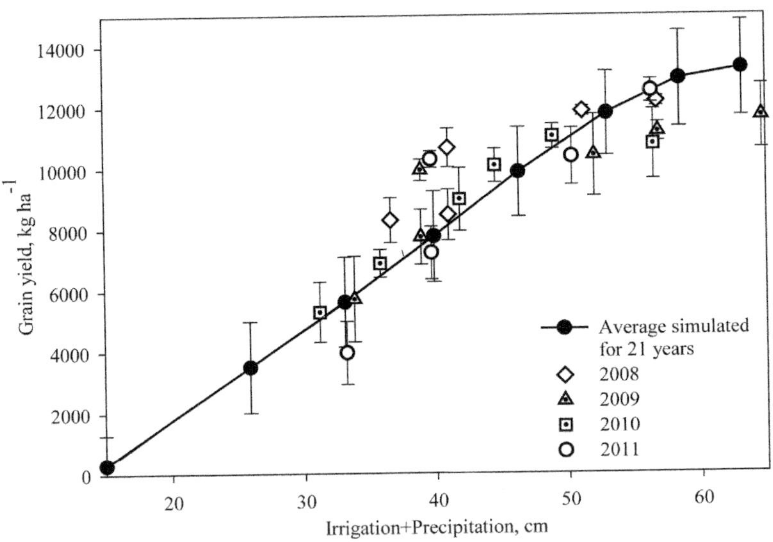

Fig. 2–2. Measured and simulated average crop water production functions in Greeley, CO. Measured corn yields were from 2008 through 2011, and simulated were the 21-yr average grain yield responses to precipitation and irrigations at 7-d intervals simulated by the Root Zone Water Quality Model. Error bars indicate one standard deviation from the mean of the long-term simulations. In addition to precipitation and applied irrigation, stored soil water at planting also contributed to the grain yields. Adapted with permission from Saseendran et al. (unpublished data, 2014).

LIRF experiments. It is important to note that in the LIRF experiments, N was applied, on the basis of soil N tests, to supply ample N to the plants for potential growth. Therefore, there were no N treatments aimed at optimizing N application rates in the limited water irrigation scenarios developed. Also, the effects of varying irrigation frequencies and initial soil water levels on corn responses were not studied.

Long-Term CWPF Responses to Frequencies of Irrigation at Various I_w Levels

In limited water irrigation management, it is imperative that the CWPF used for irrigation decision support take into account not only the amount but also the intervals of water applications. In the long-term simulations, corn yields decreased significantly ($p < 0.05$) with irrigation frequencies from 3 to 14 d (Fig. 2–3). N rates were constant at the rate of 200 kg ha^{-1} across all treatments for N stress free simulations at LIRF. In general, across low, medium, average, and high I_w levels and irrigation frequencies of 3 and 7 d, yield response to irrigation amounts from 0 to 80% ET replacement levels was approximately linear and pla-

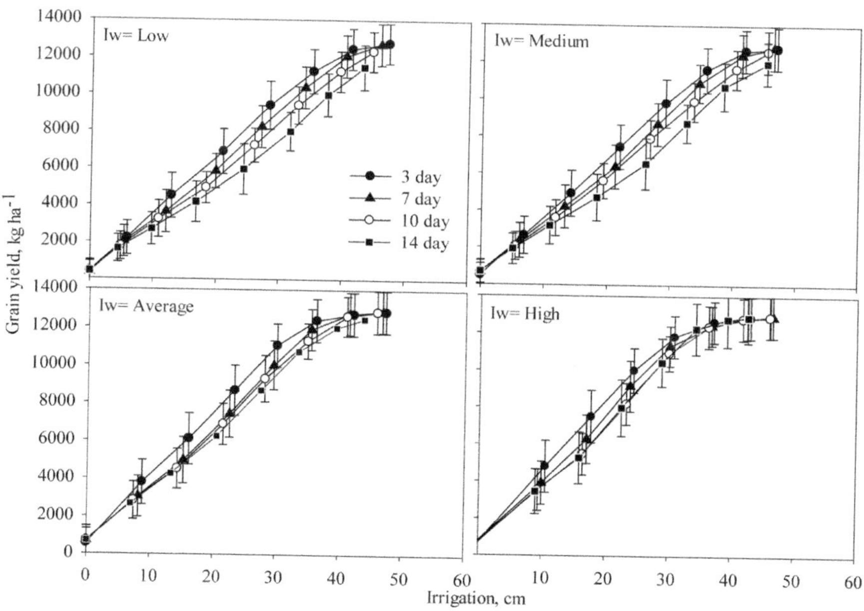

Fig. 2–3. Effect of irrigations of 3-, 7-, 10-, and 14-d frequencies and low, medium, average, and high initial soil water (I_w) levels on average (1992–2012) crop water production functions (yield vs. irrigations). Symbols plotted along the curves from left to right correspond to irrigations at 0, 40, 50, 60, 70, 80, and 100% ET replacement levels. Error bars indicate one standard deviation from the mean. N levels were constant across all the simulations of 200 kg ha^{-1}.

teaued with higher irrigations. Many experiments have shown that when other factors are not extremely limiting, the grain yield produced and the water consumed (CWPF) by a given plant species are linearly related (e.g., Hanks, 1983; Stewart and Hagan, 1973). De Wit (1958) developed the first analytical approach to formalize the relationship. It has also been observed that when only water is limiting, the grain yield response of most crops (CWPF) rises initially to a maximum then falls off with further application of water (Geerts and Raes, 2009; Stewart and Hagan, 1973). Averaged across irrigation levels amounts (at 0 to 100% ET replacement) and different initial water levels (low I_w, medium I_w, average I_w, and high I_w), the decline in grain yield with 7-, 10-, and 14-d irrigations relative to yield in response to 3-d irrigations were 5, 11, and 18%, respectively. The yield decline increased gradually from 0 to about 70% ET irrigation level, then reduced at higher irrigation levels.

In the case of the average and high I_w levels, plateauing of linear increase in yield response at about 80% ET replacement irrigation level was seen up to 14-d irrigation frequencies (Fig. 2–3). These higher I_w levels reduced the yield reduction due to lower irrigation frequencies considerably. Yield decrease due to irrigation frequencies of 10 and 14 d were not significant showing the dominance of precipitation distribution over those of irrigation at these frequencies ($p > 0.05$). Effects of irrigation frequencies on CWPF of grain yield vs. plant water supply (PWS) under different I_w levels remained more or less similar to CWPF of yield vs. irrigation (Fig. 2–4). However, effects of irrigation frequencies on CWPF of yield as a function of ET was less (Fig. 2–5) and as function of transpiration (T) were insignificant ($p > 0.05$) (Fig. 2–6), which indicates that lower irrigation frequencies reduced yields by reducing T. Looking at the overall yield performance at optimum N levels (simulated at an N level of 200 kg ha^{-1}), across irrigation frequencies simulated, the 3- and 7-d intervals promise higher yield responses.

Long-Term CWPF Responses to I_w Levels under Different Irrigation Intervals

Grain yields increased significantly ($p < 0.05$) with increasing I_w levels from low to medium, average, and high I_w under the 3- to 14-d irrigation frequencies simulated (Fig. 2–7). Averaged across the 3- to 14-d irrigation frequencies and 0 to 100% ET replacement levels, the yield increases relative to low I_w were 7, 18, and 26%, respectively, for medium, average, and high I_w. In general, the rate and pattern of yield increase with irrigation amounts remained more or less similar with irrigation frequencies of 3, 7, 10, and 14 d (Fig. 2–7). With low and medium I_w levels and irrigation frequencies of 3 and 7 d, the yield response to irrigation amounts from 0 to 80% ET replacement levels was approximately linear and plateaued

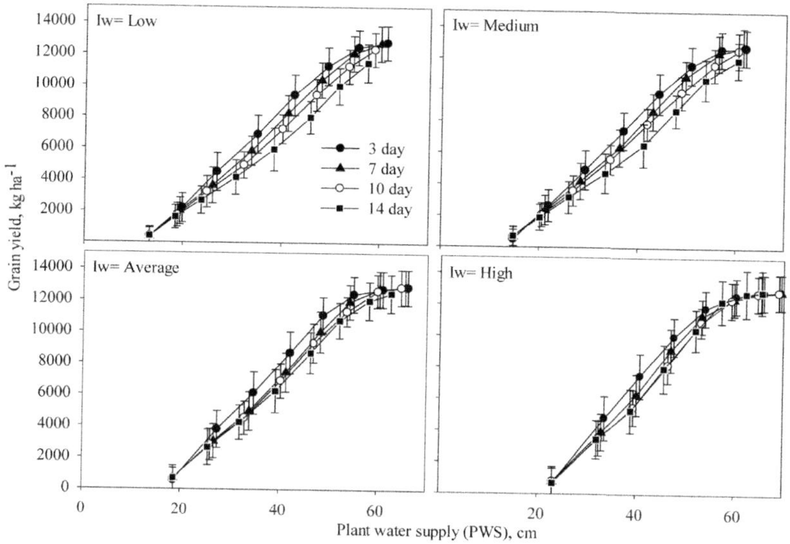

Fig. 2–4. Effect of plant water supply (PWS) that includes irrigation at 3-, 7-, 10-, and 14-d frequencies and low, medium, average, and high initial soil water (I_w) levels on average (1992–2012) crop water production functions (yield vs. PWS) for corn. Symbols plotted along the curves from left to right correspond to irrigations at 0, 40, 50, 60, 70, 80, and 100% ET replacement levels. Error bars indicate one standard deviation from the mean. N levels were constant across all the simulations of 200 kg ha^{-1}.

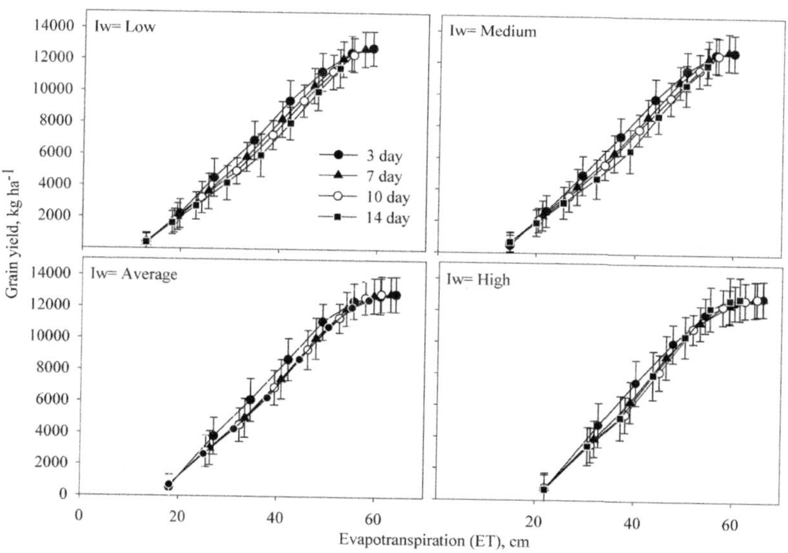

Fig. 2–5. Effect of irrigations of 3-, 7-, 10-, and 14-d frequencies and low, medium, average, and high initial soil water (I_w) levels on average (1992–2012) crop water production functions (yield vs. ET) for corn. Symbols plotted along the curves from left to right correspond to irrigations of 0, 40, 50, 60, 70, 80, and 100% ET replacement levels. N levels were constant across all the simulation at 200 kg ha^{-1}. Error bars indicate one standard deviation from the mean.

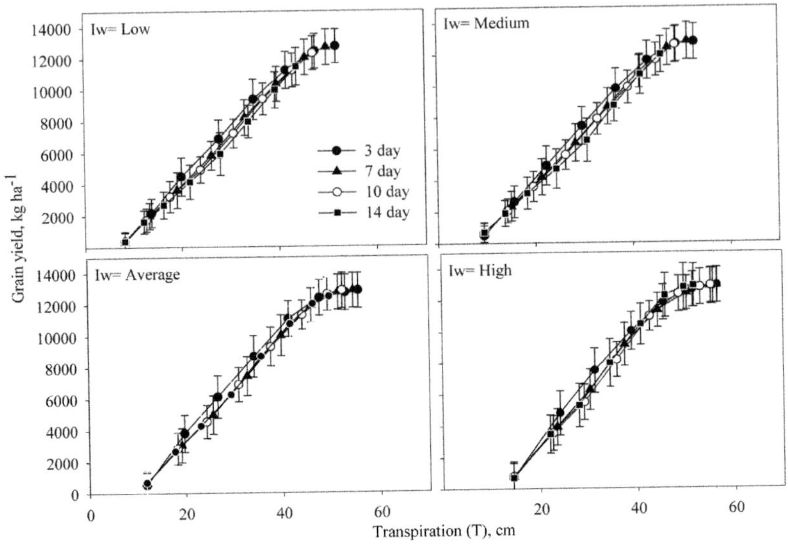

Fig. 2–6. Effect of irrigations of 3-, 7-, 10-, and 14-d frequencies and low, medium, average, and high initial soil water (I_w) levels on average (1992–2012) crop water production functions (yield vs. transpiration [T]) for corn. Symbols plotted along the curves from left to right correspond to irrigations at 0, 40, 50, 60, 70, 80, and 100% ET replacement levels. N levels were constant across all the simulation at 200 kg ha^{-1}. Error bars indicate one standard deviation from the mean.

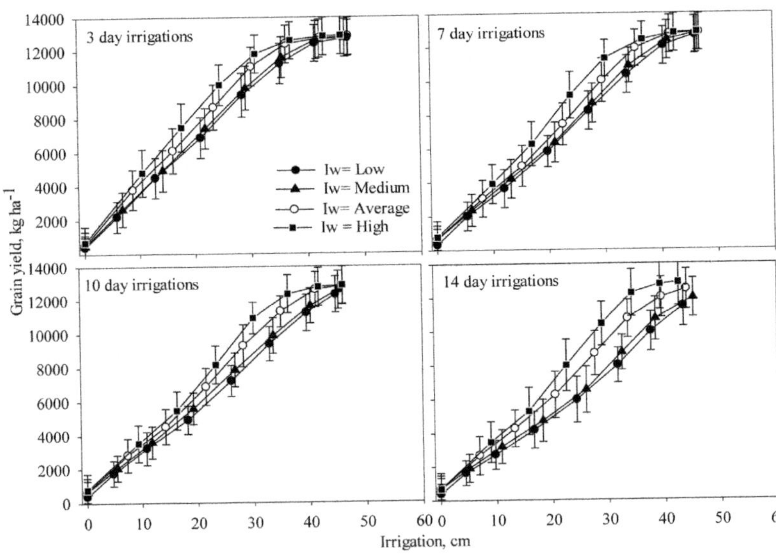

Fig. 2–7. Effect of low, medium, average, and high initial soil water (I_w) levels and 3-, 7-, 10-, and 14-d irrigation frequencies on average (1992–2012) crop water production functions (yield vs. irrigations). Symbols plotted along the curves from left to right correspond to irrigations at 0, 40, 50, 60, 70, 80, and 100% ET replacement. N levels were constant across all the simulations at 200 kg ha^{-1}. Error bars indicate one standard deviation from the mean.

with higher irrigations. With lower irrigation frequencies at 10 and 14 d under low and medium I_w levels, the yield response remained linear with irrigation amounts from 0 to 100% ET replacement (Fig. 2–7). The higher I_w values increased yields at 0 to about 80% ET irrigation levels and did not increase yields at 90 and 100% ET irrigation levels for 3- and 7-d frequencies. The yield increase was higher at higher amounts of irrigation, up to the maximum at the 70% ET irrigation level. For 10- and 14-d frequencies, the yield increase continued up to the 100% ET irrigation level. The higher I_w values obviously provided more water for growth and yield (and for ET and T) at lower irrigation levels, but the benefit was highest at the 70% ET level. However, the higher I_w levels contributed to more ET and T loss from the system and as such were less water use efficient compared with lower I_w levels (data not shown). At 90 to 100% levels of irrigation, the initial soil water can be used to cut back on the water applied.

Nitrogen Uptake Responses to I_w Levels, Irrigation Frequencies, and N Rates

In the simulations, N uptakes were found to increase approximately linearly with increasing irrigation levels from 0 to 70% ET replacement levels (Fig. 2–8). In response to N applied at 200 kg ha^{-1}, N uptake across different initial water levels, irrigation frequencies, and irrigation levels ranged between 27 kg ha^{-1} (under rainfed conditions) to 240 kg ha^{-1} (under irrigations at 100% ET). Uptakes started declining with irrigation levels above 70% ET replacement and plateaued. However, at low I_w levels the linear increase of uptakes was found to decline earlier at irrigations at 60% ET levels. By and large, N uptake increased with increasing I_w levels along with the increasing irrigations. Averaged across 0 to 70% ET level irrigations and 3- to 14-d irrigation frequencies, N uptakes increased by 12, 35, and 54%, respectively, for medium, average, and high I_w levels compared with low I_w levels. The increase in N uptakes with the increase in I_w declined with an increase in the irrigation level above 70% ET. Differences in N uptakes with different I_w levels disappear at irrigations of 100% ET replacement. Several field experiments had shown the significant effects of reduced N supply due to soil water deficit on the leaf area, plant growth, and N uptake (Martin et al., 1982; McCullough et al., 1994; Pandey et al., 1984, 2000a, 2000b).

In general, N uptake decreased with increased time between irrigations. This was a consequence of the simulated increase in the number of days with water stress associated with the increased frequency of irrigation. Xu et al. (2006) reported a water stress induced decline in N uptake and reduced crop growth in corn with decreasing frequency of irrigations. Oktem et al. (2003) reported similar observations with interval of irrigation in winter wheat. Compared with

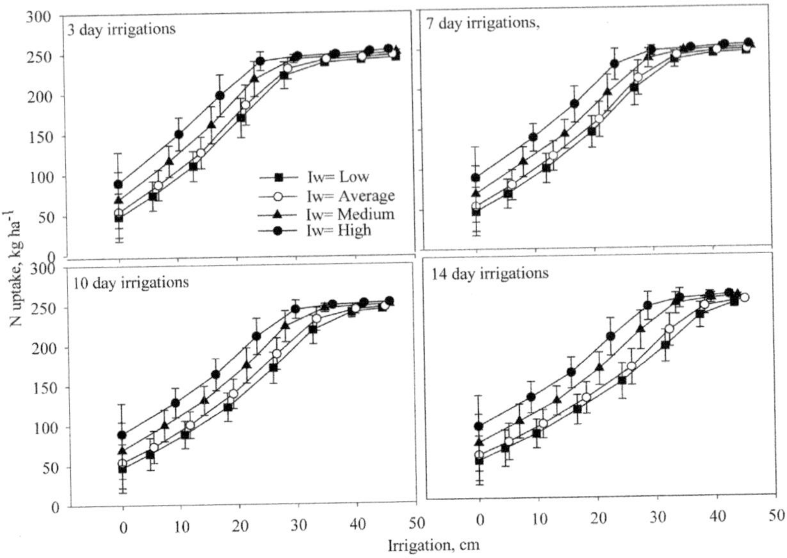

Fig. 2–8. Average (1992–2012) corn N uptake in response to low, medium, average, and high initial soil water (I_w) levels and 3-, 7-, 10-, and 14-d irrigation frequencies to replacement 0, 40, 50, 60, 70, 80, 90, and 100% ET. Symbols plotted along the curves from left to right correspond to irrigations at 0, 40, 50, 60, 70, 80, and 100% ET replacement levels. N levels were constant across all the simulations at 200 kg N ha^{-1}. Error bars indicate one standard deviation from the mean.

the 3-d irrigation frequency, the decreases in N uptakes due to irrigations at 7-, 10-, and 14-d frequencies were 8, 16, and 22%, respectively. Also, the simulated increase in N uptake was more dependent on the applied fertilizer than the irrigation water level (Fig. 2–9). The increase in N uptake at N rates > 200 kg ha^{-1} was substantial only at higher irrigations at 80% ET replacement and above. N uptake decreased with N rates < 200 kg ha^{-1} in a consistent manner, e.g., at the average I_w level, in response to N rates of 50, 100, and 150 kg ha^{-1}, N uptake peaked with irrigation at 40, 50, and 60% ET, respectively.

N recovered from the soil in terms of crop uptake expressed as NRE increased quadratically (equation not shown) with the irrigation amount (Fig. 2–10). On average, the maximum NRE attained with different I_w levels and irrigation frequencies was less than 90%, and this was achieved at lower irrigation levels with lower N rates. For example, at average I_w level with seven irrigation frequencies, simulated NRE reached 90% at irrigation levels of 60, 60, 70, 80, and 90% ET replacements for N rates of 50, 100, 150, 200, and 250 kg ha^{-1}, respectively (Fig. 2–10). However, for the 300 kg N ha^{-1} rate, NRE was on average 80% of those achieved with N rates of 50, 100, 150, 200, and 250 kg ha^{-1} and irrigations at 100% ET replacement.

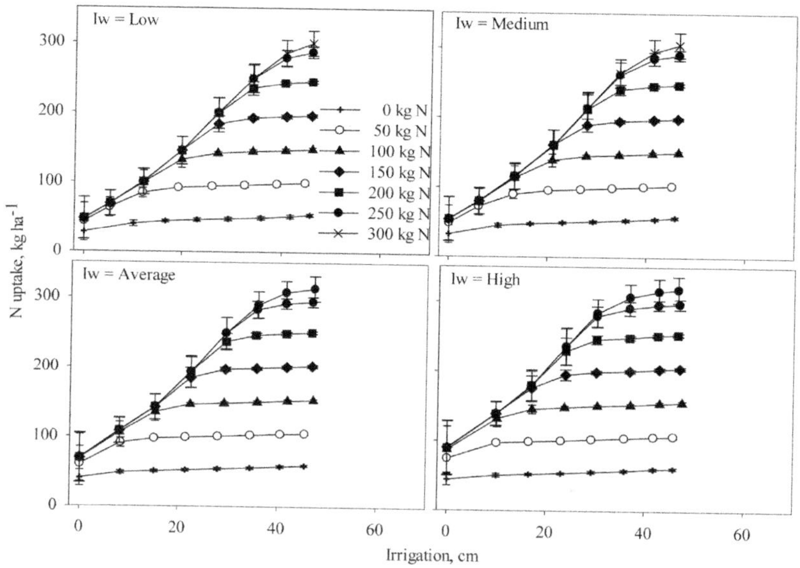

Fig. 2–9. Average (1992–2012) crop N uptake in response to 0, 50, 100, 150, 200, 250, and 300 kg ha^{-1} N applications and 7-d irrigations to replacement 40, 50, 60, 70, 80, 90, and 100% ET under low, medium, average, and high initial soil water (I_w). Error bars indicate one standard deviation from the mean.

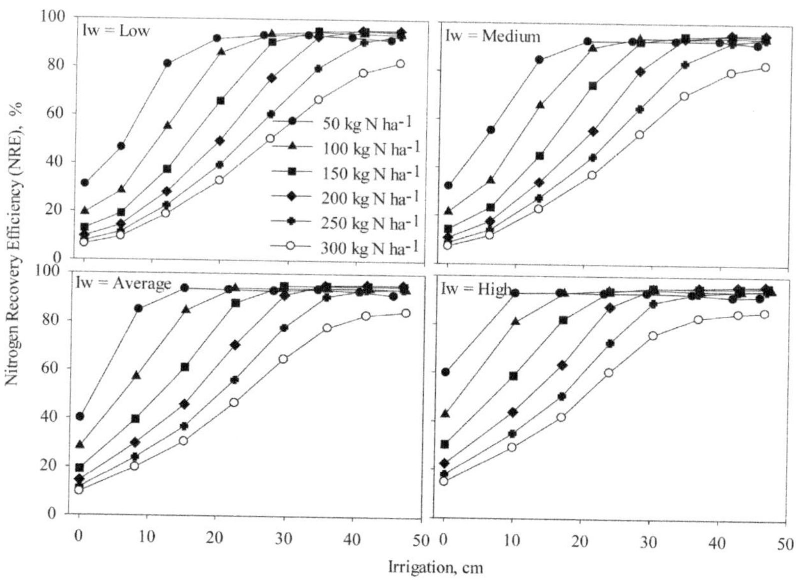

Fig. 2–10. Average (1992–2012) corn N recovery efficiency (NRE) in response to N levels of 0, 50, 100, 150, 200, 250, and 300 kg ha^{-1} and 7-d irrigations to replacement 40, 50, 60, 70, 80, 90, and 100% ET (symbols along the curves from left to right) under low, medium, average, and high initial soil water (I_w).

Residual N Responses to I_w Levels, Irrigation Frequencies, and N Rates

Plant water supply in the soil in terms of the initial water in the soil profile at planting and the precipitation and irrigation inputs during the crop growth season modify both N availability and the uptake by the plant. As a result, more N remaining in the soil at crop harvest in response to deficit irrigations was reported (e.g., Fapohunda and Hussain, 1990; Pandey et al., 2000a, 2000b). In the simulations, in response to an N rate of 200 kg ha^{-1}, N leached out of the 200-cm soil profile was found to be of negligible magnitude (<0.5 kg ha^{-1}) (data not shown). This was mainly due to the low precipitation received during the crop season and irrigation water applied only to meet the ET demand of the crop in the semiarid climate of the location. In this regard, in general, the amount of N leached from agricultural soils was reported to depend mainly on the precipitation regime when irrigation was applied only to supplement the rainwater to meet crop demand (Poch-Massegú et al., 2014). In the 21-yr simulation period in this study, the average growing season precipitation was 149 mm. Residual N in the soil at harvest increased consequent to N input levels above the crop need for growth under water stress associated with low irrigation levels (Fig. 2–11). Residual N in the soil also increased significantly ($p << 0.01$) with water content in the soil at planting from high I_w to medium I_w levels (Fig. 2–11). The effect of soil water at planting on residual N became considerably less significant from average I_w to low I_w levels because the N uptake differences at these levels also narrowed (Fig. 2–8). The pattern of the amount of residual N in response to low I_w at N application rates of 50, 100, 150, 250, and 300 kg ha^{-1} (Fig. 2–12) followed that of 200 kg N ha^{-1} (Fig. 2–11).

The amount of residual N increased consistently with N rates from 50 to 300 kg N ha^{-1}. However, the amount of residual N decreased with irrigation rates due to higher uptakes simulated by more accurate crop growth (due to less water stress), e.g., averaged across I_w levels and 3- to 14-d irrigation intervals, 21 kg N ha^{-1} out of applied 50 kg N ha^{-1} appeared as residual N in the soil at harvest at an irrigation level of 40% ET (Fig. 2–12). On the other hand, the applied N was inadequate to meet the crop demands at irrigation levels > 40% ET replacement (termed "critical irrigation level"). Similarly, the N rate at 100 kg ha^{-1} was in excess of crop need (resulted in residual N in the soil at harvest) for irrigation rates up to 50% ET replacement and inadequate for higher irrigation rates. Such critical irrigation levels for 150, 200, 250, and 300 kg ha^{-1} N application rates were, respectively, 60, 70, 80, and 90% ET replacement. In summary, the critical N application levels for avoiding residual N in the soil at harvest are 50 kg ha^{-1} for rainfed and irrigation at 40% ET replacement. For irrigations at 50, 60, 70, 80, 90, and 100% ET replace-

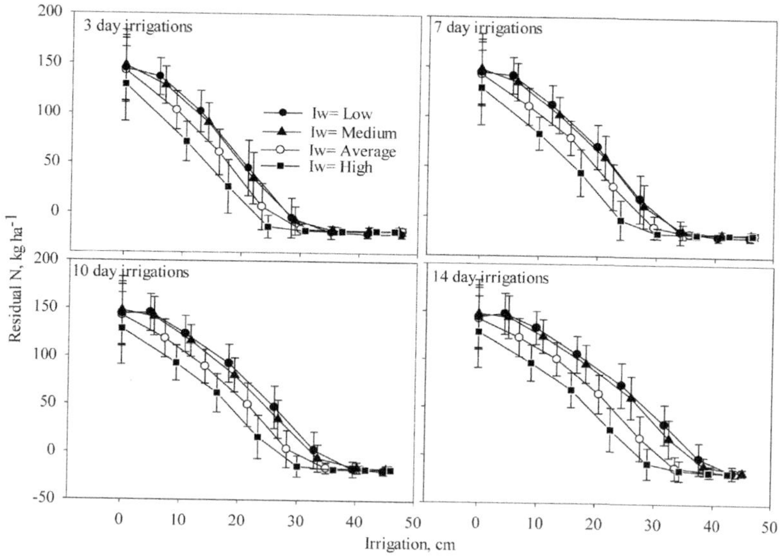

Fig. 2–11. Average (1992–2012) residual N in the soil at harvest under low, medium, average, and high initial soil water (I_w) levels under 3-, 7-, 10-, and 14-d irrigation frequencies. Symbols along the curves from left to right correspond to irrigations at 0, 40, 50, 60, 70, 80, and 100% ET replacement. N levels were constant across all the simulations at 200 kg ha^{-1}. Error bars indicate one standard deviation from the mean.

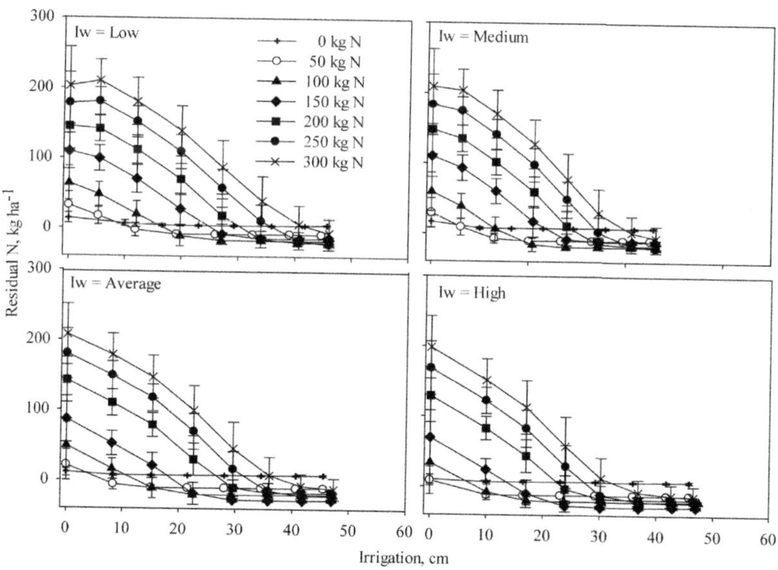

Fig. 2–12. Average (1992–2012) residual N in the soil at harvest in response to N levels of 0, 50, 100, 150, 200, 250, and 300 kg ha^{-1} and 7-d irrigations to replacement 40, 50, 60, 70, 80, 90, and 100% ET (symbols plotted along the curves from left to right) under low, medium, average, and high initial soil water (I_w). The I_w levels were for a soil profile depth of 200 cm. Error bars indicate one standard deviation from the mean. The negative value of residual N means that the amount of N in the soil at harvest is less than the amount at planting. With no N application (0 kg N ha^{-1}), plant growth was minimum and as such N uptake was less than other N application levels, hence soil residual N at harvest remained more or less constant with different water applications.

ments, N rates should be within 100, 150, 200, 250, and 250 kg ha^{-1}, respectively, to avoid residual soil N.

Corn Yield Responses to N (CW-NPF) and Water Interactions

Field experiments of the crop yield response to N reported quadratic relationships with rapid yield increase with increased applied N at the low end of N application rates followed by a decrease in the rate of increase with larger N application rates until the response reaches a plateau with minor to negligible changes (e.g., Vanotti and Bundy, 1994; Shipley et al., 1992). Grain yields simulated in this study also showed quadratic responses to N rates under all I_w and irrigation amounts and frequency levels (data not shown). However, grain yields did not show any responses to N rates > 50 kg ha^{-1} under rainfed conditions in the semiarid climate of the location, as the simulated water stress limited crop responses to higher N rates. Grain yields under rainfed conditions ranged from 398 to 700 kg ha^{-1} with an average N uptake of 49 kg ha^{-1}. These uptakes were mostly met from the initial N available in the soil at planting and N mineralized during the crop season (Fig. 2–9).

In general, grain yield increases with N were confined only up to 50, 100, 100, 150, 200, 200, and 200 kg N ha^{-1} when irrigations were at 40, 50, 60, 70, 80, 90, and 100% ET replacement, respectively (Fig. 2–13). Across different irrigation frequencies and I_w levels, grain yield increased with N but plateaued for N rates between 200 and 250 kg ha^{-1}. The simulated grain yield responses did not differ significantly ($p > 0.05$) between irrigation frequencies and I_w levels (Fig. 2–13). Grain yields simulated in response to different N levels and irrigation levels were between 385 (rainfed at 0 kg N ha^{-1}) and 13,346 kg ha^{-1} (irrigated at 3-d intervals with high I_w and 300 kg N ha^{-1}).

Grain yields did not respond to irrigations above the 90% ET level and N rates above 200 kg N ha^{-1} significantly ($p > 0.05$) for irrigation frequencies up to 7 d. Yield responses to N rates above 200 kg ha^{-1} and irrigation levels above 90% ET replacement simulated in response to irrigation frequencies above 7 d were small (Fig. 2–13). Weighing the cost and other practical difficulties that may be involved in irrigating the crop at 3-d intervals against the yield advantages, irrigation at 7-d intervals with 200 N kg ha^{-1} appears to be the best management choice when resources are not limiting.

Under the 7-d irrigation frequency, the long-term CW-NPF (yield vs. irrigation or ET) are presented in Fig. 2–14 and 2–15. As reflected in the discussions above, these CW-NPF also were profoundly influenced by the N rate, in addition to the irrigation levels, because of the high level of interactions between these resources. These CW-NPF also showed responses to water up to 90% ET replacement and

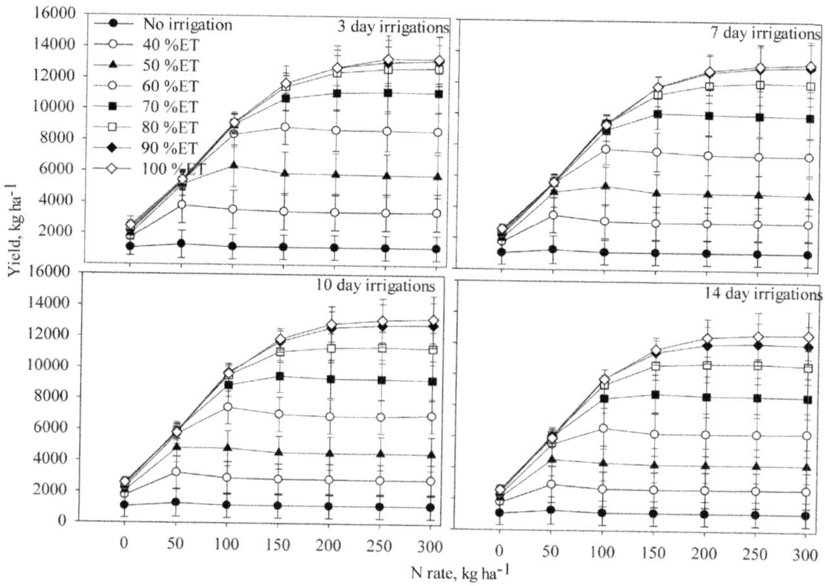

Fig. 2–13. Average (1992–2012) grain yields in response to N levels of 0, 50, 100, 150, 200, 250, and 300 kg ha^{-1} and 3-, 7-, 10-, and 14-d irrigations to replacement 40, 50, 60, 70, 80, 90, and 100% ET (symbols along the curves from left to right) constantly under average initial soil water (I_w) level. Error bars indicate one standard deviation from the mean.

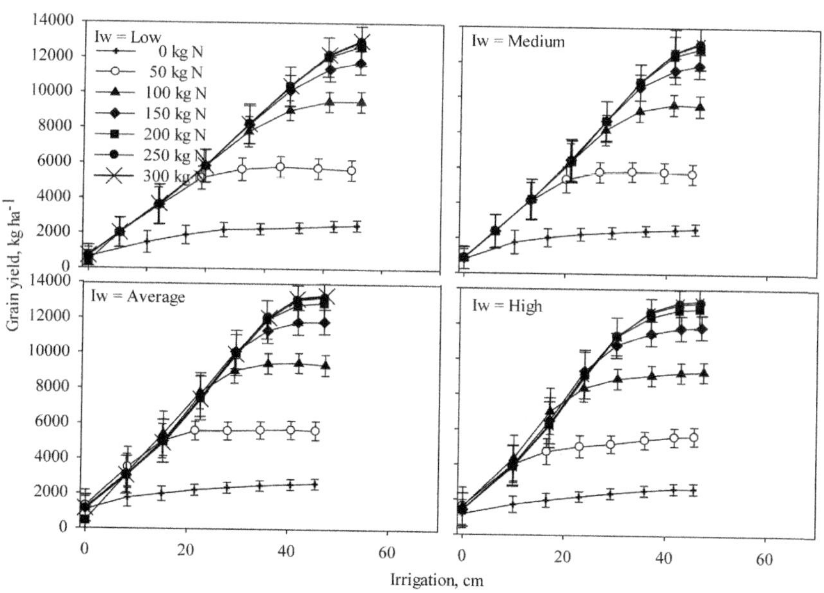

Fig. 2–14. Effect of N levels of 0, 50, 100, 150, 200, 250, and 300 kg ha^{-1} and 7-d irrigations to replacement 40, 50, 60, 70, 80, 90, and 100% ET (symbols along the curves from left to right) under low, medium, average, and high initial soil water (I_w) on average (1992–2012) crop water and nitrogen production functions (yield vs. irrigations at 7-d intervals) for corn. Error bars indicate one standard deviation from the mean.

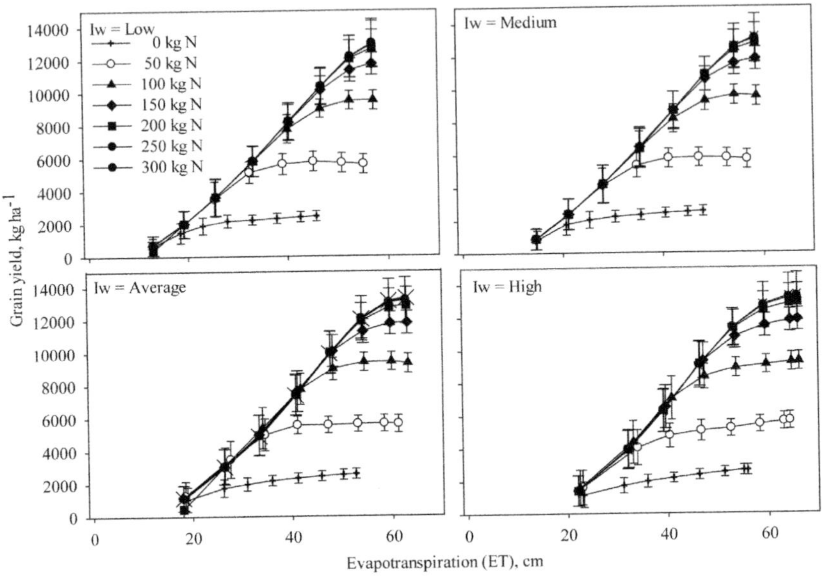

Fig. 2–15. Effect of N levels of 0, 50, 100, 150, 200, 250, and 300 kg ha^{-1} and irrigations to replacement 40, 50, 60, 70, 80, 90, and 100% ET (symbols along the curves from left to right) under low, medium, average, and high initial soil water (I_w) on average (1992–2012) crop water and nitrogen production functions (yield vs. ET) for corn. Irrigation frequency was constant at 7-d intervals. Error bars indicate one standard deviation from the mean.

N rates up to about 200 kg ha^{-1}, above which yield responses to N and irrigation levels were not significant.

Long-Term NUE and IWUE Responses and Interactions

Normally, WUE in plants increases when subjected to mild water stress (Eck, 1984, 1986; Howell, 2001; Howell et al., 1998; Pandey et al., 2000, 2000b). Also, under mild to moderate soil water deficiencies, the application of N enhanced root development into deeper soil layers, thereby enhancing the amount of water extracted and the overall IWUE in crop production (Eck, 1984; Ogola et al., 2002). Nevertheless, in corn, when water deficiency dominated over N deficiency, water stress limited crops yield more than N stress (Eck., 1984; Gholamhoseini et al., 2013). Both N and water deficits applied concurrently or otherwise were reported to adversely affect corn growth and development (Sinclair et al., 1975; Wolfe et al., 1988). In the current simulations, in general, for a given N rate, IWUE increased initially with irrigation, reaching a maximum and then declining with higher irrigation rates (Fig. 2–16). At the 7-d irrigation frequency and average I_w, the IWUE simulated with the 50 kg N ha^{-1} rate were 246, 246, 199, 153, 126, 107, and 95 kg ha^{-1} cm $^{-1}$, respectively, for irrigations of 40, 50, 60, 70, 80, 90, and 100% ET replacement. The maximum IWUE was simulated with an N rate of 200 kg ha^{-1}, after which it decreased with increasing N rate. With N rates of 50, 100, 150, and 200 kg ha^{-1}, IWUE reached a maximum level with irrigations at 40, 50, 60, and 70% ET replacement, respectively. Although the yield continued to increase with increasing irrigations above 70% ET replacement (with N above 150 kg ha^{-1}) (Fig. 2–3 and 2–13), IWUE declined drastically (Fig. 2–16).

Any abiotic or biotic factor limiting N supply can have a negative effect on NUE. For example, Gholamhoseini et al. (2013) observed that in low irrigation treatments, root access to the mass flow of N was reduced, resulting in low N uptake and the consequent decrease in NUE. In the current simulations, NUE increased with water stress abatement with irrigations (Fig. 2–17). With a lower N level at 50 kg N ha^{-1}, NUE initially increased in response to irrigations up to 60% ET replacement and thereafter plateaued and declined slightly with higher irrigation amounts. The decline in yields with higher irrigations at this N rate was caused by the leaching of the small amount of N available out of the shallow plant root zone at the beginning of the season affecting crop growth.

Field experiments also reported lower NUE with increased N rates due to the low capability of the crops to utilize the entire applied N before it is lost from

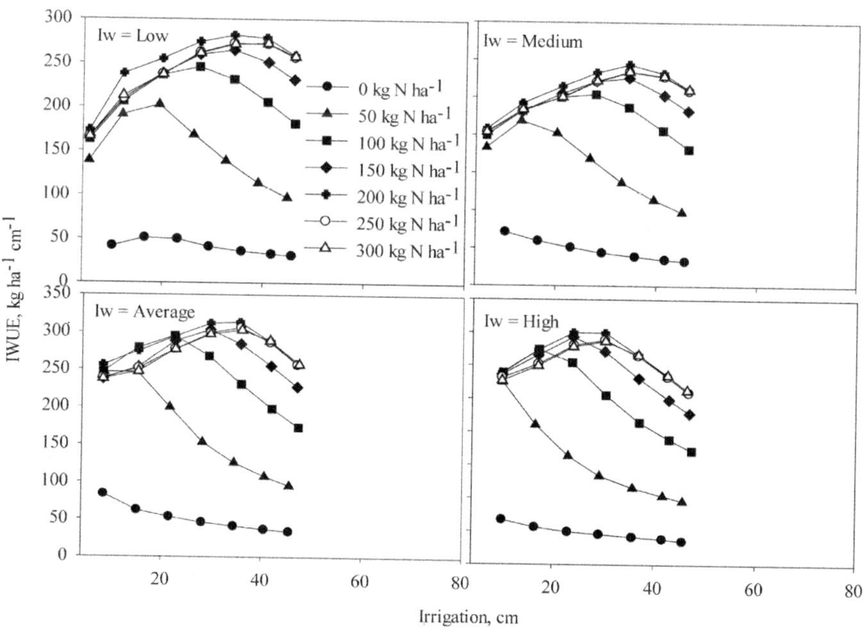

Fig. 2–16. Irrigation water use efficiency (IWUE, yield per unit irrigation) responses to N levels of 0, 50, 100, 150, 200, 250, and 300 kg ha^{-1} and irrigations to replacement 40, 50, 60, 70, 80, 90, and 100% ET (symbols along the curves from left to right) at 7-d intervals under low, medium, average, and high initial soil water (I_w) levels.

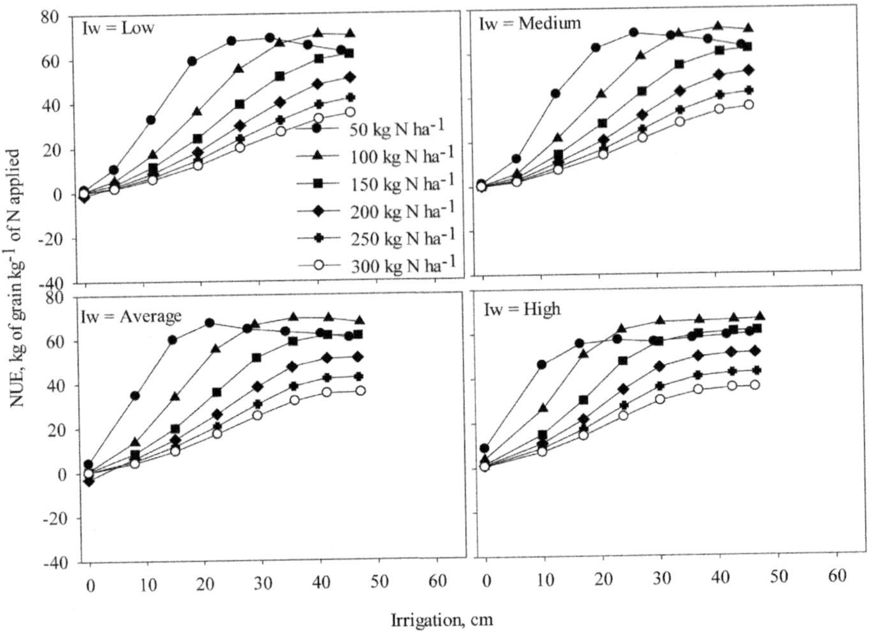

Fig. 2–17. Corn nitrogen use efficiency (NUE) responses to N levels of 0, 50, 100, 150, 200, 250, and 300 kg ha^{-1} and irrigations to replacement 40, 50, 60, 70, 80, 90, and 100% ET (symbols along the curves from left to right) at 7-d intervals under low, medium, average, and high initial soil water (I_w) levels.

the system (Dawson et al., 2008). Similar patterns in NUE in response to higher N application rates were simulated. In the current simulations, the highest NUE was achieved with 50 kg N ha^{-1}, and it decreased thereafter in response to N rates up to 300 kg ha^{-1} (Fig. 2–17). At the 50 kg N ha^{-1} rate, the maximum NUE achieved was 67.5 kg grains kg^{-1} of N applied, with irrigations of 60% ET replacement and 7-d intervals (Fig. 2–17). The maximum NUE achieved for N rates of 100, 150, 200, 250, and 300 kg ha^{-1} were 69.4, 61.5, 51.3, 42.3, and 35.7 kg grains kg^{-1} of N applied, respectively, at irrigation levels of 80, 90, 100, 100, and 100% ET, respectively. In summary, for managing limited water with optimum IWUE and NUE, it is necessary to derive CW-NPF with their interaction represented adequately (Fig. 2–14 and 2–15).

Conclusions

In the semiarid climate of Colorado and its surrounding regions, farmers rely on the irrigation water supply for dependable returns from corn farming, and the pressures on freshwater supplies keep rising from the expanding needs of many other competing human enterprises. In this context, to enhance water use efficiency in corn production, we developed the site-specific and long-term average CW-NPF that take into account the change in water supply and demand due to varying weather across crop seasons, initial water stored in the soil at planting, irrigation frequency, and N rates in Greeley, CO. Among the intervals of irrigations simulated, 3-d frequencies were best for higher yield responses and N uptake efficiencies and minimum residual N in the soil at harvest in the limited water irrigation scenario, but the 7-d frequencies gave equally good results near full irrigation. Simulated higher initial water in the soil at planting were found to increase crop responses to limited irrigation in the climate of the location. Hence, soil–crop–residue–water management practices to improve soil water storage in the soil during the fallow periods can help increase corn production with limited water irrigation. Taking into account the yield responses to N under different irrigation scenarios, optimum N rates for optimum corn yield for irrigations levels of 40, 50, 60, 70, 80, 90, and 100% ET replacement were 50, 100, 100, 150, 200, 200, and 200 kg ha^{-1}, respectively. N applications above these N rates were found to have lower IWUE and NUE and the potential for higher residual soil N at harvest that can pollute the environment and natural resources later. NUE increased in response to irrigations up to 60% ET replacement, but plateaued thereafter and declined gradually with further irrigations. In this study, regardless of irrigation level, NUE decreased and residual soil N at harvest increased in response to N rates from 50 to 300 kg ha^{-1} as observed in field experiments. Actual amount of irrigation applied for meeting a specific ET replacement level in a given crop season can vary with the amount of precipitation received between the last irrigation event and the day of irrigation. If the rainfall received exceeds the targeted ET replacement level, additional N applications will help to optimize yield with the increased water inputs. Therefore, the RZWQM2 model used in the study can be used for the development of crop responses to water and N and their interactions in different soils and climates across the world.

References

Ahuja, L.R., K.W. Rojas, J.D. Hanson, M.J. Shafer, and L. Ma, editors. 2000. Root Zone Water Quality Model. Modeling management effects on water quality and crop production. Water Resour. Publ., Highlands Ranch, CO.

Bausch, W., T.J. Trout, and G. Buchleiter. 2011. Evapotranspiration adjustments for deficit-irrigated corn using canopy temperature: A concept. Irrig. Drain. 60:682–693. doi:10.1002/ird.601

Brooks, R.H., and A.T. Corey. 1964. Hydraulic properties of porous media. Hydrol. Pap. No. 3. Colorado State Univ., Fort Collins.

Cornic, C., and A. Massacci. 1996. Leaf photosynthesis under drought stress. In: N.R. Baker, editor, Photosynthesis and environment. Kluwer Acad., Dordrecht, the Netherlands. p. 347–366.

Dawson, J.C., D.R. Huggins, and S.S. Jones. 2008. Characterizing nitrogen use efficiency in natural and agricultural ecosystems to improve the performance of cereal crops in low-input and organic agricultural systems. Field Crops Res. 107:89–101. doi:10.1016/j.fcr.2008.01.001

De Wit, C. T. 1958. Transpiration and crop yield. Versl. Landbouwkd. Onderz. No. 64, Instituut voor Biologisch en Scheikusdig Ondergoek van Landvouwgewassen, Wageningen, the Netherlands.

Eck, H.V. 1984. Irrigated corn yield response to nitrogen and water. Agron. J. 76:421–428. doi:10.2134/agronj1984.00021962007600030014x

Eck, H.V. 1986. Effects of water deficits on yield, yield components, and water use efficiency of irrigated corn. Agron. J. 78:1035–1040. doi:10.2134/agronj1986.00021962007800060020x

Falkenmark, M., and M. Lannerstad. 2005. Consumptive water use to feed humanity-curing a blind spot. Hydrol. Earth Syst. Sci. 9:15–28. doi:10.5194/hess-9-15-2005

Fapohunda, H.O., and M.M. Hussain. 1990. Water and fertilizer interrelations with irrigated corn. Agric. Water Manage. 18:49–61. doi:10.1016/0378-3774(90)90035-W

Farahani, H.J., and L.R. Ahuja. 1996. Evapotranspiration modeling of partial canopy/residue covered fields. Trans. ASAE 39:2051–2064. doi:10.13031/2013.27708

Farahani, H.J., and D.G. DeCoursey. 2000. Potential evapotranspiration processes in the soil-crop-residue system. In: L.R. Ahuja, K.W. Rojas, J.D. Hanson, M.J. Shafer, and L. Ma, editors, Root Zone Water Quality Model. Modeling management effects on water quality and crop production. Water Resour. Publ., Highlands Ranch, CO. p. 51–80.

Ferguson, R.B., C.A. Shapiro, G.W. Hergert, W.L. Kranz, N.L. Klocke, and D.H. Krull. 1991. Nitrogen and irrigation management practices to minimize nitrate leaching from irrigated corn. J. Prod. Agric. 4:186–192. doi:10.2134/jpa1991.0186

Frank, A., and D. Carlson. 1999. Colorado's net irrigation requirements for agriculture, 1995. Colorado Department of Agriculture. http://cospl.coalliance.org/fedora/repository/co:3072/ag92ir71999internet.pdf (accessed 18 Feb. 2014).

Frink, C.R., P.E. Waggoner, and J.H. Ausubel. 1999. Nitrogen fertilizer: Retrospect and prospect. Proc. Natl. Acad. Sci. USA 96:1175–1180. doi:10.1073/pnas.96.4.1175

Geerts, S., and D. Raes. 2009. Deficit irrigation as an on-farm strategy to maximize crop water productivity in dry areas. Agric. Water Manage. 96:1275–1284. doi:10.1016/j.agwat.2009.04.009

Gholamhoseini, M., M. Agha Alikhani, S.A.M. Modarres Sanavy, and S.M. Mirlatifi. 2013. Interactions of irrigation, weed and nitrogen on corn yield, nitrogen use efficiency and nitrate leaching. Agric. Water Manage. 126:9–18. doi:10.1016/j.agwat.2013.05.002

Gleick, P. 2003. 'Soft path' solution to 21st-century water needs. Science 302:1524–1528. doi:10.1126/science.1089967

Gordon, L.J., C.M. Finlayson, and M. Falkenmark. 2010. Managing water in agriculture for food production and other ecosystem services. Aric. Water Manage. 97:512–519. doi:10.1016/j.agwat.2009.03.017

Grassini, P., and G. Cassman. 2012. High-yield maize with large net energy yield and small global warming intensity. Proc. Natl. Acad. Sci. USA 109(4):1074–1079. doi:10.1073/pnas.1116364109

Halvorson, A.D., D.C. Nielsen, and C.A. Reule. 2004. Nitrogen fertilization and rotation effects on no-till dryland wheat production. Agron. J. 96:1196–1201. doi:10.2134/agronj2004.1196

Hanks, R.J. 1983. Yield and water-used relationships: An overview. In: H.M. Taylor, W.R. Jordan, and T.R Sinclair, editors, Limitations to efficient water use in crop production. ASA, Madison, WI. p. 393–411.

Hoogenboom, G., J.W. Jones, and K.J. Boote. 1991. A decision support system for prediction of crop yield, evapotranspiration, and irrigation management. In: W.F. Ritter, editor, Irrigation and drainage. Proceedings of the 1991 National Conference, Honolulu, HI. 22–26 July 1991. ASCE, Reston, VA. p. 198–204.

Howell, T.A. 2001. Enhancing water use efficiency in irrigated agriculture. Agron. J. 93:281–289. doi:10.2134/agronj2001.932281x

Howell, T.A., J.A. Tolk, A.D. Scheider, and S.R. Evett. 1998. Evapotranspiration, yield, and water use efficiency of corn hybrids differing in maturity. Agron. J. 90:3–9. doi:10.2134/agronj1998.00021962009000010002x

Jones, J.W., G. Hoogenboom, C.H. Porter, K.J. Boote, W.D. Batchelor, L.A. Hunt, P.W. Wilkens, U. Singh, A.J. Gijsman, and J.T. Ritchie. 2003. The DSSAT cropping system model. Eur. J. Agron. 18:235–265. doi:10.1016/S1161-0301(02)00107-7

Klocke, N.L., J.P. Schneekloth, S.R. Melvin, R.T. Clark, and J.O. Payero. 2004. Field scale limited irrigation scenarios for water policy strategies. App. Eng. Agric. 20:623–631.

Knisel, W.G., and E. Turtola. 2000. GLEAMS model application on a heavy clay soil in Finland. Agric. Water Manage. 43:285–309. doi:10.1016/S0378-3774(99)00067-0

Ma, L., L.R. Ahuja, S.A. Saseendran, R.W. Malone, T.R. Green, B.T. Nolan, P.N.S. Bartling, G.N. Flerchinger, K.J. Boote, and G. Hoogenboom. 2011. A protocol for parameterization and calibration of RZWQM2 in field research. In: L.R. Ahuja and L. Ma, editors, Methods of introducing system models into agricultural research. Vol. 2. ASA, CSSA, and SSSA, Madison, WI. p. 1–64.

Ma, L., G. Hoogenboom, L.R. Ahuja, J.C. Ascough, II, and S.A. Saseendran. 2006. Evaluation of the RZWQM-CERES-Maize hybrid model for maize production. Agric. Syst. 87:274–295. doi:10.1016/j.agsy.2005.02.001

Ma, L., G. Hoogenboom, L.R. Ahuja, D.C. Nielsen, and J.C. Ascough, II. 2005. Evaluation of the RZWQM-CROPGRO hybrid model for soybean production. Agron. J. 97:1172–1182. doi:10.2134/agronj2003.0314

Ma, L., G. Hoogenboom, A.A. Saseendran, P.N.S. Bartling, L.R. Ahuja, and T.R. Green. 2009. Estimates of soil hydraulic properties and root growth factor on soil water balance and crop production. Agron. J. 101:572–583. doi:10.2134/agronj2008.0206x

Ma, L., D.C. Nielsen, L.R. Ahuja, R.W. Malone, S.A. Saseendran, K.W. Rojas, J.D. Hanson, and J.G. Benjamin. 2003. Evaluation of RZWQM under varying irrigation levels in eastern Colorado. Trans. ASAE 46:39–49.

Ma, L., T.J. Trout, L.R. Ahuja, W.C. Bausch, S.A. Saseendran, R.W. Malone, and D.C. Nielsen. 2012. Calibrating RZWQM2 model for maize responses to deficit irrigation. Agric. Water Manage. 103:140–149. doi:10.1016/j.agwat.2011.11.005

Martin, D.L., D.G. Watts, L.N. Mielke, K.D. Frank, and D.E. Eisenhauer. 1982. Evaluation of nitrogen and irrigation management for corn production using water high in nitrate. Soil Sci. Soc. Am. J. 46:1056–1062. doi:10.2136/sssaj1982.03615995004600050034x

Matthews, R., W. Stephens, T. Hess, T. Middleton, and A. Graves. 2002. Applications of crop/soil simulation models in tropical agricultural systems. Adv. Agron. 76:31–124. doi:10.1016/S0065-2113(02)76003-3

McCullough, D.E., P. Giradin, M. Mihajlovic, A.A. Guilera, and M. Tollenaar. 1994. Influence of N supply on development and dry matter accumulation of an old and new maize hybrid. Can. J. Plant Sci. 74:471–477. doi:10.4141/cjps94-087

Mwanamwenge, J., S.P. Loss, K.H.M. Siddique, and P.S. Cocks. 1999. Effect of water stress during floral initiation, flowering and podding on the growth and yield of faba bean (*Vicia faba* L.). Eur. J. Agron. 11:1–11. doi:10.1016/S1161-0301(99)00003-9

Nimah, M.N., and J.R. Hanks. 1973. Model for estimating soil water, plant and atmospheric interrelations: I. description and sensitivity. Soil Sci. Soc. Am. Proc. 37:522–527. doi:10.2136/sssaj1973.03615995003700040018x

Ogola, J.B.O., T.R. Wheeler, and P.M. Harris. 2002. Effects if nitrogen and irrigation on water use of maize crops. Field Crops Res. 78:105–117. doi:10.1016/S0378-4290(02)00116-8

Oktem, A., M. Siesek, and G. Oktem. 2003. Deficit irrigation effects on sweet corn (*Zea mays saccharata* Sturt) with drip irrigation system in a semi-arid region. I: Water-yield relationship. Agric. Water Manage. 61:63–74. doi:10.1016/S0378-3774(02)00161-0

Pandey, R.K., W.A.T. Herrera, and J.W. Pendelton. 1984. Drought response of grain legume under irrigation gradient. Yield and yield components. Agron. J. 76:549–553. doi:10.2134/agronj1984.00021962007600040009x

Pandey, R.K., J.W. Maranville, and A. Admou. 2000a. Deficit irrigation and nitrogen effects on maize in a Sahelian environment I. Grain yield and yield components. Agric. Water Manage. 46:1–13. doi:10.1016/S0378-3774(00)00073-1

Pandey, R.K., J.W. Maranville, and M.M. Chetima. 2000b. Deficit irrigation and nitrogen effects on maize in a Sahelian environment II. Shoot growth, nitrogen uptake and water extraction. Agric. Water Manage. 46:15–27. doi:10.1016/S0378-3774(00)00074-3

Poch-Massegú, R., J. Jiménez-Martínez, K.J. Wallis, F. Ramírez de Cartagena, and L. Candela. 2014. Irrigation return flow and nitrate leaching under different crops and irrigation methods in Western Mediterranean weather conditions. Aric. Water Manage. 134:1–13. doi:10.1016/j.agwat.2013.11.017

Saseendran, S.A., L.R. Ahuja, L. Ma, T.J. Trout, G.S. McMaster, A.A. Andales, J. Chavez, and J. Ham. 2014. Enhancing RZWQM2 for water stress responses of corn (*Zea mays*. L.). Agron. J. 106(1):81–94. doi:10.2134/agronj2013.0300

Saseendran, S.A., L.R. Ahuja, D.C. Nielsen, T.J. Trout, and L. Ma. 2008. Use of crop simulation models to evaluate limited irrigation management options for corn in a semiarid environment. Water Resour. Res. 44:W00E02. doi:10.1029/2007WR006181

Saseendran, S.A., L. Ma, D.C. Nielsen, M.F. Vigil, and L.R. Ahuja. 2005. Simulating planting date effects on corn production using RZWQM and CERES-Maize. Agron. J. 97:58–71. doi:10.2134/agronj2005.0058

Saseendran, S.A., D.C. Nielsen, D.J. Lyon, L. Ma, D.G. Felter, D.D. Baltensperger, G. Hoogenboom, and L.R. Ahuja. 2009. Modeling responses of dryland spring triticale, proso millet and foxtail millet to initial soil water in the High Plains. Field Crops Res. 113:48–63. doi:10.1016/j.fcr.2009.04.008

Scanlon, B.R., I. Jolly, M. Sophocleous, and L. Zhang. 2007. Global impacts of conversions from natural to agricultural ecosystems on water resources: Quantity versus quality. Water Resour. Res. 43:W03437. doi:10.1029/2006WR005486

Shipley, P.R., J.J. Meisinger, and A.M. Decker. 1992. Conserving residual corn fertilizer nitrogen with winter cover crops. Agron. J. 84:869–876. doi:10.2134/agronj1992.00021962008400050020x

Sinclair, T.R., G.E. Bingham, E.R. Lemon, and L.H. Allen, Jr. 1975. Water use efficiency of field-grown maize during moisture stress. Plant Physiol. 56:245–249. doi:10.1104/pp.56.2.245

Stewart, J.I., and R.M. Hagan. 1973. Functions to predict effects of crop water deficits. J. Irrig. Drain. Div. 99:421–439.

Trout, T.J., W.C. Bausch, and G.W. Buchleiter. 2010. Water production functions for Central Plains crops. 5th National Decennial Irrigation Conference Proceedings, Phoenix, AZ. 5–8 Dec. 2010. ASABE, St. Joseph, MI.

United Nations Educational, Scientific and Cultural Organization. 2012. Managing water under uncertainty and risk. U.N. World Water Dev. Rep. 4. UNESCO, Paris.

Vanotti, M.B., and L.G. Bundy. 1994. Frequency of nitrogen fertilizer carryover in the humid Midwest. Agron. J. 86:881–886. doi:10.2134/agronj1994.00021962008600050023x

Wolfe, D.W., D.W. Henderson, T.C. Hsiao, and A. Alvino. 1988. Interactive water and nitrogen effects on senescence of maize. I. Leaf area duration, nitrogen distribution, and yield. Agron. J. 80:859–864. doi:10.2134/agronj1988.00021962008000060004x

Xu, Z., Z.W. Yu, and D. Wang. 2006. Nitrogen translocation in wheat plants under soil water deficit. Plant Soil 280:291–303. doi:10.1007/s11104-005-3276-2

A Mobile Irrigation Water Management System Using a Collaborative GIS and Weather Station Networks

A.A. Andales,* T.A. Bauder, and M. Arabi

Abstract

Improved irrigation water management (IWM) in 22.2 million hectares of irrigated U.S. farmland can play a role to conserve water, prevent pollution, and enhance crop production. Our objectives were to develop and evaluate a scalable device-independent mobile IWM system for Colorado and promote adoption by irrigators. A prototype IWM system with input from crop producers, researchers, and conservation and commodity representatives was developed for Colorado, accessible through the Internet. The tool is hosted on the collaborative environmental Risk Assessment and Management System (eRAMS) GIS with online modeling services on the Cloud Services Innovation Platform. The tool gives irrigators near-real-time estimates of daily soil water deficit (D_c) of fields as an aid to schedule irrigations. It accesses weather data from several Colorado networks, providing daily rainfall and reference evapotranspiration (ET_r) at 84 locations. Apps for handheld devices display IWM information and synchronize with the eRAMS tool. The IWM system was used for three growing seasons (2010–2012) to calculate daily root zone D_c for a sprinkler-irrigated corn field in Colorado. Daily calculated D_c had relative errors ranging from 2 to 31% compared with measured daily D_c. The average RMSE was 16 mm of water, an acceptable amount as it could be compensated by a single irrigation event. Accuracy of calculated D_c can be improved if midseason corrections based on measured profile soil water content are made. The IWM system was demonstrated at workshops across Colorado and adoption will lead to increased irrigation application efficiencies and improved water quality.

Abbreviations: AWC, available water capacity; CoAgMet, Colorado Agricultural Meteorological Network; CSIP, Cloud Services Innovation Platform; DP, deep percolation; eRAMS, Environmental Risk Assessment and Management System; FC, field capacity; GDD, growing degree days; IWM, irrigation water management; MAD, management allowed depletion; MBE, mean bias error; M&I, municipal and industrial; NCWCD, Northern Colorado Water Conservancy District; PAW, plant-available water; PWP, permanent wilting point; RE, relative error; SRO, surface runoff; SSURGO, USDA Soil Survey Geographic Database.

A.A. Andales and T.A. Bauder (troy.bauder@colostate.edu), Dep. of Soil and Crop Sciences, Colorado State University, 1170 Campus Delivery, Fort Collins, CO 80523-1170. *Corresponding author (allan.andales@colostate.edu).

M. Arabi, Dep. of Civil and Environmental Engineering, Colorado State University, 1372 Campus Delivery, Fort Collins, CO 80523-1372 (mazdak.arabi@colosate.edu).

doi:10.2134/advagricsystmodel5.c3

Practical Applications of Agricultural System Models to Optimize the Use of Limited Water
Lajpat R. Ahuja, Liwang Ma, and Robert J. Lascano, Editors
Advances in Agricultural Systems Modeling, Volume 5. Lajpat R. Ahuja, Series Editor

USDA-ARS

In a recent appraisal for the Soil and Water Resources Conservation Act, the USDA identified irrigation water conservation as a national need (USDA-NRCS, 2011). The appraisal highlighted that "improvements in irrigation water management (IWM) also have the potential to reduce water applications, and improved management is usually required to capture the potential in the improved technology." Given that 22.2 million hectares (54.9 million acres) of farmland are irrigated in the United States (USDA-NASS, 2008), widespread adoption of improved IWM can have a significant impact on water conservation, prevention of water pollution, and water productivity (crop yield). Sprinkler systems irrigate around 56% of this total irrigated area, while 40% is irrigated by gravity systems and the remaining is irrigated by micro- or sub-irrigation methods.

In the United States, there is a low adoption rate for precise irrigation scheduling methods such as soil water sensing (8.6% of farms), plant water sensing (1.7%), commercial or government scheduling services (8.0%), and daily crop ET reports (9.1%) (USDA-NASS, 2008). The top two irrigation scheduling methods in farms are "looking at crop condition" (77.7%) and "feel of soil" (42.6%). Total percentages exceed 100% because farms use more than one method to schedule irrigations (USDA-NASS, 2008). In a survey of Colorado irrigators, it was found that 89% of irrigators still rely on imprecise methods of irrigation scheduling such as using past experience or relying on crop appearance (Bauder and Waskom, 2005). These methods, which are not based on actual consumptive water use or soil water content, can result in significant overirrigation or under-irrigation. Consequently, overirrigation leads to losses of water and agricultural chemicals via surface runoff (SRO) or deep percolation (DP) (Waskom, 1994); and under-irrigation leads to crop water stress and yield reductions. More accurate irrigation scheduling methods based on soil water and ET tracking have had low adoption rates because of lack of economic incentive (e.g., low value crops), low energy costs (e.g., gravity irrigation), use of less efficient irrigation methods (e.g., furrow), or lack of technical support (Leib et al., 2002). In general, growers using more sophisticated irrigation systems (associated with higher energy or pumping costs) also used more sophisticated scheduling methods (Bauder and Waskom, 2005). Water conservation, high crop yield, fertilizer savings, and nonpoint pollution reduction were considered only secondary benefits of irrigation scheduling (Leib et al., 2002). Therefore, there is a need to provide low-cost irrigation scheduling tools that are "smart" (minimizing the technical knowledge needed by irrigators). There is also a need for continual education regarding the benefits of IWM for water conservation and prevention of nonpoint-source pollution.

Irrigation scheduling approaches can be classified into three classes: (i) soil water measurement, including various methods of measuring soil water potential or soil water content; (ii) soil water balance calculations, typically using a computer model; and (iii) plant stress sensing, including monitoring of tissue water status, physiological responses, or growth rate (Jones, 2004). The focus of this chapter is on the second approach, where calculations of the soil water balance are used to determine amount and timing of irrigations. In the United States, Jensen et al. (1971) were one of the early users of this approach, integrating advancements in ET theory, meteorology, physics, and agronomy into a mathematical computer model for estimating soil water depletion. Jensen and Wright (1978) recognized the need for accurate ET models for irrigation scheduling, and ET modeling is a key feature of current water balance models. The Penman–Monteith method for calculating reference ET (Allen et al., 1998) along with crop coefficients for calculating crop ET (ET_c) have been widely adapted for irrigation scheduling (Joshi et al., 1995, Ma et al., 2013).

Before widespread availability of the Internet and online data sources, water balance models for irrigation scheduling were standalone software tools that ran on individual computers (Jensen et al., 1971; Broner and Lambert, 1989a, 1989b; Cahoon et al., 1990; Joshi et al., 1995; Bailey and Spockman, 1996; Sammis et al., 2012; Steele et al., 2010). More computationally intensive cropping system models have also been used as strategic tools for designing or recommending irrigation schedules (Bergez et al., 2001; Endale and Fipps, 2001; He et al., 2013; Tojo Soler et al., 2013). These models calculated one-dimensional (vertical) movement of water for individual fields. The GISAREG (Fortes et al., 2005) is an example of a simulation model that provides GIS capabilities for scheduling irrigations for multiple fields or at the project scale. In general, these standalone software tools required technical expertise and access to input data (weather, soils) that made them impractical for use by irrigators for real-time irrigation scheduling. Some have attempted to overcome this limitation by using the models with long-term weather data to develop simple graphical irrigation scheduling calendars that can be widely distributed in print form (Hill and Allen, 1996; Geerts et al., 2010).

Increased accessibility of the Internet and online data sources for automated soil water balance calculations has allowed the development of web-based irrigation scheduling tools. Recent examples of web-based tools are AQUAMAN (Chauhan et al., 2013) for irrigation scheduling of peanuts in Australia and the Florida Automated Weather Network (FAWN) Interactive Irrigation Tool for lawn turf (Dobbs et al., 2013). Web-based irrigation tools feature automated inputs of weather data from online weather networks. Irrigation scheduling advice can also be delivered to mobile devices, such as the Italian IRRINET expert system that sends information on ET_c, next irrigation date, and calculated irrigation amount (water depth) via short message service (SMS) (Mannini et al., 2013) or a similar system developed for South African sugarcane(*Saccharum* spp.) farms (Singels and Smith, 2006). Irrigation system manufacturers have also started providing smartphone apps to monitor system operation (Johnson, 2010). For more examples of irrigation scheduling tools, the reader is referred to an online list provided by the Agricultural Water Conservation Clearinghouse (http://www.agwaterconservation.colostate.edu/ToolsCalculators.aspx [accessed 20 June 2014]).

This chapter focuses on a water balance-based irrigation scheduling tool for Colorado. The soil water balance method was chosen because it takes advantage of publicly available online weather and soils data sources that could be used across the state. The approach was readily adapted for Internet and mobile device access and made irrigation scheduling advice widely available to irrigators. This is in contrast to soil water measurement or plant stress sensing methods that require capital investment in soil water content or plant water status sensors. The objectives of the project were to (i) develop a scalable device-independent mobile IWM system for Colorado, (ii) evaluate the performance of the IWM system for selected irrigated fields, and (iii) promote adoption of the IWM system by irrigators. Colorado is an ideal location for this project because water conservation is a requirement for survival of irrigated agriculture and meeting water demands from population growth. There is extreme pressure in eastern Colorado to transfer significant amounts of water from agriculture to municipal and industrial (M&I) uses because of rapid population growth along the Front Range. In typical, traditional agricultural transfers, farmland is usually "dried up" or no longer irrigated and the water historically consumed during irrigation of this land is used for meeting M&I needs. Generally only the portion of the water historically consumptively used (ET) can be transferred to protect other appropriators who rely on return flows, which are composed of SRO and DP that return to streams and aquifers from upstream users (Colorado Water Conservation Board, 2004). Therefore, alternatives to traditional agricultural transfers such as transferring saved ET from deficit irrigation or rotational fallowing, and water leasing, are actively being pursued in eastern Colorado (Colorado Water Conservation Board, 2010). These practices will require precise irrigation scheduling to achieve necessary savings in ET.

Materials and Methods

The Mobile IWM System

The IWM system leverages three key technologies available at Colorado State University: (i) the Colorado Agricultural Meteorological Network (CoAgMet) of 60+ automatic weather stations around Colorado (Andales et al., 2009a; http://ccc.atmos.colostate.edu/~coagmet/ [accessed 25 May 2014]); (ii) the eRAMS (Arabi, 2011; https://www.erams.com/) that provides a web-based GIS and environmental modeling tools; and (iii) the Cloud Services Innovation Platform (CSIP) (David et al., 2013; Lloyd et al., 2012) developed in collaboration with USDA to provide cloud-based parallel computing and virtualization services for multiple users over a network or across the Internet. A functional prototype of the IWM system has been developed and is currently being tested on several irrigated fields in northeast Colorado.

The mobile IWM system provides two user interfaces: (i) a full-featured web-based interface available on eRAMS via a web browser and (ii) a mobile app that provides basic irrigation scheduling information and input capabilities via smartphones or tablets. A conceptual diagram of the system is shown in Fig. 3–1. The mobile IWM system benefits from and builds on the collaborative eRAMS

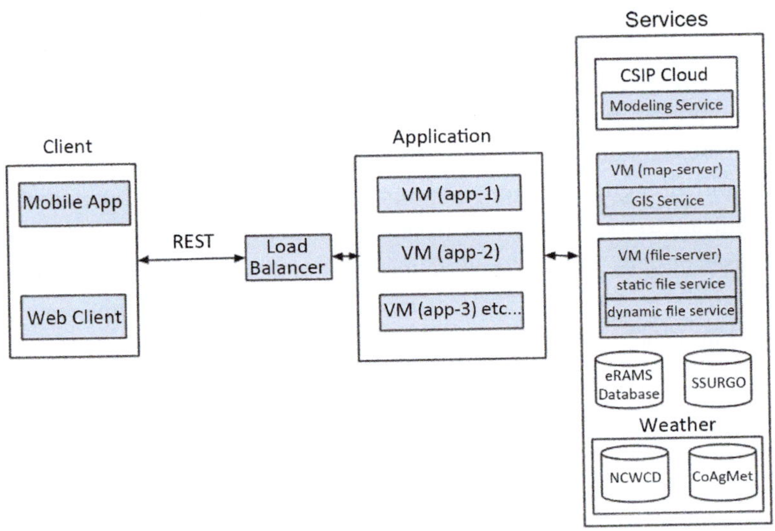

Fig. 3–1. Conceptual diagram of the mobile irrigation water management system implemented within the Risk Assessment and Management System (eRAMS) and Cloud Services Innovation Platform infrastructure (CSIP). Arrows indicate the flow of information between components. Abbreviations: CoAgMet, Colorado Agricultural Meteorological Network; NCWCD, Northern Colorado Water Conservancy District; REST, representational state transfer distributed-computing specifications for web services; SSURGO, USDA Soil Survey Geographic Database; VM, virtual machine.

online GIS platform. As an online, open-source, web-based collaborative platform, eRAMS avoids the need for expensive software licensing by end users. Equipped with a powerful array of online tools for collection and organization of geospatial data, eRAMS requires minimal training for use by stakeholders and other interested parties. The eRAMS platform includes digitization capabilities that allow users to draw features on high-resolution aerial photos (Google/Bing maps and NASA imager) and enter attributes. For example, irrigators can draw polygon features on high-resolution aerial photos to delineate field boundaries and specify field attributes such as irrigation system characteristics, soil hydraulic properties, and other information. The eRAMS platform is also equipped with state-of-the-art computational capabilities for collection, organization, and sharing of heterogeneous geospatial data, visualization, modeling, scenario analysis, and optimization. The web-based interface on eRAMS is platform independent and can be accessed by stakeholders using any web browser on a device (e.g., personal computer, tablet) with Internet connectivity.

A mobile app for smartphones communicates with eRAMS to display summary information of the soil profile water status of each field. Actual gross irrigation or locally measured precipitation can also be entered by the user. Yesterday's weather can also be viewed for weather stations preselected by the user.

The key function of the IWM system is the daily calculation of soil water content of the managed soil profile and near-real-time delivery of field-specific irrigation requirement (based on soil water deficit and application efficiency). The "smart" feature of the IWM system is the automated tracking of daily soil water status in the field based on estimation of components of the soil water balance. For the following sections, in addition to SI units note that English units are also given, consistent with how the IWM system presents information. These units are customary to most farmers and irrigators in the United States.

Estimating the Water Demand: Water Balance Approach to Schedule Irrigation

As the crop grows and extracts water from the soil to satisfy its ET_c requirement, the stored soil water is gradually depleted. In general, the net irrigation requirement is the amount of water required to refill the root zone soil water content to field capacity (FC). This amount, which is the difference between FC and current soil water content, corresponds to the soil water deficit (*D*). For each field specified by the user, the IWM system keeps track of *D*, which gives the net amount of irrigation water to apply. On a daily basis, *D* can be estimated using the following accounting equation for the soil root zone (Andales et al., 2011):

$$D_c = D_p + ET_c - P - Irr - U + SRO + DP, \quad [1]$$

where D_c is the soil water deficit (net irrigation requirement, mm) in the root zone on the current day, D_p is the soil water deficit on the previous day (mm), ET_c is the crop ET rate for the current day (mm), P is the gross precipitation for the current day (mm), Irr is net irrigation amount for the current day (mm), U is upflux of shallow ground water into the root zone (mm), SRO is surface runoff (mm), and DP is deep percolation or drainage (mm). Net irrigation amount is estimated by multiplying gross irrigation amount by irrigation application efficiency (mm mm^{-1} [in in^{-1}]).

The last three variables in Eq. [1] (U, SRO, DP) are difficult to estimate in the field. In many situations, the water table is significantly deeper than the root zone and U is zero. Currently the IWM system assumes that U is zero, which limits applicability of the tool to fields that do not have a shallow water table. When natural precipitation occurs, the SRO component is estimated using the curve number approach (USDA-NRCS, 2004). In field studies of irrigation scheduling in eastern Colorado (Gleason, 2013), it was found that DP can be accounted for in a simple way by setting D_c to zero whenever P – SRO + Irr are greater than $D_p + ET_c$. Using these assumptions, Eq. [1] can be simplified to

$$D_c = D_p + ET_c - P - Irr + SRO \text{ (if } D_c < 0\text{, then } D_c = 0.0). \quad [2]$$

The D_c value is set equal to zero if its value becomes negative. This will occur if effective precipitation and/or net irrigation exceed ($D_p + ET_c$) and means that the soil is already at FC. This method does not account for DP directly, but simply assumes that any water that exceeds FC is lost to DP. Knowledge of Irr or P rates and soil infiltration rate can be used to improve the estimates of D_c. The IWM system allows a user to correct or override the calculated D_c with actual measured D in the field (from soil water content readings). This is consistent with recommended irrigation scheduling practice using the water balance approach (Robinson, 2009).

The soil profile is discretized into 2.5-mm (1-in) layers, and D_c is tracked for each layer. Water infiltration is simulated using a "tipping bucket" approach. Effective P + Irr is routed from the top layer, setting the layer's $D_c = 0$ if effective P + Irr exceeds D_c for that layer. Any excess P + Irr is moved to the next lower layer, and this continues down until all the effective P + Irr is distributed. Upward movement of water caused by soil water potential gradients is not simulated.

The available water capacity (AWC) of the soil is the amount of water between FC and permanent wilting point (PWP) and is assumed to be the maximum amount of soil water that can be used by the plants. The AWC of soil is typically expressed in terms of mm (in) of water per mm (in) of soil depth or in terms of m^3 of water per m^3 of soil, and values for specific soils are obtained from the USDA Soil Survey Geographic (SSURGO) Database.

In irrigation practice, only a percentage of AWC is allowed to be depleted because plants start to experience water stress even before soil water is depleted down to PWP. Therefore, a management allowed depletion (MAD, %) of the AWC must be specified by the user, or default values will be assumed. The rooting depth and MAD for a crop will change with developmental stage. The MAD is expressed in terms of depth of water (d_{MAD}; mm [in] of water) using the following equation:

$$d_{\mathrm{MAD}} = (\mathrm{MAD}/100) \times (\mathrm{AWC} \times D_{\mathrm{rz}}), \quad [3]$$

where MAD is management allowed depletion (%) and D_{rz} is depth of root zone [mm (in)]. Currently, the IWM system uses one value of MAD for the entire growing season, so changes in d_{MAD} are only caused by changes in D_{rz} (Eq. [7]). Variable MAD values for different crop growth phases will be included in a future version of the IWM system.

In the IWM system, the value of d_{MAD} is used as a guide for deciding when to irrigate. Typically, irrigation will be recommended when D_c approaches d_{MAD} or when $D_c > d_{MAD}$. To minimize water stress on the crop, the magnitude of D_c should be kept smaller than d_{MAD}. The user of the IWM system is given a summary of the profile soil water deficit (net irrigation requirement) for the current day. The decision to irrigate (gross amount and date) is left to the irrigator. Once the user applies a certain amount of gross irrigation on a specific day, he/she needs to input the amount in the IWM system to update the soil water balance.

Estimating Crop ET

The ET_c, in mm d^{-1} (in d^{-1}), is estimated as (Allen et al., 1998)

$$\mathrm{ET_c} = \mathrm{ET_r} \times K_{\mathrm{cr}} \times K_{\mathrm{s}}, \quad [4]$$

where ET_r is the evapotranspiration rate, mm d^{-1} (in d^{-1}), from a tall reference crop (i.e., alfalfa [*Medicago sativa* L.]), K_{cr} is a crop coefficient (alfalfa based) that varies by crop development stage (ranges from 0 to 1), and K_s is a water stress coefficient (ranges from 0 to 1). Recent studies have shown that ET_r can be underestimated by the commonly used Penman–Monteith equation and that a recursive method that solves for temperature of the evaporating surface can improve accuracy of ET_r calculations (Lascano and van Bavel, 2007; Lascano et al., 2010). Given the possible underestimation of ET_r using the Penman–Monteith equation, it is possible to obtain K_{cr} values that may exceed 1.0. The K_{cr} values incorporate the effects of crop development on ET_c. The general shape of the seasonal K_{cr} curve is similar to the one given by Allen et al. (1998, 2007b). However, growing degree days (GDD) are used as the time unit instead of days after planting, and users can modify the default values to better represent their crop variety. If D_c remains smaller than d_{MAD}, K_s is assumed to be 1 (no water stress). The CoAgMet is the primary source of daily alfalfa reference ET_r calculated using the hourly version of the American Society of Civil Engineers (ASCE) Standardized Penman–Monteith Equation (Allen et al., 2005) by summing 24 hourly values. The hourly version of the Standardized Equation is recommended when significant shifts in wind and humidity occur hourly (Allen et al., 2005), which is common in the semiarid irrigated areas of Colorado (Andales et al., 2009b). The Standardized Equation has been shown to match diurnal changes in ET_r rates in semiarid areas (Evett et al., 2000) and was successfully validated at 49 climatically diverse sites across the United States, including 10 sites in Colorado (Itenfisu et al., 2003). The CoAgMet collects hourly weather data at all stations. In addition, the IWM system also has access to daily weather data from a network of 19 automatic weather stations owned by Northern Colorado Water Conservancy District (NCWCD) (http://www.northernwater.org/WaterConservation/WeatherandETData.aspx). These stations are located in the South Platte River Basin of northeast Colorado. The IWM system calculates K_{cr} internally, based on emergence date and accumulated GDD.

In cases when water availability is limited (e.g., lack of precipitation or irrigation; $D_c > d_{MAD}$), then K_s will be < 1, and crop ET_c will not occur at the potential (non-water-limited) rate. The K_s is calculated by (Allen et al., 1998)

$$K_s = (\mathrm{TAW} - D)/[(1 - \mathrm{MAD})\mathrm{TAW}] \quad (K_s = 1 \text{ if } D < d_{MAD}), \qquad [5]$$

where TAW is total available water in the soil root zone mm (in), D is the soil water deficit mm (in), and MAD is management allowed depletion (decimal fraction). The value of TAW is calculated from

$$\mathrm{TAW} = \mathrm{AWC} \times D_{rz}, \qquad [6]$$

where AWC is available water capacity of the root zone (mm of water/mm of soil [in of water/in of soil] or m^3 of water/m^3 of soil), and D_{rz} is the total depth of the root zone, mm (in). The nonstressed ET_c obtained from ET_r and K_{cr} is multiplied by K_s to estimate an "actual" (water-stressed) ET_c value that is used in Eq. [2].

Equations [2]–[6] have been incorporated into both the web-based and mobile IWM modules, which track daily values of P, Irr, SRO, D_{rz}, TAW, K_{cr}, K_s, ET_c, and D_c. The above algorithms and the graphical user interface were written as a module in eRAMS using Python and Javascript. Daily values of ET_r and P are obtained from CoAgMet or NCWCD. However, users of the IWM system are given the option of providing field-specific P values from their own rain gauges.

Root Growth and Soil Water Extraction

A simple root growth model conceptualized by the senior author is used to estimate plant available water at different phases of crop development. Initially, the root zone is assumed to be 152 mm (6.0 in) and remains so until emergence. The managed depth is assumed to be where most (e.g., 70%) of the roots are. The modeled root zone depth is assumed to reach the management depth once the crop reaches full canopy (i.e., once peak K_{cr} is reached). For any day between crop emergence and attainment of full canopy, the root zone is estimated as

$$D_{rz} = \frac{D_{max} - 152}{GDD_M \times cutoff2} \times GDD_{cum} + 152, \qquad [7]$$

where D_{rz} is depth of root zone (mm), D_{max} is the user-specified management depth (mm), GDD_M is the total cumulative GDDs from emergence required to reach crop maturity (°C d), cutoff2 is the fraction of GDD_M corresponding to the start of peak K_{cr} (when crop reaches full canopy), and GDD_{cum} is accumulated GDD on the current day (°C d). For any day, the soil water deficit is only calculated for the D_{rz}. The D_{rz} is assumed to be constant ($D_{rz} = D_{max}$) from attainment of full canopy until crop maturity.

For any D_{rz}, daily root water extraction that corresponds to daily ET_c is assumed to be partitioned as follows: 40% from the top quarter of D_{rz}, 30% from the second quarter of D_{rz}, 20% from the third quarter of D_{rz}, and 10% from the bottom quarter of D_{rz}. This root water extraction pattern is typical of most crops (USDA-SCS, 1991).

Setup and Operation of the IWM System

The online tool can be accessed in eRAMS (https://erams.com/) using a web browser and an Internet connection. A user needs to create an eRAMS account in which all associated field data are stored. An individual field having a single crop and served by a specific irrigation system (e.g., center pivot sprinkler system) is the management unit used in the IWM system. A user who manages several fields or several crops can simply go through the setup process for each field, which is saved as a separate "Irrigation" project in eRAMS. The tool has capabilities for locating a specific field on an aerial map and drawing field boundaries (Fig. 3–2). Once the field boundaries are drawn, the tool extracts soil properties of the field from the SSURGO database. The water holding capacity of the soil in the field is the primary piece of information used by the tool for estimating soil water content. Nine crops including alfalfa, barley (*Hordeum vulgare* L.), corn (*Zea mays* L.), dry beans (*Phaseolus vulgaris* L.), potato (*Solanum tuberosum* L.), sorghum (*Sorghum* L.), sugarbeet (*Beta vulgaris* L.), sunflower (*Helianthus* L.), and winter wheat (*Triticum aestivum* L.) are available from the crop type menu.

The tool automatically locates the online weather stations closest to the selected field, according to a user-specified search radius. Nearby weather stations are displayed on the map, and the user can select the weather station(s) that will be used to estimate P and ET_r on the selected field. If more than one weather station is selected, the weather variables are averaged using inverse distance weighting. Charts of recent weather data can also be viewed (Fig. 3–3).

To completely set up a field for irrigation scheduling, the user also has to input crop information: type, planting date, emergence or green-up date, and managed root depth; irrigation system information: type, application efficiency,

Fig. 3–2. View of the online irrigation scheduler showing tools for drawing field boundaries on a map.

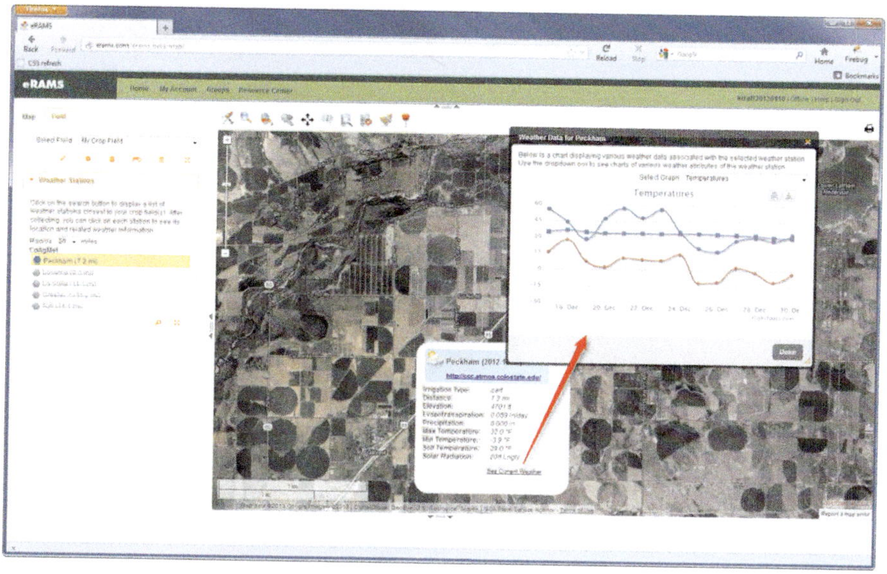

Fig. 3–3. Online irrigation scheduler showing nearby weather stations and charts of recent weather data.

capacity, and typical irrigation frequency; and soil information: initial soil water content at planting. Once a crop type is selected, default values of K_{cr} (used to estimate ET_c from Eq. [4]) are provided. An example of a K_{cr} curve is provided in a later section.

After the user provides all the field information, the tool can begin estimating the daily soil water content of the managed root zone. Gross amounts of irrigation applied to the field must be inputted whenever irrigations occur. The daily water balance of the root zone can be viewed as a summary for the current day (Fig. 3–4), as a daily spreadsheet (image not shown), or as a seasonal graph (Fig. 3–5).

Mobile App for Smartphones

A prototype iPhone app has been developed to synchronize information with eRAMS and give mobile access to a field's current water status. At the time of writing, the app is still in the process of being submitted to the Apple app store. A water gauge or "bucket" diagram is used to show the amount of water available to the crop and the recommended amount of irrigation (Fig. 3–6). Options for inputting actual gross irrigation or locally measured precipitation are also available. Yesterday's weather can also be viewed for the preselected weather station. A platform-independent mobile application that functions on any mobile

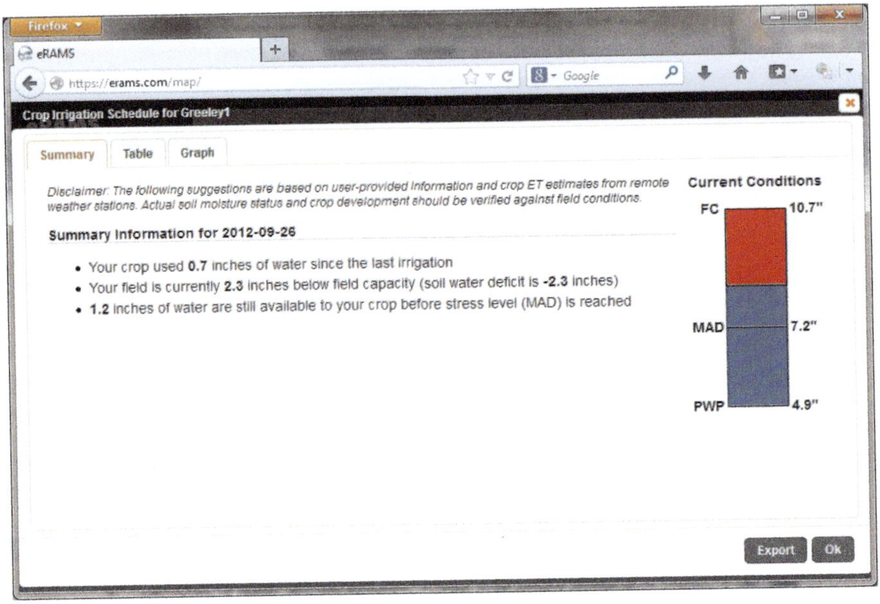

Fig. 3–4. Summary screen for a field showing soil profile water status using a "water bucket" diagram of total soil water deficit (red bar; 58 mm [2.3 in]) relative to field capacity (FC; 272 mm [10.7 in]), management allowed depletion (MAD; 183 mm [7.2 in]), and permanent wilting point (PWP; 124 mm [4.9 in]). Summary information is also given to the left of the diagram (18 mm [0.7 in] of water used since last irrigation; 30 mm [1.2 in] of water available before MAD is reached).

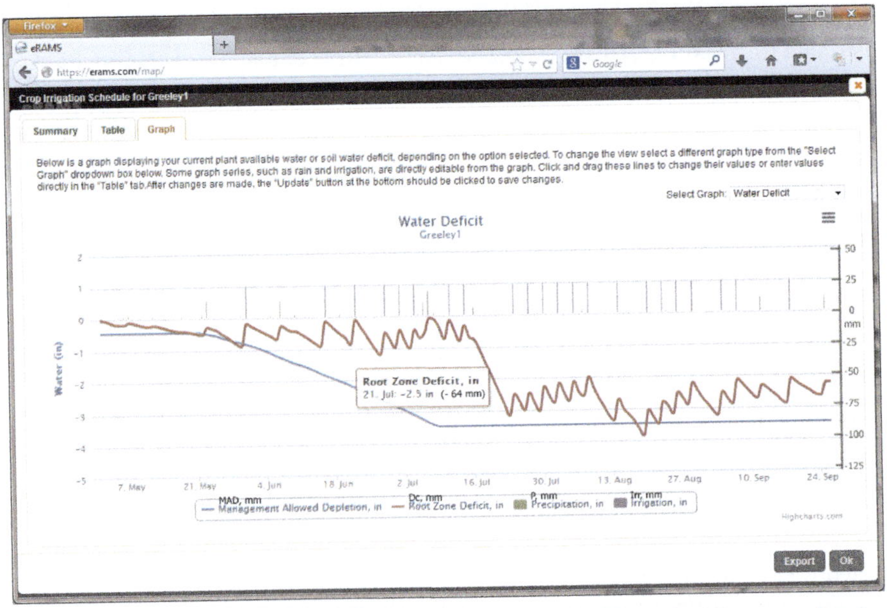

Fig. 3–5. Example graph showing daily water deficits in the root zone (red line; mm of water shown on right *y* axis) in relation to management allowed depletion (MAD, blue line). Irrigation is recommended when the red line approaches or falls below the MAD. Daily precipitation and irrigation are also shown as vertical bars in the graph. The actual web interface shows water depths in inches (left *y* axis).

device (e.g., iPhone, Android smartphones) having a web browser was also developed using HTML5 Internet markup language standards. The functionality and appearance of the HTML5 version of the mobile app are similar to the iPhone app (Fig. 3–6).

Evaluation for Selected Center Pivot Sprinkler-Irrigated Fields

The IWM system was used for a center pivot-irrigated corn field near Greeley, CO (40.460545° N, 104.57894° W; 1429 m above mean sea level). The study was conducted from 2010 to 2012. The dominant soil series in the field is Olney fine sandy loam (fine-loamy, mixed, superactive, mesic Ustollic Haplargids) (Soil Survey Staff, 2013).

Soil bulk density in the corn field was determined using a Madera probe (Precision Machine Company Inc.) and methods described by Evett (2008). Soil bulk density was not a required input for the IWM system, but was used to convert measured gravimetric soil water content to volumetric values. Two samples were taken from each of two sampling points in the field at the end of the 2011 growing season at depths of 0 to 15, 15 to 30, 30 to 45, 45 to 60, 60 to 90, and 90 to 105 cm. The PWP for each soil layer was determined using a WP4-T Dewpoint PotentiaMeter (Decagon Devices, Inc.). A soil water release curve was created to obtain gravimetric water content at 1.5 MPa of tension. Volumetric water content at PWP was then calculated using methods described by Evett (2008). The FC for each soil layer was determined in situ by gravimetric sampling 24 h after several deep irrigation or precipitation events in 2011. Values for soil properties used in the IWM system evaluation are summarized in Table 3–1.

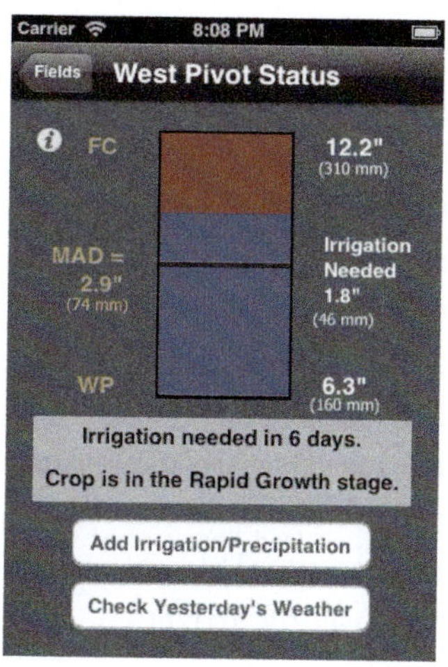

Fig. 3–6. Prototype iPhone app showing a "water bucket" representation of soil water status of a field. The actual app shows water amounts in inches ("), so the figure was annotated with mm amounts given in parentheses. Field capacity (FC) and wilting point (WP) show the upper and lower limits of plant available water (inches of water) in the root zone, respectively. The red bar shows the estimated amount of deficit or depletion (irrigation needed) relative to management allowed depletion (MAD).

Table 3–1. Corn field soil properties used in the evaluation of the irrigation water management system near Greeley, CO.

Sampling point	Depth	Bulk density	Field capacity	Wilting point	Available water capacity
	cm	g cm^{-3}		m^3 m^{-3}	
North point	0–15	1.16	0.33	0.17	0.16
	15–30	1.13	0.27	0.17	0.10
	30–45	1.33	0.28	0.14	0.14
	45–60	1.28	0.26	0.14	0.13
	60–90	1.30	0.26	0.11	0.15
	90–105	1.24	0.27	0.15	0.12
South point	0–15	1.06	0.33	0.15	0.18
	15–30	1.31	0.32	0.18	0.14
	30–45	1.25	0.27	0.14	0.13
	45–60	1.22	0.23	0.13	0.10
	60–90	1.40	0.25	0.12	0.13
	90–105	1.28	0.28	0.16	0.12

The crop and irrigation system was maintained and managed by a cooperating grower who allowed weekly monitoring of soil water content and crop development in his field. Actual gross irrigation amounts and dates were provided by the grower. Irrigation application efficiency of the system was estimated to be 90%. The K_{cr} curve used for corn was similar to curves reported by Allen et al. (2007b), with adjustments made to fit the canopy development of corn grown in the 2010 growing season and days after planting (x axis) converted to percentage of GDDs to maturity (Fig. 3–7). Daily weather data for use in the IWM system were taken from the GLY04 (40.4487° N, 104.638° W, 1427 m above mean sea level) CoAgMet weather station located 5.2 km west of the corn field.

The performance of the irrigation scheduling tool in estimating current total soil water deficit (D_c) in the managed root zone was tested by comparing the scheduler's calculated D_c to measured D_c on measurement dates (once a week during the growing season). This evaluation method is similar to the one used by Jensen et al. (1971). Measured D_c was derived from measured gravimetric soil water content of the managed root zone (Gleason, 2013). The management depth used in the evaluation of the irrigation scheduler was set at 1.05 m (41 in). In each year, the tool was initialized with the first measured D_c at each sampling point.

In the 2013 growing season, the IWM system was also tested on four center pivot sprinkler-irrigated sugarbeet fields in northeast Colorado. This initial year of testing for sugarbeet was primarily done to gather base line irrigation, crop

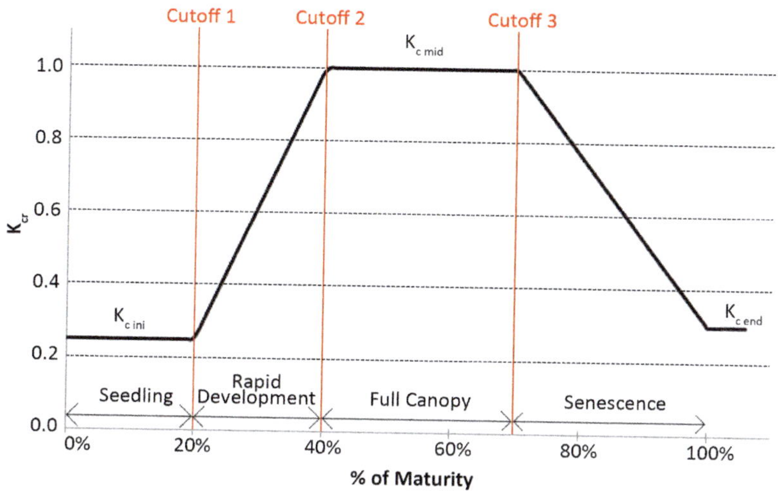

Fig. 3–7. K_{cr} curve used to calculate corn ET (ET_c) in the irrigation scheduler with crop development phases and initial ($K_{cr\,ini}$ = 0.25), peak or midpoint ($K_{cr\,mid}$ = 1.0), and ending ($K_{cr\,end}$ = 0.30) values of the K_{cr} curve indicated. Cutoffs (curve inflections) are displayed in red and represent the percentage of crop maturity based on growing degree days.

development, and management information to improve on a seasonal K_{cr} curve for sugarbeet that was initially obtained from Allen et al. (2007b).

Promotion of the IWM System

After an online prototype of the IWM system was developed, a stakeholder committee was formed by inviting progressive crop producers and advisers, researchers, USDA-NRCS agency personnel, farm managers, and crop commodity group representatives to a 2-h winter workshop demonstrating the prototype. A total of 10 individuals were part of the committee. Each member was provided an online computer and was given an opportunity to set up a field and simulate an irrigation schedule using online CoAgMet weather data from the previous growing season. The last half hour of the workshop was spent collecting comments and suggestions from the committee regarding the web-based and smartphone interfaces, soil water balance calculations and assumptions, and type of irrigation scheduling information displayed. A list of changes to the IWM system was compiled and implemented by the computer programmers.

After the workshop, several committee members volunteered to test the IWM system on some of their irrigated fields during the upcoming 2013 growing season. Once the changes and improvements were made to the IWM system, project team members gave online demonstrations at more than 10 producer or water manager conferences and workshops across the state. These led to additional volunteers who began testing the tool for their irrigated fields. The volunteers were able to identify some errors in the calculations and gave suggestions for improvement of the tool. The project team corrected the errors and implemented the improvements.

Following the first growing season with the stakeholder group, a second workshop was held to learn about any problems the group had during the previous summer, demonstrate new features, get additional suggestions for improvement, and make plans for the 2014 growing season. Following this workshop, a short survey was circulated to allow users to provide more confidential feedback on the IWM system and learn about its potential for more widespread adoption.

Results and Discussion

Evaluation for Selected Center Pivot Sprinkler-Irrigated Fields

An example of daily D_c for corn in 2010 is shown in Fig. 3–8. Performance statistics (calculated versus measured D_c on measurement dates) in 2010 to 2012 are given in Table 3–2. Recall that the calculated D_c from the IWM system was from a simplified daily soil water balance model (Eq. [2]). The IWM system was initialized with the measured D_c for the profile at the start of each growing season. No corrections were made to the daily calculated D_c throughout each growing season. Nevertheless, overall relative error (RE) was relatively low at 13.6%. The RMSE, with the exception of 2012 at the north sampling point, was less than what could be applied by the center pivot in a single irrigation event (<19 mm). The RMSE was deemed acceptable because just one irrigation application could eas-

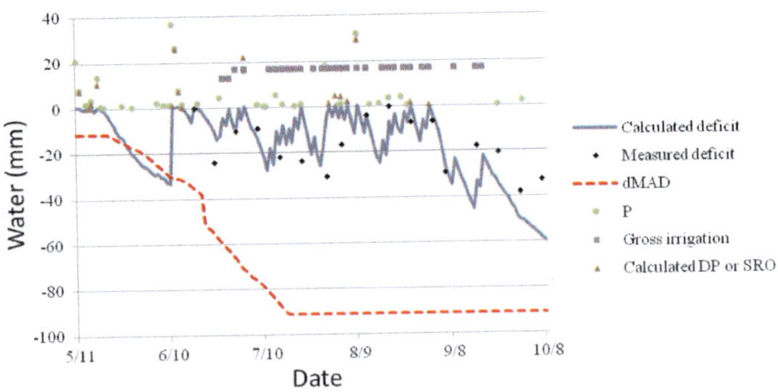

Fig. 3–8. Comparison of calculated deficit (blue line) and measured deficit (black diamonds) from the north sampling point during the 2010 growing season. Also indicated is the depth of management allowed depletion (d_{MAD}; dashed red line), gross precipitation (*P*), gross irrigation, estimated deep percolation (DP), and surface runoff (SRO). All variables are represented as depth of water in the root zone (mm).

Table 3–2. Comparison of irrigation water management system daily calculated and measured deficits during the 2010–2012 corn growing seasons. Measured deficits were obtained manually through gravimetric sampling.

Site and year	*n*†	RMSE	MBE‡	MAE§	RE¶
		mm			%
North 2010	16	13.0	−0.3	10.6	1.8
South 2010	16	16.4	−1.5	13.1	8.6
North 2011	16	12.1	−1.8	10.8	11.3
South 2011	16	15.7	−1.6	12.6	11.3
North 2012	15	22.9	−12.9	18.0	30.9
South 2012	15	13.4	−2.9	10.8	6.5
All	94	15.9	−3.4	12.6	13.6

† *n*, number of measurements.

‡ MBE, mean bias error.

§ MAE, mean absolute error.

¶ RE, relative error.

ily compensate for it. The positive RE values (Table 3–2) mean that the magnitude of calculated D_c was overestimated, which indicated that the IWM system was estimating a drier soil profile than actual conditions. Calculated and measured D_c values were both represented as negative values (lack of water in the profile) in the calculation of mean bias error (MBE). Thus, the negative MBE values in Table 3–2 indicated that calculated D_c values were larger in magnitude than measured D_c values, consistent with the positive RE values. This bias would indicate the need for slightly more irrigation water than actually required, but would not increase the risk of water stress and yield reductions.

To demonstrate potential water savings through irrigation scheduling using the IWM system, Gleason (2013) compared actual (farmer-managed) gross irrigation amounts in 2011 for the abovementioned corn field with hypothetical gross irrigation amounts that would result from scheduling irrigations based on D_c estimated by the IWM system. Precipitation in late May and the first half of June allowed the farmer to delay irrigations until 24 June 2011. Afterward, the center pivot ran on a fixed schedule, applying 13 to 17 mm of gross irrigation every 2 d. The center pivot was turned off for only one irrigation cycle in mid-July when significant rainfall occurred (15 mm on 11–13 July 2011). Gross irrigations were increased to 19 mm every 2 d beginning 28 July 2011 (reproductive growth phase) until 5 Sept. 2011. An irrigation cycle was skipped only when significant rainfall occurred (28 mm on 2–4 Aug. 2011) or was forecasted.

In contrast, hypothetical irrigations were applied only when the calculated D_c approached MAD, taking into consideration the system capacity of the center pivot. Calculated water balance components were compared during the period 13 June 2011 to 10 Oct. 2011 (Table 3–3). The actual gross irrigation for the period was 511 mm. In contrast, the recommended (hypothetical) gross irrigation amount was 372 mm using the IWM system for irrigation scheduling. This represented a 27% (139 mm) reduction in gross irrigation, which is a significant savings in water and electrical energy (energy savings not calculated). The number of irrigation applications was also reduced from 30 down to 20, which would have reduced electricity costs for running the center pivot. Runoff and deep percolation calculated by the IWM system was 146 mm for the season with the actual irrigation amounts, while it was reduced to 37 mm (75% reduction) using the IWM system for irrigation scheduling. The reduced water losses could have potentially reduced nutrient and chemical transport from the field, which is a benefit for water quality (actual water quality not measured). Following the recommended irrigations from the IWM system would have stored more water in the profile (ΔS = 42 mm; 250% increase) than actual irrigations (ΔS = 12 mm).

For the above comparison, corn grain yield would not have been reduced if recommended irrigations were applied because there was no reduction in ET_c (Table 3–3). Corn grain yield has a linear response to cumulative ET_c (Tolk et al., 1998), so the same ET_c would result in the same yield. This study showed that for the 13 June 2011 to 10 Oct. 2011 period, the corn crop was overirrigated by approximately 139 mm.

Figure 3–9 shows an example graphical output of the IWM system for a sugarbeet field in 2013. In contrast to Fig. 3–5 and Fig. 3–8, which show soil water status in terms of water deficits, Fig. 3–9 shows soil water status in terms of plant-available water (PAW) in the root zone. The IWM system has the capability of presenting soil water status in either mode, depending on the preference of the user. Daily estimated PAW is shown relative to AWC and MAD. Net irrigation, effective rainfall, and DP (if any) are also shown on the graph. As a criterion for

Table 3–3. Calculated water balance components using actual and irrigation water management system recommended irrigations for center pivot irrigated corn at Greeley, CO in 2011 (13 June–10 October).

Water balance component†	With actual irrigations	With recommended irrigations
	mm	
ET_c (mm)	501	501
Gross Irr (mm)	511	372
P (mm)	125	125
DP + SRO (mm)	146	37
ΔS (mm)	12	42

† ET_c, crop ET; Gross Irr, gross irrigation; *P*, gross precipitation; DP, deep percolation; SRO, surface runoff; ΔS, change in soil water storage in managed root zone (1050 mm).

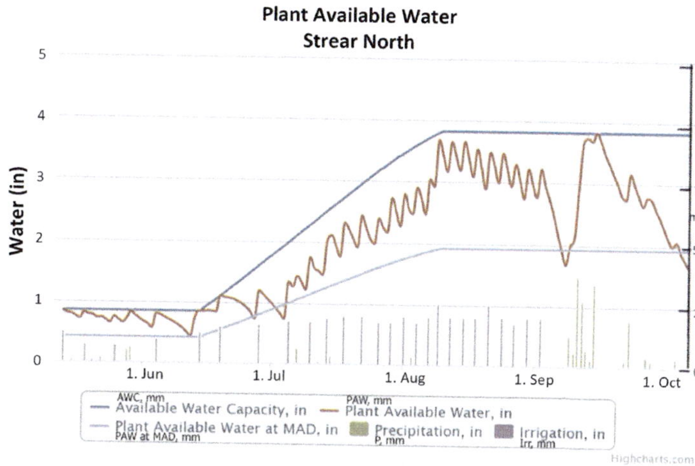

Fig. 3–9. Seasonal plant available water (mm of water shown on right *y* axis) calculated by the online irrigation scheduler for a center pivot-irrigated sugarbeet field in northeast Colorado during the 2013 growing season. The actual web interface shows water depths in inches (left *y* axis).

irrigation scheduling, the PAW is ideally kept in the region between AWC and MAD when deciding irrigation amount and timing. A second growing season is still needed to fully evaluate the effectiveness of the IWM system for scheduling irrigations for sugarbeet.

Factors Affecting Accuracy

The estimation of ET_c, which determines crop water requirements, is one of the main sources of uncertainty in irrigation scheduling using the soil water balance approach (Rhenals and Bras, 1981). The IWM system estimates ET_c using Eq. [4], which has three variables that may affect the accuracy of calculated ET_c. The above evaluation study for corn revealed that daily D_c (Eq. [2]) was generally overestimated. From Eq. [2], it may be deduced that corn ET_c was overestimated or P was underestimated. Gleason (2013) found that overestimation of D_c mostly occurred in the latter part of the corn growing season—in August and September (Fig. 3–8). In all three corn growing seasons (2010–2012), measured and calculated D_c remained smaller than d_{MAD} such that $K_s = 1.0$. This suggested that corn K_{cr} values during this period should be decreased. However, actual corn ET_c in this study was not directly measured, so this theory needs further investigation with measurements of actual ET_c.

Measured P was obtained from a CoAgMet weather station 5.2 km west of the corn field. Spatial variability of precipitation between the weather station and the corn field may have introduced errors in the daily calculation of D_c. For instance, errors in P can induce errors in calculated SRO (Troutman, 1983). In an analysis of spatial precipitation variability in northeast Colorado (May–September growing season for 1939–1959 and 1971–2006), Augustine (2010) found a mean spatial variation of 30 mm across distances of 6 to 8 km. Spatial precipitation variability increased linearly across distances of 10 to 160 km (up to 86-mm average spatial variation across distances > 120 km). He observed that spatial variation in growing season precipitation was low across distances of < 10 km (6.2 mi), but suggested that rain gauges should be within 2 km (1.2 mi) of any site to accurately represent local P in eastern Colorado. He also observed that the magnitude of spatial variation was significantly higher in wet years than in dry years. Therefore, it is recommended that an onsite rain gauge be used for irrigation scheduling of fields if the nearest CoAgMet weather station is > 2 km (1.2 mi) away. Measured P from rain gauges between 2 and 10 km (6.2 mi) may still be acceptable but used with caution if an onsite rain gauge is not available.

The above discussion highlights the importance of selecting a CoAgMet weather station that represents the weather conditions at the irrigated field. For calculation of the ASCE standardized ET_{rs}, the station should ideally be above well-watered clipped grass vegetation and extending at least 100 m in all directions from the weather station (Allen et al., 2005). The IWM system classifies the nearby weather stations in terms of surrounding conditions as either fully irrigated, partially irrigated, or dryland. It is recommended that fully irrigated or partially irrigated weather stations be selected. The use of a dryland weather station will result in overestimation of ET_{rs} because of elevated temperatures and increased advection that are typical of dryland sites. The CoAgMet weather station used in this study (GLY04) was above a fully irrigated grass vegetation and satisfied the recommendations given by Allen et al. (2005).

Errors in estimating D_c can accumulate through time. In the evaluation study for center pivot irrigated corn, calculated D_c was not corrected using midseason measurements of actual D_c. However, the IWM system allows users to input measured D_c for any date to correct any cumulative errors in calculated D_c that may have occurred up to that date. Therefore, occasional measurements of actual D_c in the field are recommended for use in correcting the calculated deficit. Frequent measurements of profile D_c may not be practical, but even one or two measurements during midseason can greatly improve the accuracy of the IWM system in calculating net irrigation requirements. For example, one midseason correction in 2011 (north sampling point) using measured D_c on 15 July 2011 reduced overall RE of daily calculated D_c from 11.3 to 6.5% and slightly reduced RMSE from 12.1 to 11.7 mm.

The AWC and corresponding TAW (Eq. [6]) for the managed root zone are the most important soil properties for irrigation scheduling (Prats and Picó, 2010). These parameters determine the water storage capacity of the root zone and greatly influence the calculations of D_c, SRO, and DP. Therefore, efforts should be made to obtain accurate estimates of these parameters for the IWM system. The measured AWC values shown in Table 3–1 were within the ranges of AWC given by the Web Soil Survey (Soil Survey Staff, 2013) for Olney fine sandy loam soil. However, the measured bulk densities were relatively low, especially for the 0- to 15-cm top soil layer (1.06–1.16 g cm^{-3}). These low values were likely due to the measurement of bulk density shortly after conventional tillage of the corn field before planting, as tillage may have enhanced the porosity. Such values have been observed before, as reported by Dam et al. (2005), who found lower bulk densities in conventionally tilled soil (as low as 0.96–1.12 g cm^{-3}) than in no-tilled or ridge-tilled sandy loam soil.

Stakeholder Involvement and Outreach to Achieve Improved Adoption

Adoption of IWM systems for irrigation scheduling is currently very low in Colorado and in the United States. For example, USDA-NASS (2009) reported that <1% of 12,788 respondents in Colorado used "computer simulation models" for deciding when to irrigate. There are many likely explanations for this low level of adoption including availability, usability, and accuracy for irrigation scheduling. An obvious process to improve the usability and utility of an IWM system is to involve potential users in a process that allows for stakeholders to use and provide feedback on the system early in the development process.

Our experiences have shown that the timing of when stakeholders are brought into the discussions of how an irrigation scheduling system should look, function, and perform is critical to its ultimate adoption. Developers and outreach professionals need to find the balance between introducing a nearly finished product with little room for stakeholder influence or input and demonstrating a system with multiple glitches and errors. Both of these situations can hurt the ultimate adoption of the IWM system. Another key step is to clearly communicate the current status of the system development and the expectations of the stakeholder committee to minimize any frustration from stakeholder users, especially if they experience multiple problems early in the development process. Researchers and system developers should keep technical jargon to a minimum with communications about IWM system development. For example, many agricultural practitioners may not understand what a beta version of a software product really means in terms of functionality and reliability. It is also crucial to emphasize that while individual users may have good ideas for improvement of the system for their particular situation, it may not be feasible to customize it for every user.

Getting the right composition and size of a committee is also important to maximize the utility of IWM system stakeholder input. If an individual expresses that they would never use an IWM system for irrigation scheduling or other farm management function, they are not going to be useful to the process even if they are influential in the agricultural community. Alternatively, stakeholders that are likely users but are reluctant to provide honest, constructive criticism of the system are also not helpful to improve the program. Individuals that have had experiences with previous IWM systems, especially irrigation scheduling, are useful in providing input on what they did and did not appreciate about those programs. Ultimately, stakeholders that will spend some time attempting to use the IWM system between meetings and provide honest feedback regarding their experiences will be most useful to improve the product.

The stakeholder committee for the irrigation scheduler described in this chapter includes crop producers and advisers, researchers, USDA-NRCS agency personnel, farm managers, and crop commodity group representatives. This mix allows for input from actual users at multiple levels and advocacy for adoption by individuals who have invested time and experiences in developing and improving the product.

In our experience, presentations of the IWM system for irrigation scheduling utility and functionality can occur at several scales and use a variety of methods. We have found that workshops with stakeholders where potential users can get hands-on experience with the tool are the most useful for finding glitches and errors or circumstances that developers may have never encountered. These sessions can identify and address questions or problems that are shared by several members of the community. However, sufficient numbers of knowledgeable staff should be on hand to help stakeholders work through examples and issues that arise. Having problems with user login and creating accounts can distract from showing the utility of the IWM system for improved water management and getting feedback for improvement. Ensuring that the workshop site has adequate Internet speed and reliability are also crucial to having a productive meeting. However, if the IWM system will operate slower on a home or rural Internet, this should be made clear to the group.

Presentations to larger potential user groups relatively early in the outreach process can be useful to generate interest. The authors experienced this scenario when a demonstration to a large group of crop consultants sparked interest from agronomists working for a sugarbeet cooperative. This interest resulted in a field demonstration and validation project of the tool to four of their grower clients.

We included cooperating stakeholders from farms where the irrigation scheduling system was field tested. These stakeholders could follow the field water balance online and enter irrigation events, while data was collected to validate the IWM system. This process demonstrates that the tool is validated with sound data and recommendations are not from a "black box" process that the user does not understand.

For presentations to larger groups other than the stakeholder committee (>15), we have found it more useful to have the presenter walk through the primary features of the IWM system live, while allowing for time for questions or to explore the minor detailed features of the program as interest arises. For these larger venues, we have used presentations with screenshots of the irrigation scheduler as backup for Internet failure or computer glitches. These problems can quickly ruin an otherwise productive presentation. Having prepared questionnaires or other feedback mechanisms on hand can help record input for future consideration.

Utilizing survey instruments can be one method to improve IWM system usability and understand potential users' needs. These can be conducted online for larger groups or by paper copy for small stakeholder committees. Table 3–4 provides summarized results from a short survey we utilized to direct our efforts in improving the IWM system discussed here. Response rate from our committee was 87%. From this effort we learned how our committee felt about the IWM system development progress and how they preferred to get user support. Most of the committee felt the system was close to being ready for use, and online video was the most popular resource for support. We also learned that time to learn and set up the IWM system could be a critical impediment to adoption. This result shows how crucial it is to develop systems that are simple and intuitive to set up. Positive feedback from these results is that 100% of the committee planned to use the IWM system in the future and would also recommend the system to a neighbor or colleague.

The best IWM system for irrigation scheduling will be reasonably intuitive or direct first-time users to setup fields with a minimum of support and tutoring. Most users today will be unwilling to spend much time reading and understanding a user manual. Fortunately, new technologies allow alternative approaches for teaching users to understand how to work through an IWM system. For example, screen recording programs are available that allow you to record how-to videos to demonstrate basic functionality of the IWM system. These can be made into short YouTube videos or provided within the tool. We have also utilized technologies such as online webinars and "Lifesize" rooms to successfully demonstrate and teach the scheduling tool to remote audiences. Eventually, the most useful skill to have when developing a user-friendly IWM system for irrigation

Table 3–4. Summary of stakeholder committee post-workshop survey results from second meeting. Percentages may not add to 100% when respondents chose more than one answer (n = 14).

Survey question	Responses
How do you feel about the current development of this product?	ready to go with minor adjustment: 50% needs a little work: 29% would like the app: 14% N/A: 7%
What would be the most useful resource(s) to help users understand how to use this program initially or when they encounter problems or questions?	online video: 50% written: 35% webinar/workshop: 14% email or phone: 29% all: 7%
Did you use the irrigation scheduler this past summer?	yes: 50% no: 43% N/A: 7%
Do you plan to use this tool in the future?	yes: 100% no: 0%
What do you think are the primary reasons that could keep producers or others from using this tool?	time to setup and/or learn: 57% understanding of ET scheduling: 28% familiarity and ease of operation: 28% other: 14%
What do you think would need to be changed for *you* to use this tool more?	phone app (Android and Apple): 29% time to learn: 14% upload speed: 7% adapt to USDA-NRCS standards: 7% other: 29%
Would you recommend the tool to a neighbor/colleague?	yes: 100% no: 0%
As we look toward revenue for future upkeep and improvement, would you rather pay for advanced features of the scheduler, iPhone and Android apps, or technical support?	would pay for apps or advanced features: 43% maybe: 43% would not pay: 14%
Would a cost share program (e.g., USDA-NRCS) encourage use of this tool?	yes: 64% no: 14% not sure: 7% irrigation water management cost share is available: 7% N/A: 7%

scheduling is active listening. Cultivate a sincere, interested stakeholder group, allow them to provide feedback, earnestly listen to their needs and concerns, and make changes accordingly. This model has worked well for the authors in the improvement and adoption of their IWM system.

Future Work

Localized K_{cr} curves for other major irrigated crops are required to expand the use of the IWM system in Colorado and beyond. Therefore, validated K_{cr} curves will continually be added to the IWM system. It is anticipated that a database of

K_{cr} will be incorporated in the IWM system to include curves for various crop varieties (e.g., early- vs. late-maturing varieties of the same crop) and geographic regions. Given that seed companies are continually releasing new crop varieties, the process of updating the K_{cr} curves will continue as the need arises. New technologies such as satellite-based remote sensing of ET_c have made it possible to develop regional K_{cr} curves that account for spatial variability of ET_c (Garcia et al., 2013).

The IWM system can potentially be used for other regions that have access to automatic weather station networks. The spatial analysis capabilities of eRAMS can be extended to other regions, as long as the appropriate soils and weather data are available. This could lead to expanded availability of the IWM system beyond the boundaries of Colorado. For future versions of the IWM system, atmometers for estimating local reference ET_r may be used to improve local ET_r estimates at sites with no accessible weather station (Gleason et al., 2013).

A novel feature that will be tested in a future version of the IWM system is the capability of using ET_c estimates from remotely sensed images (e.g., from satellites). The energy balance algorithm called "Remote Sensing of ET" or "ReSET" (Elhaddad et al., 2011) will be used to estimate daily values of ET_c for use in calculating the soil water balance at the field level. The ReSET model estimates the actual ET_c at the time of a satellite image. The model can be applied to large areas (180 by 180 km) yet retaining a resolution of 30 by 30 m that allows the model to determine ET_c for small parcels. The ReSET model is built on the same theoretical basis of its two predecessors METRIC (Allen et al., 2007a) and SEBAL (Bastiaanssen et al., 1998a, 1998b), with the additional ability to handle data from multiple weather stations. This capability enhances regional ET_c estimates by taking into consideration the spatial variability of weather conditions through data acquired from different weather stations (across the area covered by the remote sensing system or imagery).

The ReSET model can add better spatial coverage of ET_c (field specific) to the IWM system, to fill in gaps between weather stations. Actual ET_c from ReSET can be used for fields that are far from a CoAgMet weather station. It estimates actual ET_c, so it can also be used to validate ET_c values estimated using crop coefficients (Eq. [4]). The infrequent availability (e.g., every 2 wk) of satellite imagery can be overcome by interpolating daily ET_c values between limited remotely sensed ET_c using an algorithm that accounts for temporal variations in local weather during the interpolation periods. The ReSET model does this by using daily wind and reference ET_r values from weather stations (Elhaddad et al., 2011), such as those in the CoAgMet network. Several other techniques may have practical utility for irrigation scheduling in the near future. Accuracy of short-term weather forecasts are improving and may also be used for forecasting ET_c and irrigation requirements (Gowing and Ejieji, 2001) or rainfall (Cai et al., 2011). Also, soil water content could be estimated from crop water stress index derived from infrared thermometry (Taghvaeian et al., 2012). Advances in these techniques will eventually improve the accuracy as well as the ease of calculating irrigation requirements.

Conclusions

In the past, the use of computer-based IWM systems has been hindered by lack of accessibility, lack of input data, and system complexity. Recently, the convergence of Internet accessibility, availability of automatic weather station networks, development of cloud services, and widespread use of mobile devices (e.g., smartphones) has provided new opportunities in making IWM systems more accessible and user-friendly. This chapter described one such IWM system developed for Colorado.

The IWM system is accessible through either a web browser or a mobile app. The system takes initial user inputs (field boundaries, crop information, selected weather station[s], irrigation system characteristics, initial profile soil water content) for individual fields and automates the calculation of daily D_c in the profile. The calculated D_c can then be used to guide irrigation decisions in terms of timing and amount. A 3-yr study using the tool for a center pivot sprinkler-irrigated corn field in northeast Colorado showed that daily calculated D_c had errors ranging from 2 to 31% compared with measured daily D_c on 15 to 16 measurement dates per season. The average RMSE was 16 mm of water, which was deemed acceptable because it could easily be compensated by just one irrigation application (center pivot system could apply 19 mm). Accuracy of the calculated D_c can be improved by making midseason corrections based on measured profile soil water content.

Widespread adoption of the IWM system for irrigation scheduling will only occur if potential users are satisfied with its functionality and there is an economic benefit. Stakeholder feedback regarding an early functional prototype of the IWM system helped the project team significantly in developing a tool that is both user-friendly and sufficiently accurate. Future improvements to the IWM system will include the addition of more crop coefficients for major irrigated crops, the addition of other weather station networks to expand its geographic applicability, and the capability of using remotely sensed ET_c in the water balance calculations. It is anticipated that widespread adoption of the tool will lead to increased irrigation application efficiencies and improvements in environmental water quality.

Acknowledgements

This project is funded by the USDA-NIFA, the Colorado Water Conservation Board, and the Colorado Agricultural Experiment Station. Computer programming was done by Kyle Traff, Dave Patterson, and Andrew Bartlett. Olaf David provided expertise on CSIP. Aymn Elhaddad provided descriptions of the ReSET algorithm. Erik Wardle and Andrew Bartlett are coordinating the demonstration and field testing of the IWM tool. Funding was provided by Western Sugar Cooperative to test the IWM tool on sugarbeet fields. David Gleason evaluated an early version of the tool for center pivot sprinkler-irrigated corn.

References

Allen, R.G., L.S. Pereira, D. Raes, and M. Smith. 1998. Crop evapotranspiration—Guidelines for computing crop water requirements. Irrig. Drain. Pap. 56. FAO, Rome.

Allen, R.G., M. Tasumi, and R. Trezza. 2007a. Satellite-based energy balance for mapping evapotranspiration with internalized calibration (METRIC)—Model. J. Irrig. Drain. Eng. 133:380–394.

Allen, R.G., I.A. Walter, R. Elliott, T. Howell, D. Itenfisu, and M. Jensen, editors. 2005. The ASCE standardized reference evapotranspiration equation. Environ. and Water Resour. Inst. of the Am. Soc. of Civ. Eng., Reston, VA.

Allen, R.G., J.L. Wright, W.O. Pruitt, L.S. Pereira, and M.E. Jensen. 2007b. Water requirements. In: G.J. Hoffman et al., editors, Design and operation of farm irrigation systems. 2nd ed. Chap. 8. ASAE, St. Joseph, MI. p. 208–288.

Andales, A.A., T.L. Bauder, and N.J. Doesken. 2009a. The Colorado Agricultural Meteorological Network (CoAgMet) and crop ET reports. Fact Sheet No. 4.723. Colorado State Univ. Ext. http://www.ext.colostate.edu/pubs/crops/04723.html (accessed 29 May 2014).

Andales, A.A., J.L. Chávez, and T.L. Bauder. 2011. Irrigation scheduling: The water balance approach. Fact Sheet No. 4.707. Colorado State Univ. Ext. http://www.ext.colostate.edu/pubs/crops/04707.html (accessed 29 May 2014).

Andales, A.A., D. Straw, T. Ley, and A. Berrada. 2009b. Alfalfa reference ET from a weighing lysimeter and estimates from the ASCE Standardized Reference ET Equation in the Arkansas Valley of Colorado. In: S. Starrett, editor, Proceedings of the World Environmental and Water Resources Congress 2009: Great Rivers, Kansas City, MO. 17–21 May 2009. Am. Soc. of Civ. Eng., Reston, VA. p. 4093–4101.

Arabi, M. 2011. Environmental Risk Assessment and Management System. Colo. Water 28(1):15–17. http://wsnet.colostate.edu/cwis31/ColoradoWater/Images/Newsletters/2011/CW_28_1.pdf (accessed 29 May 2014).

Augustine, D.J. 2010. Spatial versus temporal variation in precipitation in a semiarid ecosystem. Landscape Ecol. 25:913–925. doi:10.1007/s10980-010-9469-y

Bailey, R.J., and E. Spockman. 1996. A model for estimating soil moisture changes as an aid to irrigation scheduling and crop water use studies. I. Operational details and description. Soil Use Manage. 12:122–128. doi:10.1111/j.1475-2743.1996.tb00532.x

Bastiaanssen, W.G.M., M. Menenti, R.A. Feddes, and A.A.M. Holtslag. 1998a. Remote sensing surface energy balance algorithm for land (SEBAL): 1. Formulation. J. Hydrol. 212–213(1–4):198–212. doi:10.1016/S0022-1694(98)00253-4

Bastiaanssen, W.G.M., H. Pelgrum, J. Wang, Y. Ma, J.F. Moreno, G.J. Roerink, and T. van der Wal. 1998b. Remote sensing surface energy balance algorithm for land (SEBAL): 2. Validation. J. Hydrol. 212–213(1–4):213–229. doi:10.1016/S0022-1694(98)00254-6

Bauder, T.A., and R.M. Waskom. 2005. Survey of irrigation, nutrient and pesticide management practices in Colorado. Tech. Rep. 05-07. Colorado State Univ., Agric. Exp. Stn., Fort Collins.

Bergez, J.-E., P. Debaeke, J.-M. Deumier, B. Lacroix, D. Leenhardt, P. Leroy, and D. Wallach. 2001. MODERATO: An object-oriented decision tool for designing maize irrigation schedules. Ecol. Modell. 137:43–60. doi:10.1016/S0304-3800(00)00431-2

Broner, I., and J. Lambert. 1989a. Optimal scheduling of irrigation machines. I: Model development. J. Irrig. Drain. Eng. 115:862–879. doi:10.1061/(ASCE)0733-9437(1989)115:5(862)

Broner, I., and J. Lambert. 1989b. Optimal scheduling of irrigation machines. II: Applications and results. J. Irrig. Drain. Eng. 115:880–893. doi:10.1061/(ASCE)0733-9437(1989)115:5(880)

Cahoon, J., J. Ferguson, D. Edwards, and P. Tacker. 1990. A microcomputer-based irrigation scheduler for the humid mid-south region. Appl. Eng. Agric. 6(3):289–295.

Cai, X., M.I. Hejazi, and D. Wang. 2011. The value of probabilistic weather forecasts—An assessment by real-time optimization of irrigation scheduling. J. Water Resour. Plan. Manage. 137:391–403.

Chauhan, Y.S., G.C. Wright, D. Holzworth, R.C.N. Rachaputi, and J.O. Payero. 2013. AQUAMAN: A web-based decision support system for irrigation scheduling in peanuts. Irrig. Sci. 31:271–283. doi:10.1007/s00271-011-0296-y

Colorado Water Conservation Board. 2004. Statewide water supply initiative, prepared for Colorado Department of Natural Resources by CDM. Colorado Water Conservation Board. http://cwcb.state.co.us/public-information/publications/Pages/StudiesReports.aspx (accessed 29 May 2014).

Colorado Water Conservation Board. 2010. Colorado's water supply future—Statewide water supply initiative, prepared for Colorado Department of Natural Resources by CDM. Colorado Water Conservation Board. http://cwcb.state.co.us/water-management/water-supply-planning/Documents/SWSI2010/SWSI2010.pdf (accessed 29 May 2014).

Dam, R.F., B.B. Mehdi, M.S.E. Burgess, C.A. Madramootoo, G.R. Mehuys, and I.R. Callum. 2005. Soil bulk density and crop yield under eleven consecutive years of corn with different tillage and residue practices in a sandy loam soil in central Canada. Soil Tillage Res. 84:41–53. doi:10.1016/j.still.2004.08.006

David, O., J.C. Ascough, II, W. Lloyd, T.R. Green, K.W. Rojas, G.H. Leavesley, and L.R. Ahuja. 2013. A software engineering perspective on environmental modeling framework design: The Object Modeling System. Environ. Model. Software 39:201–213. doi:10.1016/j.envsoft.2012.03.006

Dobbs, N.A., K.W. Migliaccio, M.D. Dukes, K.T. Morgan, and Y.C. Li. 2013. Interactive irrigation tool for simulating smart irrigation technologies in lawn turf. J. Irrig. Drain. Eng. 139:747–754. doi:10.1061/(ASCE)IR.1943-4774.0000612

Elhaddad, A., L.A. Garcia, and J.L. Chávez. 2011. Using a surface energy balance model to calculate spatially distributed actual evapotranspiration. J. Irrig. Drain. Eng. 137(1):17–26. doi:10.1061/(ASCE)IR.1943-4774.0000276

Endale, D.M., and G. Fipps. 2001. Simulation-based irrigation scheduling as a water management tool in developing countries. Irrig. Drain. 50:249–257. doi:10.1002/ird.23

Evett, S. 2008. Gravimetric and volumetric direct measurements of soil water content. In: S.R. Evett et al., editors, Field estimation of soil water content: A practical guide to methods, instrumentation, and sensor technology. IAEA-TCS-30, 1018–5518. IAEA, Vienna, Austria. p. 23–37. http://www-pub.iaea.org/mtcd/publications/pdf/tcs-30_web.pdf (accessed 29 May 2014).

Evett, S.R., T.A. Howell, R.W. Todd, A.D. Schneider, and J.A. Tolk. 2000. Alfalfa reference ET measurement and prediction. In: Proceedings of the 4th National Irrigation Symposium, Phoenix, AZ. 14–16 Nov. 2000. ASAE, St. Joseph, MI. p. 266–272.

Fortes, P.S., A.E. Platonov, and L.S. Pereira. 2005. CISAREG—A GIS based irrigation scheduling simulation model to support improved water use. Agric. Water Manage. 77:159–179. doi:10.1016/j.agwat.2004.09.042

Garcia, L.A., A. Elhaddad, J. Altenhofen, and M. Hattendorf. 2013. Developing corn regional crop coefficients using a satellite-based energy balance model (ReSET-Raster) in the South Platte River Basin of Colorado. J. Irrig. Drain. Eng. 139:821–832.

Geerts, S., D. Raes, and M. Garcia. 2010. Using AquaCrop to derive deficit irrigation schedules. Agric. Water Manage. 98:213–216. doi:10.1016/j.agwat.2010.07.003

Gleason, D.J. 2013. Evapotranspiration-based irrigation scheduling tools for use in eastern Colorado. M.S. thesis, Colorado State Univ., Fort Collins.

Gleason, D.J., A.A. Andales, T. Bauder, and J.L. Chávez. 2013. Performance of atmometers in estimating reference evapotranspiration in a semi-arid environment. Agric. Water Manage. 130:27–35. doi:10.1016/j.agwat.2013.08.008

Gowing, J.W., and C.J. Ejieji. 2001. Real-time scheduling of supplemental irrigation for potatoes using a decision model and short-term weather forecasts. Agric. Water Manage. 47:137–153. doi:10.1016/S0378-3774(00)00101-3

He, J., H. Cai, and J. Bai. 2013. Irrigation scheduling based on CERES-wheat model for spring wheat production in the Minqin Oasis in Northwest China. Agric. Water Manage. 128:19–31. doi:10.1016/j.agwat.2013.06.010

Hill, R.W., and G.R. Allen. 1996. Simple irrigation scheduling calendars. J. Irrig. Drain. Eng. 122:107–111. doi:10.1061/(ASCE)0733-9437(1996)122:2(107)

Itenfisu, D., R.L. Elliott, R.G. Allen, and I.A. Walter. 2003. Comparison of reference evapotranspiration calculations as part of the ASCE standardization effort. J. Irrig. Drain. Eng. 129:440–448. doi:10.1061/(ASCE)0733-9437(2003)129:6(440)

Jensen, M.E., and J.L. Wright. 1978. The role of evapotranspiration models in irrigation scheduling. Trans. ASAE 21:82–87. doi:10.13031/2013.35254

Jensen, M.E., J.L. Wright, and B.J. Pratt. 1971. Estimating soil moisture depletion from climate, crop, and soil data. Trans. ASAE 14:954–959. doi:10.13031/2013.38430

Johnson, A. 2010. Growers can monitor irrigation via smartphone. Farm & Ranch Guide. http://www.farmandranchguide.com/news/agri-tech/growers-can-monitor-irrigation-via-smartphone/article_e4bf838a-5d3b-5240-baf3-dda195d5ff76.html (accessed 29 May 2014).

Jones, H.G. 2004. Irrigation scheduling: Advantages and pitfalls of plant-based methods. J. Exp. Bot. 55:2427–2436. doi:10.1093/jxb/erh213

Joshi, M.B., J.S.R. Murthy, and M.M. Shah. 1995. CROSOWAT: A decision tool for irrigation schedule. Agric. Water Manage. 27:203–223. doi:10.1016/0378-3774(95)01158-F

Lascano, R.J., and C.H.M. van Bavel. 2007. Explicit and recursive calculation of potential and actual evapotranspiration. Agron. J. 99:585–590. doi:10.2134/agronj2006.0159

Lascano, R.J., C.H.M. van Bavel, and S.R. Evett. 2010. A field test of recursive calculation of crop evapotranspiration. Trans. ASABE 53:1117–1126. doi:10.13031/2013.32601

Leib, B.G., M. Hattendorf, T. Elliott, and G. Matthews. 2002. Adoption and adaptation of scientific irrigation scheduling: Trends from Washington, USA as of 1998. Agric. Water Manage. 55:105–120. doi:10.1016/S0378-3774(01)00191-3

Lloyd, W., O. David, J. Lyon, K.W. Rojas, J.C. Ascough, II, T.R. Green, and J.R. Carlson. 2012. The Cloud Services Innovation Platform—Enabling service-based environmental modelling using infrastructure-as-a-service cloud computing. In: R. Seppelt et al., editors, Proceedings of the 2012 International Congress on Environmental Modelling and Software, Leipzig, Germany. 1–5 July 2012. Int. Environ. Modell. and Software Soc., Manno, Switzerland. http://www.iemss.org/sites/iemss2012//proceedings/D3_1_0905_David_et_al.pdf (accessed 29 May 2014).

Ma, Y., S. Feng, and X. Song. 2013. A root zone model for estimating soil water balance and crop yield responses to deficit irrigation in the North China Plain. Agric. Water Manage. 127:13–24. doi:10.1016/j.agwat.2013.05.011

Mannini, P., R. Genovesi, and T. Letterio. 2013. IRRINET: Large scale DSS application for on-farm irrigation scheduling. Procedia Environ. Sci. 19:823–829. doi:10.1016/j.proenv.2013.06.091

Prats, A.G., and S.G. Picó. 2010. Performance evaluation and uncertainty measurement in irrigation scheduling soil-water balance approach. J. Irrig. Drain. Eng. 136:732–743. doi:10.1061/(ASCE)IR.1943-4774.0000245

Rhenals, A.E., and R.L. Bras. 1981. The irrigation scheduling problem and evapotranspiration uncertainty. Water Resour. Res. 17:1328–1338. doi:10.1029/WR017i005p01328

Robinson, P.M. 2009. The next generation of irrigation management programs. Proceedings of the World Environmental and Water Resources Congress: Great Rivers, Kansas City, MO. 17–21 May 2009. Am. Soc. of Civ. Eng., Reston, VA.

Sammis, T., P. Sharma, M.K. Shukla, J. Wang, and D. Miller. 2012. A water-balance drip-irrigation scheduling model. Agric. Water Manage. 113:30–37. doi:10.1016/j.agwat.2012.06.012

Singels, A., and M.T. Smith. 2006. Provision of irrigation scheduling advice to small-scale sugarcane farmers using a web-based crop model and cellular technology: A south-African case study. Irrig. Drain. 55:363–372. doi:10.1002/ird.231

Soil Survey Staff. 2013. Web Soil Survey. USDA-NRCS. http://websoilsurvey.sc.egov.usda.gov/App/HomePage.htm (accessed 25 May 2014).

Steele, D.D., T.F. Scherer, D.G. Hopkins, S.R. Tuscherer, and J. Wright. 2010. Spreadsheet implementation of irrigation scheduling by the checkbook method for North Dakota and Minnesota. Appl. Eng. Agric. 26:983–995. doi:10.13031/2013.35914

Taghvaeian, S., J.L. Chávez, and N.C. Hansen. 2012. Infrared thermometry to estimate crop water stress index and water use of irrigated maize in northeastern Colorado. Remote Sens. 4:3619–3637. doi:10.3390/rs4113619

Tojo Soler, C.M., A. Suleiman, J. Anothai, I. Flitcroft, and G. Hoogenboom. 2013. Scheduling irrigation with a dynamic crop growth model and determining the relation between simulated drought stress and yield for peanut. Irrig. Sci. 31:889–901. doi:10.1007/s00271-012-0366-9

Tolk, J.A., T.A. Howell, and S.R. Evett. 1998. Evapotranspiration and yield of corn grown on three high plains soils. Agron. J. 90:447–454. doi:10.2134/agronj1998.00021962009000040001x

Troutman, B.M. 1983. Runoff prediction errors and bias in parameter estimation induced by spatial variability of precipitation. Water Resour. Res. 19:791–810. doi:10.1029/WR019i003p00791

USDA-NASS. 2008. 2007 Census of agriculture—Farm and ranch irrigation survey (2008). Vol. 3. Special studies. Part 1. USDA-NASS, Washington, DC.

USDA-NASS. 2009. 2007 Census of agriculture—Farm and ranch irrigation survey (2008). Vol. 3. Special studies. Part 1. AC-07-SS-1. USDA-NASS, Washington, DC.

USDA-NRCS. 2004. Estimation of direct runoff from storm rainfall. USDA-NRCS Agric. Handb. 630. U.S. Gov. Print. Office, Washington, DC. ftp://ftp.wcc.nrcs.usda.gov/wntsc/H&H/NEHhydrology/ch10.pdf (accessed 29 May 2014).

USDA-NRCS. 2011. Soil and Water Resources Conservation Act appraisal. USDA-NRCS, Washington, DC.

USDA-SCS. 1991. Soil–plant–water relationships. 2nd ed. USDA-SCS Agric. Handb. 15.1. U.S. Gov. Print. Office, Washington, DC.

Waskom, R.M. 1994. Best management practices for irrigation management. Bull. XCM-173. Colorado State Univ. Ext., Fort Collins.

Circular Planting to Enhance Rainfall Capture in Dryland Cropping Systems at a Landscape Scale: Measurement and Simulation

Robert J. Lascano* and J. Randall Nelson

Abstract

Declining availability of irrigation water from the Ogallala Aquifer in the Southern High Plains (SHP) of Texas has increased dryland production and the need to increase rainfall use efficiency with minimal expense to producers. We hypothesized that planting dryland crops in a circular pattern would increase surface roughness and storage volume for precipitation and therefore increase rainfall infiltration when compared with conventional straight rows. The objective was to use the Precision Agricultural Landscape Modeling System (PALMS) to simulate a circular planting in a dryland cropping system and evaluate the impact on the surface water balance, mainly runoff and infiltration, across the landscape. Field studies on predominantly sandy soils were conducted during 2007, 2008, and 2009 on four commercial dryland fields, ranging in size from 26 to 64 ha, in Dawson County, TX to measure soil water content and crop response to circular planting patterns. Measured soil water content from all locations and years yielded 421 values that when compared with corresponding simulated values produced a linear regression with a slope of 0.88, r^2 of 0.86, and RMSE of 0.038 $m^3\ m^{-3}$. This agreement was used to infer that simulated components of the surface water balance were correct. Values of soil water content indicated less runoff from fields planted in circular patterns occurring during large, intense rainstorms. The PALMS simulations indicated a longer ponding time and increased infiltration in fields planted in a circle, most notably in locations with significant topographic relief. We conclude that circular planting is a viable option under dryland production.

Abbreviations: IBIS, Integrated Biosphere Simulator; PALMS, Precision Agricultural Landscape Modeling System; SHP, Southern High Plains; VWC, volumetric water content.

R.J. Lascano, Wind Erosion and Water Conservation Research Unit, Cropping Systems Research Laboratory, USDA-ARS, 3810 4th St., Lubbock, TX 79415. *Corresponding author (robert.lascano@ars.usda.gov).

J.R. Nelson, North Central Kansas Experiment Field, Department of Agronomy, Kansas State University, 1300 60 Rd., Courtland, KS 66939.

doi:10.2134/advagricsystmodel5.c4

Practical Applications of Agricultural System Models to Optimize the Use of Limited Water
Lajpat R. Ahuja, Liwang Ma, and Robert J. Lascano, Editors
Advances in Agricultural Systems Modeling, Volume 5. Lajpat R. Ahuja, Series Editor

USDA-ARS

Profitability of dryland cropping systems is important in the SHP of Texas because of the reduced availability of irrigation water from the Ogallala Aquifer (Mapp, 1988). Efficient capture and use of the limited precipitation in this semiarid region is significant to the success of dryland crop production (Nielsen et al., 2005) Of the 3 million hectares of cropland in the SHP of Texas, approximately two-thirds is under dryland production (Llano Estacado Regional Water Planning Group, 2006). Although dryland crop production only accounts for 25% of the total farm sales value of the SHP (Llano Estacado Regional Water Planning Group, 2006), it is projected that declining groundwater availability for irrigation could result in a significant increase in dryland area within the next 30 yr (Mapp, 1988; Colaizzi et al., 2009). A similar trend has also been documented in other regions of the Great Plains (Norwood and Dumler, 2002).

It is estimated that 80 to 85% of precipitation in the SHP can be lost as evaporation from the soil and runoff of water (Jones, 1975; Unger, 1994). For example, Jones (1975) found that although runoff only accounted for 15% of mean annual precipitation, management and use of runoff significantly increased grain yields in dryland cropping systems. At Bushland, TX, runoff volume from row cropped sorghum (*Sorghum bicolor* L. Moench) was twice as great as runoff from wheat (*Triticum aestivum* L.) fields (Jones et al., 1985). In the same study, it was found that <1% of storms, between 1960 and 1979, produced precipitation events >51 mm. However, these storms were 10% of the total precipitation and yielded 36 to 41% of runoff from cultivated watersheds.

While progress is being made in the SHP toward reduced tillage for soil water conservation, further innovations are necessary to increase the effectiveness of rainfall in dryland cropping systems. For example, the conventionally tilled cotton (*Gossypium hirsutum* L.) monoculture cropping system common to the SHP introduces several problems for soil water management in this semiarid environment (Lascano and Baumhardt, 1996). When compared with wheat and sorghum, cotton produces too little aboveground residue to effectively reduce soil water evaporation (Unger and Parker, 1976; Lascano et al., 1994) and may not affect infiltration (Baumhardt et al., 1993). Furthermore, multiple tillage events further reduce residue amounts on the surface. It has been shown that mean cumulative infiltration is lowest on bare soils and increases curvilinearly with increasing residue amounts because of raindrop impact interception (Baumhardt and Lascano, 1996).

There are several physical means to reduce runoff in row crop systems including terracing and several tillage practices, that is, conservation, contour, and basin (Wiedemann and Smallacombe, 1989; Jones and Stewart, 1990; Troeh et al., 1991; Bosch et al., 2005). Terracing is the practice of reducing slope length of a field by modifying the terrain across a hillside. This is an ancient technique used for both soil erosion and water management and was successfully implemented by the South American Incas (Resck et al., 2006). In addition, conservation tillage (Harman et al., 1989; Bosch et al., 2005), contour tillage, and furrow diking (Lyle and Dixon, 1977; Jones and Clark, 1987; Jones and Stewart, 1990; Howell et al., 2002; Jones and Baumhardt, 2003; Nuti et al., 2009) have all demonstrated effectiveness for reducing runoff and/or increasing soil water content. Level ridge-type terraces are level on the contour of the land and have blocked outlets to retain storm water until infiltration occurs. Level terraces are generally intended for water conservation but can also assist in soil erosion control. Alternatively, conservation bench terraces consist of several level benches combined with runoff contributing watershed areas (Troeh et al., 1991). These terraces are designed to direct runoff from watershed areas for spreading across the benches and increase infiltration. Most terraces in arid regions are broad-based to allow cropping on the ridge tops and channels, as well as the area between terraces. Drawbacks to terracing include the installation cost, as well as complicating the cropping process for the grower. Terraces are expensive to install and provide some production management challenges, but contour tillage can provide some of the same benefits and at a reduced cost. Ridge-forming tillage that follows the contour of the land can produce a similar effect to that of terraces. Contour tillage can improve water capture and reduce runoff for moderate storms, but depending on soil type and ridge size, they may not effectively control runoff for severe storms (Unger, 1996). Contour practices in a wheat–fallow system using plow tillage have improved cumulative infiltration by 60% (Dickey et al., 1983). In minimum tillage treatments with adequate residue cover, however, no significant differences in infiltration were measured between contour and standard tillage treatments. Furrow dikes are essentially small basins created across the furrows of a ridge–furrow tillage system. Similar to terracing, the practice of furrow diking is to increase the volume to store precipitation and potentially reducing storm runoff and increasing soil water storage and availability to crops and therefore increase the "opportunity time" for infiltration (Jones and Stewart, 1990). These basins allow rainfall to be retained in the furrow until infiltration occurs, rather than losing it to runoff along the furrow. Additional information on different techniques to improve the efficiency of rainwater use is given by Stroosnijder (2009) and by Oweis et al. (2012). A review of different rainwater harvesting techniques in the southeastern United States is given by Jones and Hunt (2010) and in African drylands by Vohland and Barry (2009) and by Biazin et al. (2012).

The response of terracing and conservation tillage practices on crop yield is variable. Hauser and Cox (1962) reported that grain yields could be increased in areas using conservation bench terraces, as compared with fields with level terraces or no terraces present. Haas et al. (1966) reported that conservation bench terraces improved crop yields through increased snow capture, heavy rainfall runoff control, and reduced erosion. The response of furrow diking to crop yield can be inconsistent because of fluctuations in weather patterns. For example, in wet years, mechanical weed control can be hindered because of excessive water storage in the furrow (Lawes, 1963). However, corn (*Zea mays* L.) and sorghum grain yield increases due to furrow diking were measured throughout the Great Plains in years with typical rainfall (Jones and Stewart, 1990). As furrow diking and contour tillage have limitations in water capture and runoff control, the possibility of combining the two practices may alleviate some of the problems encountered when incorporated alone (Unger et al., 2006).

The evaluation of both short- and long-term effects of conservation practices on water runoff, soil erosion, and crop yield is complicated by both spatial and temporal variables. For example, the topography and soil properties such as texture and infiltration vary across the landscape (Li et al., 2001) and typically during a growing season rainfall events vary both in quantity and in frequency. This complexity contributed to the development of simulation models to evaluate the effect of tillage practices on soil erosion and crop yield (e.g., Diekkrüger et al., 1995; Connolly, 1998; Takken et al., 2001 Batchelor et al., 2002; DeJonge et al., 2007; Lascano and Booker, 2013). A model also developed for this purpose was PALMS (Rodgers et al., 2000; Morgan et al., 2003; Zhu et al., 2004; Molling et al., 2005). The PALMS is a biophysical mechanistic model that calculates the flow of water, energy, and mass through the air–plant–soil system on each grid cell for the simulated field. Input data of topography and three-dimensional soil properties can be specified on a 5-, 10-, or 20-m grid. Hourly weather input data of wind, short-wave irradiance, air temperature and humidity, and rain provide the temporal scale that drives the simulation over one or more growing season (Molling et al., 2005). The PALMS model was designed as a precision rainfall–runoff landscape model and has been used to simulate crop growth of corn and soybean [*Glycine max* (L.) Merr.] in the Midwest (Morgan et al., 2003; Molling et al., 2005) and erosion processes (Bonilla et al., 2007, 2008). We evaluated PALMS to simulate soil water content throughout the growing season for several years and for three major soil series of the SHP (Alvarez-Acosta, 2009; Nelson, 2010; Nelson et al., 2013), and recently PALMS was combined with the Cotton2K model (Marani, 2004) to produce PALMScot (Booker, 2013; Booker et al., 2014).

In recent years, precision agriculture principles as a means to manage agronomic inputs in response to cropping system variability has been implemented by producers (Stafford, 2000; Zhang et al., 2002; McBratney et al., 2005; Heege, 2013; Lascano and Booker, 2013). Advances in the availability of the required input data to operate models such as PALMS are key to their continued adoption. For example, in the United States, soil data are provided by the USDA, elevation data are provided by the USGS, and weather data are available from weather mesonets (networks). The combination of crop-specific models with geographic information data of soils and elevation, real-time weather, and management information systems provide a framework to manage crops at spatial and temporal scales that capture the inherent variability of soils across the landscape.

A key factor for the continued success and implementation of dryland cropping systems in the Texas SHP is the development of practices that are both economical and effective to decrease runoff from rainfall. Thus, the overall goal of our research is to increase rainfall use efficiency, as described by Stroosnijder (2009), with minimal additional expense to producers. We hypothesized that planting dryland crops in a circular pattern would increase surface roughness and ponding time and therefore increase rainfall infiltration compared with conventional straight row configurations. In the SHP, circular planting is common under center pivot irrigations and the availability of tractor-guidance systems makes the implementation of this operation cost effective. Therefore, the objective of this study was to simulate with the PALMS model a circular planting configuration under dryland conditions for a sandy soil series in the Texas High Plains at a landscape scale. These simulations were done on quarter section–size (160 acres $\approx$ 65 ha) fields for three growing seasons and to provide an initial assessment on the effect of a circular pattern on the surface water balance across the landscape. This is an ongoing effort, and our intent was to use the mechanistic simulation model PALMS to evaluate an agronomic practice at a landscape scale.

Materials and Methods

PALMS Model Description

The PALMS is a precision rainfall–runoff landscape model designed for agricultural applications. It represents the combination of two models: (i) a two-dimensional diffusive wave, runoff model with ponding and (ii) a one-dimensional point column, land process, or biophysical model known as the Integrated Biosphere Simulator (IBIS) (Foley et al., 1996). The model PALMS is structured on a three-dimensional grid, with the horizontal dimensions (easting and northing) consisting of a uniform fixed grid and a third dimension consisting of vertical soil layers (Morgan et al., 2003). Agronomic crop submodels in PALMS are limited to corn and soybean, but other models exist for forest, shrubland, and grassland. The PALMS model operates on a grid cell size of 5 to 20 m. The IBIS model is executed at each grid point until precipitation occurs in excess of infiltration and detention storage. At this point, the diffusive wave model is activated and rainfall is simultaneously routed over the landscape and infiltrated into the soil (Morgan et al., 2003; Molling et al., 2005). The PALMS model includes ponding and reinfiltration on a field, which is excluded in most existing models. The system has demonstrated the ability to reliably simulate tillage effects on runoff and the effects of soil type and topography on the spatial distribution of soil water and crop yield (Rodgers et al., 2000; Zhu et al., 2004). A flow diagram of the major components of PALMS is given in Fig. 4–1 (Morgan et al., 2003). Inputs to the model are soil texture and topography, values related to tillage type and timing, crop cultivar and planting date, and hourly weather or shorter time

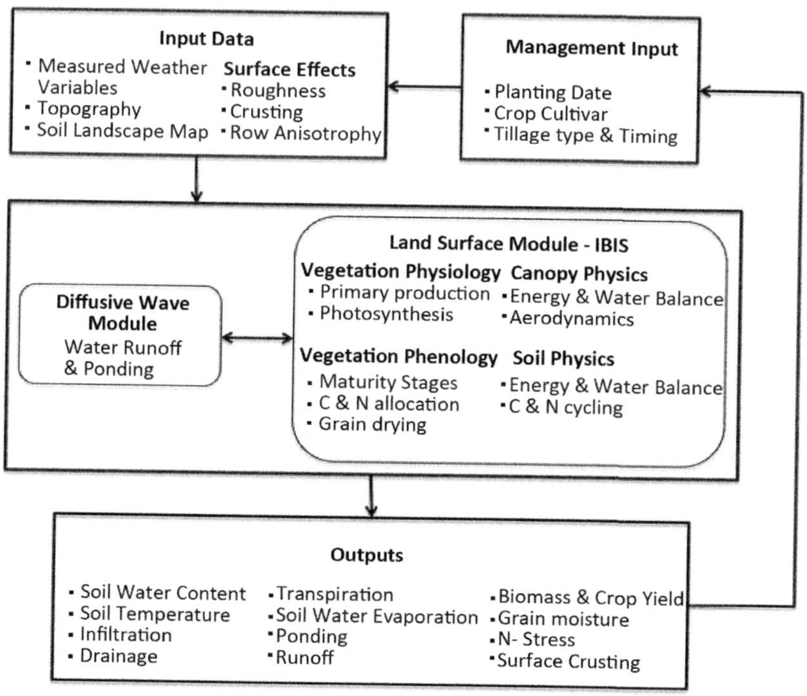

Fig. 4–1. Flowchart showing the major components of the Precision Agricultural Landscape Modeling System representing model data and management inputs, the diffusive wave model used for surface hydrology, the Integrated Biosphere Simulator (IBIS), and model outputs. Adapted from Morgan et al. (2003).

interval if available. As output, PALMS calculates hourly and daily values of soil water content and temperature in three dimensions—surface ponding, values of the water (evaporation, drainage, runoff, and infiltration) and energy (latent and sensible heat, net radiation, and soil heat flux) balance—and of the crop status on each grid cell on the field. Therefore, PALMS is a simulation model capable of simulating varying soil temperature and water content caused by the spatial variability of the topography and of the soil on a single field. This model provides a tool to effectively simulate the effects that different management practices have on the retention of rainfall and the resulting changes in soil water available for crop production. Additional information of PALMS is given by Molling et al. (2005), Bonilla et al. (2007, 2008), Nelson (2010), Nelson et al. (2013), Booker (2013), and Booker et al. (2014).

Site Characteristics and Soils

In 3 yr (2007–2009), field studies were conducted in Dawson County, TX at four privately owned dryland fields. Hereafter these fields are designated as F1, F2, F3, and F4. A summary of site characteristics for each field is given in Table 4–1, and elevation maps for the four fields are shown in Fig. 4–2. In 2007, all four fields were used, and in 2008 and 2009 we only used fields F3 and F4. Fields F1 and F2 were located ~14 km northeast of Lamesa, TX. Field F1 was planted in a conventional straight row pattern in a north–south orientation. Field F2 was planted in a circular pattern similar to some crops planted under center pivot irrigation systems. Soils at F1 are classified as Amarillo fine sandy loam (a fine-loamy, mixed, superactive, thermic Aridic Paleustalf) and Posey fine sandy loam (a fine-loamy, mixed, superactive, thermic Calcidic Paleustalf). Soils at F2 are classified as Amarillo fine sandy loam. Fields F1 and F2 were adjacent to each other and were managed by

Table 4–1. Location, elevation, slope, size, soil series, and planting pattern of the four fields in Dawson County, TX used in our studies.

Field no.	Latitude (N)	Longitude (W)	Elevation	Slope range	Size	Soil series	Planting pattern
			m	%	ha		
F1	32°50′26.8″	101°53′0.9″	920	0–6	62	Amarillo and Posey	straight (N–S)
F2	32°50′32.7″	101°52′30.9″	921	0–6	62	Amarillo	circular
F3	32°33′54.3″	101°45′15.2″	862	0–1	26	Acuff and Portales	straight (E–W)
F4	32°32′7.5″	101°46′7.5″	858	0–1.5	64	Acuff and Sharvana	circular

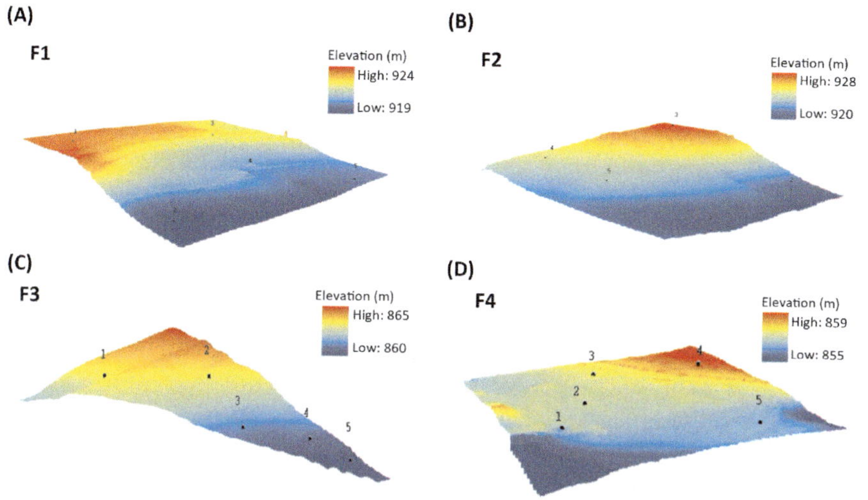

Fig. 4–2. Soil water content measurement locations in 2007 at four fields: (a) field F1, (b) field F2, (c) field F3, and (d) field F4. Sample locations are indicated by numbers overlaid on elevation maps.

the same producer. Both fields were ~63 ha and generally sloped from north to south toward a playa lake, which are naturally occurring ephemeral lakes along the landscape (Osterkamp and Wood, 1987; Zartman et al., 1994). The slope of both locations ranged from 0 to 6%, with a pronounced area of 4 to 6% slopes on the east edge of F1 and on the west edge of F2.

Fields F3 and F4 were nonadjacent fields managed by the same producer and located ~5 km northwest of Ackerly, TX. Field F3 was planted in a conventional straight pattern with row orientation in the east–west direction. Field F4 was planted in a circular pattern similar to F2. Soils at F3 are classified as a combination of Acuff sandy clay loam (a fine-loamy, mixed, superactive, thermic Aridic Paleustoll) and Portales loam (a fine-loamy, mixed, superactive, thermic Aridic Calciustoll). Soils at F4 are classified predominately as Acuff sandy clay loam with a small area of Sharvana fine sandy loam (a loamy, siliceous, superactive, thermic, shallow Ustic Petroargid). Field F3 was ~26 ha, declining slightly in elevation from south to north. Slopes at this location ranged from 0 to 1%. Field F4 was ~64 ha, declining in elevation from the northeast to the southwest. Slopes at this location range from 0 to 1.5%. The small areas of 1.5% slope can be attributed to unmaintained terraces in the northeast corner of the field. The remainder of the field was generally flat.

Data Collection and Cultural Practices

Fields at all four sites were managed by the producers according to conventional cultural practices. In 2007, fields F1 and F2 were planted in 1-m rows with cotton cultivar FM 958 (FiberMax Cotton, Bayer Crop Science).[1] Fields F3 and F4 were planted in 1-m rows with grain sorghum hybrid Pioneer 85G01 (Sorghum Pioneer, DuPont). Following harvest in 2007, access to fields F1 and F2 was denied, and the study was continued only at fields F3 and F4. In 2008, fields F3 and F4 were planted in 1-m rows with grain sorghum hybrid Pioneer 85G01. In 2009, field F3 was initially planted with cotton cultivar FM 958. One month following planting, the crop failed because of dry conditions, and grain sorghum hybrid Pioneer 85G01 was planted as a catch crop. Field F4 was planted to cotton cultivar FM 958 in 2009 and progressed well into the fruiting stage before a hailstorm destroyed the crop in late August.

In 2007, soil water content was monitored weekly using soil water sensors (Model 30KTCD-NL, Watermark, Irrometer Company) buried at ~0.2 m soil depth at five locations in each field. These sensors are useful to monitor soil water in commercial fields because they do not interfere with normal field operations and are relatively inexpensive and low maintenance. Measurement locations are shown in Fig. 4–2. In 2008, soil water content was measured weekly using the sensors at eight locations (0.2-m depth) in fields F3 and F4 as shown in Fig. 4–3a and 4–3b. In 2009, soil water content was measured every 2 wk at 12 locations in field F3 (Fig. 4–3c) and at 18 locations in field F4 (Fig. 4–3d). The output of the sensors used to measure soil water content was in terms of soil water potential (i.e., kPa) and thus provided a relative measurement of soil water content. Sensor readings were converted to corresponding soil volumetric water content (VWC, $m^3\ m^{-3}$) using the Mualem–van Genuchten equation (van Genuchten, 1980) and estimated parameters from Baumhardt et al. (1995), Alvarez-Acosta (2009), and Alvarez-Acosta et al. (2012). This conversion from water potential to soil VWC is not a reliable absolute measurement of soil water content, but it does provide a practical method for presenting relative values of soil VWC for comparison of fields planted in circular and straight planting patterns. In addition, 15 soil water sensors used in our field measurements were randomly selected and used to compare with measurements of soil water potential measured using pressure plates (Model 1500, Soil Moisture Equipment) at 10, 100, and 1500 kPa using methods given by Klute (1986). For this purpose, we used undisturbed soil samples, from a 0.1 to 0.2 m depth, taken with brass rings from fields F2 and F3. These sensor-specific calibrations were combined with the conversion obtained from estimated

[1] Mention of this or other proprietary products is for the convenience of the readers only and does not constitute endorsement or preferential treatment of these products by USDA-ARS.

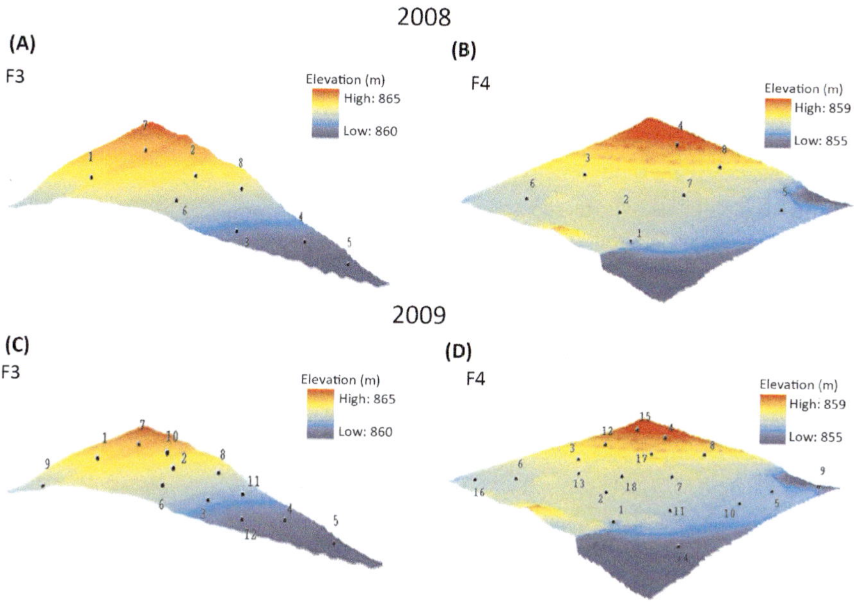

Fig. 4–3. Soil water content measurement locations at (a) field F3 and (b) field F4 in 2008 and locations on (c) field F3 and (d) field F4 in 2009 overlaid on elevation maps of the respective fields.

parameters and thereafter used to convert all sensor readings from soil water potential to soil VWC.

Soil samples were taken to a depth of 0.9 m at each site in increments of 0.0 to 0.15 m, 0.15 to 0.3 m, 0.3 to 0.6 m, and 0.6 to 0.9 m for soil texture determination using the hydrometer method (Gee and Bauder, 1986). A total of 36 samples were collected at field F1, 43 samples at F2, 17 samples at F3, and 40 samples at field F4. The Amarillo series similar to that at fields F1 and F2 were analyzed for soil physical and hydraulic properties by Baumhardt et al. (1995).

Elevation surveys were conducted at all sites using a survey grade GPS (Model 4700 Dual Channel RTK, Trimble) with an accuracy of <0.05 m, and values obtained were used to generate elevation maps for the four fields, shown in Fig. 4–2 and 4–3. Transects ~20 m apart were covered with the GPS receiver mounted to a utility vehicle. Hand samples for crop yield were taken at the soil sampling locations at each site. Samples were collected in a two-row 4-m^2 area at each point.

PALMS Inputs

The PALMS model requires as input four layers of landscape information: (i) gridded topography of the field; (ii) gridded soil texture in three dimensions (x, y, and soil depth) on the same grid as the topography; (iii) a map of vegetation type and management practices that include planting, tillage, etc.; and (iv) a land mask to indicate the grid cells that are in and/or out of the area of interest for the field (Molling et al., 2005). All landscape data were created on 10-m grids corresponding to the highest resolution for the available computing resources. The model PALMS can accommodate up to 23 soil layers down to a depth of 2.5 m at thicknesses from 0.02 m near the surface to 0.25 m at the bottom of the profile. At all sites, soil texture inputs were created from the core sampling previously described. Point samples of soil texture for each depth were interpolated using the inverse distance weighted function, power four in ArcGIS (ESRI). Elevation points from the GPS surveys were interpolated using the same procedure for soil texture data to create the required digital elevation model. Because the model PALMS includes only corn and soybean crop models, soybean was used as a substitute as the only C3 crop similar to cotton, and corn in similar fashion for C4 grain sorghum.

Atmospheric weather data used as input to the PALMS model were collected from a West Texas Mesonet (http://www.mesonet.ttu.edu) station located (32°42′36.7″ N, 101°56′13.2″ W; 891-m elevation) ~3 km southeast of Lamesa, TX. Air temperature and relative humidity, solar irradiance, rainfall, and wind speed were all measured at a screen height of 2.0 m with methods similar to those described by Lascano (2000). At each site, rainfall data were collected using a duplicate tipping bucket rain gauge with an event data logger (Model RG3, Onset Computer) and a similar rain gauge (Model TE525-L, Texas Electronics) attached to a data logger (Model CR1000, Campbell Scientific). The model was executed with weather data on 15-min time steps.

The PALMS management input file for each simulation, which includes tillage, fertilizer, planting date and population, variety information, and initial soil conditions, was created on the basis of actual management practices or the best estimate of common practices if not all information was available (Table 4–2). Information collected from soil data measurements taken at the weather stations previously mentioned were used to estimate initial soil VWC and temperature values and used as input to the model.

Surface Hydrology and Soil Water Response to Planting Patterns in PALMS

The PALMS model was executed for each site year in which field data were collected, with management inputs similar to actual field conditions and the corresponding weather data. Tillage treatment weather data corresponded to the actual year. Tillage treatment and in particular straight vs. circular planting pattern for these simulations corresponded with field conditions (Tables 4–1

Table 4–2. General timing and description of field operations for all simulation years and locations.

Month	Field operation
Mar.–Apr.	lister
May–June	planting
June	rotary hoe
June	cultivator
July	cultivator
Sept.–Oct.	harvest

and 4–2). The output of calculated soil VWC was compared with corresponding measured values.

Archived weather data from the West Texas Mesonet were used to determine the effect of variable weather conditions on PALMS simulations. Weather information from 2007 to 2009 was used to determine the effects of different planting patterns on soil water and surface hydrology. For each of the large (~62 ha) fields (F1, F2, and F4), PALMS was run with simulated straight and circular planting patterns for the years of available weather data (2007–2009). Simulated hourly, daily, and annual rates for various processes were compared across planting treatments and cropping and/or tillage scenarios. For every year of simulation, a range of storm intensities were chosen to evaluate cumulative hourly and daily values for ponding depth and time, infiltration, runoff, and erosion for each of the scenarios and planting patterns.

Simulated values of soil VWC, runoff, and net runoff were extracted from sample points located at different slope positions at each of the fields. At field F1 (Fig. 4–4a) and field F2 (Fig. 4–4b), 14 and 15 points, respectively, were chosen to represent summit slope, side slope, and toe slope positions (Table 4–3). At field F4 (Fig. 4–4c), 12 points were distributed randomly across each field because of the lack of topographic relief. In a similar fashion, values for runoff and net runoff were extracted from each simulation on the days where runoff-inducing rainfall occurred. Values of soil VWC were extracted 2 d following runoff-inducing rainfall to allow time for redistribution and reinfiltration of precipitation.

Results and Discussion

An important component in the development of simulation models is their verification and validation (Kleijnen, 1995; Thacker et al., 2004; Sargent; 2005; Johnson, 2011). While there is no consensus on how this is accomplished, an initial step is to compare calculated values obtained with the model to experimental data and then use statistics to quantify differences between measured and calculated values. Examples of these statistics are regression analysis and RMSE (Kobayashi

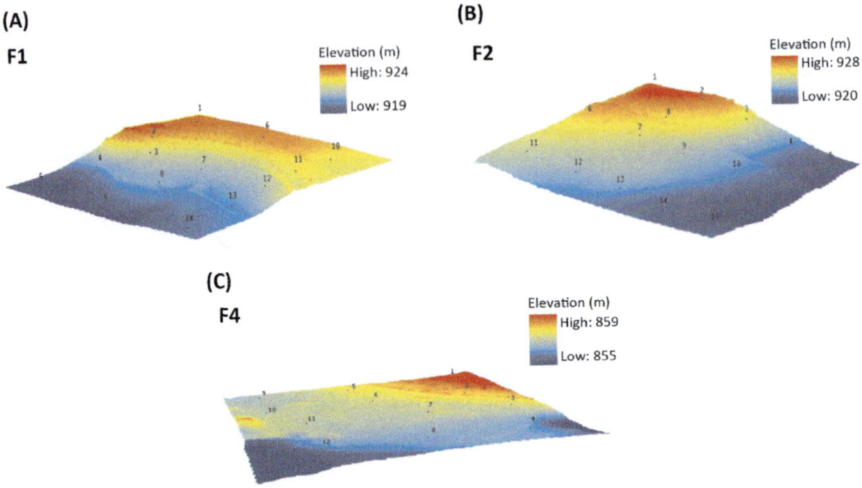

Fig. 4–4. Locations of extracted values of soil water and net runoff taken from Precision Agricultural Landscape Modeling System simulations at (a) field F1, (b) field F2, and (c) field F4 overlaid on elevation maps of the respective fields.

Table 4–3. Slope position for each sample number at fields F1 and F2.

Field F1		Field F2	
Sample number	Slope position	Sample number	Slope position
1	summit	1	summit
2	side	2	side
3	side	3	side
4	side	4	toe
5	toe	5	toe
6	summit	6	summit
7	side	7	side
8	side	8	side
9	toe	9	side
10	summit	10	toe
11	side	11	side
12	side	12	side
13	side	13	side
14	toe	14	toe
–	–	15	toe

and Salam, 2000), coefficient of efficiency (Legates and McCabe, 1999; 2013), the Nash–Sutcliffe efficiency, percent bias, and ratio of the RMSE to the standard deviation of measured data (Moriasi et al., 2007). An added difficulty with a complex biophysical model such as PALMS, particularly when applied to simulate the water, carbon, and energy balance of a 65-ha field, is that a large number of variables can be tested with experimental data. However, the fact that the field used in the simulation was divided into grid cells of 100 m^2, the calculated values obtained with the model were extracted to correspond to the georeferenced location where the data were measured. Our experimental objective was to compare measured and calculated values of soil water content from all locations and years using regression analysis and RMSE to quantify differences. The values of soil water content were selected as surrogates to infer the calculated values of the surface soil water balance (Basso et al., 2001; Gauch et al., 2003; Faivre et al., 2004).

The simulation of a crop growing across the landscape and throughout the growing season is complex and involves processes that mainly exchange water, nitrogen, carbon, and energy between the soil, plant, and atmosphere. In general, the boundary conditions of an agronomic field are set at 2 m above ground level, that is, screen height (x dimension) where weather variables are measured, and at a 2-m soil depth (z dimension), which represents the root zone where water and nutrient uptake take place. Many crop simulation models, for example, cotton (Baker et al., 1983; Jackson et al., 1988; Marani, 2004) and sorghum (Arkin et al., 1976; Rosenthal et al., 1989), are considered column models in that they operate in the spatial domain defined by the x and y boundaries and normally represent one plant per unit land area (m^2). These models operate under the assumption that all input values are the same and thus do not include the spatial variability that occurs across the landscape (y dimension). This is not the case with the PALMS model where the landscape is divided into grids and it is possible to capture the spatial variability of the field provided the input of soil physical and chemical properties are sampled at different locations across the field. With the PALMS model, we used a 10-m grid to characterize the landscape input information. For example, fields F1 and F2, each 62 ha, were thus represented by 6200 cells and PALMS calculated values of all variables of the water, energy carbon, and nitrogen balance, including crop yield for each cell. This avoids having to assume that a single value of crop yield that is calculated with a column model applies to the entire field as done by Nair et al. (2013) using the Cotton2K model (Marani, 2004) to evaluate the effect of different irrigation amounts across a 0.4-km center pivot irrigated field. Furthermore, the extrapolation of a single crop yield across the field is also avoided as suggested by Dugas et al. (1983) and Trouslard-Kerdiles and Grondona (1997). The ability to calculate the crop yield across the landscape position is an obvious advantage when using a model such as PALMS.

Of interest was the comparison between measured and calculated values of crop yield across the landscape; however, the uncertainty of using substitute crops, that is, corn for sorghum and soybean for cotton, in the PALMS model precluded a direct comparison. Furthermore, the measured cotton lint yield and grain sorghum yield CV across the fields exceeded 50%, and this variability was due to factors such as pests (weeds, nematodes, bird damage) that were not captured by the PALMS model. Nevertheless, the comparison of measured and calculated crop yield is important but was beyond the scope of this work and is the subject of a forthcoming study and paper.

Crop Response to Planting Patterns—Field Observations

In years where crop yield was measured (i.e., 2007 and 2008), there was no significant response of crop yield to either the straight or the circular pattern at either location. At fields F1 and F2 in 2007, average cotton lint yield across the entire field was ~30% higher for field F1 (straight) than for field F2 (circular) (data not shown). However, the CV of cotton lint yield across both fields was >50%. Field F2 was converted from straight to circular planting 2 wk before planting, and no crop had been cultivated in the previous 3 yr because of the lack of adequate rainfall. As a result, field F2 was overtaken by weeds and had no history of fertilization, and any stored soil water from rain was likely used by weeds and not available to the planted crop. At fields F3 and F4 in 2007, there was no difference in average grain sorghum yield between planting treatments (data not shown). Again, the CV of the measured grain sorghum yield was >70%. In 2008, field F3 was divided between cotton and grain sorghum, thus crop yield comparisons were not made for grain sorghum between fields F3 and F4.

Measured values of soil VWC at each field are shown in Fig. 4–5 and 4–6. In general, soil VWC values throughout the growing season were slightly higher in field F1 than in field F2, but the increase in soil VWC did not correspond to a yield increase. It is possible that a nematode problem on a large portion of field F1 existed where water stood early in the growing season, which may have negatively affected cotton lint yield. At soil VWC measurement site 1, there were significant fluctuations in soil VWC throughout the season (Fig. 4–5a). Although the fluctuation patterns are similar to those of the other sensors, the magnitude of change at this site is much greater. This location is an example of the effect of a circular tillage pattern reducing runoff from a toe slope position. At this point, the tillage pattern ran diagonally across the side slope. This created significant ponding and could be the reason for the significant changes in soil VWC when compared with other sample points. Several large precipitation events (i.e., >15 mm) occurred before planting time, and significant runoff was evident on the adjacent field planted in a straight configuration. At fields F3 and F4, soil VWC values were similar throughout the 2007 growing season (Fig. 4–5c and 4–5d). It was assumed that soil VWC response to the planting tillage pattern was negligible in these locations, as the lack of topographic relief only allows significant surface water movement during large

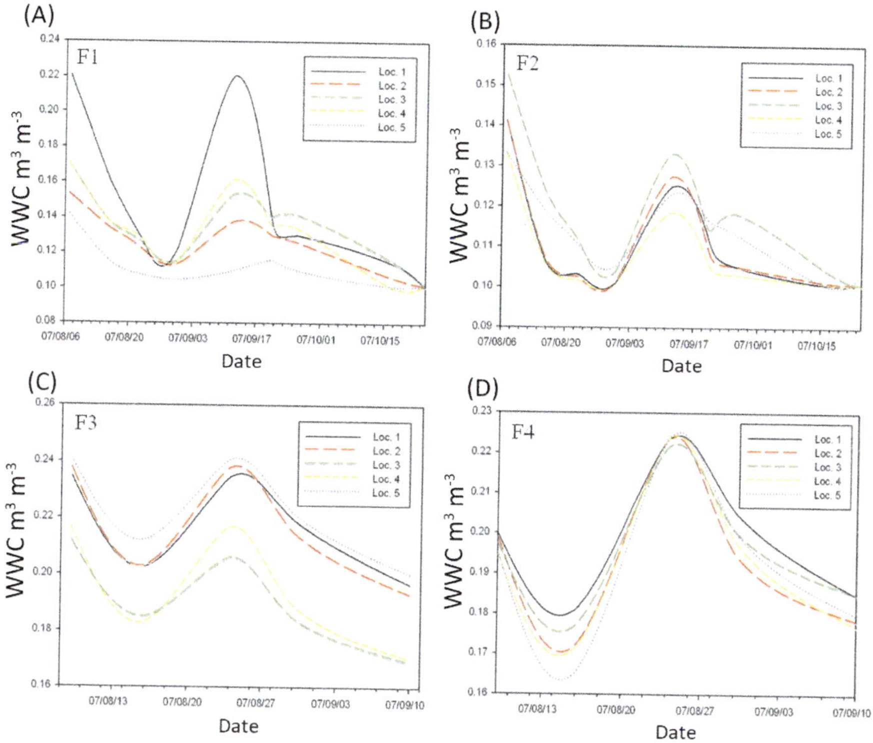

Fig. 4–5. Measured values of volumetric soil water content (VWC, m³ m⁻³) as a function of time at a depth of 0.2 m taken at (a) field F1, (b) field F2, (c) field F3, and (d) field F4 in 2007.

storms, for example, >15 mm h^{-1} (Jones et al., 1985). Soil VWC did not affect crop yield at either location, as average grain sorghum yield was the same at both locations (data not shown). Results in 2008 were similar to those in 2007. Seasonal soil VWC was not different between tillage treatment locations (Fig. 4–6a and 4–6b). In 2009, seasonal fluxes of soil VWC were different between locations. At field F4, a more noticeable range of values was measured at some sample points (Fig. 4–6b). This could be explained by the greater number of sample points capturing more of the spatial variability of soil VWC at each site. It is difficult to draw any conclusions from measurements taken at fields F3 and F4 for several reasons. Conditions at field F3 were not ideal for monitoring soil VWC. The field was located below a terraced dryland field and received run-on during moderate precipitation events. Furthermore, cropping practices were not always similar at both locations, as noted above. Therefore, it is more useful to make inferences from simulated results.

We compared all measured values of soil water content from the four fields and 3 yr to corresponding calculated values obtained with the PALMS model.

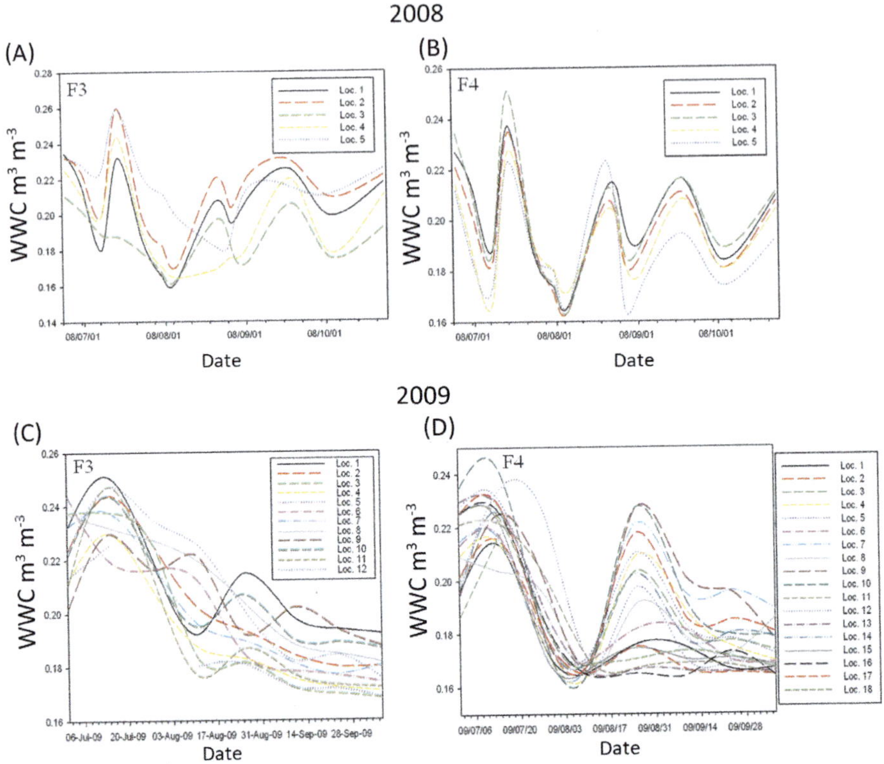

Fig. 4–6. Measured values of volumetric soil water content (VWC, m³ m⁻³) as a function of time at a depth of 0.2 m taken at (a) field F3 and (b) field F4 in 2008 and (c) field F3 and (d) field F4 in 2009.

For this purpose, we extracted the calculated value of soil water content to the corresponding location and depth (~0.2 m) and the time of measurement. This comparison yielded 421 sets of measured (y) and calculated values (x) of soil VWC. The resulting linear regression between y and x was $y = 0.12 + 0.88x$, with coefficient of determination $r^2 = 0.86$ and RMSE = 0.038 $m^3\ m^{-3}$. A similar result was reported by Molling et al. (2005), where the average daily soil water content was calculated within 5% of measured values, except during episodes of extended runoff and ponding. On first analysis and in general, this result shows that the calculated values of soil VWC obtained with the PALMS model are accurate; however, this result is misleading. For example, some calculated values of soil VWC were underestimated by as much as 0.10 $m^3\ m^{-3}$, particularly in the higher elevations (i.e., summit positions in fields F1 and F3). These positions tend to be eroded and have plants with a relative shallow root system as shown by Li et al. (2001) and by Bronson et al. (2003).

Soil Water and Runoff Response to Planting Patterns—Simulations

Calculated values of runoff, net runoff, and surface soil VWC indicated that a slight response to planting tillage pattern exists in the surface hydrology of some dryland fields. The response was more pronounced at fields F1 and F2, where significant topographic relief exists. This result is to be expected as shown by Julien et al. (1995) and by Li et al. (2001). Field F4 is a flat field (Table 4–1), and appreciable runoff only occurs during relatively large rainfall events. Although some differences in runoff and net runoff did exist between the two planting tillage patterns, there was very seldom a noticeable difference in soil water content following precipitation events. At fields F1 and F2, only two significant runoff events occurred during the 2007 growing season. On 28 May, 36 mm of rain fell over a 6-h period. Net runoff values at field F2 were slightly higher for the straight planting configuration at points 1, 2, 5, 10, and 14 (Fig. 4–7a). These locations were generally in areas where the circular tillage direction was across the slope. Therefore, the circular tillage was able to control runoff. This resulted in slightly higher

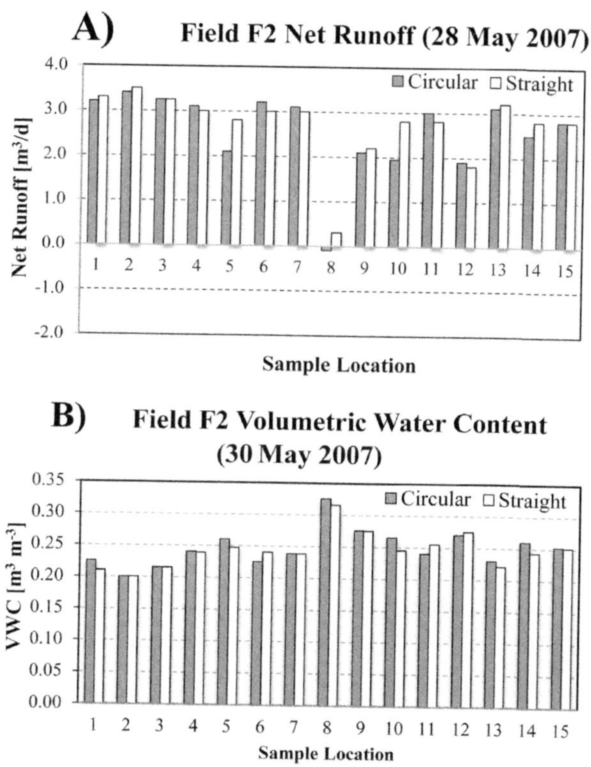

Fig. 4–7. (a) Precision Agricultural Landscape Modeling System simulated net runoff and (b) resulting calculated soil volumetric water content from a rain event at field F2 on 28 May 2007.

values of soil VWC at all locations when compared with the straight planting tillage configuration (Fig. 4–7b). Further evidence of this result is shown in Fig. 4–8. The difference in ponding between the straight and circular tillage pattern indicated that ponding was generally increased near the toe slope area by the circular pattern. Regardless of this information, changes in soil VWC were likely not significant enough to affect plant growth. A storm of 41 mm occurred over 2 h on 29 May 2008. We emphasize that runoff values were dissimilar at field F1 for the same storm. This field has remnants of some unmaintained contour terraces evident in Fig. 4–2a that reduced the calculated runoff. Net runoff was noticeably reduced by the circular planting tillage pattern at points 6 and 11 (Fig. 4–9a). Although a storm of this magnitude likely created rills through some of the tillage rows, there were some slight increases in soil VWC at several sample points on the circular planting tillage pattern (Fig. 4–9b). A similar result was reported by DeCoursey (1980) showing that land slope and tillage operations affected runoff and by Li et al. (2001) and Bronson et al. (2003) showing the importance of position across the landscape.

In most cases, runoff was generally not significantly affected by the planting tillage pattern at field F4. This was expected because of the lack of topographic relief (i.e., 0–1.5% slope) at this location. Net runoff was reduced slightly by the circular tillage pattern at field F4 during a very large storm on 12 Sept. 2008 (Fig. 4–10a). However, a storm of this magnitude (15 mm h^{-1}) usually only occurs 1 to 2 times during the typical growing season and generally runs off regardless of tillage or management (Lindstrom et al., 1998; Takken et al., 2001; DeLaune and Sij, 2012). This is demonstrated by the fact that soil VWC was not affected at all by tillage treatment (Fig. 4–10b). Further evidence is shown by the lack of difference in ponding between the two tillage treatments (data not shown).

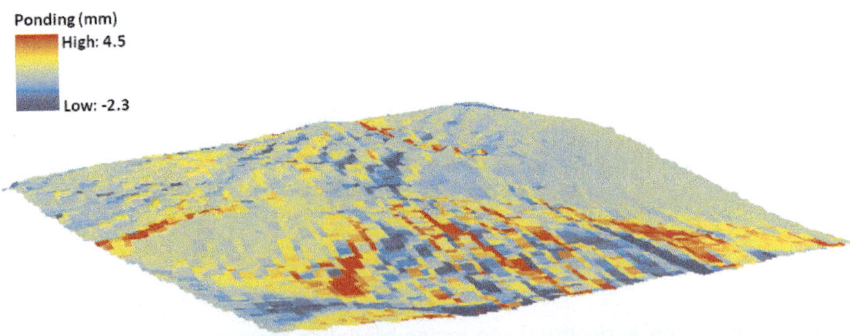

Fig. 4–8. Calculated difference in ponding values (mm) between the circular and straight tillage pattern for a single precipitation event at field F2 on 28 May 2007.

As an approximation, in the version of the PALMS model (Molling et al., 2005) used in our simulations we substituted the built-in soybean crop model to simulate a cotton crop. The main purpose was to illustrate the use of a circular planting pattern using a large-scale three-dimensional model to capture the interaction of rainfall, with infiltration and ponding across the landscape. We have now merged a cotton-specific model, Cotton2K, with PALMS and the result is PALMScot (Booker, 2013; Booker et al., 2014). This represents a next-generation cotton simulation model (Thorp et al., 2014) that provides a tool to evaluate how agronomic inputs could be managed to maximize lint yield across the landscape.

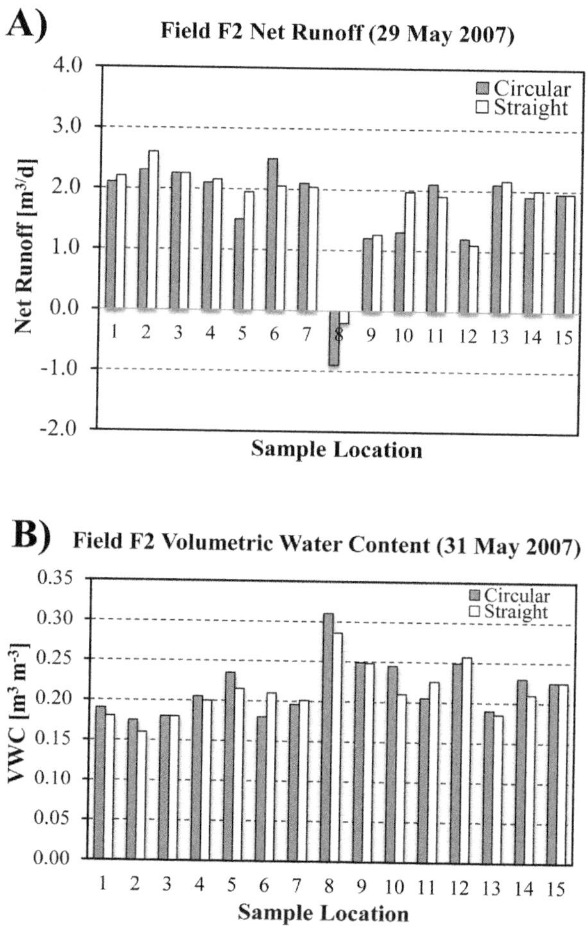

Fig. 4–9. (a) Precision Agricultural Landscape Modeling System simulated net runoff and (b) resulting calculated soil volumetric water content from a single rain event at field F2 on 29 May 2008.

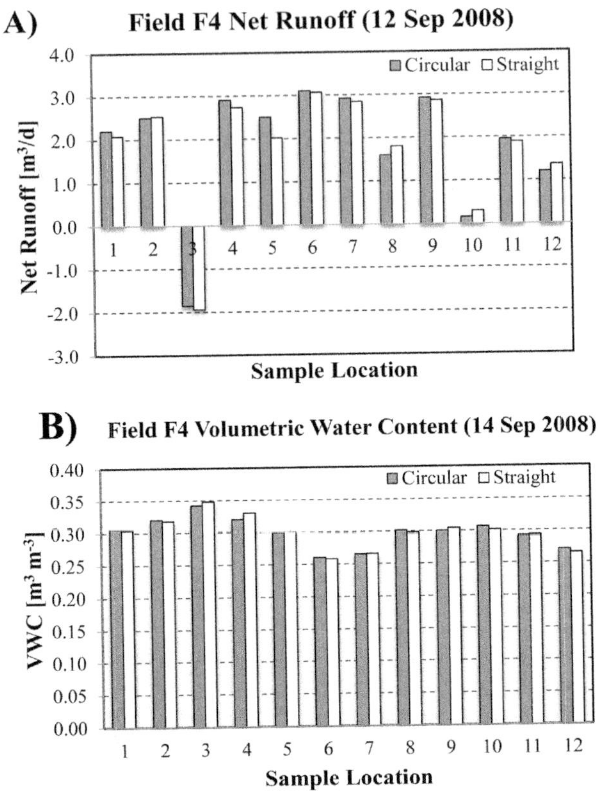

Fig. 4–10. (a) Precision Agricultural Landscape Modeling System simulated net runoff and (b) resulting calculated soil volumetric water content from a single rain event at field F4 on 14 Sept. 2008.

Conclusions

Since the 1970s two important concepts were introduced that impacted production agriculture, that is, simulation models as tools for crop management and precision agriculture to manage the crop according to specific needs across the landscape. These two concepts merged, and with advances in computer hardware and software, electronics, equipment technology, and expert systems and with increased availability of soil, elevation, and weather data (McBratney et al., 2005; Lascano and Booker, 2013), it is now possible to manage a crop in real time using a three-dimensional model such as PALMS. These concepts brought to the forefront the importance to manage agronomic inputs in response to cropping system variability. For example, instead of a blanket application of an agronomic input across the field, each input is managed according to specific needs across the landscape. The use of simulation models allows us to, for example, modify cultural practices and evaluate their impact on crop yield. In this paper, we used the PALMS simulation model to evaluate the effect of using a circular planting pattern in dryland cropping systems in a semiarid environment.

The effect of circular planting in dryland cropping systems on the SHP was compared with the traditional linear planting and the results varied across fields and years. In some cases, soil VWC was not affected, and in other cases it increased. For example, at fields F1 and F2, calculated soil VWC values increased following some runoff-inducing precipitation events. At field F4, there was no significant response because of the planting pattern, due in most part to the lack of topographic relief (i.e., <1.5% slope). In very large storms, the calculated runoff was slightly reduced by the circular tillage pattern, but it had no effect on available soil water. Measurements at fields F1 and F2 indicated some slight differences in soil VWC between planting patterns. However, crop yield differences were not noticeable, and yield reductions on the circular tilled field were most likely not related to soil water deficit. From this study and as expected, it is apparent that the success of a circular planting pattern for enhancing rainfall capture depends on the topographic characteristics of the field. The circular pattern would be most effective in fields with noticeable topographic relief (e.g., >3% slope) that are prone to runoff losses during intense rainfall events. Thus, the system is not ideal for all fields, but it provides the possibility of improving in-season soil water availability in certain dryland cropping systems with very little risk and cost to the producer.

Acknowledgments

This research was supported in part by the Ogallala Aquifer Program, a consortium between USDA-Agricultural Research Service, Kansas State University, Texas A&M AgriLife Research, Texas A&M AgriLife Extension Service, Texas Tech University, and West Texas A&M University. The support of Cotton Incorporated is also appreciated.

References

Alvarez-Acosta, C.A. 2009. Use of Precision Agricultural Landscape Modeling (PALMS) for estimating the effects of cultural practices on infiltration, runoff and erosion. M.Sc. thesis, Wageningen Univ., Wageningen, the Netherlands.

Alvarez-Acosta, C.A., R.J. Lascano, and L. Stroosnijder. 2012. Test of the Rosetta pedotransfer function for saturated hydraulic conductivity. Open J. Soil. Sci. 2:203–212. doi:10.4236/ojss.2012.23025

Arkin, G.F., R.L. Vanderlip, and J.T. Ritchie. 1976. A dynamic grain sorghum growth model. Trans. ASABE 19:622–630. doi:10.13031/2013.36082

Baker, D.N., J.R. Lambert, and J.M. McKinion. 1983. GOSSYM: A simulator of cotton crop growth and yield. S.C. Agric. Exp. Stn. Bull. 1089. Clemson Univ., Clemson, SC.

Basso, B., J.T. Ritchie, F.J. Pierce, R.P. Braga, and J.W. Jones. 2001. Spatial validation of crop models for precision agriculture. Agric. Syst. 68:97–112. doi:10.1016/S0308-521X(00)00063-9

Batchelor, W.D., B. Basso, and J.O. Paz. 2002. Examples of strategies to analyze spatial and temporal yield variability using crop models. Eur. J. Agron. 18:141–158. doi:10.1016/S1161-0301(02)00101-6

Baumhardt, R.L., and R.J. Lascano. 1996. Rain Infiltration as affected by wheat residue amount and distribution in ridged tillage. Soil Sci. Soc. Am. J. 60:1908–1913. doi:10.2136/sssaj1996.03615995006000060041x

Baumhardt, R.L., R.J. Lascano, and D.R. Krieg. 1995. Physical and hydraulic properties of a Pullman and Amarillo soil on the Texas South Plains. Tech. Rep. 95-1. Texas Agric. Exp. Stn., Lubbock.

Baumhardt, R.L., C.W. Wendt, and J.W. Keeling. 1993. Tillage and residue effects on infiltration into soils cropped to cotton. Agron. J. 85:379–383. doi:10.2134/agronj1993.00021962008500020038x

Biazin, B., G. Sterk, M. Temesgen, A. Abdulkedir, and L. Stroosnijder. 2012. Rainwater harvesting and management in rainfed agricultural systems in sub-Saharan Africa—A review. Phys. Chem. Earth 47–48:139–151. doi:10.1016/j.pce.2011.08.015

Booker, J.D. 2013. Modeling landscape-scale water balance in irrigated cotton systems—Pullman and Amarillo soil series. Ph.D. dissertation, Dep. of Plant and Soil Sciences, Texas Tech Univ., Lubbock.

Booker, J.D., R.J. Lascano, S.R. Evett, and R.E. Zartman. 2014. Evaluation of a landscape-scale approach to cotton modeling. Agron. J. 106:1–17. doi:102134/agronj14.0202

Bonilla, C.A., J.M. Norman, and C.C. Molling. 2007. Water erosion estimation in topographically complex landscapes: Model description and first verifications. Soil Sci. Soc. Am. J. 71:1524–1537. doi:10.2136/sssaj2006.0302

Bonilla, C.A., J.M. Norman, C.C. Molling, K.G. Karthikeyan, and P.S. Miller. 2008. Testing a grid-based soil erosion model across topographically complex landscapes. Soil Sci. Soc. Am. J. 72:1745–1755. doi:10.2136/sssaj2007.0310

Bosch, D.D., T. Potter, C. Truman, C. Bednarz, and T. Strickland. 2005. Surface runoff and lateral subsurface flow as a response to conservation tillage and soil-water conditions. Trans. ASABE 48:2137–2144. doi:10.13031/2013.20099

Bronson, K., J.W. Keeling, J.D. Booker, T.T. Chua, T.A. Wheeler, R.K. Boman, and R.J. Lascano. 2003. Influence of landscape position, soil series, and phosphorus fertilizer on cotton lint yield. Agron. J. 95:949–957. doi:10.2134/agronj2003.0949

Colaizzi, P.D., P.H. Gowda, T.H. Marek, and D.O. Porter. 2009. Irrigation in the Texas High Plains: A brief history and potential reductions in demand. Irrig. Drain. 58:257–274. doi:10.1002/ird.418

Connolly, R.D. 1998. Modelling effects of soil structure on the water balance of soil-crop systems: A review. Soil Tillage Res. 48:1–19. doi:10.1016/S0167-1987(98)00128-7

DeCoursey, D.G. 1980. Runoff, erosion, and crop yield simulation for land use management. Trans. ASABE. 31:379–386. doi:10.13031/2013.34589

DeJonge, K.C., A.L. Kaleita, and K.R. Thorp. 2007. Simulating the effects of spatially variable irrigation on corn yields, costs, and revenue in Iowa. Agric. Water Manage. 92:99–109. doi:10.1016/j.agwat.2007.05.008

DeLaune, P.B., and J.W. Sij. 2012. Impact of tillage on runoff in long term no-till wheat systems. Soil Tillage Res. 124:32–35. doi:10.1016/j.still.2012.04.009

Dickey, E.C., C.R. Fenster, J.M. Laflen, and R.H. Mickelson. 1983. Effects of tillage on soil erosion in a wheat-fallow rotation. Trans. ASABE 26:814–820. doi:10.13031/2013.34029

Diekkrüger, B., D. Söndgerath, K.C. Kersebaum, and C.W. McVoy. 1995. Validity of agroecosystem models: A comparison of results of different models applied to the same data set. Ecol. Modell. 81:3–29. doi:10.1016/0304-3800(94)00157-D

Dugas, W.A., G.F. Arkin, and B.S. Jackson. 1983. Factors affecting simulated crop yield spatial extrapolation. Trans. ASABE 26:1440–1444. doi:10.13031/2013.34147

Faivre, R., D. Leenhardt, M. Voltz, M. Benoit, F. Pappy, G. Dedieu, and D. Wallace. 2004. Spatialising crop models. Agronomie 24:205–217. doi:10.1051/agro:2004016

Foley, J.A., I.C. Prentice, N. Ramankutty, S. Levis, D. Pollard, S. Sitch, and A. Haxeltine. 1996. An integrated biosphere model of land surface processes, terrestrial carbon balance, and vegetation dynamics. Global Biogeochem. Cycles 10:603–628. doi:10.1029/96GB02692

Gauch, H.G., Jr., J.T. Gene Hwang, and G.W. Fick. 2003. Model evaluation by comparison of model-based predictions and measured value. Agron. J. 95:1442–1446. doi:10.2134/agronj2003.1442

Gee, G.W., and J.W. Bauder. 1986. Particle-size analysis. In: A. Klute, editor, Methods of soil analysis. Part 1. 2nd ed. Agron. Monogr. 9. ASA and SSSA, Madison, WI. p. 383–423.

Haas, H.J., W.O. Willis, and G.O. Boatwright. 1966. Moisture storage and spring wheat yields on level-bench terraces as influenced by contributing area cover and evaporation control. Agron. J. 58:297–299. doi:10.2134/agronj1966.00021962005800030015x

Harman, W.L., G.J. Michels, and A.F. Wiese. 1989. A conservation tillage system for profitable cotton production in the Central Texas High Plains. Agron. J. 81:615–618. doi:10.2134/agronj1989.00021962008100040013x

Hauser, V.L., and M.B. Cox. 1962. Evaluation of Zingg conservation bench terraces. Agric. Eng. 43:462–464.

Heege, H.J., editor. 2013. Precision in crop farming. Site specific concepts and sensing methods: applications and results. Springer, Dordrecht, the Netherlands.

Howell, T.A., A.D. Schneider, and D.A. Dusek. 2002. Effects of furrow diking on corn response to limited and full sprinkler irrigation. Soil Sci. Soc. Am. J. 66:222–227. doi:10.2136/sssaj2002.0222

Jackson, B.S., G.F. Arkin, and A.B. Hearn. 1988. The cotton simulation model "COTTMAN": Fruiting model calibration and testing. Trans. ASABE 31:846–854. doi:10.13031/2013.30790

Johnson, I.R. 2011. Testing and evaluating large-scale agricultural simulation models. In: F. Chan, D. Marinova, and R.S. Anderssen, editors, MODSIM2011, 19th International Congress on Modelling and Simulation, Perth, Western Australia. 12–16 Dec. 2011. Modell. and Simul. Soc. of Australia and New Zealand, Perth, Western Australia. p. 84–96.

Jones, M.P., and W.F. Hunt. 2010. Performance of rainwater harvesting systems in the southeastern United States. Resour. Conserv. Recycl. 54:623–629. doi:10.1016/j.resconrec.2009.11.002

Jones, O.R. 1975. Yields and water-use efficiencies of dryland winter wheat and grain sorghum production systems in the Southern High Plains. Soil Sci. Soc. Am. J. 39:98–103. doi:10.2136/sssaj1975.03615995003900010027x

Jones, O.R., and R.L. Baumhardt. 2003. Furrow dikes. In: B.A. Stewart and T.A. Howell, editors, Encyclopedia of water science. Vol. 1. Marcel Dekker, New York. p. 317–320.

Jones, O.R., and R. Clark. 1987. Effects of furrow dikes on water conservation and dryland crop yields. Soil Sci. Soc. Am. J. 51:1307–1314. doi:10.2136/sssaj1987.03615995005100050039x

Jones, O.R., H.V. Eck, S.J. Smith, G.A. Coleman, and V.L. Hauser. 1985. Runoff, soil, and nutrient losses from rangeland and dry-farmed cropland in the Southern High Plains. J. Soil Water Conserv. 40:161–164.

Jones, O.R., and B.A. Stewart. 1990. Basin tillage. Soil Tillage Res. 18:249–265. doi:10.1016/0167-1987(90)90064-K

Julien, P.Y., B. Saghafian, and F.L. Ogden. 1995. Raster-based hydrologic modeling of spatially-varied surface runoff. Water Resour. Bull. 31:523–536. doi:10.1111/j.1752-1688.1995.tb04039.x

Kleijnen, J.P.C. 1995. Verification and validation of simulation models. Eur. J. Oper. Res. 82:145–162. doi:10.1016/0377-2217(94)00016-6

Klute, A. 1986. Water retention: Laboratory methods. In: A. Klute, editor, Methods of soil analysis. Part 1. 2nd ed. Agron. Monogr. 9. ASA and SSSA, Madison, WI. p. 635–686.

Kobayashi, K., and M.U. Salam. 2000. Comparing simulated and measured values using mean squared deviation and its components. Agron. J. 92:345–352. doi:10.2134/agronj2000.922345x

Lascano, R.J. 2000. A general system to measure and calculate daily crop water use. Agron. J. 92:821–832. doi:10.2134/agronj2000.925821x

Lascano, R.J., and R.L. Baumhardt. 1996. Effects of crop residue on soil and plant water evaporation in a dryland cotton system. Theor. Appl. Climatol. 54:69–84. doi:10.1007/BF00863560

Lascano, R.J., R.L. Baumhardt, S.K. Hicks, and J.L. Heilman. 1994. Soil and plant water evaporation from strip-tilled cotton: Measurement and simulation. Agron. J. 86:987–994. doi:10.2134/agronj1994.00021962008600060011x

Lascano, R.J., and J.D. Booker. 2013. Precision agriculture: Water and nutrient management. In: S.E. Jørgensen, editor, Encyclopedia of environmental management. 1st ed. Vol. III. Taylor & Francis, Oxford, UK. p. 2217–2223.

Lawes, D.A. 1963. A new cultivation technique in tropical Africa. Nature 198:1328. doi:10.1038/1981328a0

Legates, D.R., and G.J. McCabe, Jr. 1999. Evaluating the use of "goodness-of-fit" measures in hydrologic and hydroclimatic model validation. Water Resour. Res. 35:233–241. doi:10.1029/1998WR900018

Legates, D.R., and G.J. McCabe. 2013. A refined index of model performance: A rejoinder. Int. J. Climatol. 33:1053–1056. doi:10.1002/joc.3487

Li, H., R.J. Lascano, J. Booker, L.T. Wilson, and K.F. Bronson. 2001. Cotton lint yield variability in a heterogeneous soil at a landscape level. Soil Tillage Res. 58:245–258. doi:10.1016/S0167-1987(00)00172-0

Lindstrom, M.J., T.E. Schumacher, N.P. Cogo, and M.L. Blecha. 1998. Tillage effects on water runoff and soil erosion after sod. J. Soil Water Conserv. 53:59–63.

Llano Estacado Regional Water Planning Group. 2006. Regional water plan. High Plains Underground Water Conservation District No. 1, Lubbock, TX.

Lyle, W.M., and D. Dixon. 1977. Basin tillage for rainfall retention. Trans. ASABE 20:1013–1017. doi:10.13031/2013.35693

Mapp, H.P. 1988. Irrigated agriculture on the High Plains: An uncertain future. West. J. Agric. Econ. 13:339–347.

Marani, A. 2004. Cotton2K Model version 4.0. Hebrew Univ. of Jerusalem. http://departments.agri.huji.ac.il/plantscience/cotton/Cotton2KModelDoc.htm (accessed 4 Feb. 2014).

McBratney, A., B. Whelan, T. Ancev, and J. Bouma. 2005. Future directions of precision agriculture. Precis. Agric. 6:7–23. doi:10.1007/s11119-005-0681-8

Molling, C.C., J.C. Strikwerda, J.M. Norman, C.A. Rodgers, R. Wayne, C.L.S. Morgan, G.R. Diak, and J.R. Mecikalski. 2005. Distributed runoff formulation designed for a Precision Agricultural Landscape Modeling System. J. Am. Water Resour. Assoc. 41:1289–1313. doi:10.1111/j.1752-1688.2005.tb03801.x

Morgan, C.L.S., J.M. Norman, C.C. Moiling, K. McSweeney, and B. Lowery. 2003. Evaluating soil data from several sources using a landscape model. In: Y.A. Pachepsky, D.E. Radcliffe, and H. Magdi Salim, editors, Scaling methods in soil physics. CRC Press, New York. p. 243–260.

Moriasi, D.N., J.G. Arnold, M.W. Van Liew, R.L. Bingner, R.D. Harmel, and T.L. Veith. 2007. Model evaluation guidelines for systematic quantification of accuracy in watershed simulations. Trans. ASABE 50:885–900. doi:10.13031/2013.23153

Nair, S., S. Maas, C. Wang, and S. Mauget. 2013. Optimal field partitioning for center-pivot-irrigated cotton in the Texas High Plains. Agron. J. 105:124–133. doi:10.2134/agronj2012.0219

Nelson, J.R. 2010. Modification of surface hydrology to enhance rainfall capture and infiltration in dryland cropping systems. Ph.D. dissertation, Texas Tech Univ., Lubbock.

Nelson, J.R., R.J. Lascano, J.D. Booker, R.E. Zartman, and T.S. Goebel. 2013. Evaluation of the Precision Agricultural Landscape Modeling System (PALMS) in the semiarid Texas Southern High Plains. Open J. Soil Sci. 3:169–181. doi:10.4236/ojss.2013.34020

Nielsen, D.C., P.W. Unger, and P.R. Miller. 2005. Efficient water use in dryland cropping systems in the Great Plains. Agron. J. 97:364–372. doi:10.2134/agronj2005.0364

Norwood, C.A., and T.J. Dumler. 2002. Transition to dryland agriculture: Limited irrigated vs. dryland corn. Agron. J. 94:310–320. doi:10.2134/agronj2002.0310

Nuti, R.C., M.C. Lamb, R.B. Sorensen, and C.C. Truman. 2009. Agronomic and economic response to furrow diking tillage in irrigated and non-irrigated cotton (*Gossypium hirsutum L.*). Agric. Water Manage. 96:1078–1084. doi:10.1016/j.agwat.2009.03.006

Osterkamp, W.R., and W.W. Wood. 1987. Playa-lake basins on the Southern High Plains of Texas and New Mexico: Part I. Hydrologic, geomorphic, and geologic evidence for their development. Geol. Soc. Am. Bull. 99:215–223. doi:10.1130/0016-7606(1987)99<215:PBOTSH>2.0.CO;2

Oweis, T.Y., D. Prinz, and A.Y. Hachum. 2012. Water harvesting for agriculture in the dry areas. Taylor & Francis, London.

Resck, D.V.S., J.E. da Silva, A.S. Lopes, and L.M. da Costa. 2006. Management systems in northern South America. In: G.A. Peterson, P.W. Unger, and W.A. Payne, editors, Dryland agriculture. 2nd ed. Agron. Monogr. 23. ASA, CSSA, and SSSA, Madison, WI. p. 427–525.

Rodgers, C., J. Mecikalski, C. Molling, J. Norman, C. Kucharik, and C. Morgan. 2000. Spatially distributed hydrologic-biophysical modeling: Applications in precision agriculture. In: 24th Conference on Agricultural and Forest Meteorology, Davis, CA. 14–18 Aug. 2000. Am. Meteorol. Soc., Boston, MA. p. 86–87

Rosenthal, W.D., R.L.Vanderlip, G.F. Arkin, and B.S. Jackson. 1989. SORKAM: A grain sorghum crop growth model. Misc. Publ. MP-1699. Texas Agric. Exp. Stn., College Station.

Sargent, R.G. 2005. Verification and validation of simulation models. In: M.E. Kuhl et al., editors, Proceedings of the 2005 Winter Simulation Conference, Orlando, FL. 4–7 Dec. 2005. IEEE, Piscataway, NJ. p. 130–143.

Stafford, J.V. 2000. Implementing precision agriculture in the 21st century. J. Agric. Eng. Res. 76:267–275. doi:10.1006/jaer.2000.0577

Stroosnijder, L. 2009. Modifying land management in order to improve efficiency of rainwater use in the African highlands. Soil Tillage Res. 103:247–256. doi:10.1016/j.still.2008.05.019

Takken, I., G. Govers, V. Jetten, J. Nachtergaele, A. Steegen, and J. Poesen. 2001. Effects of tillage on runoff and erosion patterns. Soil Tillage Res. 61:55–60. doi:10.1016/S0167-1987(01)00178-7

Thacker, B.H., S.W. Doebling, F.M. Hemez, M.C. Anderson, J.E. Pepin, and E.A. Rodriguez. 2004. Concepts of model verification and validation. Tech. Rep. LA-14167-MS. Los Alamos Natl. Lab., Los Alamos, NM.

Thorp, K.R., S. Ale, M.P. Bange, E.M. Barnes, G. Hoogenboom, R.J. Lascano, A.C. McCarthy, S. Nair, J.O. Paz, N. Rajan, K.R. Reddy, G.W. Wall, and J.W. White. 2014. Development and application of process-based simulation models for cotton production: A review of past, present, and future directions. J. Cotton Sci. 18:1–38.

Troeh, F.R., J.A. Hobbs, and R.L. Donahue. 1991. Soil and water conservation. Prentice Hall, Englewood Cliffs, NJ.

Trouslard-Kerdiles, V., and M.O. Grondona. 1997. A case of combined use of crop simulation models and general linear models. Ecol. Modell. 99:71–85. doi:10.1016/S0304-3800(96)01941-2

Unger, P.W. 1994. Tillage effects on dryland wheat and sorghum production in the Southern Great Plains. Agron. J. 86:310–314. doi:10.2134/agronj1994.00021962008600020019x

Unger, P.W. 1996. Common soil and water conservation practices. In: M. Agassi, editor, Soil erosion, conservation and rehabilitation. Marcel Dekker, New York. p. 147–161.

Unger, P.W., and J.J. Parker. 1976. Evaporation reduction from soil with wheat, sorghum, and cotton residues. Soil Sci. Soc. Am. J. 40:938–942. doi:10.2136/sssaj1976.03615995004000060035x

Unger, P.W., W.A. Payne, and G.A. Peterson. 2006. Water conservation and efficient use. In: G.A. Peterson, P.W. Unger, and W.A. Payne, editors, Dryland agriculture. 2nd ed. Agron. Monogr. 23. ASA, CSSA, and SSSA, Madison, WI. p. 39–85.

van Genuchten, M.Th. 1980. A closed-form equation for predicting the hydraulic conductivity of unsaturated soils. Soil Sci. Soc. Am. J. 44:892–898. doi:10.2136/sssaj1980.03615995004400050002x

Vohland, K., and B. Barry. 2009. A review of *in situ* rainwater harvesting (RWH) practices modifying landscape functions in African drylands. Agric. Ecosyst. Environ. 131:119–127. doi:10.1016/j.agee.2009.01.010

Wiedemann, H.T., and B.A. Smallacombe. 1989. Chain diker—A new tool to reduce runoff. Agric. Eng. 70:12–15.

Zartman, R.E., P.W. Evans, and R.H. Ramsey. 1994. Playa lakes on the Southern High Plains in Texas: Reevaluating infiltration. J. Soil Water Conserv. 49:299–301.

Zhang, N., M. Wang, and N. Wang. 2002. Precision agriculture—A worldwide overview. Comput. Electron. Agric. 36:113–132. doi:10.1016/S0168-1699(02)00096-0

Zhu, J., C.L.S. Morgan, J.M. Norman, W. Yue, and B. Lowery. 2004. Combined mapping of soil properties using a multi-scale tree-structured spatial model. Geoderma 118:321–334. doi:10.1016/S0016-7061(03)00217-9

Quantifying and Managing Corn Water Use Efficiencies under Irrigated and Rainfed Conditions in Nebraska Using the Hybrid-Maize Simulation Model

Haishun Yang* and Patricio Grassini

Abstract

This study used the Hybrid-Maize model to quantify corn (*Zea mays* L.) water use and water use efficiencies (WUEs) under irrigated and rainfed conditions at Mead and North Platte in Nebraska. Water use efficiency is quantified using both evapotranspiration (ET) and transpiration (T), respectively, based on 20 yr of simulations. Across-year variations of crop total ET and T and WUE are greater in rainfed fields than irrigated fields. When crop growth is limited by water, crop biomass and grain yield depend strongly on seasonal total ET, while the dependence becomes much weaker or even insignificant once water limitation is eliminated. Seasonal total T is about 81% of total ET under irrigated conditions and 76% under rainfed conditions. Seasonal dynamics of ET and T follows the pattern of crop biomass growth. Evapotranspiration-based WUE is low in the spring when plants are small and the canopy is open, while WUE reaches its plateau shortly before silking. Evapotranspiration-based WUE for total biomass at a plateau ranges from 45 to 55 kg ha^{-1} mm^{-1}, while T-based WUE ranges from 55 to 65 kg ha^{-1} mm^{-1} in Nebraska. Evapotranspiration-based WUE for grain filling ranges from 50 to 70 kg ha^{-1} mm^{-1}. Eliminating non-water-related crop growth constraints is critical to the improvement of WUE, and meeting crop water demand during the reproductive stage is important for maximizing grain yield in deficit irrigation.

Abbreviations: GDD, growing degree day; LAI, leaf area index; T, transpiration; WUE, water use efficiency.

H. Yang (hyang2@unl.edu) and P. Grassini (pgrassini2@unl.edu) Department of Agronomy and Horticulture, University of Nebraska, 377M, PLSH, Lincoln, NE 68583. *Corresponding author.

doi:10.2134/advagricsystmodel5.c5

Practical Applications of Agricultural System Models to Optimize the Use of Limited Water
Lajpat R. Ahuja, Liwang Ma, and Robert J. Lascano, Editors
Advances in Agricultural Systems Modeling, Volume 5. Lajpat R. Ahuja, Series Editor

USDA-ARS

Crop production must increase steadily to meet the increasing demand for food, fiber, and fuel due to growth of the global population and the change in diet in booming economies (Godfray et al., 2010). In many parts of the world, a further increase of crop production depends largely on water availability for irrigation (Grassini et al., 2011a, 2011b; Payero et al., 2006; Steduto et al., 2009). At the same time, however, water resources for irrigation have been shrinking at local, regional, and global levels (Vorosmarty et al., 2000). Hence, crop irrigation management and crop WUE must be further improved to irrigate more arable land using the limited or even shrinking irritation water resources (Howell, 2001). To achieve this goal, a good understanding of the relationship of crop growth and yield formation with water consumption is required. Achieving this goal also requires a good quantification of crop WUE under current crop management practices as well as WUE that can be achieved using optimal and water-saving management practices. The gap between current WUE and optimal WUE is the measure of potential improvement that can be made through adjustment and changes in crop management, including irrigation, fertilization, tillage, and crop residues management.

Crop water efficiency (WUE) can be generally defined as the yield of a harvest product (e.g., grain for cereal crops) per unit of seasonal total water consumption. As transpiration is the true water consumption by the crop, de Wit (1958) defined WUE based on crop total transpiration (WUE_T) as

$$WUE_T = Y/T \qquad [1]$$

where Y is product yield (kg ha^{-1}), T is seasonal total transpiration (mm), and WUE_T is in kg ha^{-1} mm^{-1}. Although transpiration is the true part of water consumption by a crop, it is difficult to separate it from total ET, which is the sum of transpiration and soil evaporation. As a result, seasonal total ET is more often used a measure of crop water consumption (Howell and Hiler, 1975; Grassini et al., 2009). Hence, WUE based on ET (WUE_{ET}) can be defined as

$$WUE_{ET} = Y/ET \qquad [2]$$

where WUE_{ET} has the same unit as WUE_T. To estimate actual WUE_T or WUE_{ET}, actual yield and actual total transpiration or ET should be used, while crop yield potential and corresponding transpiration or ET should be used when estimating WUE_T or WUE_{ET} under optimal crop management (Grassini et al., 2009).

The WUE of many common crops has been measured and quantified in experimental studies. In their reviews, Tanner and Sinclair (1983) and Steduto et al. (2007) found that WUE_{ET} and WUE_T, particularly for crop total biomass, are rather conservative in the same environment or across different environments if climatic conditions are normalized. Steduto et al. (2007) further suggests that a significant alteration in WUE may come only from a genetic breakthrough that brings in fundamental changes in plant biosynthetic capacities. At the same time, studies have demonstrated that crop economic yield (e.g., grains of cereal crops) can vary significantly depending on the timing and amount of water supply in relation to the crop growth stage and yield sensitivity to water stress (Payero et al., 2009; Perry et al., 2009). For instance, Howell (2000) shows that elimination of water stress during corn pollination can increase grain yield WUE by as much as 48% in a surface irrigation experiment.

Most studies on crop WUE quantified WUE_{ET} or WUE_T at the end of a crop growth cycle when a crop matures. As seasonal total crop water consumption and dry matter production or yield is a result of daily accumulation, it is important to understand the dynamic nature of water consumption, dry matter accumulation, and WUE in the course of crop development. In addition, separating T from ET and the resulting separate WUE_T and WUE_{ET} may lead to a better understanding of the dynamics of crop water consumption at different stages of crop development and possible effects of crop management (e.g., tillage, fertilization, mulching by crop residues or plastic film) on crop yield and WUE.

Mechanistic crop simulation integrates a quantitative understanding of processes of crop growth and development in response to weather conditions, soil properties, and crop management. Under optimal crop management for nutrients and control of diseases, insects, and pests, a crop simulation model can estimate yield potentials for irrigated or rainfed conditions. If crop water consumption can be estimated in terms of either ET or T, WUE_{ET} or WUE_T can then be quantified and responses of WUE_{ET} or WUE_T to crop management can be studied.

The objective of this chapter was to use a crop simulation model to quantify corn water use and WUEs in two cropping environments in Nebraska, in terms of WUE_{ET} and WUE_T, for combinations of aboveground biomass as well as grain yield, at the end of the season as well as seasonal dynamics, under fully irrigated and rainfed conditions. The study used Hybrid-Maize, a corn simulation model (Yang et al., 2004, 2006, 2013), and 20 yr of historical weather data from 1993 to 2012 at each site to effectively capture weather variability while avoiding possible confounding effects from technological advancement over time.

MATERIALS AND METHODS

Hybrid-Maize Model

The Hybrid-Maize model is a computer simulation model that simulates daily growth of a corn crop under nonlimiting (yield potential) or water-limited yield potential (rainfed or partially irrigated) conditions (Yang et al., 2004, 2006, 2013; http://hybridmaize.unl.edu/). It combines the strength of two maize simulation approaches represented by Wageningen models, including World Food Studies (WOFOST) (van Diepen et al., 1989) and INTERCOM (Lindquist, 2001), and by the Crop Environment Resource Synthesis-Maize (CERES-Maize) model (Jones and Kiniry, 1986; Kiniry et al., 1997). The model can be used to (i) assess yield potential and its variability at a given location based on historical weather data, (ii) evaluate changes in yield potential using combinations of planting date, hybrid maturity, and plant density, (iii) explore options for irrigation management, and (iv) forecast final yield scenarios in the middle of a growing season on the basis of real-time weather data and historical weather records. Hybrid-Maize does not account for yield losses due to suboptimal nutrient management, weeds, insects, diseases, lodging, and other stresses. Hybrid-Maize has been validated for irrigated as well as rainfed corn systems in the U.S. Western Corn Belt (Yang et al., 2004; Grassini et al., 2009), North China (Chen et al., 2013), and South Asia (Timsina et al., 2010) to assess maize yield potential and yield gaps and to evaluate management options (Chen et al., 2011; Grassini et al., 2011a), impact of climate change (Chen et al., 2013; Lobell et al., 2009), water productivity (Grassini et al., 2009, 2011b), and nutrient management (Meng et al., 2012) across diverse maize systems and production environments.

Hybrid-Maize Simulation of Evaporation, Transpiration, and Water Stress

The Hybrid-Maize model simulates daily soil evaporation, T, and crop water stress. Required input data include reference evapotranspiration (ET_0) from weather data, water input from precipitation and irrigation and simulated soil moisture content of the previous day, crop root distribution, and crop leaf area index (LAI). Required field and soil properties include field slope and drainage class for estimation of surface runoff and soil texture and bulk density for pedotransfer function for conversion between volumetric water content and water potential.

Simulation of soil evaporation and T are based on ET_0 from the weather data. As a well-watered corn field at LAI greater than 4 typically has greater ET than that of grass, ET_0 is adjusted to reflect ET of a corn field using a modified method of Allen et al. (1998)

$$\text{if LAI} \leq 3.5, \text{ then Coef}_{\text{adj}} = 1$$
$$\text{else Coef}_{\text{adj}} = 1 + (\text{LAI} - 3.5) \times (1.2 - 1) / (4.5 - 3.5)$$
$$\text{if Coef}_{\text{adj}} > 1.2, \text{ then Coef}_{\text{adj}} = 1.2$$
$$\text{ET}_{0\text{adj}} = \text{ET}_0 \times \text{Coef}_{\text{adj}}$$

where ET_{0adj} is the ET_0 adjusted for corn canopy.

ET_{0adj} is then split into maximum transpiration ($Transp_{max}$) and potential evaporation ($Evap_{pot}$) on the basis of LAI

$$\text{Transp}_{\text{max}} = \text{ET}_{0\text{adj}} \times [1 - \exp(-\text{LAI} \times k)]$$
$$\text{Evap}_{\text{pot}} = \text{ET}_{0\text{adj}} - \text{Transp}_{\text{max}}$$

where k is the light extinction coefficient.

The maximum evaporation ($Evap_{max}$) is estimated by accounting for the effect of soil coverage by crop residues (Rosenberg et al., 1983)

$$\text{Evap}_{\text{max}} = \text{Evap}_{\text{pot}} \times \exp(-\text{soilCover})$$

where soilCover is the fraction of soil surface that is covered by crop residues. Actual soil evaporation ($Evap_{act}$) is estimated using the two-step evaporation scheme as in FAO Report 56 (Allen et al., 1998). Soil evaporation is assumed to occur in the top 10-cm depth, and evaporation rate is at maximum when soil is wet (i.e., Step 1)

$$\text{Evap}_{\text{act}} = \text{Evap}_{\text{max}}$$

The evaporation rate starts to decrease, and continuously so, when soil water content drops below a threshold (Step 2). This threshold (step2EvapWater) is estimated as

$$\text{EvapWater}_{\text{max}} = (\theta_{\text{FC}} - 0.5 \times \theta_{\text{PWP}}) \times 10$$
$$\text{step2EvapWater} = \text{step2threshold} \times \text{EvapWater}_{\text{max}}$$
$$\text{waterForEvap} = (\theta - 0.5 \times \theta_{\text{PWP}}) \times 10$$
$$\text{Evap}_{\text{act}} = \text{Evap}_{\text{max}} \times \text{waterForEvap} / \text{step2EvapWater}$$

where waterForEvap is the amount of water available for evaporation, $EvapWater_{max}$ is the maximum amount of water that can evaporate, θ, θ_{FC}, and θ_{PWP} are the topsoil volumetric water content (in fraction) and its values at field capacity and permanent wilting point, respectively, 10 is the depth (in cm) of the evaporating soil, and step2threshold is a fraction typically ranging around 0.5 ~ 0.7.

Actual transpiration ($Transp_{act}$) is estimated from the maximum water uptake by roots from soil rooting depth, which is divided into layers of 10 cm, relative to maximum transpiration ($Transp_{max}$). The smaller one of the two is taken as $Transp_{act}$.

Crop water stress index (WSI) is estimated as

$$WSI = 1 - Transp_{act}/Transp_{max}$$

where WSI (dimensionless) is scaled from 0 (no stress) to 1 (full stress). In the Hybrid-Maize model, direct crop responses to water stress include proportional reduction of photosynthesis and gross dry matter production, reduced daily canopy expansion rate if water stress occurs before silking, or greater canopy senescence if water stress occurs after silking. Other impacts of water stress include reduced kernel setting if water stress occurs during the kernel setting window, which is about 4 wk centered on silking, and increased respiration losses due to translocation of carbohydrate reserve for grain filling if water stress occurs during grain filling (Yang et al., 2004). Besides the ability of simulating corn growth and yield under irrigated and rainfed conditions, Hybrid-Maize also has the option of estimating irrigation water requirements (both for timing and amount) for achieving stress-free crop growth. The model outputs include daily values of LAI; crop dry matter accumulation; respiration losses; ET demand and actual ET; and soil water balance at depths of 30 cm, 30 to 60 cm, and below to crop rooting depth.

Selected Locations, Simulation Setup, and Weather Data

Two locations with contrast cropping environments were selected for the simulation study (Table 5–1): North Platte, NE (41.14° N, 100.76° W) and Mead, NE (41.23° N, 96.49° W). North Platte, located in western Nebraska, is a typical irrigated corn production environment because of its inadequate growing season rainfall and strong ET demand. Mead, located in eastern Nebraska, has both irrigated and rainfed corn production because of its relatively abundant rainfall during the growing season while having a significant yield gain when irrigated. Soil at both locations are typical Mollisols with a rooting depth at least 1.5 m or deeper.

Table 5–1. Growing season weather characteristics and water balance at Mead (41.23° N, 96.49° W) and North Platte (41.14° N, 100.76° W) in Nebraska. The distance between the two locations is 360 km. The growing season is defined from a common planting date of 1 May to the average date when a maize crop reaches physiological maturity for common hybrids. All values are means of 20 growing seasons from 1993 to 2002.

	Growing days	Mean temperature†			Total rainfall	Mean daily ET demand‡	Total ET demand‡	Required irrigation§	Corn yield ratio of rainfed to irrigated¶
		T_{min}	T_{max}	T_{mean}					
	d	°C			mm				
Mead	136	14.7	27.8	21.3	400	3.85	523	101	0.68
North Platte	144	12.5	27.7	20.1	322	4.19	603	180	0.39

† T_{min}, minimum temperature; T_{max}, maximum temperature; T_{mean}, mean temperature.

‡ Crop reference evapotranspiration adjusted for corn canopy size dynamics throughout the season as simulated by the Hybrid-Maize model.

§ Simulated by the Hybrid-Maize model for growth of water stress free. Soil available water is calculated to a rooting depth of 1.5 m with an initial water content set at 75% of available water for the top 30 cm and at field capacity for the subsoil.

¶ Based on NASS county yield data (http://quickstats.nass.usda.gov/) of 1993–2012 for the two locations (Saunders County for Mead and Lincoln County for North Platte).

Weather Data

Hybrid-Maize requires daily weather data for simulation. Weather variables for simulation of fully irrigated yield potentials include maximum and minimum temperature and solar radiation. For rainfed production or fully irrigated conditions with the option of estimating irrigation water requirements, extra weather variables are required, including precipitation, mean humidity, and reference ET (ET_0). Daily weather data of 1993 to 2012 for the two locations were obtained from the High Plains Regional Climate Center (http://www.hprcc.unl.edu/).

Simulation Setup

Two sets of simulations were conducted for both locations: one set for fully irrigated conditions with estimation of irrigation requirements (both timing and amount) and the other set for rainfed conditions. Hybrid-Maize requires crop information, including (i) hybrid maturity, either by total growing degree days (GDDs) or by relative maturity or actual date of physiological maturity, (ii) planting date, and (iii) plant population. For rainfed fields or irrigated fields but with estimation of irrigation water requirements, basic soil properties are required, including soil rooting depth, texture of the top 30 cm and subsoil, and soil water content at planting time.

Input settings used for this study include a corn hybrid of relative maturity of 114 d for Mead and 110 d for North Platte (Klein and Lyon, 2011), a common planting date of 1 May and plant population of 80,000 ha^{-1}, and silt clay loam soil with a rooting depth of 1.5 m at both locations. The field capacity and permanent wilting point of the soil are set at 0.32 and 0.14 soil volumetric fraction, respectively (Driessen and Konijn, 1992). Bulk density of the top 30 cm soil was set at 1.3 g cm^{-3}, while subsoil bulk density was estimated internally in the model by using soil texture information. Soil water content at planting was assumed to be 75% of available water for the top 30 cm and 100% (i.e., fully recharged) for subsoil. For both sites, 20 individual year simulations were conducted using the same settings for each year, including the soil moisture status at planting. Although the soil moisture status at planting varies from year to year, the assumed 75 and 100% of recharging rate for the top 30 cm and below, respectively, are generally valid in this region (Grassini et al., 2010), and simulation results are not very sensitive to the settings because of good rainfall after planting and in early summer.

Simulation Output

The Hybrid-Maize model outputs daily simulation results of dry matter accumulation of stover (sum of leaves, stem, and cob), grain, and roots, ET demand and actual ET, water stress index (WSI), and soil water content of 0 to 30 cm, 30 to 60 cm, and from 60 cm to the end of rooting depth, along with GDD accumulation, LAI, photosynthesis, and respiration. For this study, the model was revised to also output daily actual T to estimate WUE_T using Eq. [1].

Results

Results and discussion are based on simulation results of 20 yr from 1993 to 2012 of corn total aboveground biomass, grain yield, water use by ET as well as T, and ET- and T-based WUEs for total biomass and grain yield. Both biomass and grain yield refer to dry matter.

Grain Yield and Total Biomass

At both locations of Mead and North Platte, simulated corn grain yield and total aboveground biomass are much higher under irrigated conditions than those under rainfed conditions (Table 5–2, Fig. 5–1). Under rainfed conditions, grain yield and total biomass vary more from year to year than corresponding irrigated values (Fig. 5–1). Under irrigated conditions, grain yield and total biomass are determined largely by temperature and solar radiation, while grain yield and biomass under rainfed conditions are further limited by seasonal rainfall. At Mead, the 20-yr average grain yield is 12.6 and 9.3 Mg ha^{-1} under irrigated and rainfed conditions, respectively, while at North Platte, the average yield is 12.4 and 6.7 Mg ha^{-1} under irrigated and rainfed conditions, respectively (Table 5–2). The lower

Table 5–2. Means and coefficient of variation (CV in parentheses) of 20 yr from 1993 to 2012 of model simulated corn grain yield, total aboveground biomass (both in dry matter), seasonal total evapotranspiration (ET), transpiration (*T*), and ET- and T-based water use efficiencies for biomass ($WUE_{biomass,\ ET}$, $WUE_{biomass,\ T}$) and grain yield ($WUE_{grain,\ ET}$, $WUE_{grain,\ T}$).

	Grain yield	Total biomass	ET	*T*	$WUE_{grain,\ ET}$	$WUE_{grain,\ T}$	$WUE_{biomass,\ ET}$	$WUE_{biomass,\ T}$
	—— Mg ha^{-1} ——		—— mm ——		—— kg ha^{-1} mm^{-1} ——			
Mead, irrigated	12.6 (9%)	26.7 (10%)	523 (8%)	422 (9%)	24.2 (12%)	30.0 (12%)	51.1 (8%)	63.4 (8%)
Mead, rainfed	9.3 (39%)	20.9 (23%)	440 (13%)	340 (17%)	20.6 (32%)	26.5 (30%)	46.9 (14%)	60.7 (10%)
North Platte, irrigated	12.4 (13%)	28.4 (6%)	603 (5%)	490 (6%)	20.5 (12%)	25.3 (12%)	47.2 (6%)	58.1 (6%)
North Platte, rainfed	6.7 (46%)	17.8 (29%)	439 (18%)	335 (21%)	14.6 (30%)	19.1 (27%)	39.9 (12%)	52.5 (9%)

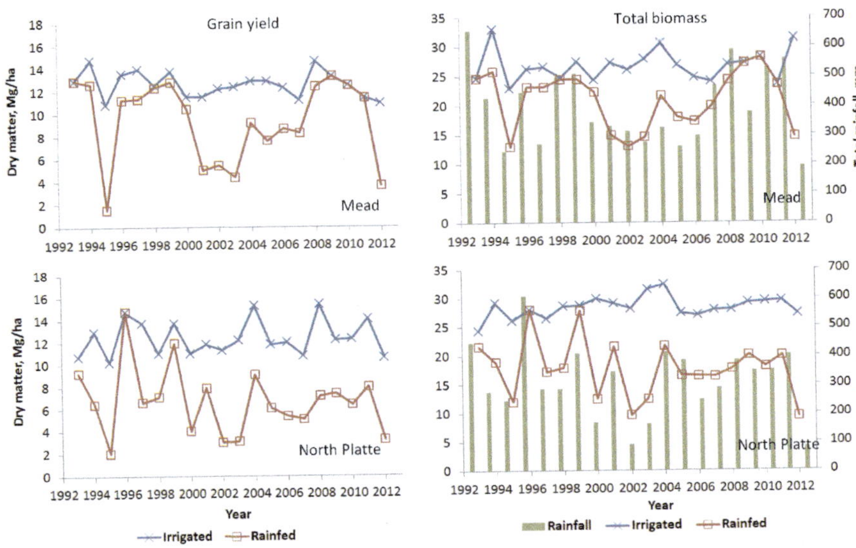

Fig. 5–1. Simulated corn grain yield (left panels) and aboveground biomass (right panels) of 20 yr from 1992 to 2012 under irrigated and rainfed conditions at Mead and North Platte in Nebraska. The vertical bars in the right panels are total rainfall from planting to physiological maturity of each year.

rainfed grain yield at North Platte is largely due to the a greater water deficit than Mead (Table 5–1). The coefficient of variation (CV) of irrigated grain yield is 9 and 13% at Mead and North Platte, respectively, while the CV of rainfed grain yield is 39 and 46% at Mead and North Platte, respectively. For total biomass, the CV is mostly lower than those of grain yield.

Simulated grain yields and total biomass are on average higher than farmers' reported values because of the assumption in simulation of optimal crop management (under irrigated conditions) except water shortage (under rainfed conditions). But the CV of grain yield is close to that of county yield statistics (http://quickstats.nass.usda.gov/), which is 14 and 11%, respectively, for irrigated and rainfed corn yield of Saunders County (for Mead), and 11 and 38%, respectively, for irrigated and rainfed corn yield of Lincoln County (for North Platte).

Total Evapotranspiration and Total Transpiration

Daily crop ET simulated by the Hybrid-Maize model is based on daily reference ET, canopy size (i.e., LAI), and soil water supply from the entire rooting zone. Seasonal crop ET is the total of daily values from emergence to physiological maturity. Simulated crop ET under irrigated conditions is higher than that under rainfed conditions and varies less across years than the latter (Fig. 5–2). On average over 20 yr, crop ET is 523 mm under irrigated conditions and 440 mm under rainfed conditions at Mead, while the corresponding values at North Platte are 603 mm and 439 mm, respectively, for irrigated and rainfed conditions. These values are slightly lower than the reported long-term values of 610 mm for the Mead region and 710 mm for the North Platte region (Kranz et al., 2008). The lower total ET under rainfed conditions is attributed to two factors: poor water supply when

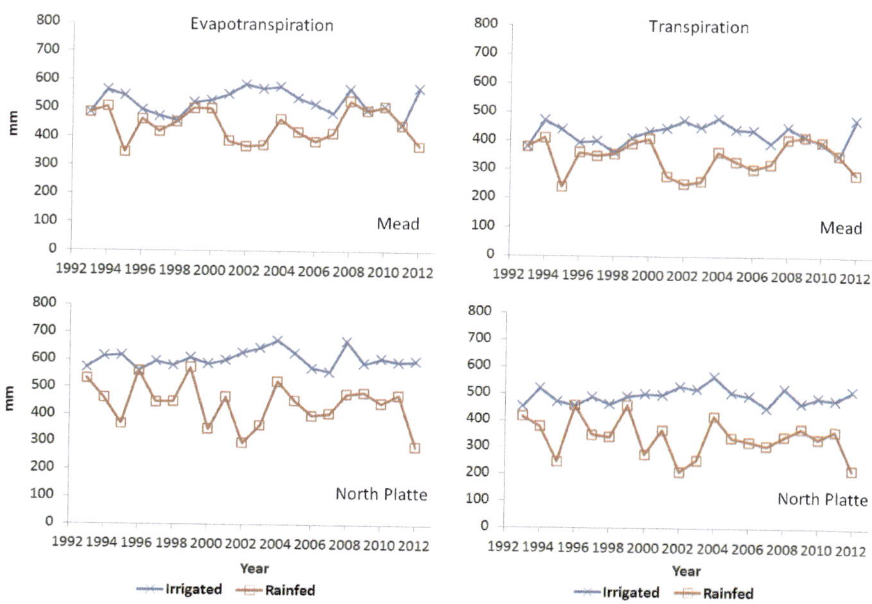

Fig. 5–2. Simulated seasonal total evapotranspiration (left panels) and transpiration (right panels) of 20 yr from 1993 to 2012 under irrigated and rainfed conditions at Mead and North Platte.

soil becomes dry and smaller plant canopy size as a result of water stress. Under irrigated conditions, water supply in soil is supposed to satisfy the demand for ET. As a result, irrigated crop ET will be higher if weather conditions are favorable to water evaporation and T (e.g., the year 2012 at Mead in Fig. 5–2). For rainfed crops, however, greater ET demand can lead to earlier and more severe soil water depletion and reduced plant size, leading to reduction in crop ET (French and Schultz, 1984; Howell et al., 1995).

Transpiration is the part in ET that represents direct water consumption through plants and is directly responsible for crop physiological functions. At Mead, the average seasonal total T of the 20 yr is 422 mm under irrigated conditions and 340 mm under rainfed conditions (Table 5–2). The corresponding total T at North Platte is 490 mm and 335 mm, respectively, under irrigated and rainfed conditions. The higher total T at North Platte than at Mead is due primarily to the higher overall ET demand at North Platte than at Mead, as shown in Table 5–1. At both locations, seasonal total T represents about 81% of total ET under irrigated conditions and 77% of total ET under rainfed conditions. These values are well within the range of normal crop fields (Allen et al., 1998; Kranz et al., 2008).

Relationship of Total Biomass and Grain Yield with Evapotranspiration

Under irrigated conditions, total biomass and grain yield have a weak relationship with seasonal total ET, whereas the relationship under rainfed conditions is much stronger (Fig. 5–3). For irrigated conditions, the R^2 of correlation between total ET and total biomass is 0.32 and 0.22, respectively, at Mead and North Platte, and the corresponding R^2 of ET with grain yield is 0.002 and 0.189, respectively, at Mead and North Platte. For rainfed conditions, the R^2 of correlation of ET with total biomass is 0.84 and 0.92, respectively, at Mead and North Platte, and the corresponding R^2 of ET with grain yield is 0.79 and 0.83, respectively, at Mead and North Platte. Under irrigated conditions, soil water is supposed to be at levels that avoid water stress, and crop growth is determined by weather factors only, especially temperature and solar radiation. Under rainfed conditions, seasonal water supply by precipitation cannot always meet the ET demand of water at both locations (Table 5–1). As a result, crop biomass production and gain yield depend closely on the amount of soil water available for ET. In addition, the relationship of total biomass with ET is stronger than that of grain yield with ET (Fig. 5–3), because total ET is for the entire season, which corresponds to the total biomass, while grain yield is formed only during the second half of the growth cycle.

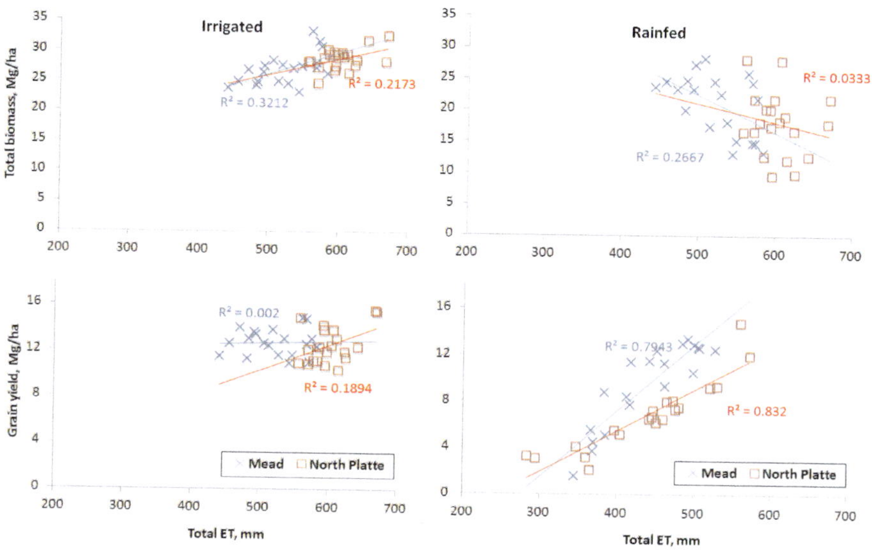

Fig. 5–3. Relationship of total aboveground biomass (upper panels) and grain yield (lower panels) with evapotranspiration (ET) under irrigated (left panels) and rainfed conditions (right panels) for 20 yr from 1992 to 2012 at Mead and North Platte.

Water Use Efficiency of Total Biomass

Evapotranspiration-based WUE of total aboveground biomass ($WUE_{biomass, ET}$) is higher under irrigated conditions than that under rainfed conditions (Table 5–2). At Mead, the 20-yr mean $WUE_{biomass, ET}$ is 51.1 kg ha^{-1} mm^{-1} under irrigated conditions and 46.9 kg ha^{-1} mm^{-1} under rainfed conditions. At North Platte, $WUE_{biomass, ET}$ is 47.2 kg ha^{-1} mm^{-1} under irrigated conditions and 39.9 kg ha^{-1} mm^{-1} under rainfed conditions. Because T is always lower than ET, T-based WUE should be higher than ET-based WUE. At mead, T-based WUE of total biomass ($WUE_{biomass, T}$) is 63.4 kg ha^{-1} mm^{-1} under irrigated conditions and 60.7 kg ha^{-1} mm^{-1} under rainfed conditions, while $WUE_{biomass, T}$ at North Platte is 58.1 kg ha^{-1} mm^{-1} under irrigated conditions and 52.5 kg ha^{-1} mm^{-1} under rainfed conditions (Table 5–2). Between the two locations, the slightly lower $WUE_{biomass, ET}$ at North Platte than at Mead under both irrigated and rainfed conditions could be the result of higher ET demand at North Platte, which is on average 4.19 mm d^{-1}, than at Mead, which is 3.85 mm d^{-1} (Table 5–1).

Across-year variations of $WUE_{biomass, ET}$ and $WUE_{biomass, T}$ are smaller than that of total biomass (Fig. 5–1, Table 5–2), especially under rainfed conditions. At Mead, the 20-yr CV of $WUE_{biomass, ET}$ is 8 and 14%, respectively, for irrigated and rainfed conditions, while the corresponding value for total biomass is 10 and 23%, respectively. At North Platte, the 20-yr CV of $WUE_{biomass, ET}$ is 6 and 12%, respectively, for irrigated and rainfed conditions, while the corresponding value for total biomass is 6 and 29%, respectively.

Water Use Efficiency of Grain Yield

Evapotranspiration-based WUE of grain yield ($WUE_{grain, ET}$) is higher under irrigated conditions than that under rainfed conditions (Table 5–2). At Mead, the 20-yr mean $WUE_{grain, ET}$ is 24.2 kg ha^{-1} mm^{-1} under irrigated conditions and 20.6 kg ha^{-1} mm^{-1} under rainfed conditions, while $WUE_{grain, ET}$ at North Platte is 20.5 kg ha^{-1} mm^{-1} under irrigated conditions and 14.6 kg ha^{-1} mm^{-1} under rainfed conditions. For T-based WUE of grain yield ($WUE_{grain, T}$), the value is 30.0 kg ha^{-1} mm^{-1} under irrigated conditions and 26.5 kg ha^{-1} mm^{-1} under rainfed conditions at Mead, while $WUE_{grain, T}$ at North Platte is 25.3 kg ha^{-1} mm^{-1} under irrigated conditions and 19.1 kg ha^{-1} mm^{-1} under rainfed conditions (Table 5–2). The values of $WUE_{grain, ET}$ derived from the simulations are well within the range of 16 to 27 kg ha^{-1} mm^{-1} reported in literature for corn production of the U.S. Corn Belt and West Coast (Tanner and Sinclair, 1983; Al-Kaisi and Yin, 2003; Grassini et al, 2011b; Djaman and Irmak, 2012) and are at the higher end of a wider range from 10 countries as reviewed by Zwart and Bastiaanssen (2004).

Across-year variations of $WUE_{grain, ET}$ and $WUE_{grain, T}$ are greater than $WUE_{biomass, ET}$ and $WUE_{biomass, T}$ (Fig. 5–4 and Fig. 5–5, Table 5–2), especially under rainfed

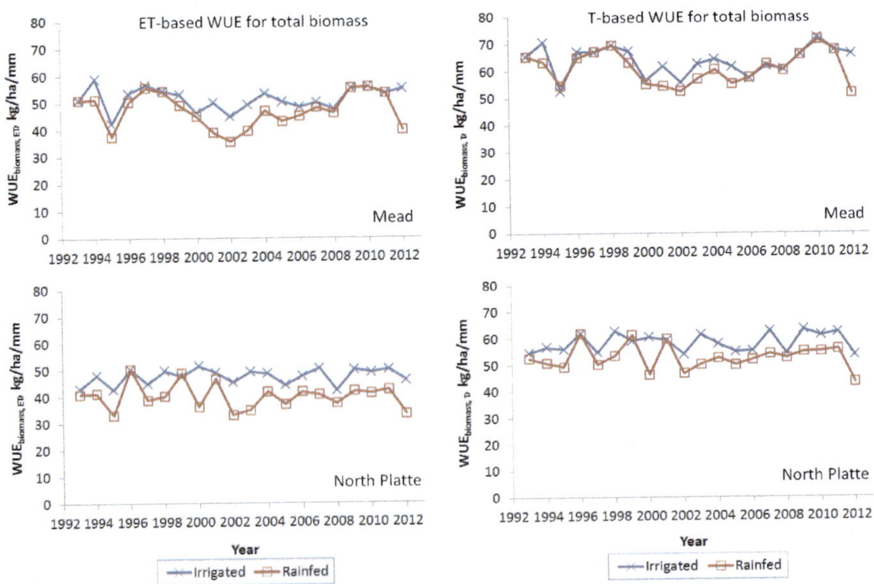

Fig. 5–4. Water use efficiency of total biomass based on evapotranspiration ($WUE_{biomass, ET}$, left panels) and transpiration ($WUE_{biomass, T}$, right panels) of 20 yr from 1993 to 2012 at Mead and North Platte.

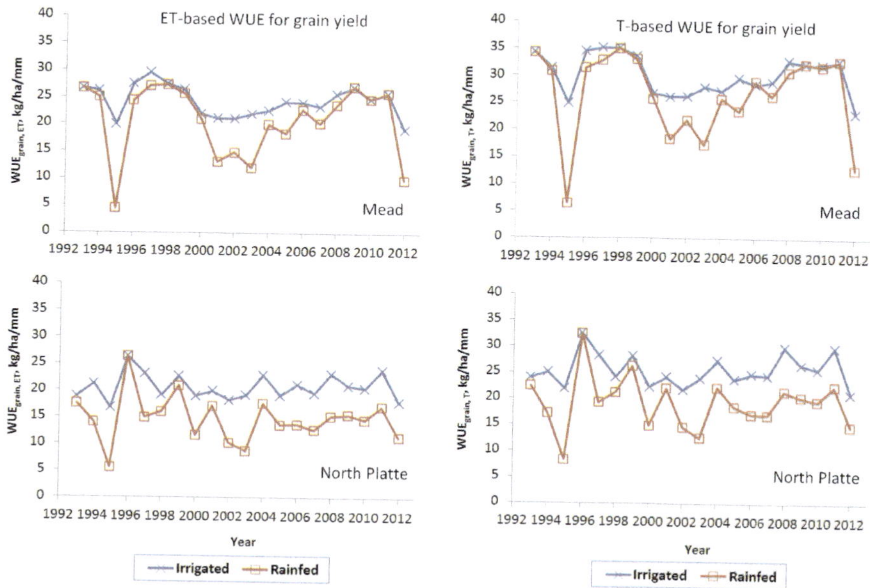

Fig. 5–5. Water use efficiency of grain yield based on evapotranspiration ($WUE_{grain, ET}$, left panels) and transpiration ($WUE_{grain, T}$, right panels) of 20 yr from 1993 to 2012 at Mead and North Platte.

conditions. At Mead, the 20-yr CV of $WUE_{grain, ET}$ is 12 and 32%, respectively, for irrigated and rainfed conditions, while the corresponding value at North Platte is 12 and 30%, respectively, for irrigated and rainfed conditions.

Relationship of Total Biomass and Grain Yield with Water Use Efficiency

Under rainfed conditions, total biomass and grain yield show a strong, positive, and linear correlation with their respective ET- or T-based WUE at both locations, while the correlation under irrigated conditions is relatively weak (Fig. 5–6 and 5–7). Under rainfed conditions, WUE at low total biomass or grain yield is similar at Mead and North Platte, but WUE increases faster with increase of total biomass or grain yield at Mead than at North Platte. This may reflect the overall lower ET demand at Mead than that at North Platte during growing season (Table 5–1).

Seasonal Dynamics of Crop Growth

Although the general pattern of growth dynamics looks similar under irrigated and rainfed conditions (Fig. 5–8), there are several differences. Overall, crop growth under rainfed conditions becomes more restricted after silking than that under irrigated conditions. For LAI, its peak value is lower and LAI declines more quickly under rainfed conditions than under irrigated conditions. Total biomass accumulation after silking is less under rainfed conditions than under

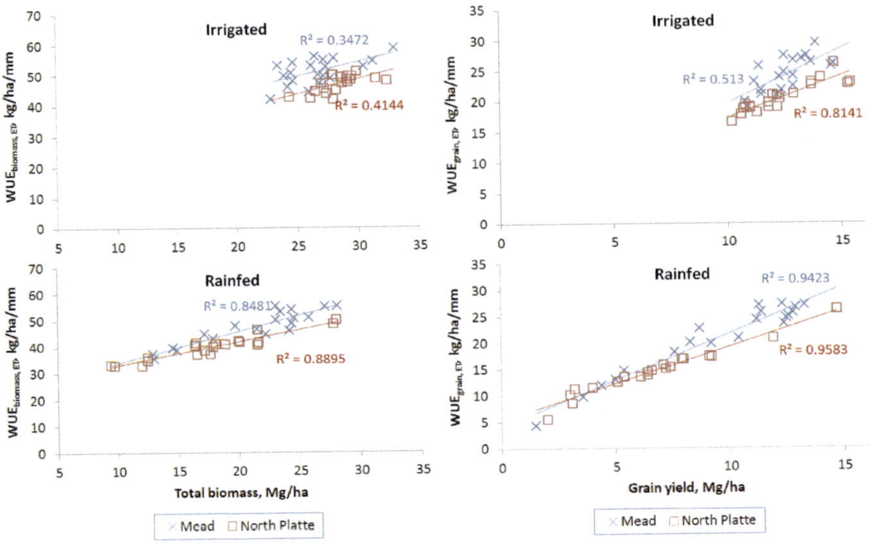

Fig. 5–6. Relationship of total biomass and grain yield with their evapotranspiration-based water use efficiency $WUE_{biomass,\ ET}$ and $WUE_{grain,\ ET}$ for 20 yr from 1993 to 2012 at Mead and North Platte.

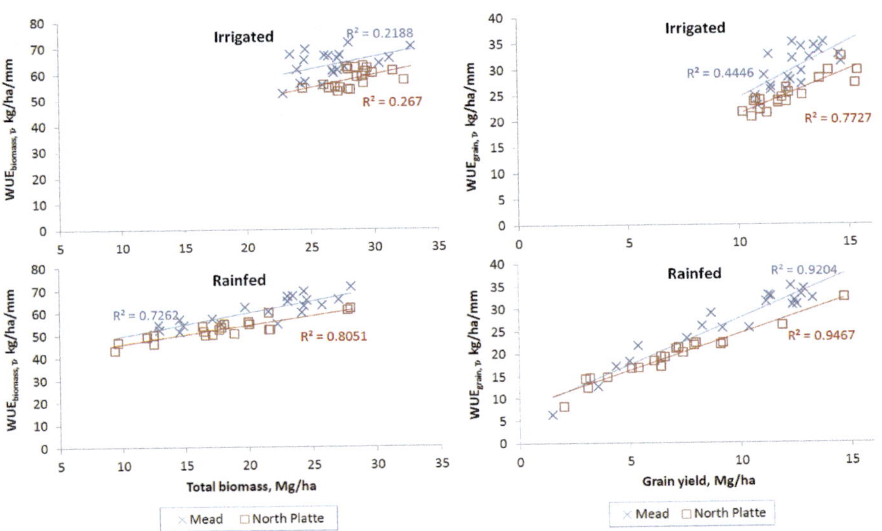

Fig. 5–7. Relationship of total biomass and grain yield with their transpiration-based water use efficiency $WUE_{biomass,\ T}$ and $WUE_{grain,\ T}$ for 20 yr from 1993 to 2012 at Mead and North Platte.

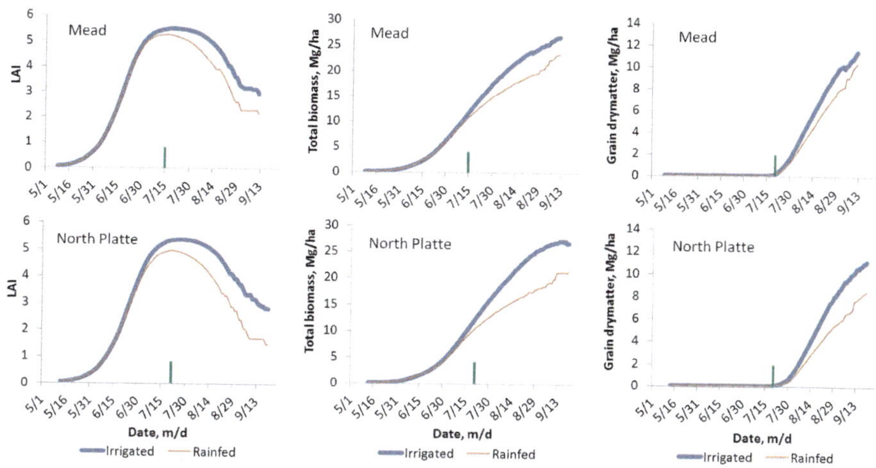

Fig. 5–8. Simulated corn growth dynamic of leaf area index (LAI, left two panels) and accumulation of aboveground total biomass (middle panels) and grain dry matter (right panels) under irrigated and rainfed conditions at Mead and North Platte. For each line, the daily value is the mean of 20 yr from 1993 to 2012. The starting and ending date of the plots are the 20-yr mean date of emergence and physiological maturity, respectively. The vertical bar on the *x* axis in each graph indicates the mean date of silking.

irrigated conditions, because of restricted growth under rainfed conditions due to the water stress starting from silking under rainfed conditions.

Seasonal Dynamics of Evapotranspiration and Transpiration

Reference ET is slightly higher during the early part of the season than the later part (Fig. 5–9). Crop ET, however, starts low in the beginning of the season when plants are small as indicated by LAI in Fig. 5–8 and peaks when the plant canopy reaches its maximum size. Under irrigated conditions, crop ET becomes close to reference ET during the period when the canopy is closed (i.e., LAI > 5), and soil water supply can always meet the ET demand. Under rainfed conditions, crop ET becomes significantly lower than reference ET as well as irrigated crop ET for the second half of the growing season on average over 20 yr at both locations. The lower crop ET under rainfed conditions is because of the average water deficit of 101 mm at Mead and 180 mm at North Platte (Table 5–1) to achieve full ET and drought-free corn growth. Between the two locations, reference ET and crop ET under irrigated conditions are higher at North Platte than at Mead.

As a part of crop ET, the proportion of T is very small at the seedling stage, but increases quickly with expansion of the crop canopy (Fig. 5–8 and 5–10). Transpiration completely dominates crop ET when LAI reaches 4 or higher, or about 2 to 3 wk before silking.

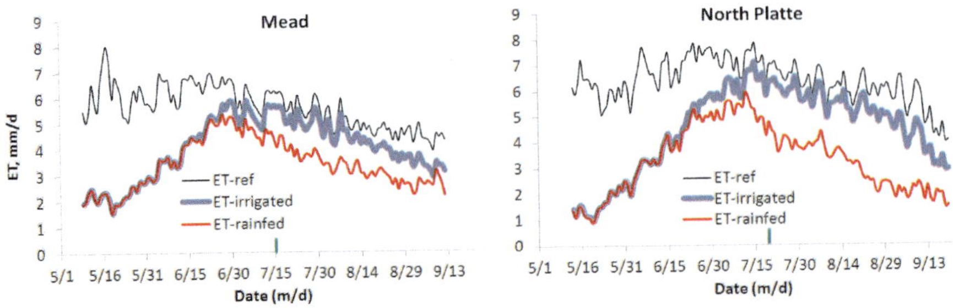

Fig. 5–9. Seasonal dynamics of daily reference evapotranspiration (ET-ref, alfalfa based) and simulated crop ET under irrigated (ET-irrigated) and rainfed conditions (ET-rainfed). For each line, the daily value is the mean of 20 yr from 1993 to 2012. The starting and ending date are the mean date of emergence physiological maturity, respectively. The vertical bar on the *x* axis in each graph indicates the mean date of silking.

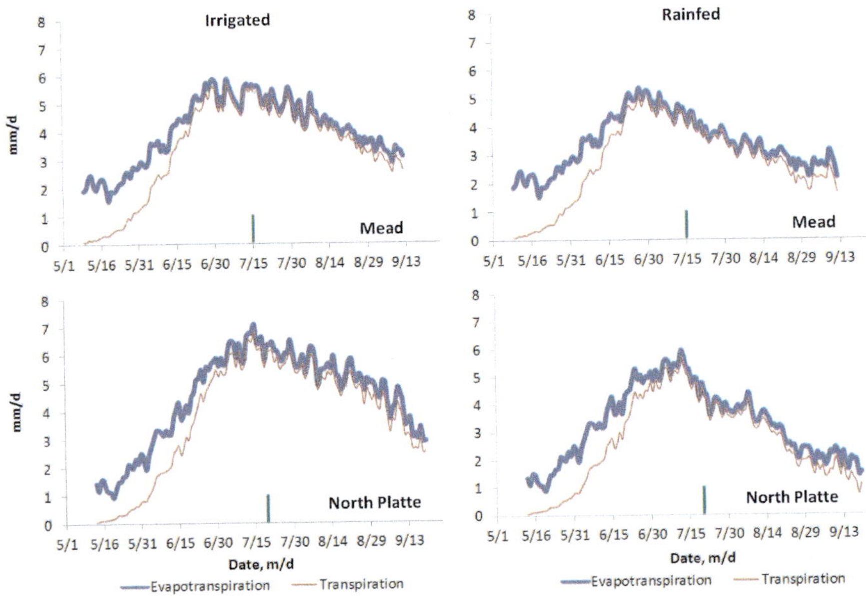

Fig. 5–10. Simulated seasonal dynamics of corn daily evapotranspiration and transpiration under irrigation and rainfed conditions at Mead and North Platte. For each line, the daily value is the mean of 20 yr from 1993 to 2012. The starting and ending date is the mean date of emergence and physiological maturity, respectively. The vertical bar on the *x* axis in each graph indicates the mean date of silking.

Crop Water Stress under Rainfed Conditions

Under rainfed conditions, simulated crop water stress starts from late June on average of the 20 yr at both locations (Fig. 5–11). Between the two locations, drought is more severe at North Platte than Mead, which is consistent with the difference in the gap between irrigated ET and rainfed ET in Fig. 5–9 as well as the overall seasonal water deficit shown in Table 5–1. Because drought occurs mainly during the

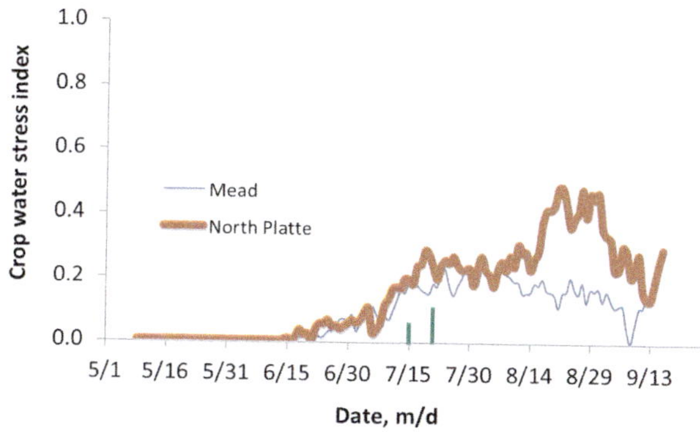

Fig. 5–11. Simulated crop water stress index under rainfed conditions at Mead and North Platte. For each line, the daily value is the mean of 20 yr from 1993 to 2012. The starting and ending date is the mean date of emergence and physiological maturity, respectively. The water stress index ranges from 0 (stress free) to 1 (complete closure of leaf stomata and stop of photosynthesis). The shorter and longer vertical bars indicate the mean silking date at Mead and North Platte, respectively.

second half of the growing season when grain filling takes place at both locations, the impact of drought would be reflected mainly on grain yield.

Dynamics of Water Use Efficiency

Evapotranspiration-based WUE of biomass accumulation ($WUE_{biomass, ET}$) is low at the early of the growing season and increases fast until silking (Fig. 5–12). $WUE_{biomass, ET}$ reaches a plateau shortly after silking, and the peak value maintains to crop maturity. The trend is the same under irrigated and rainfed conditions at both locations, although the plateau value is higher under irrigated conditions than rainfed conditions and higher at Mead than at North Platte. The low values of $WUE_{biomass, ET}$ during the early part of the season are largely because a big part of crop ET is water loss through soil evaporation before the canopy is closed, which does not contribute to crop biomass production (Allen et al., 1998).

Transpiration-based WUE of biomass accumulation ($WUE_{biomass, T}$) is much higher than $WUE_{biomass, ET}$ during the early part of the season and further increases to a plateau shortly after silking (Fig. 12). The lower values of $WUE_{biomass, T}$ during the early part of the season are likely due to the lower temperature in the spring and, as a result, a lower photosynthetic efficiency. Similar to $WUE_{biomass, ET}$, the plateau values of $WUE_{biomass, T}$ under irrigated and rainfed conditions at Mead are also slightly higher than the respective values at North Platte. Compared with $WUE_{biomass, ET}$, $WUE_{biomass, T}$ is significantly and consistently higher and is about 10 kg ha^{-1} mm^{-1} higher during the period when WUE is at a plateau under both irrigated and rainfed conditions at both locations.

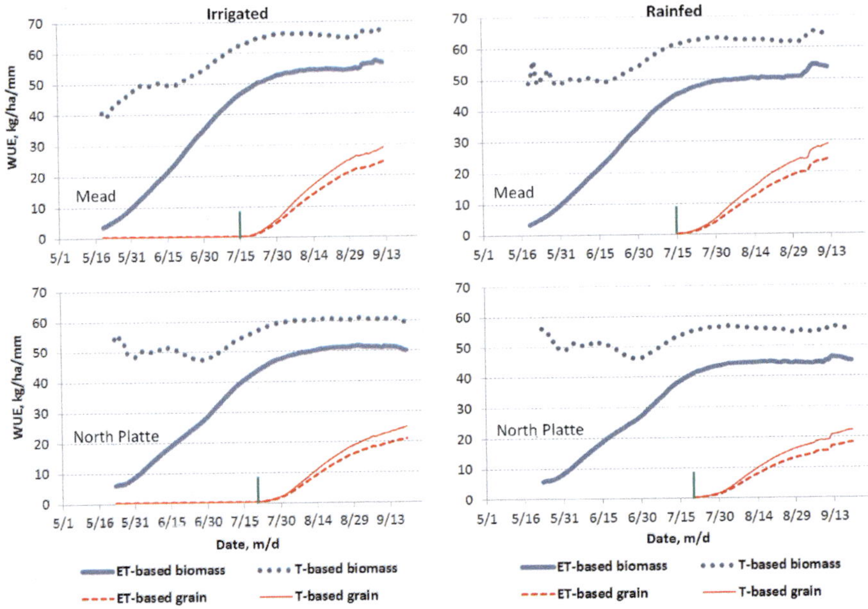

Fig. 5–12. Seasonal dynamics of evapotranspiration-based (ET-based) and transpiration-based (T-based) water use efficiency of total biomass accumulation and grain dry matter accumulation under irrigated and rainfed conditions at Mead and North Platte. For each line, the daily value is the mean of 20 yr from 1993 to 2012. The first 10 d after the mean date of emergence are not plotted because a significant part of the seedling biomass is derived from seeds instead of photosynthesis. The ending date of the plots is the mean date of physiological maturity of the 20 yr. The vertical bar on the *x* axis in each graph indicates the mean date of silking.

Evapotranspiration-based WUE of grain yield accumulation ($WUE_{grain, ET}$) resembles very much the growth curves of grain dry matter accumulation in Fig. 5–8. It begins low at the onset of grain filling and increases fast thereafter. Unlike $WUE_{biomass, ET}$, $WUE_{grain, ET}$ does not have an apparent plateau throughout the season. This is because the ET part in the calculation of $WUE_{grain, ET}$ includes ET before start of grain filling. If the amount of ET before the start of grain filling is excluded, ET-based WUE of grain filling ($WUE_{grainfilling, ET}$) is much higher (Fig. 5–13) and resembles the part of $WUE_{biomass, ET}$ during the grain-filling period in Fig. 5–12. In addition, $WUE_{grainfilling, ET}$ can be exceptionally high on some days, especially in the early stage of grain filling (e.g., at North Platte). This seemingly high WUE is due to the simulated use of carbohydrate reserve in plant stem and leaves for grain filling when daily photosynthesis cannot provide sufficient dry matter for grain filling (Yang et al., 2004; Kiniry et al., 1997). This phenomenon would disappear once the carbohydrate reserve is exhausted. Compared with irrigated conditions, $WUE_{grainfilling, ET}$ under rainfed conditions is slightly higher

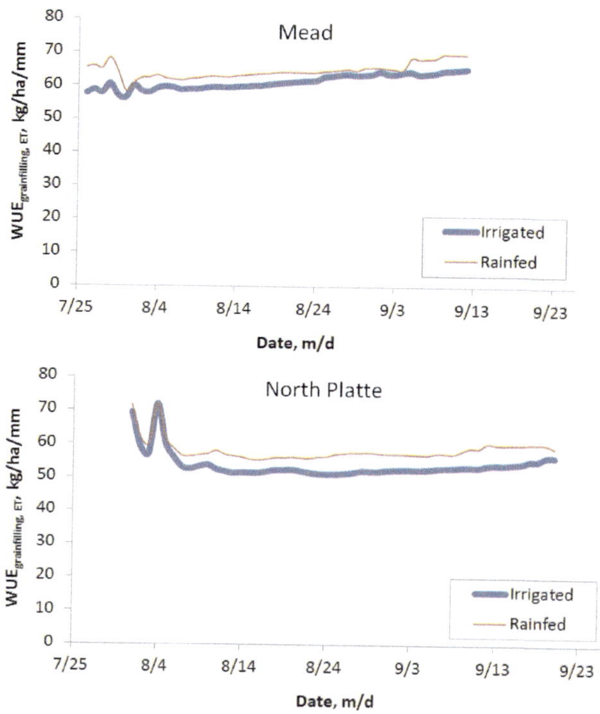

Fig. 5–13. Seasonal dynamics of ET-based water use efficiency of grain filling ($WUE_{grainfilling, ET}$) under irrigated and rainfed conditions at Mead and North Platte. In the calculation of $WUE_{grainfilling, ET}$, only the total during the grain-filling period is included. For each line, the daily value is the mean of 20 yr from 1993 to 2012. The starting and ending date is the mean date of grain filling and physiological maturity, respectively.

because of more frequent use of the carbohydrate reserve for grain filling under water stress conditions.

Discussion and Conclusions

A better understanding and quantification of crop water use and efficiencies are the foundation for improvement of crop water management (Allen et al., 1998). Although conventional experimental studies can provide first-hand evidence, mechanistic simulation models have the potential of providing more insights about crop responses to environmental conditions and interactions among processes. Simulation results of the Hybrid-Maize show that across-year variations of corn yield and total biomass are greater under rainfed conditions than under irrigated conditions at both locations, because water limitation under rainfed conditions imposes constraints for crop growth besides other growing conditions including nutrients, temperature, and solar radiation. The same variations are reflected on crop total ET and T and their corresponding WUE. Such great variations of corn yield and WUE under rainfed conditions are typical of the environment of the U.S. Western Corn Belt (Grassini et al., 2009).

Many studies have shown the strong and linear dependence of total crop biomass and grain yield on seasonal total ET in rainfed systems (Tanner and Sinclair, 1983; Howell and Hiler, 1975), but few have shown the lack of correlation of the two under irrigated conditions when water stress is eliminated. When crop growth is mainly limited by water availability like under rainfed conditions, improvement in water supply and thus crop ET would translate directly into crop biomass gain through improvement of photosynthetic rate and overall better crop development (Farooq et al., 2009). Under irrigated conditions, however, water stress is eliminated and crop growth and yield are limited by other conditions, including nutrients and weather conditions.

Compared with seasonal total ET, seasonal total T amounts to about 81% of total ET under irrigated conditions and 76% under rainfed condition at both locations. In other words, there is about 20 to 25% of water loss through evaporation as it does not contribute to crop biomass production. Any management practices that minimize soil evaporation (e.g., no till, mulching by crop residues or plastic film) would reduce overall crop water consumption to increase ET-based WUE (Munawar et al., 1990; Bu et al., 2013). Also, an earlier canopy closure (e.g., a canopy architecture that has more flat leaves when plants are small) can also reduce soil evaporation by increasing light interception. The significance of reducing water losses, especially during the early part of growing cycle, is to make more water available to crops during the grain-filling period for maximizing grain yield.

During the early part of the growing season, crop ET is much smaller than reference ET and T is even a smaller fraction of crop ET because of the smaller plant size and open canopy. This indicates that when using reference ET to estimate water consumption of small plants, a correction such as using the crop coefficient is needed (Allen et al., 1998). Once the canopy is closed (i.e., LAI > 5), the correction becomes unnecessary. As crop ET under rainfed conditions with occurrence of drought can be significantly lower than that under irrigated conditions, it also becomes necessary to adjust soil water losses under drought conditions relative to irrigated conditions (Djaman and Irmak, 2012). Otherwise, crop water consumption will be overestimated for drought conditions.

In Nebraska and many parts of the U.S. Western Corn Belt, crop water shortage often does not occur until the middle of the growing season when soil water storage becomes depleted and precipitation becomes less. This generally coincides with corn pollination and kernel setting in this region. Water stress can further develop throughout the rest of the season if water input, including rainfall and irrigation, cannot compensate crop ET. For irrigation management in this region, meeting crop water demand during the pollination window and grain-filling period is critical to high yield.

Field studies often report WUE that represents the mean of the entire growth cycle (Grassini et al., 2011b; Tanner and Sinclair, 1983), whereas Fig. 5–12 indicates lower WUE values at the early stage of crop development before reaching a plateau when the canopy is established. The lower WUE when plants are small reflects the small proportion of T in crop ET, as well as lower photosynthetic efficiency when the temperature is low during the early part of the growing season. The plateau value of 45 to 55 kg ha^{-1} mm^{-1} ET-based WUE for total biomass can be used as a benchmark in irrigation management. The range of 55 to 65 kg ha^{-1} mm^{-1} T-based WUE can represent the celling of corn WUE with no evaporation losses under the conditions of Nebraska.

Estimation of WUE for corn grain yield deserves special attention, because grain yield does not start to form until stover biomass accumulation has completed with associated water consumption. As a result, ET-based WUE for grain yield is much lower than WUE for total biomass, but the ratio of the former to the latter is equal to the harvest index, as shown in Fig. 5–13. When using WUE for total biomass to estimate WUE for grain yield, the former needs to multiply with harvest index to obtain the latter. The true ET-based WUE for grain filling (i.e., when the total crop ET before the onset of grain filling is excluded) is much higher and stable because the major part of the grain-filling period occurs during an optimal temperature regime for photosynthesis.

To improve corn WUE, for both irrigated and rainfed systems, it is critical to identify and reduce or eliminate non-water-related constraints to crop growth (e.g., nutrients deficiency, diseases, pests, and insects). Crop yield and WUE will not be improved if yield is constrained by those factors. For full irrigation management, irrigation frequency should be minimized to shorten the period when the soil surface is wet to reduce evaporation water losses and save water for crop use at later stages. Moreover, the total irrigation amount should be always limited to the amount corresponding to the maximum yield of the field. Overirrigation not only wastes water and reduces WUE but also runs the risk of lowering yield because of poor root aeration and logging (Irmak et al., 2012). For deficit irrigation, water should be allocated more to the period around pollination and kernel setting, as well as grain filling to maximize grain yield instead of stover biomass.

Results and conclusions of this study are based on results from model simulations. Although the model has been tested and validated in the target application environment and conditions (Yang et al., 2004, Grassini et al., 2009), some assumptions used in the current model (e.g., no interactions between nutrient deficiency and water shortage) may not be fully valid in some cropping systems. Experimental studies are still needed to test and verify those results. Simulation results can serve as the first step of testing hypothesis and guide experimental research.

Acknowledgments

We thank the Water for Food Institute of the University of Nebraska, the Water, Energy and Agriculture Initiative (WEAI) of the Nebraska Center for Energy Sciences Research, and the Nebraska Corn Board for providing financial support for this study.

References

Al-Kaisi, M., and X. Yin. 2003. Effects of nitrogen rate, irrigation rate, and plant population on corn yield and water use efficiency. Agron. J. 95:1475–1482. doi:10.2134/agronj2003.1475

Allen, R.G., L.S. Pereira, D. Raes, and M. Smith. 1998. Crop evapotranspiration: Guidelines for computing crop water requirements. Irrig. Drain. Pap. 56. FAO, Rome.

Bu, L., J. Liu, L. Zhu, S. Luo, X. Chen, S. Li, H. Lee, and Y. Zhao. 2013. The effects of mulching on maize growth, yield and water use in a semi-arid region. Agric. Water Manage. 123:71–78. doi:10.1016/j.agwat.2013.03.015

Chen, X., F. Chen, Y. Chen, Q. Gao, X. Yang, L. Yuan, F. Zhang, and G. Mi. 2013. Modern maize hybrids in Northeast China exhibit increased yield potential and resource use efficiency despite adverse climate change. Glob. Change Biol. 19:923–936. doi:10.1111/gcb.12093

Chen, X.P., Z.L. Cui, P.M. Vitousek, K.G. Cassman, P.A. Matson, J.S. Bai, Q.F. Meng, P. Hou, V. Römheld, and F.S. Zhang. 2011. Integrated soil-crop system management for food security. Proc. Natl. Acad. Sci. USA 108:6399–6404. doi:10.1073/pnas.1101419108

de Wit, C.T. 1958. Transpiration and crop yields. Versl. Landbouwk. Onderz. Vol. 64.6 Institute of Biological and Chemical Research on Field Crops and Herbage, Wageningen, the Netherlands.

Djaman, K., and S. Irmak. 2012. Soil water extraction patterns and crop, irrigation, and evapotranspiration water use efficiency of maize under full and limited irrigation and rainfed settings. Trans. ASAE 55:1223–1238. doi:10.13031/2013.42262

Driessen, P.M., and N.T. Konijn. 1992. Land-use systems analysis. Wageningen Agricultural University, Wageningen, the Netherlands.

Farooq, M., A. Wahid, N. Kobayashi, D. Fujita, and S.M.A. Basra. 2009. Plant drought stress: Effects, mechanisms and management. In: E. Lichtfouse, M. Navarrete, P. Debaeke, S. Véronique, and C. Alberola, editors, Sustainable agriculture. Springer, Dordrecht, the Netherlands. p. 153–188.

French, R.J., and J.E. Schultz. 1984. Water use efficiency of wheat in a Mediterranean-type environment. I. The relation between yield, water use and climate. Aust. J. Agric. Res. 35:743–764. doi:10.1071/AR9840743

Godfray, H.C.J., J.R. Beddington, I.R. Crute, L. Haddad, D. Lawrence, J.F. Muir, J. Pretty, S. Robinson, S.M. Thomas, and C. Toulmin. 2010. Food security: The challenge of feeding 9 billion people. Science 327:812–818. doi:10.1126/science.1185383

Grassini, P., J. Thorburn, C. Burr, and K.G. Cassman. 2011a. High-yield irrigated maize in the Western U.S. Corn Belt: I. On-farm yield, yield potential, and impact of agronomic practices. Field Crops Res. 120:142–150. doi:10.1016/j.fcr.2010.09.012

Grassini, P., H.S. Yang, and K.G. Cassman. 2009. Limits to maize productivity in Western Corn-Belt: A simulation analysis for fully irrigated and rainfed conditions. Agric. For. Meteorol. 149:1254–1265. doi:10.1016/j.agrformet.2009.02.012

Grassini, P., H. Yang, S. Irmak, J. Thorburn, C. Burr, and K.G. Cassman. 2011b. High-yield irrigated maize in the Western U.S. Corn Belt: II. Irrigation management and crop water productivity. Field Crops Res. 120:133–141. doi:10.1016/j.fcr.2010.09.013

Grassini, P., J. You, K.G. Hubbard, and K.G. Cassman. 2010. Soil water recharge in a semi-arid temperate climate of the Central U.S. Great Plains. Agric. Water Manage. 97:1063–1069. doi:10.1016/j.agwat.2010.02.019

Howell, T.A. 2000. Irrigation's role in enhancing water use efficiency. In: R.G. Evans, B.L. Benham, and T.P. Trooien, editors, National Irrigation Symposium: Proceedings of the 4th Decennial Symposium, Phoenix, AZ. 14–16 Nov. 2000. ASAE, St. Joseph, MI. p. 66–80.

Howell, T.A. 2001. Enhancing water use efficiency in irrigated agriculture. Agron. J. 93:281–289. doi:10.2134/agronj2001.932281x

Howell, T.A., and E.A. Hiler. 1975. Optimization of water use efficiency under high frequency irrigation. 1. Evapotranspiration and yield relationship. Trans. ASAE 18:873–878. doi:10.13031/2013.36700

Howell, T.A., A. Yazar, A.D. Schneider, D.A. Dusek, and K.S. Copeland. 1995. Yield and water use efficiency of corn in response to LEPA irrigation. Trans. ASAE 38:1737–1747. doi:10.13031/2013.28001

Irmak, S., M.J. Murgert, H.S. Yang, K.G. Cassman, D.T. Walters, W.R. Rathje, J.O. Payero, P. Grassini, M.S. Kuzila, K.J. Brunkhorst, B. VanDeWalle, J.M. Rees, W.L. Kranz, D.J. Eisenhauer, C.A. Shapiro, and G.L. Zoubek. 2012. Large-scale on-farm implementation of soil moisture-based irrigation management strategies for increasing maize water productivity. Trans. ASAE 55:881–891. doi:10.13031/2013.41521

Jones, C.A., and J.R. Kiniry. 1986. CERES-Maize: A simulation model of maize growth and development. Texas A&M Univ. Press, College Station.

Kiniry, J.R., J.R. Williams, R.L. Vanderlip, J.D. Atwood, D.C. Reicosky, J. Mulliken, W.J. Cox, H.J. Mascagni, S.E. Hollinger, and W.J. Wiebold. 1997. Evaluation of two maize models for nine US locations. Agron. J. 89:421–426. doi:10.2134/agronj1997.00021962008900030009x

Klein, R., and D. Lyon. 2011. Recommended seeding rate and hybrid selection for rainfed (dryland) corn in Nebraska. NebGuide G2068. Univ. of Nebraska–Lincoln Extension, Lincoln.

Kranz, W.L., S. Irmak, S.M. van Donk, C.D. Yonts, and D.L. Martin. 2008. Irrigation management for corn. NebGuide G1850. Univ. of Nebraska–Lincoln Extension, Lincoln.

Lindquist, J.L. 2001. Performance of INTERCOM for predicting corn-velvetleaf interference across north-central United States. Weed Sci. 49:195–201. doi:10.1614/0043-1745(2001)049[0195:POIFPC]2.0.CO;2

Lobell, D.B., K.G. Cassman, and C.B. Field. 2009. Crop yield gaps: Their importance, magnitudes, and causes. Ann. Rev. Environ. Resour. 34:179–204.

Meng, Q., Q. Sun, X. Chen, Z. Cui, S. Yue, F. Zhang, and V. Romheld. 2012. Alternative cropping systems for sustainable water and nitrogen use in the North China Plain. Agric. Ecosyst. Environ. 146:93–102. doi:10.1016/j.agee.2011.10.015

Munawar, A., R.L. Blevins, W.W. Frye, and M.R. Saul. 1990. Tillage and cover crop management for soil water conservation. Agron. J. 82(4):773–777. doi:10.2134/agronj1990.00021962008200040024x

Payero, J.O., S.R. Melvin, S. Irmak, and D. Tarkalson. 2006. Yield response of corn to deficit irrigation in a semiarid climate. Agric. Water Manage. 84:101–112. doi:10.1016/j.agwat.2006.01.009

Payero, J.O., D.D. Tarkalson, S. Irmak, D. Davison, and J.L. Petersen. 2009. Effect of timing of a deficit-irrigation allocation on corn evapotranspiration, yield, water use efficiency and dry mass. Agric. Water Manage. 96:1387–1397. doi:10.1016/j.agwat.2009.03.022

Perry, C., P. Steduto, R. Allen, and C.M. Burt. 2009. Increasing productivity in irrigated agriculture: Agronomic constraints and hydrological realities. Agric. Water Manage. 96:1517–1524. doi:10.1016/j.agwat.2009.05.005

Rosenberg, N.J., B.L. Blad, and S.V. Verman. 1983. Microclimate: The biological environment. John Wiley & Sons, New York.

Steduto, P., T.C. Hsiao, and E. Fereres. 2007. On the conservative behavior of biomass water productivity. Irrig. Sci. 25:189–207. doi:10.1007/s00271-007-0064-1

Steduto, P., T.C. Hsiao, D. Raes, and E. Fereres. 2009. AquaCrop—The FAO crop model to simulate yield response to water: I. Concepts and underlying principles. Agron. J. 101:426–437. doi:10.2134/agronj2008.0139s

Tanner, C.B., and T.R. Sinclair. 1983. Efficient water use in crop production: Research or re-search? In: H.M. Taylor, W.A. Jordan, and T.R. Sinclair, editors, Limitations to efficient water use in crop production. ASA, Madison, WI. p. 1–27.

Timsina, J., M.L. Jat, and K. Majumdar. 2010. Rice-maize systems of South Asia: Current status, future prospects and research priorities for nutrient management. Plant Soil 335:65–82. doi:10.1007/s11104-010-0418-y

van Diepen, C.A., J. Wolf, H. van Keulen, and C. Rappoldt. 1989. WOFOST: A simulation model of crop production. Soil Use Manage. 5:16–24. doi:10.1111/j.1475-2743.1989.tb00755.x

Vorosmarty, C.J., P. Green, J. Salisbury, and R.B. Lammers. 2000. Global water resources: Vulnerability from climate change and population growth. Science 289:284–288. doi:10.1126/science.289.5477.284

Yang, H.S., A. Dobermann, K.G. Cassman, and D.T. Walters. 2006. Features, applications, and limitations of the Hybrid-Maize simulation model. Agron. J. 98:737–748. doi:10.2134/agronj2005.0162

Yang, H.S., A. Dobermann, K.G. Cassman, D.T. Walters, and P. Grassini. 2013. User manual of the Hybrid-Maize. Univ. of Nebraska– Lincoln, Lincoln. http://hybridmaize.unl.edu/ (accessed 21 Sept. 2014).

Yang, H.S., A. Dobermann, J.L. Lindquist, D.T. Walters, T.J. Arkebauer, and K.G. Cassman. 2004. Hybrid-maize- a maize simulation model that combines two crop modeling approaches. Field Crops Res. 87:131–154. doi:10.1016/j.fcr.2003.10.003

Zwart, S.J., and W.G.M. Bastiaanssen. 2004. Review of measured crop water productivity values for irrigated wheat, rice, cotton and maize. Agric. Water Manage. 69:115–133. doi:10.1016/j.agwat.2004.04.007

Modeling to Evaluate and Manage Water and Environmental Sustainability of Bioenergy Crops in the United States

James R. Kiniry,* Manyowa N. Meki, Thomas E. Schumacher, Cody J. Zilverberg, Felix B. Fritschi, and Vijaya G. Kakani

Abstract

Decisions are being made that will determine the path to be taken for biofuel production and use. The objectives of this chapter were to describe some simulation modeling tools as they have been applied to issues related to biofuel production and to describe how to use the Agricultural Land Management Alternative with Numerical Assessment Criteria (ALMANAC) model for these applications. To verify that the biomass yields are realistic, we compared mean simulated yields for four sites in the Central Great Plains to reported yields of maize (*Zea mays* L.) and switchgrass (*Panicum virgatum*). We were successful in getting mean simulated values close to the mean measured values in each case. The model was then used to simulate water use efficiency (WUE) of different production systems and biomass yield stability of these systems. This study demonstrated the usefulness of this process-based model for these portions of life cycle analyses. Further work developing parameters to simulate other candidate plant species will prove useful to extend this type of approach. Likewise, further validation of the plant yields on various soils in various rainfall zones will be invaluable as a proof of concept for this approach.

Abbreviations: Agro-BGC, Agronomy-BioGeocemical Cycles; ALMANAC, Agricultural Land Management Alternative with Numerical Assessment Criteria; BIOCRO, BIOphysical CROp; CRP, conservation reserve program; DAYCENT, Daily timestep CENtury; EPIC, Erosion-Productivity Impact Calculator; GHG, greenhouse gas; LAI, leaf area index; PotLAI, potential leaf area index; RUE, radiation use efficiency; SOC, soil organic carbon; SWAT, Soil and Water Assessment Tool; WUE, water use efficiency.

J.R. Kiniry, USDA-ARS, Grassland, Soil and Water Research Lab., 808 E. Blackland Road, Temple, TX 76502. *Corresponding author (jim.kiniry@ars.usda.gov).

M.N. Meki, Blackland Research Lab., Texas Agrilife., 720 E. Blackland Road, Temple, TX 76502 (nmeki@brc.tamus.edu).

T.E. Schumacher, Plant Science, South Dakota State University, Box 2207A, Agricultural Hall– SAG 004, Brookings, SD 57007 (thomas.schumacher@sdstate.edu).

C.J. Zilverberg, Department of Natural Resource Management, South Dakota State University, Northern Plains Biostress Laboratory (SNP), Box 2140B, Brookings, SD 57007 (czilverberg@brc.tamus.edu).

F.B. Fritschi, Division of Plant Sciences, University of Missouri, 1-33 Agriculture Building, Columbia, MO 65211 (fritschif@missouri.edu).

V.G. Kakani, Department of Plant and Soil Sciences, 368 Agricultural Hall, Oklahoma State University, Stillwater, OK 74078-6028 (v.g.kakani@okstate.edu).

doi:10.2134/advagricsystmodel5.c6

Practical Applications of Agricultural System Models to Optimize the Use of Limited Water
Lajpat R. Ahuja, Liwang Ma, and Robert J. Lascano, Editors
Advances in Agricultural Systems Modeling, Volume 5. Lajpat R. Ahuja, Series Editor

USDA-ARS

Biofuel production is currently at a critical crossroads. Decisions are being made that will determine the path to be taken for biofuel production and use. Maize has recently been the crop of choice, and the impacts of its use have been dramatic. Widespread expansion of biofuel maize acreage has been reported to impact the economics of feed production, soil erosion, and water quality downstream. Other options such as perennial grasses for cellulose, oil-seed crops, biofuel sorghum (*Sorghum bicolor* L. Moench), energycane (*Sacharum spontaneum*), or sugarcane (*Saccharum officinarum* L.) each have their own advantages as well as negative aspects.

Many of the basic questions and issues concerning biofuel production and use in the U.S. have been the same for at least the last 20 yr. A background paper for the U.S. Congress by the Office of Technology Assessment (U.S. Congress Office of Technology Assessment, 1993, foreword) outlines several of the issues we are still facing today. Impacts of expanded biofuel production on "soil quality and soil erosion, water quality, air quality, habitat for a variety of species, and the global environment" are all discussed in this publication. Growing bioenergy crops on set-aside or conservation reserve program (CRP) lands and degraded lands were and still are important concerns. The ability of herbaceous environmental crops like switchgrass or short rotation woody crops such as hybrid poplar (*Populus* spp.) to "stabilize erosive soils or perhaps filter agricultural chemicals and sediments before they reach water supplies" is discussed in this publication (U.S. Congress Office of Technology Assessment, 1993, p. 2). Even economic impacts, especially on the rural economies, are also discussed. With the current concern over expanded maize production and the negative economic impacts of its use for biofuel, it is interesting to note that this 20-yr-old publication discusses positive impacts, with the hope that expanded biofuel production might provide "alternative crops for rural agricultural communities: to provide employment, to stabilize rural incomes, and to maintain the rural infrastructure of equipment and supplies distribution and service" (U.S. Congress Office of Technology Assessment, 1993, p. 13).

Likewise, with the current rush to convert CRP lands back into production agriculture, it might be wise to consider one of the main reasons this program was originally established: to protect highly erodible soils. In the 1993 report, when discussing herbaceous environmental crops like switchgrass, authors propose that "Bioenergy crops are a potential alternative cash crop that could protect fragile soils or could be grown on lands previously idled to strengthen commodity crop prices" (U.S. Congress Office of Technology Assessment, 1993, p. 13). This could result in a win-win scenario, allowing economic production on these lands while maintaining the soil and water environmental benefits of a permanent plant cover. The report goes on to discuss impacts of bioenergy crop production on soil organic matter, nutrient cycling, soil erosion, water quality, air quality, riparian zones and wetlands, wildlife habitat, greenhouse gases (GHGs), and atmospheric concentration of CO_2, all important and frequently discussed issues today.

To address these issues, process-based models have offered and continue to offer promise as valuable tools to rapidly and efficiently assess the benefits and negative impacts of these various biofuel plant species as they are considered for large-scale production. Such crop models, once parameterized for the various plants and validated across diverse landscapes and production scenarios, are useful for calculating water requirements, impacts of limited water availability on production capacity, stability of production over years and over climatic conditions such as periodic drought, and impacts on soil erosion and water quality downstream. Before the first acre is dedicated to such production, effects on water use, soil erosion, water quality downstream, and acreage lost for crop production for food or for livestock production through grazing can be assessed. Such models can assess tradeoffs between the use of smaller acreage of prime farmland to meet biofuel production goals vs. more marginal soils or rangeland, but increased acreage to reach the same goals. Local, regional, and even national assessments are possible with these models in an efficient and scientifically defendable manner.

There has been much progress in development of such process-based models and their parameterization, validation, and application in the last 20 yr since publication of the congressional report. The first objective of this chapter was to describe some of these modeling tools as they have been applied to issues related to biofuel production. Second, we will describe one such single field scale simulation tool, the ALMANAC model (Kiniry et al., 1992), as an effective means to address some of the current issues concerning biofuel production. We will describe the basic processes simulated by the model, present examples that demonstrate its usefulness, and show some applications in typical regions that can be considered for biofuel production in the future.

Materials and Methods

Applications of Simulation Models to Address Questions Related to Biofuel Production

As discussed by Behrman et al. (2014b), simulation models of biomass production by plants such as switchgrass are being used to address many of the abovementioned issues. One type of model for bioenergy crops is the statistical approach that relies on relationships between environmental variables and empirical biomass estimates. Statistical models have been used to estimate yields across large spatial extents (Wullschleger et al., 2010). Another type of model relies on the process-based approach that simulates the actual processes governing plant growth. For this chapter, we will concentrate only on the process-based modeling approach.

Process-based simulations include detailed information on plant growth and development, climate, soil dynamics, and management (Williams et al., 1989; Kiniry et al., 1992, 1996; Arnold et al., 1998; Del Grosso et al., 2005; Di Vittorio et al., 2010; Miguez et al., 2012; Gelfand et al., 2013). These models have been used not only to estimate yields but also to analyze WUE, impacts of management practices, long-term effects on soil properties, and the impact of climate change (Kiniry et al., 1996, 2008a; Brown et al., 2000).

Process-based models of plant growth that have been used to simulate switchgrass production include Agro-BGC (Agronomy-BioGeocemical Cycles), ALMANAC, BIOCRO (BIOphysical CROp), DAYCENT (Daily timestep CENtury), Erosion-Productivity Impact Calculator (EPIC), and Soil and Water Assessment Tool (SWAT) (Williams et al., 1989; Kiniry et al., 1992, 1996; Arnold et al., 1998; Del Grosso et al., 2005; Di Vittorio et al., 2010; Miguez et al., 2012; Gelfand et al., 2013). These models have been used to track GHG emissions, soil erosion, nutrient cycling, and plant productivity. Because of the variability in modeling objectives, these models vary in their functions and amount of detail incorporated to simulate plant growth. However, each model shares the following basic functionality. First, they simulate biomass production by specifying light interception, conversion of sunlight to biomass, and partitioning of biomass into structural components. Second, they simulate soil water dynamics, which depends on precipitation, runoff, and evapotranspiration. Third, they simulate soil carbon and nitrogen dynamics. Lastly, each model simulates the effect of drought stress on plant growth. Models with more complex functions for the effects of environmental stress on plant growth, such as ALMANAC and EPIC, incorporate additional stress effects including stresses due to extreme temperatures, nutrient deficiency, salinity, low pH, aluminum toxicity, and poor soil aeration.

Recent Modeling Applications

Modeling Biomass Production

The following sections will mainly concern switchgrass simulation, because of its prevalence in discussions of potential biofuel plant species. The concepts discussed herein are directly relevant for other plant species potentially useful for biofuel.

As described by Behrman et al. (2014b), Agro-BGC, ALMANAC, BIOCRO, DAYCENT, EPIC, and SWAT are process-based models that have been used to simulate switchgrass productivity (Williams et al., 1989; Kiniry et al., 1992, 1996; Arnold et al., 1998; Del Grosso et al., 2005; Di Vittorio et al., 2010; Miguez et al., 2012). Each model keeps track of the number of growing degree days to specify the developmental rate or phenological stage. The number of growing degree days is determined by the average of the daily maximum and minimum temperature above the specified baseline temperature (Williams et al., 1984). ALMANAC, EPIC, and SWAT use a function relating radiation use efficiency (RUE) to biomass on the basis of the leaf area and the amount of light intercepted (Williams et al., 1989). For switchgrass, plant development is initiated when temperatures exceed 12°C (ALMANAC, EPIC, and SWAT). Senescence begins when plants exceed the maximum number of growing degree days (Williams et al., 1989; Kiniry et al., 1992). Biomass is partitioned into roots and shoots. The DAYCENT model uses a constant energy biomass conversion factor (Parton et al., 1998). BIOCRO uses an empirical derivation of the relationship between photosynthesis, stomatal conductance, and biomass production (Humphries and Long, 1995; Miguez et al., 2012). Agro-BGC simulates carbon uptake and assimilation (Di Vittorio et al., 2010).

Modeling Water Use Efficiency

Water use efficiency can be vitally important for predicting areas suitable for switchgrass biofuel production. To avoid competition with farmland already used for food, fiber, and feed production, the areas most likely for switchgrass production are the less productive soils where soil water and nutrients often limit production (Perlack et al., 2005). In dryland production systems, limited rainfall and/or limited capacity of soils to store moisture are important issues for switchgrass production. Plant water use and WUE also are major concerns in irrigated regions, where competition between agriculture and other water users arises. Direct measurements of WUE are important but require labor-intensive procedures involving soil water measurement with soil moisture sensors, gravimetric measurements of soil moisture from soil cores, or use of weighing lysimeters. Likewise, determinations of WUE require harvesting to quantify plant dry weights. To adequately define WUE for a range of soils, plant species, and climatic conditions requires considerable resources and time.

Water Use Efficiency of Four Switchgrass Types

Behrman et al. (2014a) parameterized the ALMANAC model and used the model to simulate WUE for the four major switchgrass types: northern upland, northern lowland, southern upland, and southern lowland in multiple locations across the

Great Plains. This was an extension of the work with these four types by Kiniry et al. (2008a) in the northern U.S. and Woli et al. (2012) in Mississippi. These locations are representative sites in the region anticipated to be the primary production area for biofuel crops in the United States. The data for comparing simulated WUE among switchgrass types and among locations using ALMANAC came from two studies (Table 6–1). The first study (Kiniry et al., 2008a) simulated plant transpiration and biomass for four sites: Stephenville, TX; Mead, NE; Columbia, MO; and Ames, IA. The second study (Kiniry et al., 2013a; Behrman et al., 2014a) involved 10 sites in the Central and Southern Great Plains. These ranged from northern Missouri at Elsberry to subtropical southern Texas at Weslaco. The model was used, with appropriate soil parameters for each site and with actual measured weather for the growing season, to calculate plant transpiration. For the first study, crop parameters were adjusted to get reasonable simulated yields of each switchgrass type, as compared with the mean of the 2 yr of measured yields. For the second study, rainfall was severely limiting because of a drought throughout much of the region. Therefore, the crop parameters were adjusted to match LAI values measured early in the growing season in an effort to minimize the impact of severe drought stress on results (Behrman et al., 2014a).

Table 6–1. Simulated water use efficiency (WUE) for four switchgrass ecotypes: southern lowland (SL), northern lowland (NL), southern upland (SU), and northern upland (NU) from Kingsville, TX to Ames, IA. The greatest WUE value at each location is in italics.

Location	Latitude	Soil type	WUE for SL	WUE for NL	WUE for SU	WUE for NU
			mg g^{-1}			
Weslaco, TX†	26.22	Hidalgo sandy clay loam	4.2	*4.3*	3.7	4.2
Kingsville, TX†	27.54	Cranel sandy clay loam	*5.3*	3.1	2.6	2.8
Temple, TX†	31.04	Houston black clay	*11.3*	4.9	3.5	3.3
Nacogdoches, TX†	31.5	Atoyac sandy loam	*5.8*	3.6	2.2	2.1
Stephenville, TX†	32.22	Brackett clay loam	*3.5*	3.3	3.2	3.2
Booneville, AR†	35.09	Leadvale silty loam	3.3	*3.5*	3.5	4.7
Fayetteville, AR†	36.09	Pinckwick gravely loam	5.0	*5.2*	4.9	4.9
Stillwater, OK†	36.12	Kirkland silt loam	3.1	2.2	2.5	*3.7*
Mt. Vernon, MO†	37.03	Gerald silt loam	5.4	*5.6*	4.4	3.5
Columbia, MO‡	38.89	Keswich silt loam	4.5	*4.6*	4.1	3.9
Columbia, MO‡	38.89	Mexico silt loam	2.3	*3.1*	1.6	2.4
Elsberry, MO‡	39.16	Menfro silt loam	3.7	*5.1*	3.2	3.2
Mead, NE‡	41.23	Yutan silty clay loam	4.1	4.0	*4.5*	3.3
Ames, IA‡	42.03	Clarion loam	5.3	*5.4*	3.9	3.6
Average			*4.8*	4.1	3.4	3.5

† Reproduced from Behrman et al. (2014b).
‡ Reproduced from Kiniry et al. (2008a).

The simulated WUE of all four switchgrass types showed values generally ranging from 3 to 6 mg g^{-1}, with a few exceptions. These values are within the range of empirical data that range widely and depend on the environmental conditions and switchgrass ecotype. Byrd and May (2000) calculated the WUE of many different potted switchgrass cultivars and reported that WUE ranges from 4.3 to 8.5 mg dry weight per gram of water used. In germplasm nurseries in Tennessee and Oklahoma, WUE of switchgrass accessions were 3.5 to 6.3 mg CH_2O g^{-1} of water transpired (McLaughlin et al., 2006). In the field in Nebraska, switchgrass WUE values were 1.0 to 5.5 mg g^{-1} (Eggemeyer et al., 2006), values similar to those demonstrated by switchgrass seedlings in a growth chamber (1.45 to 5.5 mg g^{-1}) (Xu et al., 2006).

The greatest simulated WUE values were most often for the lowland types. The southern lowland types had the highest WUE values in most of Texas, and the northern lowland types generally had the greatest WUE values at locations further north (Table 6–1). The northern lowland type had the highest WUE in more than half the cases, being greatest for 8 of the 14 locations. The southern upland's values for WUE were greatest at two sites: Booneville, AR and Mead, NE. The northern upland types only had the greatest value at Stillwater, OK. High WUE of lowland varieties allows for increases in yield in southern locations where water is often limiting.

Long-Term Environmental Impacts of Bioenergy Production Systems

Mechanistic models will be useful for comparing bioenergy production systems to more conventional agricultural crops. Meki et al. (2013) applied a version of the EPIC model, the Agricultural Policy Environmental Extender (APEX), to assess the sustainability of maize stover removal from the Upper Mississippi River Basin, based on a set of "acceptable planning criteria" used in the Conservation Effects Assessment Project analysis (USDA-NRCS, 2010), to judge whether or not a farm field needed additional conservation treatment. The "acceptable criteria" included (i) N in surface runoff < 16.8 kg ha^{-1} yr^{-1} (15.0 lb ac^{-1} yr^{-1}), (ii) N in subsurface runoff < 28.0 kg ha^{-1} yr^{-1} (25.0 lb ac^{-1} yr^{-1}), (iii) total P losses < 4.5 kg ha^{-1} yr^{-1} (4.0 lb ac^{-1} yr^{-1}), (iv) sediment loss < 4.5 Mg ha^{-1} yr^{-1} (2.0 ton ac^{-1} yr^{-1}), and (v) soil organic carbon (SOC) with a more "stringent" restriction that the annual rate of change be positive. Given the critical functions of SOC in maintaining soil quality and productivity, biomass removal can only be justified if it does not deplete the SOC pool. These "acceptable" levels represent field-level losses that are feasible to attain using traditional conservation treatment (nutrient management and soil erosion control), are agronomically feasible, and can equally be adapted to most bioenergy production systems. Scientific literature on field research and edge-of-field monitoring in the U.S. Midwest, coupled with model simulations of conservation practices effects, provided guidance for identifying these thresholds (USDA-NRCS, 2010).

Examples of ALMANAC Applications

The ALMANAC model has been described numerous times, as it has been used to simulate crops (Kiniry et al., 1997; Kiniry and Bockholt, 1998; Xie et al., 2001) and warm season grasses (Kiniry et al., 1996, 2002, 2005, 2007, 2008b, 2013b; McLaughlin et al., 2006; Schilling and Kiniry, 2007; Behrman et al., 2014b). Parameters to simulate different plants continue to be refined as new research results are reported. Basically, the model simulates the soil water balance, the soil and plant nutrient balance, and the interception of solar radiation. This model includes subroutines and functions from the EPIC model (Williams et al., 1984, 1989) with added details for plant growth. The model has a daily time step. It simulates plant growth for a wide range of species and is implemented easily.

Light Interception

ALMANAC simulates light interception by the leaf canopy with Beer's law (Monsi and Saeki, 1953) and the leaf area index (LAI). The LAI is the amount of leaf area per unit ground area, a unitless variable. With greater extinction coefficient values (k), a given LAI intercepts more light. The fraction of incoming solar radiation intercepted by the leaf canopy is

$$\text{Fraction} = 1.0 - \exp(k \times \text{LAI}) \qquad [1]$$

Leaf Area Development

An accurate prediction of light interception depends on a realistic simulation of leaf area. The model estimates leaf area production up to the point of maximum leaf area for the growing season using Eq. [2]. The sigmoid curve function for the potential leaf area index (PotLAI) production takes the form

$$F = \text{SYP} / [\text{SYP} + \exp(Y1 - Y2 \times \text{SYP})] \qquad [2]$$

where F is the factor for relative LAI, SYP is the fraction of the degree days from planting to maturity, and Y1 and Y2 are the sigmoid curve coefficients generated by ALMANAC. This curve passes through the origin and through two points, asymptotically approaching $F = 1.0$. The model calculates SYP each day. The sum of degree days is zero at planting in the establishment year and at tiller emergence in subsequent years, and it reaches its maximum value at maturity.

The model describes the loss of leaf area late in the season with the LAI decline factor. The LAI begins to decrease after a defined fraction of the seasonal degree days have accumulated.

Biomass Production and Partitioning

The model simulates biomass with an RUE value for each plant species. Values for RUE have a wide range for different crops and grasses (Kiniry et al., 1992, 1999, 2007). ALMANAC describes declining RUE in later growth stages with an identical function to the one for the decrease in LAI.

The maximum rooting depth defines the potential depth in the absence of a root-restricting soil layer. Soil cores at Temple, TX in 1994 showed that switchgrass roots extend to depths of at least 2.2 m.

Simulating Water Uptake with ALMANAC

The ALMANAC model is the only process-based plant model that has been used to analyze switchgrass WUE (Kiniry et al., 2008a; Behrman et al., 2014b). ALMANAC simulates the water balance by determining plant water use while taking into account soil properties, weather, and plant species cover. ALMANAC calculates the effects of soil water availability on plant growth by calculating the potential evaporation, potential soil water evaporation, and potential plant water transpiration on the basis of the LAI. Potential evaporation is estimated by the Penman–Monteith method (Ritchie, 1972). Potential soil evaporation and plant transpiration is estimated by

$$\mathrm{EP} = \mathrm{E0}(\mathrm{LAI}/3) \quad 0 < \mathrm{LAI} < 3.0 \qquad [3]$$

$$\mathrm{EP} = \mathrm{E0} \quad \mathrm{LAI} > 3.0 \qquad [4]$$

$$\mathrm{ES} = \text{minimum of } \left[\mathrm{E0}\exp(-0.1\mathrm{BIO}),\ \mathrm{E0} - \mathrm{EP}\right] \qquad [5]$$

where EP and ES are potential plant transpiration and soil evaporation (mm), E0 is potential evaporation (mm), and BIO is the sum of the aboveground biomass and plant residue (Mg ha^{-1}). Simulated plant leaf area and biomass are reduced when available soil water is depleted. Water use depends on soil water availability and plant water demand. Water demand is based on potential evapotranspiration and the LAI. The water stress factor is estimated and decreases daily leaf area and consequently biomass growth if current available soil water is insufficient to meet demands.

Water and Nutrient Uptake

Critical for yield and biomass simulation in water-limited conditions is the simulated water demand, as discussed above. The ALMANAC model calculates effects of soil water on crop growth and yield with similar functions. Potential evaporation is calculated first, and then potential soil water evaporation and potential plant water transpiration are derived from potential evaporation and LAI. On the basis of the soil water supply and crop water demand, the water stress factor is estimated to decrease daily crop growth and yield. Water stress factor is the ratio of water use to water demand (potential plant transpiration) in ALMANAC, and water use is a function of plant extractable water and root depth.

The nutrient balance (N and P) also allows plants to acquire sufficient nutrients to meet the demands if adequate quantities are available in the current rooting zone. Nutrient values for switchgrass were refined with N concentration data collected at Stephenville during 5 yr (M. Sanderson, unpublished data, 1995).

Base Temperature, Optimum Temperature, and Total Degree Days

Base temperature in ALMANAC is constant for all growth stages of any plant species. Base temperature constrains the initiation of leaf area growth and thus dry matter accumulation. Higher optimum temperature can allow increased plant development rate later in the season when temperatures are greater. The sum of degree days to maturity controls the duration of growth. Heat units are reset to zero after maturity each year. Heat units are calculated from daily maximum and minimum temperatures, assuming the maximum equals the optimum if it exceeds the optimum.

Plant parameters can be adjusted to get reasonable simulated yields of each switchgrass type, as compared with the mean of multiple-year measured yield data, as discussed above. Using parameters developed for Alamo switchgrass in Texas (Kiniry et al., 1996, 1999, 2007), adjusting only two parameters—degree days to maturity (base 12°C) and PotLAI—can often lead to realistic simulations for other switchgrass ecotypes. The light extinction coefficient for Beer's law (Monsi and Saeki, 1953) can be set to −0.51, the average of the two means from Kiniry et al. (1999, 2007).

The values of PotLAI and degree days varies with latitude and switchgrass types. Other parameters that can be used for a wide range of locations and switchgrass types are (i) 3.9 g MJ^{-1} intercepted photosynthetically active radiation for the RUE at low vapor pressure deficits; (ii) a 0.65 unit decrease in RUE for each 1 kPa increase in vapor pressure deficits above 1 kPa; (iii) base temperature of 12°C and optimum temperature of 25°C; (iv) linear decreases in LAI and in RUE from 70% of total degree day accumulation until maturity; (v) maximum potential rooting depth of 2.2 m; (vi) optimum nitrogen concentrations of 2.57% for plants early in the spring, 1.1% for plants near midseason, and 0.28% for plants near maturity each year; and (vii) optimum phosphorus concentrations of 0.14% for plants early in the spring, 0.10% for plants near midseason, and 0.07% for plants near maturity each year.

Four Study Locations with Simulated Yields of Switchgrass and Maize

Verification Results

Before conducting comparisons among crops and soils, it is imperative to verify that the biomass yields are realistic. This was accomplished by comparing the mean simulated yields for four sites (Table 6–2) over 17 yr (we simulated 20 yr, but ignored the first three to allow various model components to stabilize). The soils in each case were the "good soil" described below. The measured yield values were derived by various ways, either by USDA-NRCS values of maize given for that soil in that county on the USDA-NRCS Web Soil Survey, the mean for the county reported by USDA-NASS (2014), published values for the site (especially for South Dakota; Zilverberg et al., 2014), or, especially for switchgrass, yields derived from USDA-NRCS values using the regressions of Johnson et al. (2010).

We were successful in getting mean simulated values close to the mean measured values in each case (Table 6–3). This was accomplished after determining the appropriate degree days or potential heat units for duration of growth for the crops in each region and with reasonable input annual fertilizer amounts. Potential LAI was the primary variable adjusted to get simulated yields close to measured values.

Table 6–2. Four sites simulated with the ALMANAC model for maize and switchgrass. All slopes were assumed to be 1% with the exception of the better soil in South Dakota, with a slope of 4%.

Location	County	Latitude, longitude	Years run	Better soil	Poorer soil
Temple, TX	Bell	31.0508 N, −97.354 W	1983–2002	Houston Black clay	Austin silty clay
Chickasha, OK	Grady	35.082 N, −97.8853 W	1990–2009	Zaneis loam	Nash–Lucien complex
Columbia, MO	Boone	38.9008 N, −92.2087 W	1949–1968	Mexico silt loam	Armstrong loam
Colman, SD	Moody	4.0824 N, −96.7381 W	1988–2007	Wentworth–Egan silty clay loam	Baltic silty clay loam

Table 6–3. Yield verification: simulated yields (on the "better" soils, see text) and reported yields of ALMANAC (last 17 yr of 20 simulated) for switchgrass (aboveground biomass) and maize (grain only) at four different locations. Switchgrass potential leaf area index (LAI) values were 1.3, 12, 6, and 12 for the four sites in order. Maize input LAI values were 3.2, 4.1, 2.0, and 3.2 for the four sites in order.

	Switchgrass		Maize	
	Simulated	Measured	Simulated	Measured
	Mg ha^{-1}			
Colman, SD†				
Mean yield	5.9	5.3	6.7	6.8
Warm season grass mixture	3.6	3.7		
Columbia, MO‡				
Mean yield	17.4	18.1	4.9	4.7
Chickasha, OK§				
Mean yields	5.9	6.6	4.0	2.3 (5.1)
Temple, TX‡				
Mean yield	14.2	14.5	4.8	4.7

† Measured data from Zilverberg et al. (2014).
‡ Measured data derived from USDA-NRCS values for the soils near the site.
§ Measured data derived from USDA-NRCS values for the soils near the site and, in parentheses, the USDA-NASS mean reported yield for 4 yr.

Biomass Yields and Water Use Results

Decisions on which crops to use and which site to use depend not only on biomass yields but also on yield stability. To be economically viable, producers must have the assurance that even dry years will have adequate biomass production.

As production is moved to poorer marginal soils (Tables 6–4 to 6–7), not only are mean yields consistently lower as expected, but also yield variability, as shown by the CVs, usually increased. Thus, not only will it take more area of these poorer marginal soils to meet yield goals, but the increased yield variability will increase risk. Thus, this type of simulation model can be readily used to assess such risk.

For South Dakota (Table 6–4), maize yields were less stable, with greater variability in yields than switchgrass or the native seed mix as expected. In addition, mean maize yields for the two soils at this site were greater than for switchgrass or the mixture of warm season grasses on the better soil, but lower than switchgrass for the poorer soil.

Somewhat similar trends occurred for the three more southern sites (Tables 6–5, 6–6, and 6–7). Yields for switchgrass were much greater than for maize in each case. Yield stability, as indicated by the CVs, was lower for switchgrass than for maize with the two soils in Missouri and for the poorer soil in Texas. For the better soil in Texas and for both soils in Oklahoma, yield stability for maize was higher than for switchgrass.

Water use efficiency is reported in three ways: yield divided by seasonal precipitation (WUE_1), yield divided by seasonal evapotranspiration (water loss from soil evapotranspiration and plant transpiration, WUE_2), and yield divided by plant transpiration (WUE_3). Thus, WUE_1 is a general measure of water use efficiency for a particular rainfall zone. WUE_2 has water loss to deep infiltration and runoff taken out of the calculation. It therefore is more indicative of how efficient the production system is in terms of actual water in the soil profile. Finally, WUE_3 has soil evaporation taken out of the calculation also. As plants develop their LAI over the course of the season, and as evaporative demand changes, plant transpiration increases and eventually decreases to zero by the day of total leaf senescence. Thus, this final WUE value is useful for comparing production systems in terms of actual plant use of water.

There are different trends for WUE depending on the way it was calculated. In terms of total precipitation (WUE_1) for the South Dakota site (Table 6–4), values for maize were highest for the better soil, while values for switchgrass were highest for the poorer soil. WUE_1 values for switchgrass were higher for both soils at the other three sites (Tables 6–5, 6–6, and 6–7). In terms of total ET (WUE_2), maize was higher than switchgrass in South Dakota for the better soil; however, switchgrass was higher for the poorer soil. Again for the three other sites, switchgrass had the higher WUE when calculated in this manner. Finally, when WUE was calculated in terms of plant transpiration (WUE_3), results consistently showed the higher values for switchgrass than for maize in all cases.

Comparing better versus poorer soil, WUE results were highly variable. For South Dakota switchgrass and the mixture, all three WUE values were nearly the same on the two soils or slightly higher on the poorer soil. For maize, WUE values were noticeably higher on the better soil. These trends varied for the other three sites. In Missouri, switchgrass WUE values were higher on the better soil for WUE_1, higher on the poorer soil for WUE_3, and nearly the same for the two soils for WUE_2. Maize at the Missouri site had the higher WUE for the better soil for the first two WUE values and nearly the same for the two soils for WUE_3. In Oklahoma, switchgrass WUE values were higher for the better soil for WUE_1 and for the poorer soil for WUE_2 and WUE_3. Maize in Oklahoma had the highest values on the better soil for all three WUE values. Switchgrass in Texas had the WUE_1 values on the better soil, the higher WUE_3 value on the poorer soil, and nearly identical values for WUE_2 on the two soils. Maize in Texas had the highest WUE_1 and WUE_3 on the better soil and nearly identically WUE_2 values on the two soils.

Table 6–4. At Colman, SD, ALMANAC model simulated yields and water use efficiency (WUE) of switchgrass, native plant mixture (both aboveground biomass), and maize (grain only). Simulations were conducted for 17 yr on two soils. WUE_1 is the ratio of yield divided by rain, WUE_2 is yield divided by evapotranspiration (ET; water loss from soil evapotranspiration and plant transpiration), and WUE_3 is yield divided by plant transpiration (EP). The mean annual precipitation was 1433 ± 300 mm.

	Yield	ET	EP	WUE_1	WUE_2	WUE_3
	Mg ha^{-1}	mm yr^{-1}		mg g^{-1}		
Switchgrass†						
Better soil						
Mean ± SD	5.86 + 0.96	432 ± 22	158 ± 22	0.42 ± 0.13	1.35 ± 0.20	3.70 ± 0.30
CV (%)	16	5	14	32	14	8
Poorer soil						
Mean ± SD	5.85 + 0.95	394 ± 19	157 ± 22	0.42 ± 0.14	1.48 ± 0.22	3.72 ± 0.31
CV (%)	16	5	14	32	15	8
Mixture‡						
Better soil						
Mean ± SD	3.57 + 0.57	551 ± 44	306 ± 47	0.26 ± 0.08	0.64 ± 0.09	1.18 ± 0.21
CV (%)	16	8	15	30	14	18
Poorer soil						
Mean ± SD	3.55 + 0.59	486 ± 56	259 ± 54	0.25 ± 0.08	0.74 ± 0.13	1.41 ± 0.29
CV (%)	17	12	21	30	18	21
Maize§						
Better soil						
Mean ± SD	8.53 + 0.16	554 ± 41	354 ± 36	0.60 ± 0.15	1.54 ± 0.26	2.42 ± 0.44
CV (%)	18	7	10	25	17	18
Poorer soil						
Mean ± SD	5.69 + 2.51	496 ± 51	287 ± 44	0.39 ± 0.11	1.12 ± 0.38	1.93 ± 0.65
CV (%)	44	10	15	37	34	33

† Switchgrass is an upland switchgrass.
‡ Mixture is a mixture of warm season and native grasses.
§ Maize is a locally adapted hybrid.

It should be noted that these comparisons between maize and switchgrass WUE values would be different if total biomass of maize was considered instead of just grain. With a harvest index of near 0.5, such whole biomass WUE values of maize would be approximately twice as large. Similar comparisons were reported by Kiniry et al. (2008a).

Water Runoff, Soil Erosion, and N and P Simulation

The permanent cover of perennial switchgrass (and the native grass mixture in South Dakota) as compared with maize had the expected effects of reducing water runoff (Q) and reducing water-induced soil erosion (MUSL) often quite dramatically, as expected (Tables 6–8, 6–9, 6–10, and 6–11). At the South Dakota site, the mean Q values for maize were 6 and 78% greater than for switchgrass. The mixture had 1 to 2% lower Q than for switchgrass. Soil erosion (MUSL) was dramatically greater for maize than for switchgrass or for the mixture.

Table 6–5. For Columbia, MO for 17 simulated years: simulated yields and water use efficiency (WUE) of ALMANAC for switchgrass and maize (grain only) at four different locations with two soils at each location. WUE_1 is the ratio of yield divided by rain, WUE_2 is yield divided by evapotranspiration (ET; water loss from soil evapotranspiration and plant transpiration), and WUE_3 is yield divided by plant transpiration (EP). The mean annual precipitation was 1222 ± 312 mm.

	Yield	ET	EP	WUE_1	WUE_2	WUE_3
	Mg ha^{-1}	—— mm yr^{-1} ——		—— mg g^{-1} ——		
Switchgrass†						
Better soil						
Mean ± SD	13.39 + 3.33	673 ± 101	470 ± 89	1.13 ± 0.34	1.96 ± 0.33	2.82 ± 0.45
CV (%)	25	15	19	30	19	16
Poorer soil						
Mean ± SD	11.26 + 3.36	578 ± 101	360 ± 88	0.97 ± 0.34	1.94 ± 0.46	3.15 ± 0.72
CV (%)	30	17	24	35	24	23
Maize‡						
Better soil						
Mean ± SD	5.14 + 1.81	607 ± 53	344 ± 41	0.43 ± 0.13	0.84 ± 0.25	1.48 ± 0.42
CV (%)	35	9	12	30	29	29
Poorer soil						
Mean ± SD	3.72 + 1.72	543 ± 74	253 ± 60	0.31 ± 0.10	0.67 ± 0.22	1.43 ± 0.38
CV (%)	46	15	24	33	34	26

† Switchgrass is a lowland switchgrass.
‡ Maize is a locally adapted hybrid.

Table 6–6. For Chickasha, OK for 17 simulated years: simulated yields and water use efficiency (WUE) of ALMANAC for switchgrass and maize (grain only) at four different locations with two soils at each location. WUE_1 is the ratio of yield divided by rain, WUE_2 is yield divided by evapotranspiration (ET; water loss from soil evapotranspiration and plant transpiration), and WUE_3 is yield divided by plant transpiration (EP). The mean annual precipitation was 899 ± 227 mm.

	Yield	ET	EP	WUE_1	WUE_2	WUE_3
	Mg ha^{-1}	—— mm yr^{-1} ——		—— mg g^{-1} ——		
Switchgrass†						
Better soil						
Mean ± SD	7.21 + 3.36	590 ± 79	545 ± 65	0.82 ± 0.37	1.20 ± 0.47	2.04 ± 0.74
CV (%)	47	13	19	45	39	36
Poorer soil						
Mean ± SD	6.39 + 3.01	513 ± 118	281 ± 57	0.74 ± 0.37	1.43 ± 1.29	2.23 ± 0.88
CV (%)	47	23	20	51	90	39
Maize‡						
Better soil						
Mean ± SD	3.97 + 1.39	583 ± 79	265 ± 52	0.47 ± 0.22	0.67 ± 0.21	1.47 ± 0.40
CV (%)	35	14	20	47	32	27
Poorer soil						
Mean ± SD	2.42 + 1.09	524 ± 79	205 ± 51	0.29 ± 0.18	0.46 ± 0.21	1.17 ± 0.45
CV (%)	45	15	25	62	45	38

† Switchgrass is a lowland switchgrass.
‡ Maize is a locally adapted hybrid.

Table 6–7. For Temple, TX for 17 simulated years; simulated yields and water use efficiency (WUE) of ALMANAC for switchgrass and maize (grain only) at four different locations with two soils at each location. WUE_1 is the ratio of yield divided by rain, WUE_2 is yield divided by evapotranspiration (ET; water loss from soil evapotranspiration and plant transpiration), and WUE_3 is yield divided by plant transpiration (EP). The mean annual precipitation was 841 ± 193 mm.

	Yield	ET	EP	WUE_1	WUE_2	WUE_3
	$Mg\ ha^{-1}$	—— $mm\ yr^{-1}$ ——		—— $mg\ g^{-1}$ ——		
Switchgrass†						
Better soil						
Mean ± SD	13.64 + 3.52	841 ± 193	669 ± 93	1.69 ± 0.55	2.04 ± 0.48	2.86 ± 0.59
CV (%)	26	23	14	33	23	21
Poorer soil						
Mean ± SD	9.94 + 2.78	493 ± 63	340 ± 64	1.26 ± 0.53	2.03 ± 0.53	2.97 ± 0.74
CV (%)	28	13	19	42	26	25
Maize‡						
Better soil						
Mean ± SD	4.82 + 1.15	620 ± 67	358 ± 75	0.60 ± 0.22	0.77 ± 0.16	1.33 ± 0.18
CV (%)	24	11	21	36	21	13
Poorer soil						
Mean ± SD	4.05 + 1.27	515 ± 49	323 ± 59	0.51 ± 0.21	0.78 ± 0.21	1.23 ± 0.28
CV (%)	31	10	18	42	27	23

† Switchgrass is a lowland switchgrass.
‡ Maize is a locally adapted hybrid.

Table 6–8. For Colman, SD for 17 simulated years: environmental impacts, water quality, and soil erosion values simulated by the ALMANAC model. Variables are surface runoff (Q), soil loss from water erosion (MUSL), organic N loss with sediment (YON), P loss with sediment (YP), NO_3 loss in surface runoff (YNO3), and soluble P loss in runoff (YAP).

	Q	MUSL	YON	YP	YNO3	YAP
	$mm\ yr^{-1}$	$Mg\ ha^{-1}\ yr^{-1}$	—— $kg\ ha^{-1}\ yr^{-1}$ ——			
Switchgrass†						
Better soil						
Mean ± SD	386 ± 186	0.32 ± 0.16	1.3 ± 0.6	0.2 ± 0.1	3.4 ± 1.6	0.2 ± 0.14
CV (%)	49	49	43	43	48	60
Poorer soil						
Mean ± SD	596 ± 248	0.07 ± 0.03	0.4 ± 0.1	0.05 ± 0.02	5.1 ± 2.0	0.05 ± 0.03
CV (%)	42	39	24	33	38	49
Mixture‡						
Better soil						
Mean ± SD	381 ± 188	0.35 ± 0.16	1.4 ± 0.6	0.2 ± 0.1	3.2 ± 1.6	0.26 ± 0.14
CV (%)	49	47	43	44	49	30
Poorer soil						
Mean ± SD	583 ± 248	0.08 ± 0.04	0.5 ± 0.2	0.1 ± 0.0	5.0 ± 2.0	0.07 ± 0.03
CV (%)	42	45	42	42	40	44
Maize§						
Better soil						
Mean ± SD	688 ± 262	24.3 ± 9.3	70.1 ± 24.4	9.3 ± 3.2	5.7 ± 2.1	0.12 ± 0.05
CV (%)	38	38	35	34	37	43
Poorer soil						
Mean ± SD	633 ± 247	4.7 ± 1.9	25.6 ± 9.7	3.4 ± 1.3	9.6 ± 5.0	0.55 ± 0.39
CV (%)	39	40	38	38	52	72

† Switchgrass is an upland switchgrass.
‡ Mixture is a mixture of warm season and native grasses.
§ Maize is a locally adapted hybrid.

Table 6–9. For Columbia, MO for 17 simulated years: environmental impacts, water quality, and soil erosion values simulated by the ALMANAC model. Variables are surface runoff (Q), soil loss from water erosion (MUSL), organic N loss with sediment (YON), P loss with sediment (YP), NO_3 loss in surface runoff (YNO3), and soluble P loss in runoff (YAP).

	Q	MUSL	YON	YP	YNO3	YAP
	mm yr^{-1}	Mg ha^{-1} yr^{-1}	kg ha^{-1} yr^{-1}			
Switchgrass†						
Better soil						
Mean ± SD	388 ± 193	0.06 ± 0.04	0.1 ± 0.1	0.02 ± 0.01	3.1 ± 1.5	0.013 ± 0.009
CV (%)	57	60	44	45	48	66
Poorer soil						
Mean ± SD	329 ± 191	0.64 ± 0.37	1.3 ± 0.6	0.2 ± 0.1	5.2 ± 2.4	0.078 ± 0.053
CV (%)	58	58	46	46	45	68
Maize‡						
Better soil						
Mean ± SD	395 ± 206	4.7 ± 2.4	11.8 ± 5.2	1.8 ± 0.8	13.6 ± 3.9	0.60 ± 0.30
CV (%)	52	51	44	43	29	51
Poorer soil						
Mean ± SD	471 ± 237	68.5 ± 33.6	109.7 ± 49.9	18.6 ± 8.7	21.7 ± 12.4	0.84 ± 0.43
CV (%)	50	49	45	47	57	51

† Switchgrass is a lowland switchgrass.
‡ Maize is a locally adapted hybrid.

Table 6–10. For Chickasha, OK for 17 simulated years; environmental impacts, water quality, and soil erosion values simulated by the ALMANAC model. Variables are surface runoff (Q), soil loss from water erosion (MUSL), organic N loss with sediment (YON), P loss with sediment (YP), NO_3 loss in surface runoff (YNO3), and soluble P loss in runoff (YAP).

	Q	MUSL	YON	YP	YNO3	YAP
	mm yr^{-1}	Mg ha^{-1} yr^{-1}	kg ha^{-1} yr^{-1}			
Switchgrass†						
Better soil						
Mean ± SD	42 ± 38	0.007 ± 0.009	0.039 ± 0.02	0.004 ± 0.005	0.45 ± 0.36	0.014 ± 0.013
CV (%)	90	120	68	140	80	94
Poorer soil						
Mean ± SD	65 ± 48	0.07 ± 0.05	0.18 ± 0.11	0.02 ± 0.01	0.61 ± 039	0.0088 ± 0.007
CV (%)	74	73	59	62	64	79
Maize‡						
Better soil						
Mean ± SD	162 ± 77	2.5 ± 1.3	6.0 ± 2.8	1.00 ± 0.50	10.6 ± 4.3	0.320 ± 0.157
CV (%)	48	50	47	50	41	49
Poorer soil						
Mean ± SD	182 ± 89	8.9 ± 4.0	16.0 ± 7.7	2.36 ± 1.13	8.7 ± 2.4	0.019 ± 0.095
CV (%)	49	52	48	48	28	50

† Switchgrass is a lowland switchgrass.
‡ Maize is a locally adapted hybrid.

Table 6–11. For Temple, TX for 17 simulated years; environmental impacts, water quality, and soil erosion values simulated by the ALMANAC model. Variables are surface runoff (Q), soil loss from water erosion (MUSL), organic N loss with sediment (YON), P loss with sediment (YP), NO_3 loss in surface runoff (YNO3), and soluble P loss in runoff (YAP).

	Q	MUSL	YON	YP	YNO3	YAP
	mm yr^{-1}	Mg ha^{-1} yr^{-1}	———	——— kg ha^{-1} yr^{-1}	———	———
Switchgrass†						
Better soil						
Mean ± SD	138 ± 78	0.02 ± 0.01	0.10 ± 0.05	0.012 ± 0.007	1.7 ± 0.9	0.006 ± 0.006
CV (%)	56	59	47	54	50	105
Poorer soil						
Mean ± SD	116 ± 64	0.14 ± 0.08	0.34 ± 0.16	0.04 ± 0.02	1.0 ± 0.5	0.004 ± 0.006
CV (%)	55	57	46	47	54	150
Maize‡						
Better soil						
Mean ± SD	161 ± 90	1.9 ± 1.1	6.6 ± 3.2	0.95 ± 0.46	7.1 ± 4.6	0.225 ± 0.155
CV (%)	56	58	49	49	65	69
Poorer soil						
Mean ± SD	170 ± 85	12.8 ± 6.5	28.5 ± 13.0	4.3 ± 1.9	6.8 ± 3.3	0.148 ± 0.088
CV (%)	50	51	46	45	49	60

† Switchgrass is a lowland switchgrass.
‡ Maize is a locally adapted hybrid.

The other three sites showed how the increased Q of maize and reduced soil cover relative to switchgrass dramatically increased soil erosion of maize. Thus, this type of model provides quantitative values for the expected trends for the two crops. These values can show which soils should be in row crop production if reduced soil erosion is the goal.

The N and P results were highly dependent on the Q and MUSL values and on the nutrient uptake by the particular crop simulated. Fertilizer applications within the model were aimed at preventing nutrient deficiencies with the continuous long-term annual harvests simulated. Annually, applied N values simulated on the two soils were 100 kg N ha^{-1} and 50 kg P ha^{-1} in South Dakota, 200 and 0 in Missouri, and 100 and 0 in Oklahoma and Texas. This allowed direct comparisons of the various production systems within each site. Variables YON and YP are the organic N and P loss in the sediment each year. YNO_3 and YP are the NO_3 and P loss in the surface runoff each year. The two sediment values, YON and YP, showed dramatically greater values for maize than for switchgrass in each case, as expected. Similarly, the two runoff nutrient values (YNO_3 and YAP) were higher for maize than for switchgrass in all cases. Again, as discussed above, these values give quantitative estimates for known phenomena, providing useful quantitative assessments for known environmental impacts on various soils and with various crops.

Concluding remarks

The development of bioenergy systems needs thorough, scientifically defendable analyses on areas required, production stability across wet and dry years, water requirements, nutrient requirements, and impacts on soils and water quality downstream. Process-based models such as the ALMANAC model described herein offer a means to effectively evaluate these complex interacting facets of such systems. This study demonstrated the usefulness of this process-based model for these portions of life cycle analyses. Further work developing parameters to simulate other candidate plant species will prove useful to extend this type of approach. Likewise, further validation of the plant yields on various soils in various rainfall zones will be invaluable as a proof of concept for this approach.

References

Arnold, J.G., R. Srinivasan, R.S. Muttiah, and J.R. Williams. 1998. Large area hydrologic modeling and assessment Part I: Model development 1. J. Am. Water Resour. Assoc. 34:73–89. doi:10.1111/j.1752-1688.1998.tb05961.x

Behrman, K.D., T.H. Keitt, and J.R. Kiniry. 2014a. Modeling a mechanistic theory of differential growth in switchgrass cultivars across the central and southern Great Plains. BioEnergy Res. (in press).

Behrman, K.D., M.N. Meki, Y. Wu, and J.R. Kiniry. 2014b. Applications of biomass production modeling for switchgrass. In: H. Lou, Y. Wu, and C. Kole, editors, Compendium of bioenergy plants: Switchgrass. CRC Press, Boca Raton. p. 356–377 (in press).

Brown, R.A., N.J. Rosenberg, C.J. Hays, W.E. Easterling, and L.O. Mearns. 2000. Potential production and environmental effects of switchgrass and traditional crops under current and greenhouse-altered climate in the central United States: A simulation study. Agric. Ecosyst. Environ. 78:31–47.

Byrd, G.T., and P.A. May. 2000. Physiological comparisons of switchgrass cultivars differing in transpiration efficiency. Crop Sci. 40:1271–1277. doi:10.2135/cropsci2000.4051271x

Del Grosso, S.J., A.R. Mosier, W.J. Parton, and D.S. Ojima. 2005. DAYCENT model analysis of past and contemporary soil N_2O and net greenhouse gas flux for major crops in the USA. Soil Tillage Res. 83:9–24. doi:10.1016/j.still.2005.02.007

Di Vittorio, A.V., R.S. Anderson, J.D. White, N.L. Miller, and S.W. Running. 2010. Development and optimization of an Agro-BGC ecosystem model for C_4 perennial grasses. Ecol. Modell. 221:2038–2053. doi:10.1016/j.ecolmodel.2010.05.013

Eggemeyer, K.D., T. Awada, D.A. Wedin, F.E. Harvey, and X. Zhou. 2006. Ecophysiology of two native invasive woody species and two dominant warm season grasses in the semiarid grasslands of the Nebraska sandhills. Int. J. Plant Sci. 167:991–999. doi:10.1086/505797

Gelfand, I., R. Sahajpal, X. Zhang, R.C. Izaurralde, K.L. Gross, and G.P. Robertson. 2013. Sustainable bioenergy production from marginal lands in the US Midwest. Nature 493:514–520. doi:10.1038/nature11811

Humphries, S.W., and S.P. Long. 1995. WIMOVAC: A software package for modelling the dynamics of plant leaf and canopy photosynthesis. Comput. Appl. Biosci. 11:361–371.

Johnson, M.-V.V., J.R. Kiniry, H. Sanchez, H.W. Polley, and P.A. Fay. 2010. Comparing biomass yields of low-input high-diversity communities with managed monocultures across the central United States. BioEnerg. Res. 3:353–361.

Kiniry, J.R. 1998. Biomass accumulation and radiation use efficiency of honey mesquite and eastern red cedar. Biomass Bioenergy 15:467–473. doi:10.1016/S0961-9534(98)00057-9

Kiniry, J.R., L.C. Anderson, M-V.V. Johnson, K.D. Behrman, M. Brakie, D. Burner, R.L. Cordsiemon, P.A. Fay, F.B. Fritschi, J.H. Houx, III, C. Hawkes, T. Juenger, J. Kaiser, T.H. Keitt, J. Lloyd-Reilley, S. Maher, R. Raper, A. Scott, A. Shadow, C. West, Y. Wu, and L. Zibilske. 2013a. Perennial biomass grasses and the Mason-Dixon Line: Comparative productivity across latitudes in the Southern Great Plains. BioEnergy Res. 6:276–291.

Kiniry, J.R., B.L. Burson, G.W. Evers, J.R. Williams, H. Sanchez, C. Wade, J.W. Featherston, and J. Greenwade. 2007. Coastal bermudagrass, bahiagrass, and native range simulation at diverse sites in Texas. Agron. J. 99:450–461. doi:10.2134/agronj2006.0119

Kiniry, J.R., K.A. Cassida, M.A. Hussey, J.P. Muri, W.R. Ocumpaugh, J.C. Read, R.L. Reed, M.A. Sanderson, B.C. Venuto, and J.R. Williams. 2005. Switchgrass simulation by the ALMANAC model at diverse sites in the southern U.S. Biomass Bioenergy 29:419–425.

Kiniry, J.R., M.-V.V. Johnson, B.C. Venuto, and B.L. Burson. 2013b. Novel application of ALMANAC: Modelling a functional group, "exotic warm-season perennial grasses. Am. J. Exp. Agric. 3(3):631–650. doi:10.9734/AJEA/2013/4313

Kiniry, J.R., L. Lynd, N. Greene, M.V. Johnson, M. Casler, and M.S. Laser. 2008a. Biofuels and water use: Comparison of maize and switchgrass and general perspectives. In: J.H. Wright and D.A. Evans, editors, New research on biofuels. Nova Science Publ., New York. p. 17–30.

Kiniry, J.R., H. Sanchez, J. Greenwade, E. Seidensticker, J.R. Bell, F. Pringle, G. Peacock, Jr., and J. Rives. 2002. Simulating grass productivity on diverse range sites in Texas. J. Soil Water Conserv. 57:144–150.

Kiniry, J.R., M.A. Sanderson, J.R. Williams, C.R. Tischler, M.A. Hussey, W.R. Ocumpaugh, J.C. Read, G. Van Esbroeck, and R.L. Reed. 1996. Simulating Alamo switchgrass with the ALMANAC model. Agron. J. 88:602–606. doi:10.2134/agronj1996.00021962008800040018x

Kiniry, J.R., M.R. Schmer, K.P. Vogel, and R.B. Mitchell. 2008b. Switchgrass biomass simulation at diverse sites in the northern Great Plains of the US. BioEnergy Res. 1:259–264. doi:10.1007/s12155-008-9024-8

Kiniry, J.R., C.R. Tischler, and G.A. Van Esbroeck. 1999. Radiation use efficiency and leaf CO_2 exchange for diverse C_4 grasses. Biomass Bioenergy 17:95–112. doi:10.1016/S0961-9534(99)00036-7

Kiniry, J.R., J.R. Williams, P.W. Gassmann, and P. Debaeke. 1992. General, process-oriented model for two competing plant species. Trans. ASABE 35:801–810. doi:10.13031/2013.28665

Kiniry, J.R., J.R. Williams, R.L. Vanderlip, J.D. Atwood, D.C. Reicosky, J. Mulliken, W.J. Cox, H.J. Mascagni, S.E. Hollinger, and W.J. Wiebold. 1997. Evaluation of two maize models for nine U.S. locations. Agron. J. 89(3):421–426. doi:10.2134/agronj1997.00021962008900030009x

McLaughlin, S.B., J.R. Kiniry, C.M. Taliaferro, and D.D. Ugarte. 2006. Projecting yield and utilization potential of switchgrass as an energy crop. Adv. Agron. 90:267–297. doi:10.1016/S0065-2113(06)90007-8

Meki, M.N., J.D. Atwood, L.M. Norfleet, J.R. Williams, T.J. Gerik, and J.R. Kiniry. 2013. Corn residue removal effects on soybean yield and nitrogen dynamics in the Upper Mississippi River Basin. Agroecology Sustainable Food Syst. 37(3):379–400. doi:10.1080/10440046.2012.724529

Miguez, F.E., M. Maughan, G.A. Bollero, and S.P. Long. 2012. Modeling spatial and dynamic variation in growth, yield, and yield stability of the bioenergy crops *Miscanthus ' giganteus* and *Panicum virgatum* across the conterminous United States. GCB Bioenergy 4(5):509–520. doi:10.1111/j.1757-1707.2011.01150.x

Monsi, M., and T. Saeki. 1953. Uber den lichtfaktor in den pflanzengesellschaften und seine bedeutung fur die 360 stoffproduktion. Jpn. J. Bot. 14:22–52.

Parton, W.J., M. Hartman, D. Oijma, and D. Schimel. 1998. DAYCENT and its land surface submodel: Description and testing. Global Planet. Change 19:35–48.

Perlack, R.D., L.L. Wright, A.F. Turhollow, R.L. Graham, B.J. Stokes, and D.C. Erbach. 2005. Biomass as feedstock for a bioenergy and bioproducts industry: The technical feasibility of a billion-ton annual supply. Tech. Rep. ORNL/TM 2006/66. Oak Ridge National Laboratory, Oak Ridge, TN.

Ritchie, J.T. 1972. Model for predicting evaporation from a row crop with incomplete cover. Water Resour. Res. 8:1204–1213. doi:10.1029/WR008i005p01204

Schilling, K.E., and J.R. Kiniry. 2007. Estimation of evapotranspiration by reed canarygrass using field observations and model simulations. J. Hydrol. 337:356–363. doi:10.1016/j.jhydrol.2007.02.003

U.S. Congress Office of Technology Assessment. 1993. Potential environmental impacts of bioenergy crop production. Background Paper OTA-BP-E-118. U.S. Gov. Print. Office, Washington, DC.

USDA-NASS. 2014. USDA Economics and Statistics System data sets: Crops. USDA-NASS. http://www.nass.usda.gov/Charts_and_Maps/Field_Crops/ (accessed 6 Mar. 2014).

USDA-NRCS. 2010. Conservation Effects Assessment Project (CEAP). Assessment of the effects of conservation practices on cultivated cropland in the Upper Mississippi River Basin. USDA-NRCS. http://www.nrcs.usda.gov/technical/nri/ceap/ (accessed 6 Mar. 2014).

Williams, J.R., C.A. Jones, and P.T. Dyke. 1984. A modeling approach to determining the relationship between erosion and soil productivity. Trans. ASABE 27:129–144. doi:10.13031/2013.32748

Williams, J.R., C.A. Jones, J.R. Kiniry, and D.A. Spanel. 1989. The EPIC crop growth model. Trans. ASABE 32:497–511. doi:10.13031/2013.31032

Woli, P., J.O. Paz, D.J. Lang, B.S. Baldwin, and J.R. Kiniry. 2012. Soil and variety effects on the energy and carbon balances of switchgrass-derived ethanol. J. Sustainable Bioenergy Syst. 2(4):65–74. doi:10.4236/jsbs.2012.24010

Wullschleger, S.D., E.B. Davis, M.E. Borsuk, C.A. Gunderson, and L.R. Lynd. 2010. Biomass production in switchgrass across the United States: Database description and determinants of yield. Agron. J. 102:1158–1168. doi:10.2134/agronj2010.0087

Xie, Y., J.R. Kiniry, V. Nedbalek, and W.D. Rosenthal. 2001. Maize and sorghum simulations with CERES-Maize, SORKAM, and ALMANAC under water-limiting conditions. Agron. J. 93:1148–1155. doi:10.2134/agronj2001.9351148x

Xu, B., F. Li, L. Shan, Y. Ma, N. Ichizen, and J. Huang. 2006. Gas exchange, biomass partition, and water relationships of three grass seedlings under water stress. Weed Biol. Manage. 6:79–88. doi:10.1111/j.1445-6664.2006.00197.x

Zilverberg, C.J., W.C. Johnson, V. Owens, A. Boe, T. Schumacher, K. Reitsma, C.O. Hong, C. Novotny, M. Volke, and B. Werner. 2014. Biomass yield from planted mixtures and monocultures of native prairie vegetation across a heterogeneous farm landscape. Agric. Ecosyst. Environ. 186:148–159. doi:10.1016/j.agee.2014.01.027

Using a Model and Forecasted Weather to Predict Forage and Livestock Production for Making Stocking Decisions in the Coming Growing Season

Quan X. Fang,* L.R. Ahuja, Allan A. Andales, and Justin D. Derner

Abstract

Forecasting peak standing crop (PSC) for the coming grazing season can help ranchers make appropriate stocking decisions to reduce enterprise risks. Previously developed PSC predictors were based on short-term experimental data (<15 yr) and limited stocking rates (SR) without including the effect of SR on PSC explicitly. Here we used long-term (30 yr) measured data of PSC and steer weight gain (SWG), extended with the help of a model for SR effect, to develop multiple-variable regression functions for predicting PSC and SWG across a wide range of SR (0.2–1.32 steers ha^{-1} for summer grazing season, June to mid-October) on a loam soil in a northern mixed-grass prairie. April to June rainfall was the primary weather variable influencing PSC (R^2 = 0.45); inclusion of SR and soil water content on 1 April improved the accuracy in predicting PSC (R^2 = 0.64). Combining the response of PSC to SR and the response of SWG to both PSC and SR enables ranchers to explore tradeoffs between economic net return and environmental impact (land conservation) as influenced by SR and weather variations. The result was further extended from the loam soil at the experimental site to the other two soil types (loam sandy and clay loam soils) by using a simple soil influence factor. A simple spreadsheet-based decision support tool can be developed to facilitate stocking decisions by ranchers in a northern mixed-grass prairie to adaptively manage rangelands in an effort to increase economic net return and reduce land degradation associated with high weather variability and SR levels.

Abbreviations: DC, Drought Calculator; HPGRS, High Plains Grasslands Research Station; HPPSC, peak standing crop; GPFARM, Great Plains Framework for Agricultural Resource Management; PSC, peak standing crop; SR, stocking rate; SWG, steer weight gain; TDN, total intake digestible nutrient.

Quan X. Fang Agronomy College, Qingdao Agricultural University, Changcheng Road 700, Chengyang District, Qingdao, Shandong, China, 266108. *Corresponding author (fangqx@igsnrr.ac.cn)

L.R. Ahuja, USDA-ARS-NPA-ASRU 2150 Centre Ave., Bldg. D, Ste. 200 Fort Collins, CO, 80526 (laj.ahuja@ars.usda.gov)

A.A. Andales, Department of Soil and Crop Sciences, Colorado State University, 1170 Campus Delivery, Fort Collins, CO 80523-1170 (Allan.Andales@colostate.edu).

J.D. Derner, USDA-ARS, High Plains Grasslands Research Station, 8408 Hildreth Road, Cheyenne, WY 82009-8899 (Justin.Derner@ars.usda.gov)

doi:10.2134/advagricsystmodel5.c7

Practical Applications of Agricultural System Models to Optimize the Use of Limited Water
Lajpat R. Ahuja, Liwang Ma, and Robert J. Lascano, Editors
Advances in Agricultural Systems Modeling, Volume 5. Lajpat R. Ahuja, Series Editor

USDA-NRCS

Ranchers need to make stocking decisions for the upcoming grazing season before the initiation of forage growth. Weather variability (e.g., rainfall) is the most important factor influencing forage production in semi-arid areas (e.g., Milchunas et al., 1994; Biondini et al., 1998; Derner and Hart, 2007). Relationships between forage production and seasonal or spring (April–June) precipitation have been well established for many rangelands, such as the northern mixed-grass prairie of the United States (Currie and Peterson, 1966; Lauenroth and Sala 1992; Derner and Hart, 2007; Derner et al., 2008b, Smart et al., 2007; Wiles et al., 2011) and the central grassland region of the United States (Sala et al., 1988). Forecasting of spring rainfall, PSC, and cattle weight gain before the grazing season would facilitate better decision-making for reducing enterprise risk for ranchers.

In addition to spring precipitation, soil water content at the beginning of the growing season increases the robustness of predictions of PSC (Briggs and Knapp, 1995; Andales et al., 2006; Torell et al., 2011). Dahl (1963) found that soil water and depth of moist soil on 15 April were useful indices for predicting grass production by early August in rangelands of eastern Colorado. Torell et al. (2011) used measured soil water to improve predictions of forage production in rangelands of New Mexico. Andales et al. (2006) used the Great Plains Framework for Agricultural Resource Management (GPFARM)-Range model with different initial soil water levels in early April to produce improved response functions of PSC to spring precipitation in the northern mixed-grass prairie.

Stocking decisions by ranchers are dependent on PSC (Wiles et al., 2011; Dunn et al., 2013) as forage availability limits cattle intake and weight gains (e.g., Bement 1969; Redmon et al., 1995; Poppi, 1996). The aforementioned studies, however, did not account for the effect of SR on PSC and cattle weight gain. Recent studies have showed that high SR can reduce PSC, such as in the northern mixed-grass prairie of the United States (Derner and Hart, 2007), and the response of PSC or livestock weight gain to weather variables differs with different grazing levels (Reeves et al., 2013a, 2013b, 2014). In this context, integrating the effect of SR and weather variables on PSC can potentially improve the predictions of PSC compared with predictions based on the weather variables alone, thus improve stocking decisions. Additionally, the cattle weight gains were not quantified as functions of PSC and SR in the previous stocking decision support tools. Thus, this precludes economic risk analyses at the different SR levels. Additionally, land degradation due to overgrazing with high SR should also be accounted for in making SR decisions (Smart et al., 2010).

In this study, we used PSC and SWG data from a long-term (1982–2012) experiment to create model simulated data sets with much higher (50–150%) SR levels than the experimental grazing levels (Derner and Hart 2007; Derner et al., 2008a). We then used the extended dataset to (i) obtain better PSC predictions as a function of spring rainfall, soil water before grazing seasons, and SR and (ii) combine the SR effects on SWG and land degradation across different soil types in the PSC predictor and help ranchers to make better SR decisions considering both net return and land conservations.

Materials and Methods

Peak Standing Crop and Steer Weight Gain from Long-Term Field Experiments

Peak standing crop was measured annually in mid to late July from a long-term grazing experiment on the semiarid northern mixed-grass prairie at the High Plains Grasslands Research Station (HPGRS) in Cheyenne, WY (41°11′ N, 104°53′ W) (Derner and Hart, 2007; Derner et al., 2008a). Mean annual precipitation at the site is 381 mm with peaks of precipitation in April, May, and June. The soils are medium textured and well drained, largely composed of Albinas, Asclon, and Altvan loams and Cascajo gravelly loam (Stevenson et al., 1984).

Season-long (June to mid-October. 4.5 mo), continuous grazing treatments were initiated in 1982 with the following three SR levels: light [8 steers per 40 ha (0.20 steer ha^{-1}), about 35% below the USDA–NRCS recommended rate], moderate [8 steers per 24 ha (0.33 steer ha^{-1}), the USDA–NRCS recommended rate], and heavy (8 steers per 18 ha (0.44 steer ha^{-1}), 33% above the USDA–NRCS recommended rate) (Hart et al., 1988). Yearling steers were used as grazing animals, and each steer was weighed before and after the season and all experimental procedures were approved by the HPGRS Animal Care and Use Committee.

Peak Standing Crop and Steer Weight Gain from GPFARM-Range Model Simulations

The forage and cattle modules of GPFARM-Range are simplified versions of the Simulating Production and Utilization of Rangeland (SPUR) model (Hanson et al., 1988, 1992). Model details, including functions and equations for both the forage and cattle modules, have been published previously (Andales et al., 2005, 2006). Recently, the GPFARM-Range model was improved by considering the SR effect on PSC and steer weight and the revised model showed a reasonable response of

PSC and SWG to different grazing levels across long-term weather conditions as shown in Fig. 7–1 (Fang et al., 2014).

The improved model was used to simulate PSC and SWG for four SR greater than the above experimental values, in addition to the three experimental SR treatments, from 1982 to 2012. The six higher SR levels were 50 to 150% greater than the highest experimental SR of 0.44 steer ha^{-1}, including (i) 0.66 steer ha^{-1} (50% greater), (ii) 0.88 steer ha^{-1} (100% greater), and (iii) 1.10 steer ha^{-1} (150% greater). See Fang et al. (2014) for more detailed information.

Multiple Regression Analysis and Economic Analysis

On the basis of previous studies on the relationships between PSC and weather variables (e.g., Eneboe et al., 2002; Derner and Hart, 2007; Patton et al., 2007; Smart et al., 2007; Torell et al., 2011; Wiles et al., 2011), the weather variables (precipitation), SR, and soil water on 1 April were used for predicting both experimentally

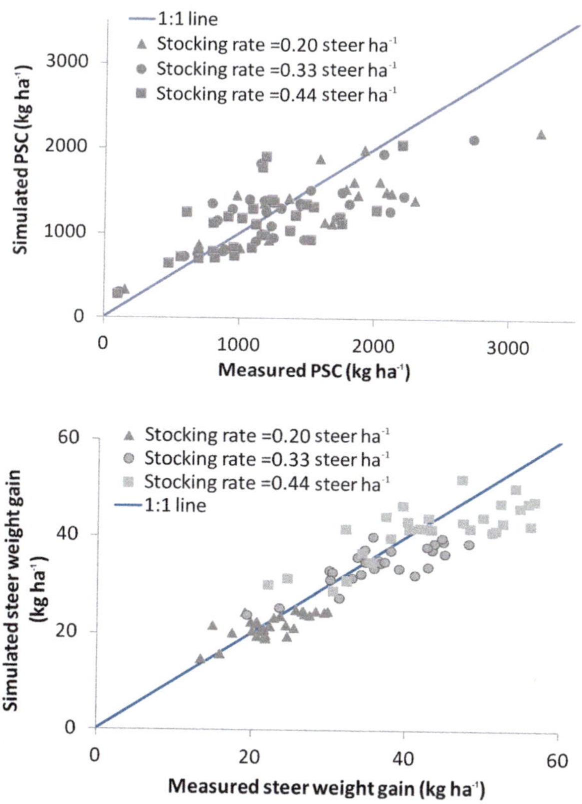

Fig. 7–1. GPFARM-Range model evaluations on simulating peak standing crop (PSC) and steer weight gain under the three stocking rate experimental treatments from 1982 to 2012 (detailed information can be found Fang et al., 2014).

observed PSC data and GPFARM model simulated PSC. Multivariate stepwise regression was used to evaluate the contribution of each of the above variables for predicting PSC. The contribution of precipitation was evaluated on a trimonthly basis from January to September according to the abovementioned studies in the region. Besides PSC, the dependent variables of harvest efficiency and SWG were also predicted on the basis of PSC and SR by using the multivariate stepwise regression method, where interactions between PSC and SR were included. The harvest efficiency, defined as the ratio of forage intake to forage production (Smart et al., 2010), reflects the grazing pressure and can be used as an important indicator for SR decisions (Galt et al., 2000). A stepwise procedure of RSREG (Response Surface Regression) in SAS (Statistical Analysis System) (Freund and Littell, 1991) was used for the variable selections ($P < 0.01$), which combines forward selection and backward elimination steps for variable selection.

Because the model does not simulate the carrying costs (such as supplement cost, salt, implants, and transportation) of a grazing enterprise, a simple method was used to estimate the economic profit based on previous studies in the region (Hart et al., 1988; Manley et al., 1997). The purchase price in March and selling price in October for the simulation period (1982–2012) in the United States were obtained from USDA–NASS (2014), and we estimated carrying costs at $US60 per steer. Predicted SWG values from the different SR levels were used to estimate economic net returns each year.

Extending the Regression Equations to Other Soils

We simulated PSC and SWG using the seven SR levels for the other two different soil types (loam sandy soil and clay loam soil) from 1982 to 2012. These simulated data were used to estimate effects of soil types on the responses of PSC and SWG to SR under the same climate conditions. To keep this tool simple, we used a simple soil factor to extend the regression equation from a loam soil (experimental site soil) to the other two soil types (clay loam soil and loam sandy soils). The soil factor was obtained based on the relationships of PSC or SWG from 1982 to 2012 between the different soils.

Results and Discussion

Identifying the Influential Factors for Yearly Variability of Peak Standing Crop or Steer Weight Gain

For predicting both observed and model simulated PSC, the trimonthly rainfall amounts from January to March or from July to September were not significant for predicting PSC across the different SR levels (Table 7–1). The April to June

Table 7–1. The most influential variables selected ($P < 0.01$) with partial R^2 values based on the multivariate regressions for both observed peak standing crop (MPSC) or steer weight gain (OSWG) and GPFARM-Range model predicted PSC (SPSC) and weight gain (SSWG) from 1982 to 2012 under different stocking rates (steer ha^{-1}) for a 4.5-mo grazing season.†

	Stocking rate for experiment				Stocking rate for long-term simulations							
Variable	0.20	0.33	0.44	Total	0.20	0.33	0.44	0.66	0.88	1.10	1.32	Total
	steer ha^{-1}				steer ha^{-1}							
	MPSC				SPSC							
$Rain_{123}$	no	no	no	no	no	no	no	no	no	no	no	no
$Rain_{456}$	0.39	0.43	0.43	0.38	0.64	0.57	0.54	0.47	0.50	0.51	0.52	0.45
$Rain_{789}$	no	no	no	no	no	no	no	no	no	no	no	no
SW	0.03	0.05	0.02	0.03	0.05	0.05	0.07	0.08	0.06	0.04	0.06	0.05
SR	–	–	–	0.08	–	–	–	–	–	–		0.14
	MSWG				SSWG							
PSC	no	no	no	no	no	no	no	no	0.79	0.74	0.68	0.04
TDN	–	–	–	–	0.87	0.85	0.81	0.74	0.09	0.08	0.08	0.37
SR	–	–	–	0.59	–	–	–	–	–	–	–	0.15

† Climatic variables: $rain_{123}$, total rainfall (mm) from January to March; $rain_{456}$, total rainfall (mm) from April to June; $rain_{789}$, total rainfall (mm) from July to September. Other variables: SW, soil water content on 1 April; SR, stocking rate; PSC, peak standing crop; TDN, total intake digestible nutrient. The word "no" means this variable was not included at $P < 0.01$ level, and "–" means this variable was not used for regression analysis.

(spring) rainfall was identified as the most important factor ($P < 0.01$, partial R^2 from 0.39 to 0.64) to both observed PSC and model simulated PSC, respectively (Table 7–1). Similar results were found with field experiments in this rangeland ecosystem (Derner and Hart, 2007; Smart et al., 2007; Derner et al., 2008b; Wiles et al., 2011). The effect of SR on PSC was significant ($P < 0.01$) but with a partial R^2 of 0.08 (observed PSC) and 0.15 (simulated PSC) (Table 7–1). Soil water affected PSC for both experimental data and model simulation results, agreeing with prior findings (Andales et al., 2006; Torell et al., 2011). Simulated soil water content on 1 April influenced PSC ($P < 0.01$) (Table 7–1). A significant positive relationship was obtained between the GPFARM simulated soil water content on 1 April with the observed PSC ($R^2 = 0.08$, $n = 93$; $P < 0.001$) or the simulated PSC ($R^2 = 0.14$, $n = 279$; $P < 0.001$) across the SR levels. This result showcases that including soil water content on 1 April improves the PSC predictions over those based on spring rainfall only. We also found a significant ($P < 0.001$) linear relationship between soil water content on 1 April and the precipitation in March of the current year and December of the previous year, which explained 41 and 18% of the variation in soil water content on 1 April across seasons, respectively (Eq. [4] in Table 7–2), and other variables of rainfall in January, October, and November of the previous year were also significant but explained less variations in soil water content on 1 April. This regression equation based on previous precipitation can be used to estimate

Table 7–2. Regression equations for GPFARM-Range simulated peak standing crop when using rainfall, stocking rate, and GPFARM-Range simulated soil water (0–50 cm) on 1 April and for GPFARM-Range model predicted forage harvest efficiency and steer weight gain using peak standing crop and stocking rate (Fig. 7–4).†

Number	Variable	Equations	Statistic results
[1]	PSC	$4.16 \times rain_{456} + 358.41$	RMSE = 358, $R^2 = 0.45$, $N = 217$
[2]	PSC	$620.36 + 4.16 \times rain_{456} - 370.04 \times SR$	RMSE = 246, $R^2 = 0.588$, $N = 217$
[3]	PSC	$-370.69 + 3.85 \times rain_{456} - 363.17 \times SR + 5800.05 \times SWP$	RMSE = 231, $R^2 = 0.637$, $N = 217$
[4]	SWP	$0.14483 - 5.61e\text{-}4 \times rain_1 + 4.14e\text{-}4 \times rain_3 + 5.79e\text{-}4 \times prain_{12} + 2.55e\text{-}4 \times prain_{10} + 2.71e\text{-}4 \times prain_{11}$	RMSE = 0.0089, $R^2 = 0.647$, $N = 217$
[5]	HE	$0.47820 - 0.00042531 \times PSC + 0.94588 \times SR$	RMSE = 0.17, $R^2 = 0.87$, $N = 217$
[8]	HE	$0.00831 + 1.63 \times SR - 6.71e\text{-}4 \times SR \times PSC$	RMSE = 0.15, $R^2 = 0.91$, $N = 217$
[9]	SWG	$-16.54 + 0.04 \times PSC + 41.15 \times SR - 1.95e5 \times PSC^2 - 45.60 \times SR^2 + 0.06 \times SR \times PSC$	RMSE = 9.1, $R^2 = 0.84$, $N = 217$

† Climatic variables: $rain_{456}$, rainfall from April to June; $rain_1$, rainfall in January; $rain_3$, rainfall in March; $prain_{10}$, monthly rainfall during October in the previous year; $prain_{11}$, monthly rainfall during November in the previous year; $prain_{12}$, monthly rainfall during December in the previous year. Other variables: PSC, peak standing crop (kg ha^{-1}); SR, stocking rate (steer ha^{-1}); SWP, GPFARM simulated soil water ($cm^3\ cm^{-3}$); HE, harvest efficiency; SWG, steer weight gain (kg ha^{-1}).

soil water content on 1 April (before forage growth begins) when the observed value is not available.

Peak standing crop was not identified as a significant ($P > 0.19$) influential factor for predicting observed SWG for the experimental SR levels (0.20–0.44 steer ha^{-1}) (Table 7–1), and similar results were also obtained for the simulated SWG. Under higher SR levels from 0.88 to 1.32 steer ha^{-1}, however, PSC became the most limiting factor for the simulated SWG with a partial R^2 of 0.79 to 0.68. Considering all the SR treatments, SR was the most important factor to limiting SWG with a partial R^2 of 0.59, and this agrees with prior research (Reeves et al., 2013b). For the simulated SWG data, however, PSC, total intake digestible nutrient (TDN) and SR were all significant in predicting SWG, and a linear relationship between PSC and TDN was also found ($R^2 = 0.50$) for these simulated data.

Improving Peak Standing Crop Predictions by Including Soil Water and Stocking Rate

Extending the PSC predictions to a wider range of weather and grazing conditions is needed to assist in decision-making for ranchers. The PSC predictions from the regression equations based on the spring rainfall (Fig. 7–2a), spring rainfall plus SR (Fig. 7–2b), or spring rainfall plus SR and predicted soil water on 1

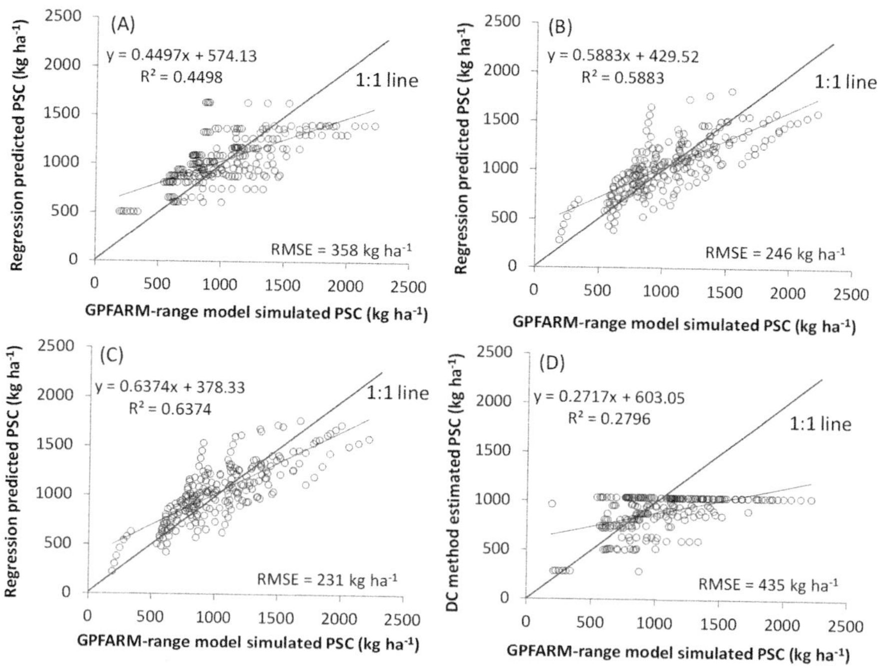

Fig. 7–2. Comparing the performances in predicting peak standing crop (PSC) for the developed regression equations based on (a) April to June rainfall (Eq. [1] in Table 7–2), (b) April to June rainfall and stocking rate (Eq. [2] in Table 7–2), (c) April to June rainfall, stocking rate, and soil water content on 1 April (Eq. [3] in Table 7–2), and (d) the Drought Calculator (DC method [Dunn et al., 2013]), with PSC values simulated by the GPFARM-Range model. The up-bound in estimated PSC by the DC method (d) is the average simulated PSC level from 1982 to 2012.

April (Fig. 7–2c) were compared with the earlier Drought Calculator (DC) method (Dunn et al., 2013) (Fig. 7–2d). As expected, using variables of spring rainfall and SR produced better PSC predictions (RMSE = 246 kg ha^{-1}; $R^2 = 0.59$) than the PSC predictions using spring rainfall only (RMSE = 358 kg ha^{-1}; $R^2 = 0.45$). Further, by including soil water on 1 April, PSC predictions were improved with an RMSE value of 231 kg ha^{-1} and R^2 of 0.64. For the DC method (Dunn et al., 2013), a better performance in predicting PSC was obtained when the PSC level was below 1000 kg ha^{-1} because of the drought stresses, and obvious under-predictions occurred when PSC level was higher than 1000 kg ha^{-1} when there was no or little drought stress (Fig. 7–2d). The regression equations improved the predictions, especially at high PSC level with little or no drought stresses, compared with the predictions from DC method (RMSE = 435 kg ha^{-1}).

Next, we compared predicted PSC (regression equations in Table 7–2 and DC method) and measured data under the three SR levels (Fig. 7–3). Both regression equations and DC methods predicted PSC well when measured PSC

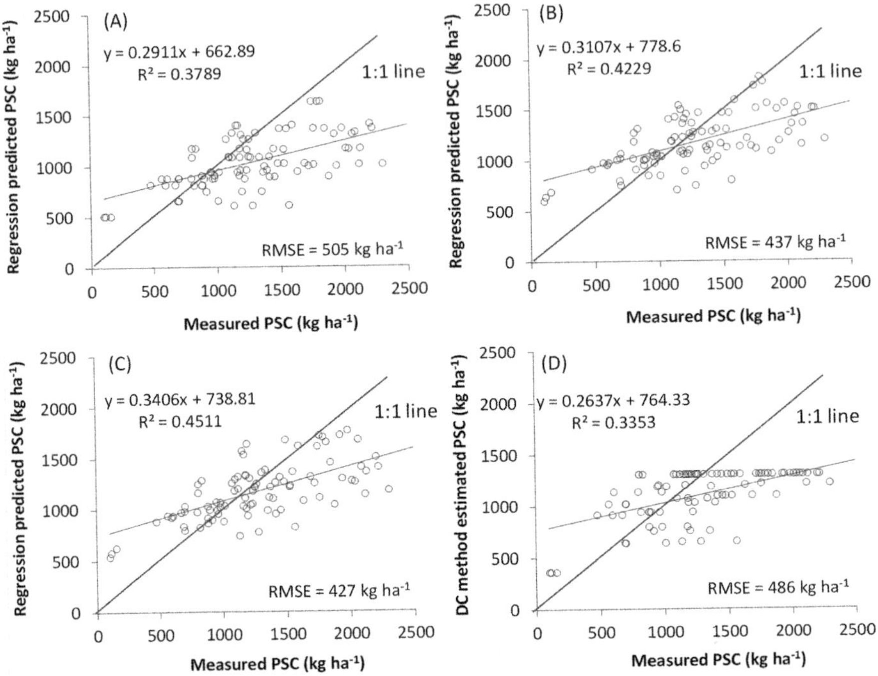

Fig. 7–3. Comparing the performances in predicting peak standing crop (PSC) for the developed regression equations based on (a) April to June rainfall (Eq. [1] in Table 7–2), (b) April to June rainfall and stocking rate (Eq. [2] in Table 7–2), (c) April to June rainfall, stocking rate, and soil water content on 1 April (Eq. [3] in Table 7–2), and (d) the Drought Calculator (DC method [Dunn et al., 2013]), with the experimental measured PSC. The up-bound in estimated PSC by the DC method (d) is the average simulated PSC level from 1982 to 2012.

was below average levels, but under-predicted PSC occurred when PSC values were above average. The regression equation based on GPFARM simulated PSC under-predicted PSC because of the GPFARM under-simulated PSC compared with measured data (Fig. 7–1). Slightly better predictions of PSC were obtained from the regression equation than from the DC method with lower RMSE values (Fig. 7–3). The inclusion of the SR effect on PSC in the regression equations was demonstrated from a field experiment study (Derner and Hart, 2007). The developed regression equations produced a comparable performance in predicting PSC to the DC method and can be applied to predict PSC across a wider range of SR and weather conditions.

Extending to Forecast Harvest Efficiency and Steer Weight Gain Using Stocking Rate and Peak Standing Crop

To assist ranchers with stocking decisions, other indicators, such as forage harvest efficiency (total forage intake/PSC as defined by Smart et al. [2010]) and SWG, are needed to account for SR effect on economic profits (SWG) and environmental impacts (land conservation). On the basis of the long-term simulations with a wider range of SR levels, the regression equations for simulated harvest efficiency or SWG based on the variables of PSC and SR are presented in Table 7–2. For predicting SWG or harvest efficiency, interactions between PSC and SR were significant ($P < 0.01$) and were included in the regression equations. This result was consistent with previous experimental studies (Derner and Hart, 2007; Smart et al., 2010). These regression equations showed good performances in predicting forage harvest efficiency (Fig. 7–4a) and SWG (Fig. 7–4b) across these seasons, with RMSE values of 0.18 and 9.3 kg ha^{-1}, respectively, which can help ranchers predict harvest efficiency and SWG based on the PSC and SR levels and make better SR decisions based on the SR effects on land degradation (low PSC level associated with high SR levels) and economic net return.

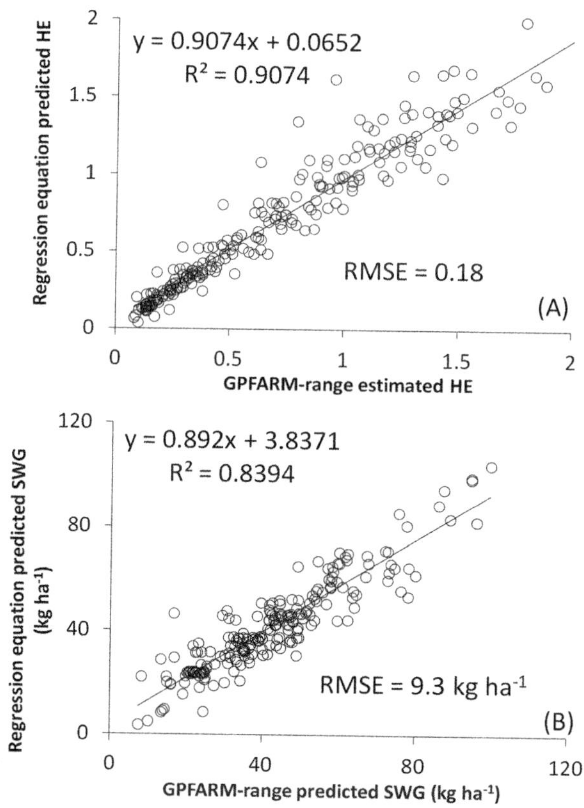

Fig. 7–4. Comparisons between regression equation predicted forage harvest efficiency (HE, Eq. [5] in Table 7–2) and steer weight gain (SWG, kg ha^{-1}, Eq. [6] in Table 7–2) with GPFARM-Range model simulated data from 1982 to 2012.

The regression equation developed for SWG in Table 7–2 (Eq. [9]) is based on simulated data for all seven SR and thus includes the effect of PSC at all SR levels. On the other hand, there was no significant relationship between SWG and PSC at the three lower SR used in the experiment for both measured and simulated data (Table 7–1), indicating that the PSC was not a limiting factor to SWG under the low SR conditions. We explored if Eq. [9] supports this effect. As shown in Fig. 7–5a, the response of the predicted SWG to different levels of PSC by the regression equation, Eq. [9], was small and apparently insignificant under the low SR level (0.20 steer ha^{-1}) compared with the high SR level (0.60 steer ha^{-1}). This result was consistent with the measured and simulated data, where no significant influence of PSC on SWG at the low SR levels was found (Table 7–1). We further compared the predicted SWG from the regression equation, Eq. [9], based on the simulated data and another equation based only on experimental measured data as a function of SR levels (Fig. 7–5b). Both equations predicted very similar SWG at the low SR level. This result indicated that the regression equation based on the simulated data (Eq. [9] in Table 7–2) was effective in predicting SWG even under the low SR levels. While at higher SR levels, the regression equation based the experimental data tended to oversimulate SWG because it did not consider the reductions in PSC caused by high SR. Therefore, the regression equation based on the simulated data with a wider range of SR was selected since it reflects the experimental data.

Extending the Regression Equations to Other Soil Types

We first compared the GPFARM simulated PSC and SWG for the different SR levels for the three soils, for example, loam soil (experimental field soil), loam sandy soil, and clay loam soil, as shown in Fig. 7–6. For the predicted PSC from

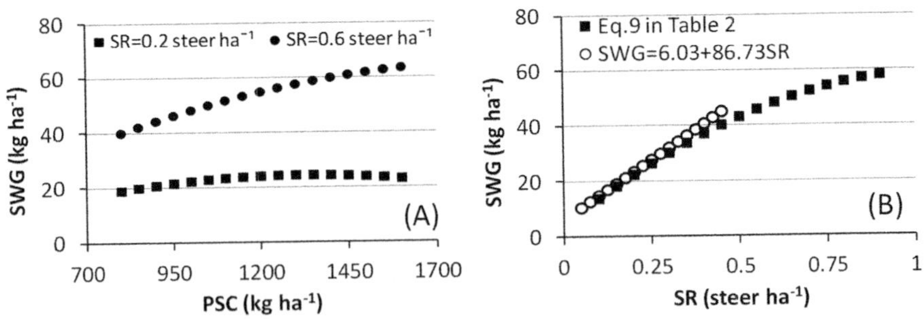

Fig. 7–5. Comparisons of (a) the response of steer weight gain (SWG) with peak standing crop (PSC) between the low (0.20 steer ha^{-1}) and high (0.60 steer ha^{-1}) stocking rates (SR) based on the regression Eq. [9] in Table 7–2 and (b) the response of SWG to SR between the regression equation based on the simulated data (Eq. [9] in Table 7–2) and the experimental data (SWG = 6.03 + 86.73 × SR based on Table 7–1).

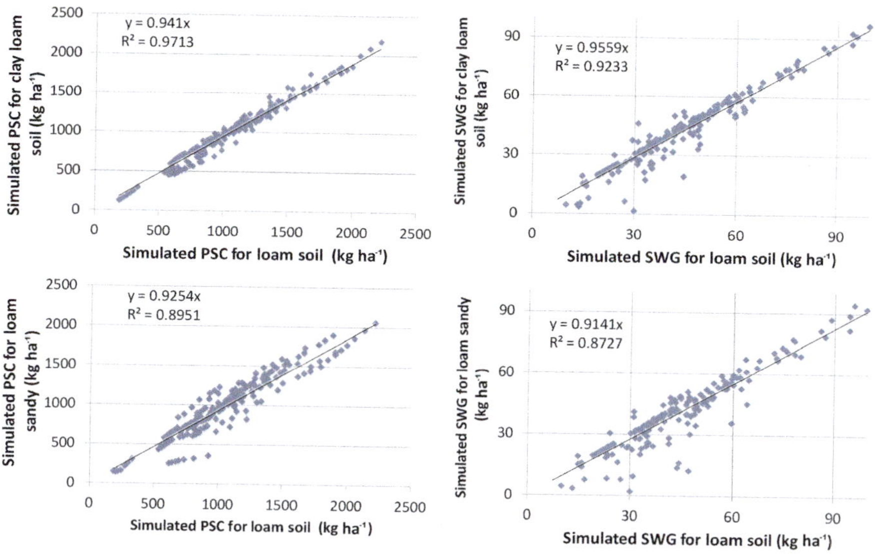

Fig. 7–6. Comparisons of GPFARM-Range simulated long-term (1982–2012) seasonal peak standing crop (PSC) and steer weight gain (SWG) on the loam soil (experimental field soil) to the clay loam or loam sandy soils.

1982 to 2012, strong linear relationships between loam and clay loam soils or between loam and loam sandy soils were found with R^2 values of 0.97 or 0.90, respectively. The long-term predicted average PSC were slightly lower for clay loam soil (slope = 0.94) and loam sandy (slope = 0.93) soils than for loam soils. The SR level showed little or no effect on abovementioned relationships between these different soils. The abovementioned relationships in PSC among these different soil types suggested that the PSC is lower for loam sandy and clay loam soils than for loam soil, and a simple soil influence factor on PSC (the slopes for the relationships) can be used to extend the regression equation of PSC for loam soil to clay loam soil (slope = 0.94) and loam sand soil (slope = 0.93). Such an extension method was very simple and was easier for ranchers to adopt.

For the predicted SWG from 1982 to 2012, linear relationships between loam and clay loam soils or between loam and loam sandy soils were also found with R^2 values of 0.92 or 0.87, respectively. The long-term predicted SWG was slightly lower for clay loam (slope = 0.96) or loam sandy (slope = 0.91) soils than for loam soils, which suggested that different soil factors should be used for the different SR levels. Specifically, when SR levels were below 0.88 steer ha^{-1}, the slopes for the SWG relationship between loam and loam sandy soils are similar (0.98–0.97), and one soil influence factor of 0.97 was used for the SR levels. When SR levels were between 1.1 and 1.32 steer ha^{-1}, the soil influence factor of 0.94 was used.

Similarly, when extending the results from loam soil to loam sandy soil, the soil influence factor of 0.95 and 0.90 was used for the SR levels below 0.88 steer ha^{-1} and above 0.88 steer ha^{-1}, respectively (Table 7–3).The GPFARM-Range predicted PSC or SWG and the estimated data from equations based on above soil influence factor were very close with a strong linear relationship (slope values between 0.98 and 1, and R^2 values between 0.87 and 0.97). This suggests that it was reasonable to extend the regression equations from the loam soil to other soils (such as clay loam and loam sandy soils) using the soil influence factor.

Better Stocking Decisions Based on Peak Standing Crop and Steer Weight Gain and Economic Analysis

On the basis of the regression equations for PSC and SWG (Table 7–2), ranchers can forecast their PSC and SWG before the grazing season, with the National Weather Service forecast of rainfall from April to June and soil water on 1 April (measured or estimated from previous rainfall based on regression Eq. [4] in Table 7–2). The forage harvest efficiency can also be predicted on the basis of the predicted PSC and SR levels (Eq. [5] in Table 7–2). These regression equations can be developed into a spreadsheet-based decision support tool and help ranchers explore the effects of SR and weather on PSC and SWG.

In Fig. 7–7, the responses of regression predicated PSC, harvest efficiency, SWG, and economic profits to SR were compared for three different weather conditions [normal seasons with average spring rainfall (160 mm), dry seasons with 25% below normal rainfall (120 mm), and wet seasons with 25% above normal rainfall (200 mm)]. As expected, PSC decreased with increasing SR in all cases. Harvest efficiency, however, increased with increasing SR, indicating

Table 7–3. The linear relationships between GPFARM-Range model simulated steer weight gain from 1982 to 2012 between loam soil and clay loam or loam sandy soil across these stocking rates.†

Stocking rate	Clay loam soil		Loam sandy soil	
steer ha^{-1}				
0.20	$y = 0.99x$	$R^2 = 0.84$	$y = 0.98x$	$R^2 = 0.70$
0.33	$y = 0.98x$	$R^2 = 0.82$	$y = 0.96x$	$R^2 = 0.60$
0.44	$y = 0.98x$	$R^2 = 0.88$	$y = 0.95x$	$R^2 = 0.70$
0.66	$y = 0.98x$	$R^2 = 0.93$	$y = 0.94x$	$R^2 = 0.68$
0.88	$y = 0.97x$	$R^2 = 0.93$	$y = 0.95x$	$R^2 = 0.87$
1.10	$y = 0.95x$	$R^2 = 0.92$	$y = 0.91x$	$R^2 = 0.83$
1.32	$y = 0.94x$	$R^2 = 0.85$	$y = 0.89x$	$R^2 = 0.81$
0.2–0.88	$y = 0.97x$	$R^2 = 0.96$	$y = 0.95x$	$R^2 = 0.88$
1.10–1.32	$y = 0.94x$	$R^2 = 0.87$	$y = 0.90x$	$R^2 = 0.82$

† The slopes for these linear relationships were used as scaling factor to extend to these different soil types under different stocking rate levels.

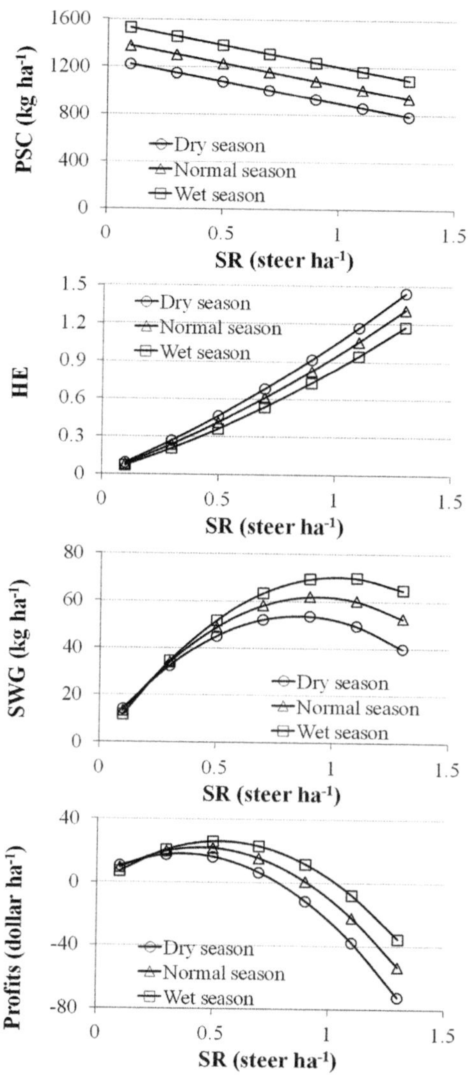

Fig. 7–7. Responses of peaking standing crop (PSC), harvest efficiency (HE), steer weight gain (SWG), and economic profits to stocking rate (SR, steer ha^{-1}) under dry (25% below average April–June rainfall, 120 mm), normal (160 mm), and wet (25% above average April–June rainfall, 200 mm) seasons. (The PSC, HE, and SWG were predicted from the equations in Table 7–2, where the soil water content was used as average value 0.18 cm^3 cm^{-3} from 1982 to 2012. The economic profits were estimated on the basis of the predicted SWG, cattle buy or sale price, and carry cost (Hart et al., 1988).

overgrazing at the high SR level (harvest efficiency > 1 indicates greater forage intake than forage production). The SWG increased quadratically with SR and showed higher values for wet seasons than for dry and normal seasons; the economic profits responded similarly to SR. In general, the biophysical optimum SR levels (that gave the highest SWG) generally increased from dry seasons to wet seasons. However, the SR values that yield the optimum economic returns were much lower than the biophysical optimums, which are associated with high harvest efficiency but also high risk of land degradation. Thus, the practical SR levels between 0.33 and 0.44 steer ha^{-1} are lower than the biophysical optimum SR and close to the experimental SR levels, avoiding negative economic profits and land degradation (overgrazing), but they can be higher for wet seasons than for dry seasons.

The soil water on 1 April showed positive influences on PSC and SWG and economic profits (Fig. 7–8). High soil water on 1 April produced higher PSC with lower harvest efficiency and produced higher SWG and economic profits than lower soil water on 1 April. The soil water effect on SWG and economic profits was pronounced with increased SR levels. The biophysical or economic optimum

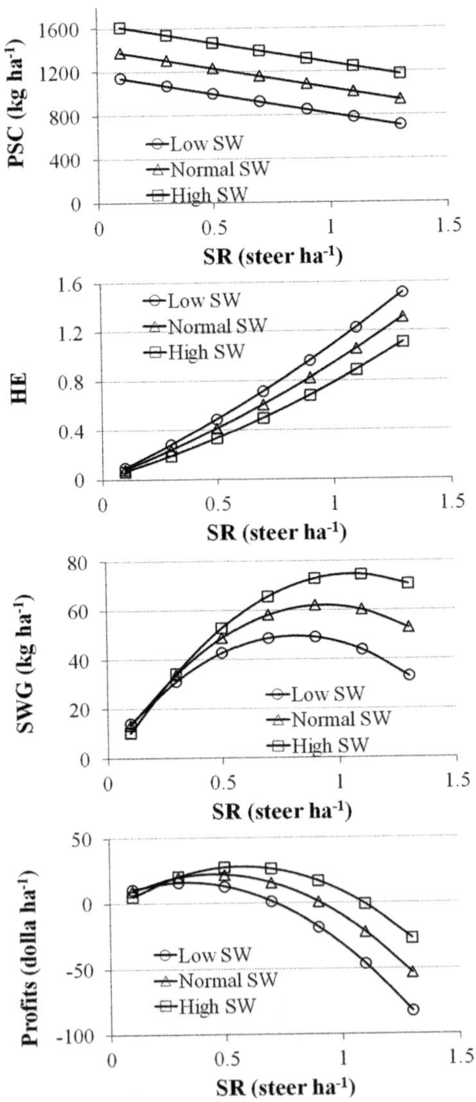

Fig. 7–8. Responses of regression predicted peaking standing crop (PSC), harvest efficiency (HE), steer weight gain per area (SWG), and economic profits to stocking rate (SR, steer ha^{-1}) as influenced by soil water conditions on 1 April (SW) (lower SW = 0.15 cm^3 cm^{-3}; normal SW = 0.18 cm^3 cm^{-3}, and high SW = 0.21 cm^3 cm^{-3}) under normal April to June rainfall conditions (160 mm). The economic profits were estimated on the basis of the predicted SWG, cattle buy or sale price, and carry cost (Hart et al., 1988).

SR levels increased from low to high soil water content on 1 April, suggesting that higher practical SR can be applied under higher soil water on 1 April than under lower soil water on 1 April.

Another influential factor to SR selections was the economic factor, such as cattle purchase or selling price. As shown in Fig. 7–9, the response of economic profits to SR as influenced by the cattle price showed that optimum SR levels were obtained at 0.3 to 0.5 steer ha^{-1} for a favorable price year and 0.1 to 0.3 steer ha^{-1} for the normal or unfavorable price year. A great difference in economic profits occurred with these different economic years (cattle price) and different weather conditions. Favorable prices for wet seasons produced high economic profits with relatively lower harvest efficiency (Fig. 7–7) when choosing a relatively high SR of 0.5 steer ha^{-1}. Unfavorable prices for dry seasons produced very low economic profits, and increasing SR can increase SWG (Fig. 7–7) but reduce the economic profits (Fig. 7–9).

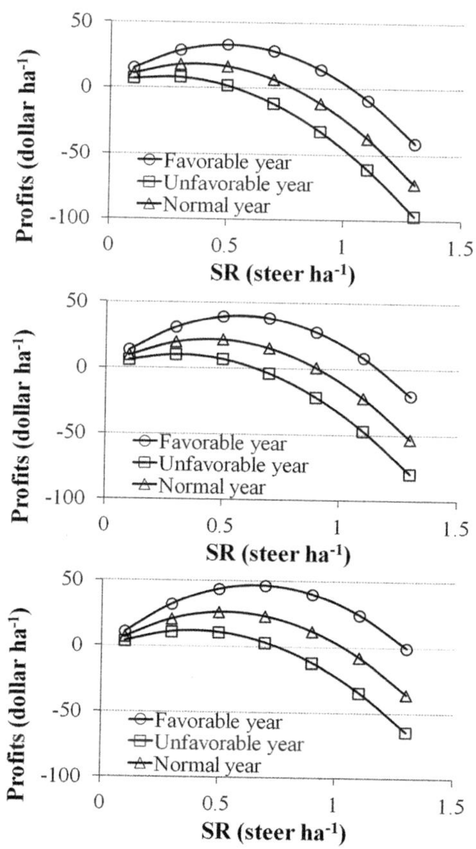

Fig. 7–9. Responses of regression predicted economic profits to stocking rate (SR, steer ha^{-1}) as influenced by cattle price (favorable price: buy at $US1.76 and sell at $US1.65; normal price: buy at $US1.61 and sell at $US1.48; unfavorable price: buy at $US1.54 and sell at $US1.32, according to Manley et al. [1997]) under normal April to June rainfall conditions (160 mm). The economic profits were estimated on the basis of the predicted SWG, cattle buy or sale price, and carry cost (Hart et al., 1988).

When extending these results to other soils, for example, clay loam and sandy loam soils, the responses of PSC, harvest efficiency, SWG, and economic profits to SR levels were similar, but with lower PSC (higher harvest efficiency), SWG, and economic profits compared with the results from loam soil, especially under high SR levels (Fig. 7–10). This result suggested that loam soil had an advantage in producing higher PSC and SWG with higher SR levels compared with clay loam and loam sandy soil. While under low SR levels, the response of harvest efficiency, SWG, and economic profits were very close among the three soil types.

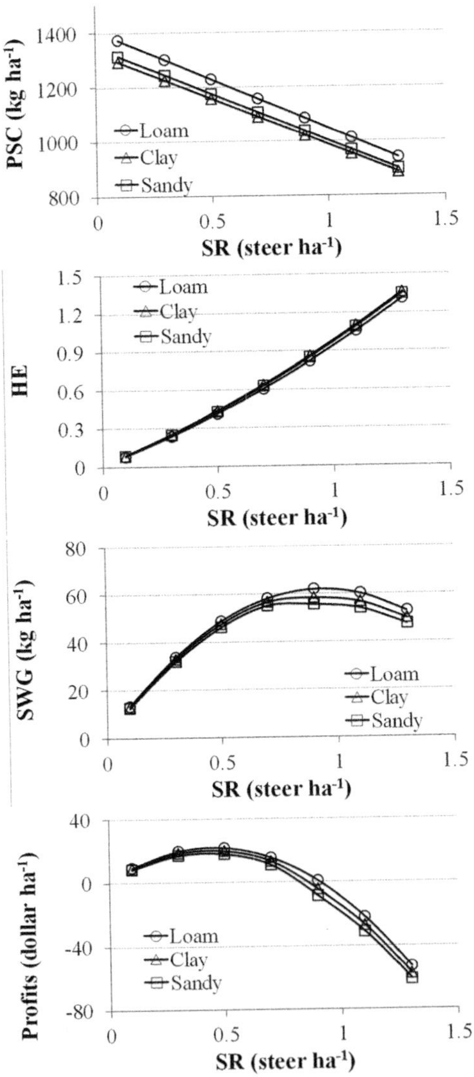

Fig. 7–10. Responses of regression predicted peaking standing crop (PSC), harvest efficiency (HE), steer weight gain per area (SWG), and economic profits to stocking rate (SR, steer ha^{-1}) as influenced soil types [the loam, clay loam, and loam sandy soils with extended regression equation predicted data from loam soil (Table 7–3)] under normal April to June rainfall conditions (160 mm). The economic profits were estimated on the basis of the predicted SWG, cattle buy or sale price, and carry cost (Hart et al., 1988).

Summary and Conclusions

On the basis of the GPFARM-Range model simulated long-term PSC data, simple regression equations can be developed in terms of rainfall from April to June, SR, and soil water content on 1 April to significantly improve the PSC predictions. Combining soil water content before the growing season and SR provides an accurate tool for ranchers to predict PSC before the grazing season and make reasonable stocking decisions in the region.

Including the predictions of PSC, SWG, and economic profits can be very helpful for ranchers in making better SR decisions, accounting for both economic return and environmental impact (land conservations). The regression equations can also be used to analyze the effects of weather and soil variations on the responses of PSC and SWG to SR levels. These regression equations are advantageous to previous methods, such as the Drought Calculator (Dunn et al., 2013), in that the regression equations consider both economic net return and environmental impacts and the coupled responses of PSC to SR and SWG to PSC and SR across various weather conditions. These equations can be used to develop a spreadsheet-based decisions support tool and help ranchers explore the tradeoffs between economic net return and environmental impacts (such as land degradation) as influenced by SR and weather variations and make better stocking decisions with reduced enterprise risks and land degradations on different soils across the northern mixed-grass prairie.

References

Andales, A.A., J.D. Derner, L.R. Ahuja, and R.H. Hart. 2006. Strategic and tactical prediction of forage production in northern mixed-grass prairie. Rangeland Ecol. Manag. 59:576–584. doi:10.2111/06-001R1.1

Andales, A.A., J.D. Derner, P.N.S. Bartling, L.R. Ahuja, G.H. Dunn, R.H. Hart, and J.D. Hanson. 2005. Evaluation of GPFARM for simulation of forage production and cow-calf weights. Rangeland Ecol. Manage. 58:247–255. doi:10.2111/1551-5028(2005)58[247:EOGFSO]2.0.CO;2

Bement, R.E. 1969. A stocking-rate guide for beef production on blue-grama range. J. Range Manage. 22:83–86. doi:10.2307/3896186

Biondini, M.E., B.D. Patton, and P.E. Nyren. 1998. Grazing intensity and ecosystem processes in a northern mixed-grass prairie. Ecol. Appl. 8:469–479. doi:10.1890/1051-0761(1998)008[0469:GIAEPI]2.0.CO;2

Briggs, J.M., and A.K. Knapp. 1995. Interannual variability in primary production in tallgrass prairie: Climate, soil moisture, topographic position, and fire as determinants of aboveground biomass. Am. J. Bot. 82:1024–1030. doi:10.2307/2446232

Currie, P.O., and G. Peterson. 1966. Using growing-season precipitation to predict crested wheatgrass yields. J. Range Manage. 19(5):284–288. doi:10.2307/3895721

Dahl, B.E. 1963. Soil moisture as predictive index of forage yield for the sandhills range type. J. Range Manage. 16:128–132. doi:10.2307/3895105

Derner, J.D., and R.H. Hart. 2007. Grazing-induced modifications to peak standing crop in northern mixed-grass prairie. Rangeland Ecol. Manage. 60:270–276. doi:10.2111/1551-5028(2007)60[270:GMTPSC]2.0.CO;2

Derner, J.D., R.H. Hart, M.A. Smith, and J.W. Waggoner. 2008a. Long-term cattle gain responses to stocking rate and grazing systems in northern mixed-grass prairie. Livest. Sci. 117:60–69. doi:10.1016/j.livsci.2007.11.011

Derner, J.D., B.W. Hess, R.A. Olson, and G.E. Schuman. 2008b. Functional group and species responses to precipitation in three semi-arid rangeland ecosystems. Arid Land Res. Manage. 22:81–92. doi:10.1080/15324980701784274

Dunn, G.H., M. Gutwein, L.J. Wiles, T.R. Green, A. Menger, and J. Printz. 2013. The Drought Calculator: Decision support tool for predicting forage growth during drought. Rangeland Ecol. Manage. 66(5):570–578. doi:10.2111/REM-D-12-00087.1

Eneboe, E.J., B.F. Sowell, R.K. Heitschmidt, M.G. Karl, and M.R. Haferkamp. 2002. Drought and grazing: IV. Blue grama and western wheatgrass. J. Range Manage. 55:73–79. doi:10.2307/4003266

Fang, Q.X., A.A. Andales, J.D. Derner, L.R. Ahuja, L. Ma, P.N.S. Bartling, J.L. Reeves, and Z. Qi. 2014. Modeling weather and stocking rate effects on forage and steer production in northern mixed-grass prairie. Agric. Syst. 129:103–114. doi:10.1016/j.agsy.2014.05.011

Freund, R.J., and R.C. Littell. 1991. SAS system for regression. 2nd ed. SAS Inst., Cary, NC.

Galt, D., F. Molinar, J. Navarro, J. Joseph, and J. Holechek. 2000. Grazing capacity and stocking rate. Rangelands 22(6):7–11.

Hanson, J.D., B.B. Baker, and R.M. Bourdon. 1992. SPUR2 Documentation and user guide. GPSR Tech. Rep. 1. USDA–ARS, Fort Collins, CO.

Hanson, J.D., J.W. Skiles, and W.J. Parton. 1988. A multispecies model for rangeland plant communities. Ecol. Modell. 44:89–123. doi:10.1016/0304-3800(88)90084-1

Hart, R.H., M.J. Samuel, P.S. Test, and M.A. Smith. 1988. Cattle, vegetation, and economic responses to grazing systems and grazing pressure. J. Range Manage. 41:282–286. doi:10.2307/3899379

Lauenroth, W.K., and O.E. Sala. 1992. Long-term forage production of north american shortgrass steppe. Ecol. Appl. 2:397–403. doi:10.2307/1941874

Manley, W.A., R.H. Hart, M.J. Samuel, M.A. Smith, J.W. Waggoner, and J.T. Manley. 1997. Vegetation, cattle, and economic responses to grazing strategies and pressures. J. Range Manage. 50:638–646. doi:10.2307/4003460

Milchunas, D.G., J.R. Forwood, and W.K. Lauenroth. 1994. Productivity of long-term grazing treatments in response to seasonal precipitation. J. Range Manage. 47:133–139. doi:10.2307/4002821

Patton, B.D., X. Dong, P.E. Nyren, and A. Nyren. 2007. Effects of grazing intensity, precipitation, and temperature on forage production. Rangeland Ecol. Manage. 60(6):656–665. doi:10.2111/07-008R2.1

Poppi, D.P. 1996. Predictions of food intake in ruminants from analyses of food composition. Aust. J. Agric. Res. 47:489–504. doi:10.1071/AR9960489

Redmon, L.A., F.T. McCollum, III, G.W. Horn, M.D. Cravey, S.A. Gunter, P.A. Beck, J.M. Mieres, and R.S. Julian. 1995. Forage intake by beef steers grazing winter wheat with varied herbage allowances. J. Range Manage. 48:198–201. doi:10.2307/4002420

Reeves, J., J. Derner, M. Sanderson, J. Hendrickson, S. Kronberg, M. Petersen, and L. Vermeire. 2014. Seasonal weather influences on yearling beef steer production in C_3-dominated Northern Great Plains rangeland. Agric. Ecosyst. Environ. 183:110–117. doi:10.1016/j.agee.2013.10.030

Reeves, J.L., J.D. Derner, M.A. Sanderson, M.K. Petersen, L.T. Vermeire, J.R. Hendrickson, and S.L. Kronberg. 2013a. Seasonal temperature and precipitation effects on cow–calf production in northern mixed-grass prairie. Livest. Sci. 155:355–363. doi:10.1016/j.livsci.2013.04.015

Reeves, J., J. Derner, M. Sanderson, M. Petersen, L. Vermeire, J. Hendrickson, and S. Kronberg. 2013b. Temperature and precipitation affect steer weight gains differentially by stocking rate in northern mixed-grass prairie. Rangeland Ecol. Manage. 66(4):438–444. doi:10.2111/REM-D-12-00157.1

Sala, O.E., W.J. Parton, L.A. Joyce, and W.K. Lauenroth. 1988. Primary production of the central grassland region of the United States. Ecology 69:40–45. doi:10.2307/1943158

Smart, A.J., J.D. Derner, J.R. Hendrickson, R.L. Gillen, B.H. Dunn, E.M. Mousel, and K.C. Olson. 2010. Effects of grazing pressure on efficiency of grazing on North American Great Plains rangelands. Rangeland Ecol. Manage. 63(4):397–406. doi:10.2111/REM-D-09-00046.1

Smart, A.J., B.H. Dunn, P.S. Johnson, L. Xu, and R.N. Gates. 2007. Using weather data to explain herbage yield on three Great Plains plant communities. Rangeland Ecol. Manage. 60:146–153. doi:10.2111/05-099R4.1

Stevenson, A., R.E. Baumgartner, and G.E. Schuman. 1984. High Plains Grasslands Research Station detailed soil survey. USDA Publ. 1–84/1C/3.62. Wyoming Agricultural Experiment Station, Laramie, WY.

Torell, L.A., K.C. McDaniel, and V. Koren. 2011. Estimating grass yield on blue grama range from seasonal rainfall and soil moisture measurements. Rangeland Ecol. Manage. 64:56–66. doi:10.2111/REM-D-09-00107.1

USDA–NASS. 2014. Economics and prices. USDA–NASS. http://www.nass.usda.gov/Statistics_by_Subject/Economics_and_Prices/index.asp (accessed 27 June 2014).

Wiles, L.J., G. Dunn, J. Printz, B. Patton, and A. Nyren. 2011. Spring precipitation as a predictor for peak standing crop of mixed-grass prairie. Rangeland Ecol. Manage. 64(2):215–222. doi:10.2111/REM-D-09-00024.1

Providing User-Oriented Uncertainty Information with a Vineyard Model Used for Irrigation Decisions

Sébastien Roux, Xavier Delpuech, Gabriel Daudin, François Brun, Jacques Wery, and Daniel Wallach*

Abstract

In southeastern France, vineyard irrigation has been increasing and now concerns about 8% of the total vineyard area in the region. Irrigation management aims to ensure both fruit quality and yield by maintaining the vineyard above a time-dependent threshold of soil water. Extension services disseminate information and advice on vineyard irrigation, based in large part on the WaLIS (Water Balance for Intercropped Systems) model described here, which computes the fraction of transpirable soil water of a vineyard with or without an intercrop. Extension bulletins currently report model results, but the extension services have indicated that adding uncertainty information would be important for informing users as to the reliability of those results. We show how uncertainty information could be calculated and presented. The major focus here is on the uncertainty due to using estimated rather than measured input variables, taking into account the way users estimate the input variables in practice. This is the source of uncertainty most relevant for users because it is an uncertainty that they can affect. However, it is also important to know how this uncertainty compares with the overall uncertainty in the predictions. We therefore also show how total uncertainty can be estimated.

Abbreviations: PLWP, predawn leaf water potential; WaLIS, Water Balance for Intercropped Systems.

Sébastien Roux, INRA, UMR System, 2 Place Pierre Viala, 34060 Montpellier Cedex 2, France (sebastien.roux@supagro.inra.fr).

Xavier Delpuech, Institut Français de la Vigne et du Vin, Pôle Rhône-Méditerranée, Domaine de Donadille

30230 Rodilhan, France (xavier.delpuech@vignevin.com).

Gabriel Daudin, Vita Consult, ZA du Pré Neuf, 19 rue du Pré Neuf, 44190 Gorges, France (gdaudin@vitaconsult.fr).

François Brun, ACTA, INRA, UMR 1248 AGIR, 24 Chemin de Borde Rouge - Auzeville, CS 52627, 31326 Castanet Tolosan, Cedex, France (francois.brun@acta.asso.fr).

Jacques Wery, Montpellier SupAgro, UMR System, 2 Place Pierre Viala, 34060 Montpellier Cedex 2, France (jacques.wery@supagro.inra.fr).

Daniel Wallach, INRA, UMR 1248 AGIR, 24 Chemin de Borde Rouge - Auzeville, CS 52627, 31326 Castanet Tolosan, Cedex, France. *Corresponding author (wallach@toulouse.inra.fr).

doi:10.2134/advagricsystmodel5.c8

Practical Applications of Agricultural System Models to Optimize the Use of Limited Water
Lajpat R. Ahuja, Liwang Ma, and Robert J. Lascano, Editors
Advances in Agricultural Systems Modeling, Volume 5. Lajpat R. Ahuja, Series Editor

USDA-ARS

The administrative regions Languedoc-Roussillon and Alpes Côtes d'Azur in southeastern France have been producing grapes (*Vitis* L.) for wine since well before the Roman conquest. Today these regions represent 41.6% of the vineyards in France, covering 326,880 ha (RGA, 2010). This area has a Mediterranean climate, with summers that can be very dry. Severe summer drought can reduce yield and affect the development of the grapes, leading to economic losses. As a result, there has been increasing use of irrigation on vineyards in this region. The irrigated area increased by 88% between 2000 and 2010, up to 25,298 ha (RGA, 2000, 2010), which is about 8% of the total vineyard area in the region and represents almost 90% of the total irrigated vineyard area in France. Irrigation will probably continue to expand if, as expected, the frequency of droughts in the Mediterranean area increases over the 21st century (Soubeyroux et al., 2011).

The goal of vineyard irrigation is to ensure both fruit quality and yield. However, yield and quality are to some extent contradictory: some water stress is required for high quality, but this reduces yield somewhat. To limit vegetative development and control the formation of the yield and quality of grapes, the water availability should be limited, and a moderate water stress should be experienced by grapevines after blooming (Ojeda, 2007). There are studies that recommend a detailed time course of water status, with different recommendations depending on the type of wine to be produced.

To implement these recommendations, it is necessary to have information about the water status of the soil–plant system. While soil moisture probes or pressure chambers for measuring plant water status are available, these are expensive and time consuming. Therefore, a water balance model was developed to simulate soil water in vineyards (Lebon et al., 2003). This model has been in use by advisory services in southwestern France since 2004, after being approved by the French Institute of Vine and Wine (IFV). Currently, the model is used by several extension services that base their irrigation advice to a significant extent on the model results.

The vineyard model as currently used provides a unique value of soil moisture at each time representing the average soil water status sensed by the root system (Pellegrino et al., 2005). There is no information about the uncertainty associated with the simulated values. However, the extension services have indicated that adding uncertainty information to the model would be useful and that they would prefer to have such information. It would allow both the extension services and the growers who use their advice to better evaluate the confidence that they should have in the model results, before making a decision about irrigation.

It seems then important to include a measure of uncertainty in model results. In a previous study, we showed how to estimate overall uncertainty for the vineyard soil water model (Roux et al., 2014). The emphasis there was on the methodology of estimating the different contributions to overall uncertainty. The purpose here is to extend that study to the case where multiple inputs to the vineyard model are approximated. The major new problems that arise are to identify the types of approximation that are used in practice, to quantify the uncertainty that each approximation entails, and to show how to evaluate the contributions of those uncertainties to overall uncertainty. This is a specific application of the general recommendation that modeling should include estimation of the reliability of the simulations (Jakeman et al., 2006; Bellocchi et al., 2010; Wallach et al., 2014; Bennett et al., 2013).

In this paper we describe the model and the way it is used as a decision support tool currently by extension services, or could be used by growers in the future, as well as the data used in the examples presented here. We then explain the methodology, with emphasis on quantifying the uncertainty in the model input variables that corresponds to the way the model is really used and evaluating the consequences of that uncertainty. Finally, we present the results of uncertainty calculations and present the discussion and conclusions.

Model, model use, and data

Vineyard Soil Water Model

The WaLIS model used in this study is a recent extension of the one used by farmer advisors and described in Lebon et al. (2003). The two models are based on the same principles, but WaLIS has additional parameters to take into account intercrop and runoff processes (Celette et al., 2010). The model extension was developed and evaluated using data from a site near Montpellier (Domaine du Chapitre) in the South of France. Additional data have been collected and used for a more extended evaluation study (Delpuech et al., 2010; Roux et al., 2014). The model is available at http://walis.vignevin.com/.

The model computes the fraction of transpirable soil water (Pellegrino et al., 2005) as a function of time [FTSW(*t*)] using a daily water balance and estimates of the main daily water fluxes (Fig. 8–1). The FTSW(*t*) is often used as the indicator of the level of water deficit experienced by a vineyard (Pellegrino et al., 2006). The model parameters are shown in Table 8–1 and the input variables in Table 8–2.

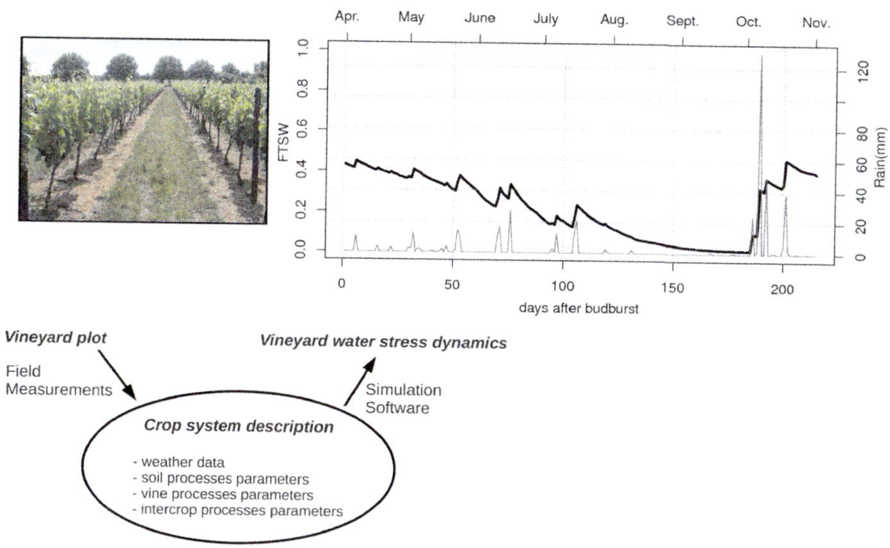

Fig. 8–1. Schematic diagram of the vineyard soil water model. The model outputs fraction transpirable soil water over time [FTSW(t)], starting at budburst.

Table 8–1. Model parameters.

Parameter name	Description	Value
b1,b2	soil evaporation parameters	14, 0.15
U	soil surface water volume	2.7
s1bare, s2bare	threshold parameters	12.7, 28
s1cover, s2cover	threshold parameters	35.6, 53
ddkmax	degree days to get maximum crop coefficient	600
WT	parameter to describe effect of water stress on transpiration	0.4
Maximum effective depth for grass roots	1	p_g
LAI0	grass initial leaf area (LAI) index	0.3
LAIrate	grass LAI growth rate parameter	0.9
LSSmin	duration of grass green leaves (DD)	700
Wig,1	parameter to describe effect of water stress on grass LAI growth	0.9
Wig,1	parameter to describe effect of water stress on grass transpiration	0.6
NNI	grass nitrogen nutrition index	0.3
ε	grass extinction coefficient	0.6
kgmax	grass maximum light interception	0.95
LAIcut	grass LAI threshold to trigger grass cover cut	3
LAIres	grass LAI after grass cut	0.3

Table 8–2. Model input variables.

Variables†	Definition
$R(t)$*, $T(t)$*, ETP(t)*	daily rainfall, mean temperature, and reference evapotranspiration
CNcover***	curve number for soil with cover
CNbare***	curve number for bare soil
kmax*	maximum vineyard light interception fraction
TTSWtot**	maximum total transpirable water
FTSW0**	fraction transpirable water at budburst
pic*	proportion of soil covered by grass
dbud*	date of budburst

† The number of stars indicates the relative difficulty of using the reference method to obtain this input (*, easiest; ***, most difficult), based on responses of extension personnel to questionnaire.

The water balance is based on the assumption that the soil of an intercropped vineyard can be represented by two reservoirs, the first with only grapevine roots and the other with both intercrop and grapevine roots. The size of the second reservoir is proportional to the proportion of soil covered by grass; if there is no intercrop, the size of the second reservoir is 0 and only one reservoir is used. The two reservoirs are characterized by their transpirable water content at each time t, ASW1(t) and ASW2(t), and their maximum transpirable water content TTSW1 and TTSW2. Total transpirable water content for the vineyard is TTSWtot = TTSW1 + TTSW2. The fraction of transpirable soil water content at time t is

$$\mathrm{FTSW}(t)=\left[\mathrm{ASW1}(t)+\mathrm{ASW2}(t)\right]/\,\mathrm{TTSWtot} \qquad [1]$$

ASW1(t) and ASW2(t) are calculated by updating the water content of the previous day using the daily water fluxes computed by submodules:

$$\begin{aligned}\mathrm{ASW1}(t)&=\max\left\{0,\min\left[\mathrm{TTSW1},\mathrm{ASW1}(t-1)+\mathrm{Sum}\left(\mathrm{Fluxes}\{t-1,t,1\}\right)\right]\right\}\\ \mathrm{ASW2}(t)&=\max\left\{0,\min\left[\mathrm{TTSW2},\mathrm{ASW2}(t-1)+\mathrm{Sum}\left(\mathrm{Fluxes}\{t-1,t,2\}\right)\right]\right\}\end{aligned} \qquad [2]$$

The fluxes involved in Eq. [2] are (i) rain, (ii) runoff (from bare soil or soil covered by intercrop), (iii) bare soil evaporation, (iv) intercrop evapotranspiration, (v) vineyard transpiration, (vi) drainage from reservoir 2 to reservoir 1, and (vii) drainage out of reservoir 1.

The detailed expressions for computing these fluxes can be found in Celette et al. (2010). Parameters for soil evaporation come from Lebon et al. (2003), and the parameters related to grass growth come from Duru et al. (2009).

The model calculates FTSW(t) but can also output predawn leaf water potential (PLWP), which is calculated from FTSW(t) (Pellegrino et al., 2005).

Model Use and Role of Uncertainty Information

The model is used by researchers, engineers of the French Vine and Wine Institute, and extension services to evaluate the water stress experiences by vineyards during the growing season (Pellegrino et al., 2006; Ripoche et al., 2011; Gaudin and Gary, 2012; Delpuech et al., 2010).

A major use of the model is as an aid to irrigation decisions. There is a recommended threshold for FTSW(*t*) or for PLWP, below which irrigation is required. The threshold depends on the growth stage of the crop, expressed as days after budburst. Simulated FTSW(*t*) or PLWP are compared with the threshold, and when the simulated values fall below the threshold, irrigation is recommended. Several threshold functions have been proposed (Ojeda, 2007; Gary et al., 2005). We use here the thresholds proposed by Gaudin and Gary (2012).

The model has been in routine use for several years by multiple extension services, including the Agricultural Bureaux (Chambres d'Agriculture) of the French departments Aude, Bouches-du-Rhône, Drôme, Gard, and Vaucluse, the center for regional agrometeorolocal information (CIRAME), and the center for wine research (CRVI) of Corsica. Consider the case of the Gard department. The model is run weekly to simulate six to eight vineyards that are representative of a variety of conditions in the region. The results of the model, and an interpretation in terms of irrigation, are in a weekly bulletin made available to growers. Figure 8–2 shows a graph from one of these bulletins. The graph shows simulated PLWP for each test vineyard, up to the current date (25 June in this case), and also

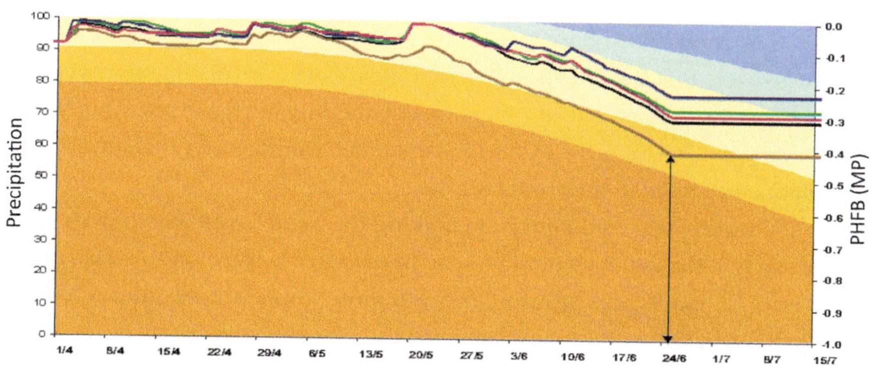

Fig. 8–2. Example of simulation results, as presented in a weekly bulletin issued by the Gard agricultural bureau. The four solid lines are simulation results from four vineyards. The model calculates FTSW(*t*), but those values have been translated into predawn leaf water potential values, which is are familiar to extension services, using the relationship established by Pellegrino et al. (2005). Values in the lowest shaded area would indicate that the threshold for acceptable water stress has been passed and irrigation is required. Note that the threshold varies (toward lower leaf water potentials) as the season advances. The lowest simulation curve has not yet crossed the threshold at the time of the bulletin, but it is quite close.

the time varying threshold below which irrigation is recommended. The PLWP is preferred here, in place of FTSW from which it is calculated, because it is more familiar to extension services and farm advisers who can also access it with direct measurement. For one of the test vineyards, the simulated PLWP at the current date is quite close to the threshold. The commentary focuses on the distance from the simulated curve to the threshold. However, the interpretation of the results is more in terms of trends than of detailed recommendations. For example, for the simulated field with the lowest FTSW, the commentary reads "even though the vineyard currently seems to still be growing or near the end of vegetative growth, the level of water reserves in the soil is likely to be low and the water situation needs to be followed closely." Furthermore, the bulletin stresses that growers should make decisions based on observations in their own fields. Thus, the model is a decision aid, an important part of the extension message, but it is not used to directly trigger irrigations.

To better understand user practices and needs, extension personnel involved with using the model were given a questionnaire that included questions about their perception of the uncertainty in the model predictions. Their estimates of uncertainty were very variable, ranging from "slight" to 75%. The perceived causes of the uncertainty included uncertainties in the model parameters, in the model equations, and in the value of initial FTSW(t) furnished to the model. When asked if they would prefer that the model produce a single value or an envelope of possible values, the clear majority indicated that they prefer an envelope. Paraphrasing one of the responses, the reason is that providing uncertainty would allow one to know how much confidence to have in the model results. In a similar vein, another response was that showing the uncertainty would make it clear to growers that the results are only approximate and that therefore the conclusions about irrigation based on the model are only approximate. This shows that the model users here, the extension personnel, would prefer that uncertainty information be added to the model results.

In the future, one could envision making the model available to individual growers so that the results would be specific to each vineyard. In this case uncertainty information would probably be even more important in order to clearly indicate the reliability of the simulated values.

Once uncertainty information becomes available, it seems reasonable to suppose that users will seek ways of reducing the uncertainty. Much of the uncertainty in model predictions depends on the model and cannot be reduced by the model user. This includes uncertainty due to uncertainty in the model equations and in the model parameters and uncertainty due to the model residual error, that is, due to the fact that the model input variables do not explain all the variability between vineyards. However, there is an additional source of uncertainty that does depend on the user—the uncertainty due to uncertainty in the model input variables.

That contribution to model uncertainty can be reduced by obtaining more accurate values for the model inputs. However, there is usually a cost involved in obtaining more accurate values. For example, one might have a single weather station for multiple vineyards, so that the weather data is not exactly the weather at each vineyard. To obtain more accurate weather data, one could install weather stations at each vineyard, but of course there is an associated cost. There is thus a trade-off between reducing uncertainty in simulated values and cost. To make a decision about investment in better input values, one would want information about the levels of uncertainty with and without the additional investment. This is a different kind of uncertainty information, concerning the average effect on uncertainty of obtaining more accurate input values.

Data

The data that were used to estimate errors here are from two locations in a Mediterranean environment (Corbane et al., 2012; Celette et al., 2009a). One location, Domaine du Chapitre (http://www.domaine-du-chapitre.com/), with data from 3 yr (2004–2006), has a clay soil with depth greater than 3 m. The second location, with data from 2 yr (2001 and 2002), has a sandy loam soil of 1.2-m depth. In each case, part of the vineyard had grass between the rows and the other part had bare soil, obtained with chemical weed control. Overall then there were 10 site-year grass cover combinations. In each of these plots, all of the input variables were measured according to the reference method indicated below, and FTSW(*t*) was measured at multiple dates using the approach of Pellegrino et al. (2005). These are the same data as used in Roux et al. (2014).

To illustrate the uncertainty calculations, we do the calculations for the Domaine du Chapitre with bare soil, for 2 yr. The 2 yr had quite different winter and spring rainfall patterns. Rainfall was high in 2004 (1110 mm between 10 Jan. 2003 and 4 Jan. 2004) and low in 2006 (435 mm over the same period).

Uncertainty Estimation

Principles

We first define more precisely what we mean by uncertainty. This is a statistical notion and is measured by the probability distribution of simulation errors, that is, of the difference between observed and simulated FTSW(*t*). When this probability distribution is wide, the difference between observed and simulated values can be large and we say that there is a large amount of uncertainty. When the probability distribution of model errors is concentrated close to 0, then the simulated and observed values are always close and we say that there is little uncertainty. Note that this implies that we have defined some target population, that is, some range of situations that are of interest. The probability distribution of errors refers to errors for vineyards in this target population. In our case the population is vineyards in southeastern France, with or without grass cover.

Following Roux et al. (2014), we can write the error as

$$\text{error}_i(t) = \text{FTSW}_i(t) - \hat{\text{F}}\text{TSW}\left(t, \hat{T}_i, X_i; \theta\right) \quad [3]$$

where $\text{FTSW}_i(t)$ is the observed value of FTSW in vineyard i at time t, and $\hat{\text{F}}\text{TSW}\left(t, \hat{T}_i, X_i; \theta\right)$ is the corresponding simulated value. The simulated value depends on $\hat{T}$, which represents those input variables that are approximated, on X, which represents those input values that are measured, and on θ, which are the model parameters.

The error can be written as the sum of two terms as follows:

$$\begin{aligned} &\text{FTSW}_i(t) - \hat{\text{F}}\text{TSW}\left(t, \hat{T}_i, X_i; \theta\right) = \varepsilon_i(t) + \tau_i(t) \\ &\varepsilon_i(t) = \left[\text{FTSW}_i(t) - \hat{\text{F}}\text{TSW}\left(t, T_i, X_i; \theta\right)\right] \\ &\tau_i(t) = \left[\hat{\text{F}}\text{TSW}\left(t, T_i, X_i; \theta\right) - \hat{\text{F}}\text{TSW}\left(t, \hat{T}_i, X_i; \theta\right)\right] \end{aligned} \quad [4]$$

The first term, $\varepsilon_i(t)$, is the error for measured inputs. It is the difference between measured FTSW and the simulated value using measured values of all input variables, $\hat{\text{F}}\text{TSW}\left(t, T_i, X_i; \theta\right)$. The error here comes from errors in the equations and in the parameter values and from model residual error but not from errors in the input variables.

We assume that the probability distribution of $\varepsilon_i(t)$, denoted by $f_E(\varepsilon)$, does not depend on t, T_i, or X_i. In general there is simply not enough past data to do better. Furthermore, in a previous study, there was no obvious dependence of the distribution of $\varepsilon_i(t)$ on t.

The distribution of errors in hindcasts is taken as our estimate of the distribution of $\varepsilon_i(t)$. Put another way, we take the distribution of past errors as our approximation to the distribution of future errors. Note however that the hindcasts must concern vineyards from the target population. Furthermore, the hindcast errors must be between observed values and simulated values where all inputs were measured, since $\varepsilon_i(t)$ is the simulation error for measured inputs.

The second term in Eq. [2], $\tau_i(t)$, is the error due to approximating inputs. It is the difference between simulated values using measured values for the inputs

T and simulated values using approximate values for the inputs T. It depends on the difference between measured and approximate values of the input variables T, denoted by η_i for vineyard i,

$$\eta_i = T_i - \hat{T}_i \quad [5]$$

and on the way errors in T propagate through the model and finally affect FTSW(t).

The distribution of η_i will depend in detail on the way the inputs are estimated (see next section). Given that distribution, one can then do a Monte Carlo calculation to obtain the distribution of $\tau_i(t)$.

We assume that the $\varepsilon_i(t)$ and η_i are all mutually independent [and therefore the $\varepsilon_i(t)$ and the $\tau_i(t)$ are mutually independent]. This assumption will not be exactly true if some of the approximations to the input variables are based on the model. In that case the errors in the input variables and in the model will be related.

Finally, one can obtain the overall uncertainty in the model outputs for vineyard i, as follows:

1. Draw a value from the distribution of η_i.
2. Run the model to obtain the simulated value $\hat{\mathrm{F}}\mathrm{TSW}\left(t,\eta_i+\hat{T}_i,X_i;\theta\right)$.
3. Add to the simulated value a value drawn at random from the distribution of $\varepsilon_i(t)$.
4. Repeat steps 1 to 3 multiple times (in our case 500).
5. The distribution of the results is an approximation to the distribution of $\mathrm{FTSW}_i(t)-\hat{\mathrm{F}}\mathrm{TSW}\left(t,\hat{T}_i,X_i;\theta\right)$. Confidence intervals of level α are obtained by removing the largest and smallest $(100-\alpha)/2\%$ of the generated values.

Application to the Vineyard Soil Moisture Model

Reference Method for Measuring Input Variables

The method of measurement for each of the input variables (the "reference" method) is explained below. This is the method that is used to obtain the measured value for each input variable T_i.

- Climate data (R, T, ETP): The reference method is to measure weather variables at the location of the vineyard.
- Curve number (CNcover, CNbare): Runoff in the model is taken into account using the curve number approach (Boughton 1989). If there is an intercrop then two values are required, one for bare soil and one for intercrop. The reference procedure is to measure runoff under various conditions and then work back to deduce the curve number.

- Maximum vineyard light interception fraction (kmax): The reference procedure here is to use a radiation model (Riou et al., 1989) that estimates the maximum light interception fraction (ratio of intercepted over incident light) of a vineyard using its geographical location, row spacing and orientation, and foliage shape (width and height). The calculation is available at http://walis.vignevin.com/kmax/.
- Maximum total transpirable water (TTSWtot): The reference method of obtaining this value (described in Pellegrino et al., 2005) is to use a moisture probe a few days after heavy rain and also after prolonged drought. The difference in soil water contents gives TTSWtot.
- Fraction transpirable water at budburst (FTSW0): The reference method here is to use a moisture probe to measure soil moisture at budburst.

Approximate Methods for Obtaining Input Variables

For each of the input variables that are difficult to measure, we define a simpler method which can be used to approximate the input.

- Climate data (R, T, ETP): The approximate procedure consists of using weather data from neighboring weather stations.
- Curve number (CNcover, CNbare): An approximate procedure is to estimate curve number from soil group and ground cover (USDA–NRCS, 2007) (see Table 8–3).
- Maximum vineyard light interception fraction (kmax): An approximation can be obtained using Table 8–4, which relates kmax to vigor and plant height. This table was obtained by running the radiation model with multiple plant heights and vigor values.
- Maximum total transpirable water (TTSWtot): The estimation procedure estimates TTSWtot from soil depth and soil texture (Gijsman et al., 2002; see Table 8–5). Many other possibilities exist (Gijsman et al., 2002).
- Fraction transpirable water at budburst (FTSW0): It is assumed that the soil is at wilting point at the end of the previous summer (i.e., after harvest and before autumn rainfalls). This is often a reasonable approximation for Mediterranean conditions. Then the model is run from the end of summer to determine soil moisture at budburst (Fig. 8–3).

The Distribution of η_i for Each Input Variable

Given the method of obtaining approximate inputs, one then needs to translate those procedures into distributions of η_i. Here we explain how this was done for the example treated here and show the resulting distributions.

- Weather data: An approximate distribution of weather sequences is obtained by assuming that the weather sequences from all neighboring weather stations are equally probable. In the example treated here, weather data from

Table 8–3. Table to obtain approximate curve numbers CNcover and CNbare and the associated uncertainty.

	Soil group†			
Situation	A	B	C	D
Bare soil	77 *U*(77 ± 5.5)	86 *U*(86 ± 4.5)	91 *U*(91 ± 3)	94 *U*(94 ± 2.5)
Bare soil and tillage	66 *U*(66 ± 8.5)	77 *U*(77 ± 5.5)	85 *U*(85 ± 4)	89 *U*(89 ± 2.5)
Grass cover	49 *U*(49 ± 10)	69 *U*(69 ± 10)	79 *U*(79 ± 5)	84 *U* (84 ± 2.5)

† The notation *U*(a ± b) signifies that the uncertainty is represented by a uniform distribution with lower limit a − b and upper limit a + b.

Table 8–4. Table to approximate vineyard maximum fraction of light interception kmax and the associated uncertainty.

	Vigor level†		
Foliage height	Low	Medium	High
m			
0.7	0.306 *U*(0.306 ± 0.034)	0.374 *U*(0.374 ± 0.034)	0.432 *U*(0.432 ± 0.029)
0.8	0.323 *U*(0.323 ± 0.036)	0.395 *U*(0.395 ± 0.036)	0.457 *U*(0.457 ± 0.031)
0.9	0.339 *U*(0.339 ± 0.038)	0.415 *U*(0.415 ± 0.038)	0.480 *U*(0.480 ± 0.032)
1.0	0.353 *U*(0.353 ± 0.040)	0.433 *U*(0.433 ± 0.040)	0.501 *U*(0.501 ± 0.034)
1.1	0.367 *U*(0.367 ± 0.042)	0.451 *U*(0.451 ± 0.042)	0.522 *U*(0.522 ± 0.035)
1.2	0.380 *U*(0.380 ± 0.043)	0.467 *U*(0.467 ± 0.043)	0.541 *U*(0.541 ± 0.037)
1.3	0.392 *U*(0.392 ± 0.045)	0.482 *U*(0.482 ± 0.045)	0.560 *U*(0.560 ± 0.039)
1.4	0.403 *U*(0.403 ± 0.047)	0.497 *U*(0.497 ± 0.047)	0.577 *U*(0.577 ± 0.040)

† The notation *U*(a ± b) signifies that the uncertainty is represented by a uniform distribution with lower limit a − b and upper limit a + b.

Table 8–5. Table to approximate maximum total transpirable water TTSWtot and the associated uncertainty.

	Soil depth†		
Soil texture	Shallow	Medium	Deep
Sand	50 *U*(50,75)‡	100 *U*(100 ± 30)	150 *U*(150 ± 40)
Medium	90 *U*(90 ± 35)	160 *U*(160 ± 35)	230 *U*(230 ± 40)
Clay	120 *U*(120 ± 30)	180 *U*(180 ± 60)	300 *U*(300 ± 60)

† The notation *U*(a ± b) signifies that the uncertainty is represented by a uniform distribution with lower limit a − b and upper limit a + b.

‡ For shallow sandy soil, the uncertainty is represented by a uniform distribution with lower and upper limits of 50 and 75, respectively.

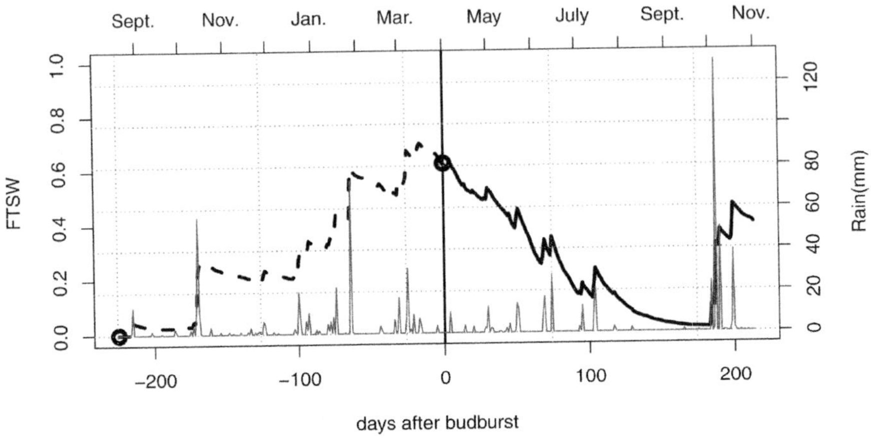

Fig. 8–3. Method for approximating the fraction transpirable soil water at budburst. It is assumed that fraction transpirable soil water is 0 at the end of the previous summer. Then the model is used to estimate the value at budburst.

two weather stations, each approximately 15 km from the vineyard in question, were used.

- Input variables approximated using more easily measured variables (kmax, CN, TTSWtot): The variable kmax will serve as an example. The approximate kmax value is given in Table 8–4 as a function of vigor level and plant height. Though each vigor–height combination is associated with a nominal kmax value, in fact the combination corresponds to a range of values. The range in each direction is estimated as half the maximum distance to an adjoining entry. For example, nominal kmax for vigor = "medium" and height = 1 m is 0.433. The neighboring values are (clockwise from top) 0.415, 0.501, 0.451, and 0.353. The greatest distance is 0.433 – 0.353 = 0.080, and so the distribution of η_i is taken as a uniform distribution with limits 0.433 – 0.040 and 0.433 + 0.040.
- The same procedure is used to estimate the distribution of η_i for CN (Table 8–4) and TTSWtot (Table 8–5).
- Fraction transpirable soil water at budburst (FTSW0): The uncertainty here is estimated as the sum of two terms, using $\eta_i = \left(\mathrm{FTSW0}_i - \tilde{\mathrm{F}}\mathrm{TSW0}_i\right) + \left(\tilde{\mathrm{F}}\mathrm{TSW0}_i - \hat{\mathrm{F}}\mathrm{TSW0}_i\right)$, where $\tilde{\mathrm{F}}\mathrm{TSW0}_i$ is an estimate of FTSW0 based on simulations that begin several years before the year in question (whereas $\hat{\mathrm{F}}\mathrm{TSW0}_i$ only starts at the end of the previous summer). It was felt that this decomposition should give a more reliable estimate of error than looking directly at $\left(\mathrm{FTSW0}_i - \hat{\mathrm{F}}\mathrm{TSW0}_i\right)$. The distribution of the first term is estimated as the empirical distribution of errors for the 10 vineyards in our data set. The distribution of the second term is estimated from multiple simulations. Note that since both $\tilde{\mathrm{F}}\mathrm{TSW0}_i$ and $\hat{\mathrm{F}}\mathrm{TSW0}_i$ are based on simulations, they both have uncertainties due to uncertainties in

any other approximated input variables. Table 8–6 shows the distribution of $FTSW0_i - \hat{FTSW0}_i$ for each scenario (combination of estimated inputs).

Results of Uncertainty Calculations with the Vineyard Model

Evaluating the Assumptions about Uncertainty in the Input Variables

We have made assumptions about the uncertainty in each approximated input variable. We can test these assumptions to some extent, using the data from the 10 vineyards in our data set where we have both approximate and measured values of the input variables. For these cases we calculate the distribution of $\eta_i + \hat{T}_i$ and compare with the measured value T_i. Ideally we should find that in x% of the cases (for any x from 0 to 100), the measured value should fall within the x percentile of the distribution. In practice we simply examine the fraction of observed values that fall within the 100% range of $\eta_i + \hat{T}_i$.

Figure 8–4 shows the distribution of $\eta_i + \hat{T}_i$ and the value of T_i for the input variables kmax, CNcover, CNbare, TTSWtot, and FTSW0, for each of the 10 vineyards in the data set used here. Some of the observed values fall outside the uncertainty range, which indicates that the range of $\eta_i + \hat{T}_i$ might be slightly underestimated in some cases. However, overall the uncertainty ranges seem reasonable.

Scenarios for Approximated Inputs

We have defined eight different scenarios, with different combinations of approximated input variables (Table 8–7). The reference scenario represents the case where all the input variables are measured. The last scenario (U7) represents the case where only the most easily obtained inputs, namely percentage grass cover and budburst date, are available. In scenario U1 only CN is approximated, and in U6 only weather is approximated. The other scenarios correspond to various combinations of approximated inputs.

Table 8–6. Uncertainty in FTSW0 for the different combinations of estimated input variables.

Lower and upper limits	Scenario†							
	Uref	U1	U2	U3	U4	U5	U6	U7
u_{min}‡	0.0	0.0	−0.20	−0.17	−0.23	−0.23	0.00	−0.23
u_{max}‡	0.00	0.00	−0.22	0.14	0.19	0.17	0.00	0.17

† Each scenario corresponds to a particular list of input variables that are estimated (see Table 8–7). In scenarios Uref, U1, and U6, FTSW0 is measured so there is no uncertainty.

‡ The uncertainty is represented by a uniform distribution with minimum and maximum values equal to u_{min} and u_{max}, respectively.

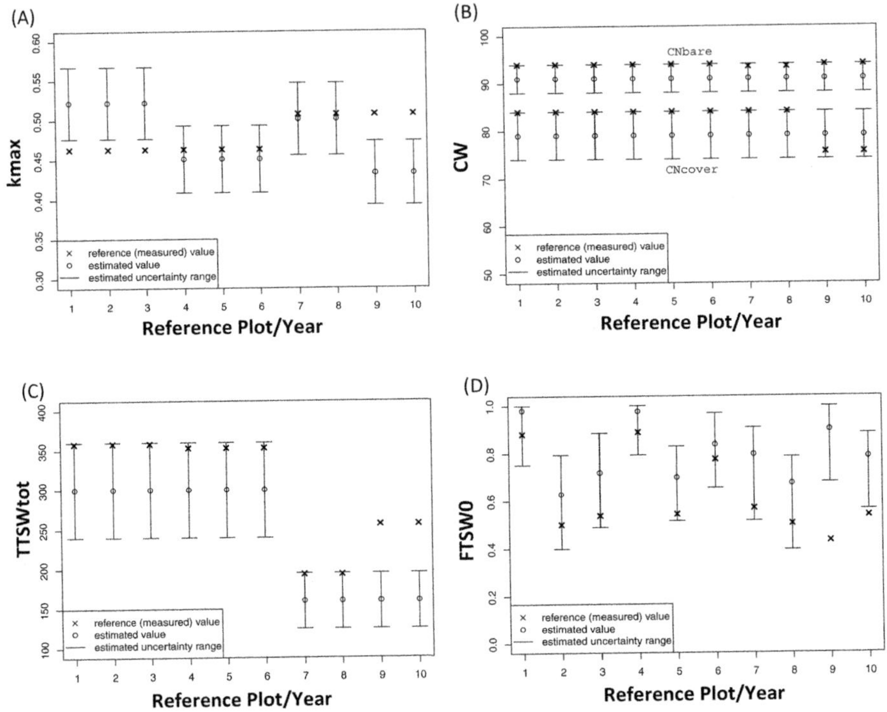

Fig. 8–4. Evaluation of the assumptions about uncertainty in the input variables (a) kmax, (b) CNcover and CNbare, (c) TTSWtot, and (d) FTSW0. For each of the 10 reference plots, the total uncertainty range is shown, together with the estimated value and the measured value. For FTSW0, the uncertainty is calculated for scenario U5.

Table 8–7. Scenarios (combinations of approximated input values).†

Scenario	pic	TTSWtot	FTSW0	dbud	CNbare and CNcover	kmax	weather
Uref	1	1	1	1	1	1	1
U1	1	1	1	1	0	1	1
U2	1	1	0	1	0	1	1
U3	1	0	0	1	1	1	1
U4	1	0	0	1	0	1	1
U5	1	0	0	1	0	0	1
U6	1	1	1	1	1	1	1
U7	1	0	0	1	0	0	1

† A value 1 (0) indicates that the input variable is measured (approximated).

Uncertainty due to Approximating Inputs

The uncertainty in simulated FTSW(*t*) due to uncertainty in the input variables was calculated for the test vineyard for both the years 2004 (Fig. 8–5) and 2006 (Fig. 8–6). The difference between each scenario and the reference scenario indicates how much uncertainty is added due to approximating those input variables that are estimated in that scenario. From a different point of view, the difference shows how much uncertainty could be reduced if the approximations were replaced by measurements.

In 2004, the high winter and spring rainfall year, simulated FTSW(*t*) is always well above the threshold for irrigation. The level of uncertainty depends on which input variables are estimated, but in any case the simulated values with uncertainty never reach the threshold.

The situation is quite different in 2006, the low rainfall year. In this case several of the scenarios have high probability of requiring irrigation. The difference between scenarios is important. Depending on the scenario, the average of the simulations can be quite biased compared with the reference scenario U0 (e.g., U4). Uncertainty in CN alone (scenario U1) leads to little uncertainty in the simulations, since this is a low winter and spring rainfall year and runoff is therefore unimportant. Uncertainty in weather alone (scenario U6) also leads to little uncertainty in the simulations. For the other scenarios the probability that irrigation is required ranges from 0.1 to 0.5.

Total Uncertainty

Figure 8–7 and Fig. 8–8 show simulated FTSW(*t*) for 2004 and 2006, respectively, first without uncertainty information, then with just uncertainty due to approximating inputs for scenario U3, and finally with total uncertainty, due to both uncertainty with known inputs and the uncertainty due to approximating inputs, again for scenario U3. The lower limit of the white area is the threshold for irrigation; irrigation is indicated if simulated FTSW(*t*) goes below that threshold.

In 2004, even taking into account total uncertainty, the simulated values indicate that irrigation is unnecessary. In 2006 on the other hand, taking into account total uncertainty rather than just the uncertainty in the inputs, increases substantially the probability that irrigation is required.

Discussion

The vineyard model presented here is used by multiple extension services as an aid to producing irrigation advice. The extension services do not present the results as a definitive recommendation for irrigation timing, but rather as a general indication of water status in vineyards in the region. Nevertheless, even in this auxiliary role, it is important to have an estimate of the uncertainty in the

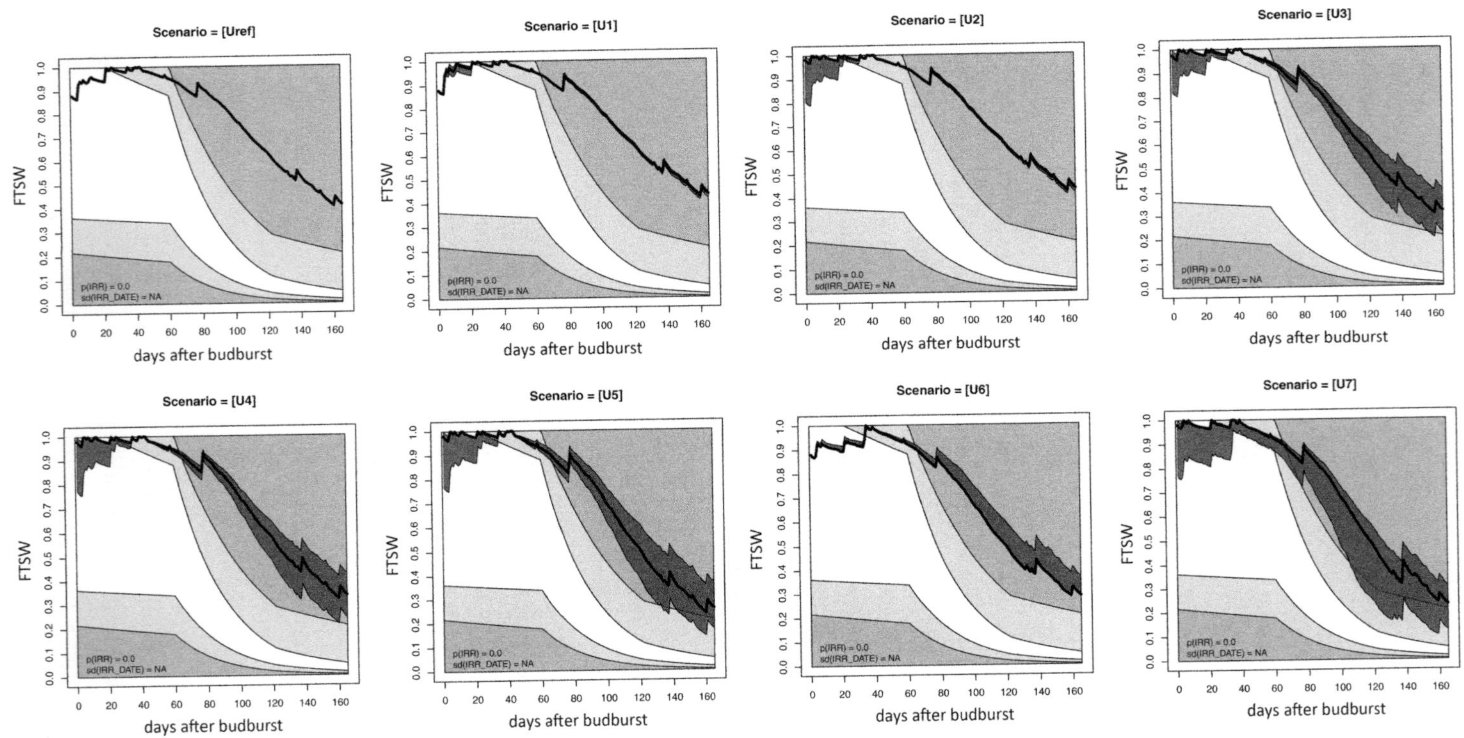

Fig. 8–5. Simulated FTSW(t) and 90% confidence interval resulting from uncertainty in input variables, for the year 2004. The scenarios (Table 8–7) correspond to different combinations of input variables that are approximated. The lower limit of the white area is the threshold for irrigation; irrigation is indicated if simulated FTSW(t) goes below that threshold; p(IRR) is the probability that irrigation is required at some time, given the uncertainty in FTSW(t); and sd(IRR_DATE) is the standard deviation of the first date when the probability of irrigation is > 0. If the probability of irrigation is 0, then sd(IRR_DATE) = NA. The histogram gives the probabilities of first irrigation date.

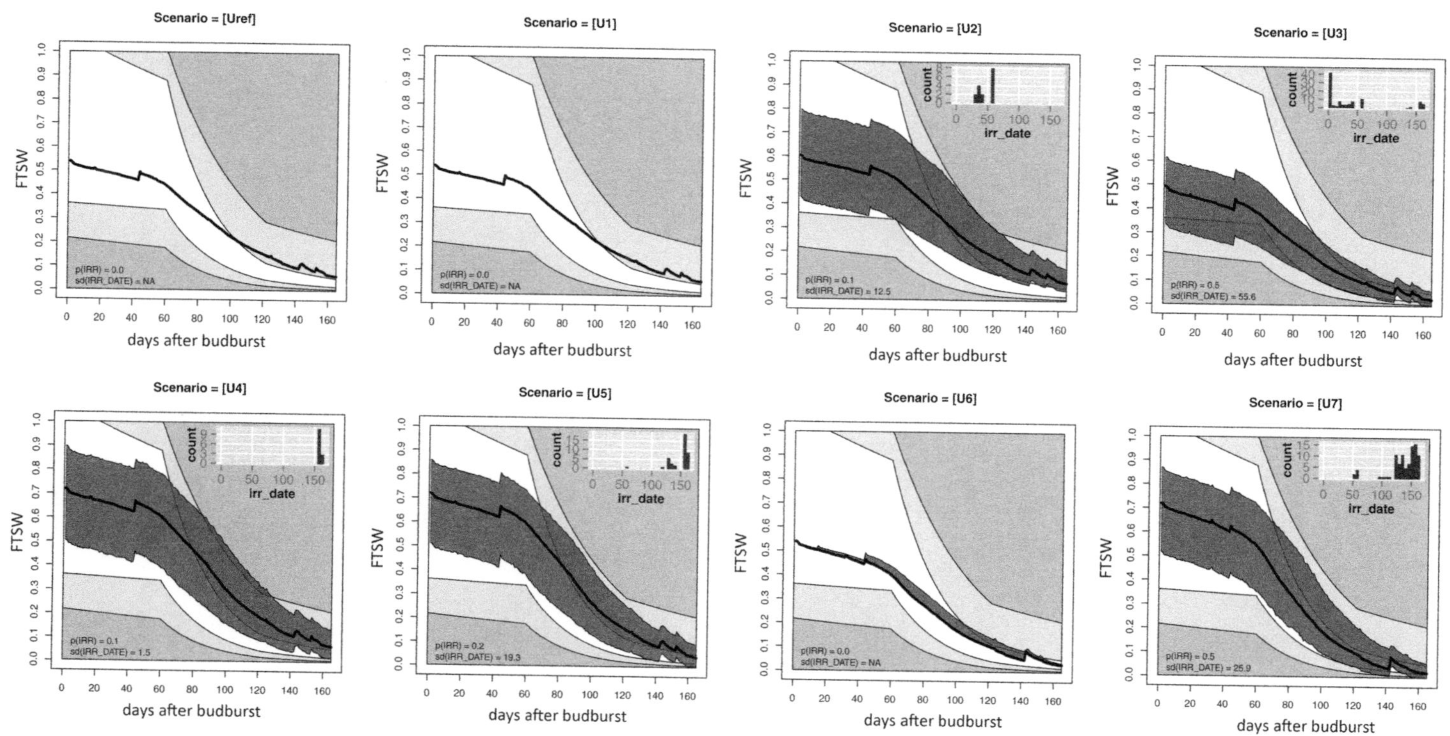

Fig. 8–6. Simulated FTSW(t) and 90% confidence interval resulting from uncertainty in input variables, for the year 2006. The scenarios (Table 8–7) correspond to different combinations of input variables that are approximated. The lower limit of the white area is the threshold for irrigation; irrigation is indicated if simulated FTSW(t) goes below that threshold; p(IRR) is the probability that irrigation is required at some time, given the uncertainty in FTSW(t); and sd(IRR_DATE) is the standard deviation of the first date when the probability of irrigation is > 0. If the probability of irrigation is 0, then sd(IRR_DATE) = NA. The histogram gives the probabilities of first irrigation date.

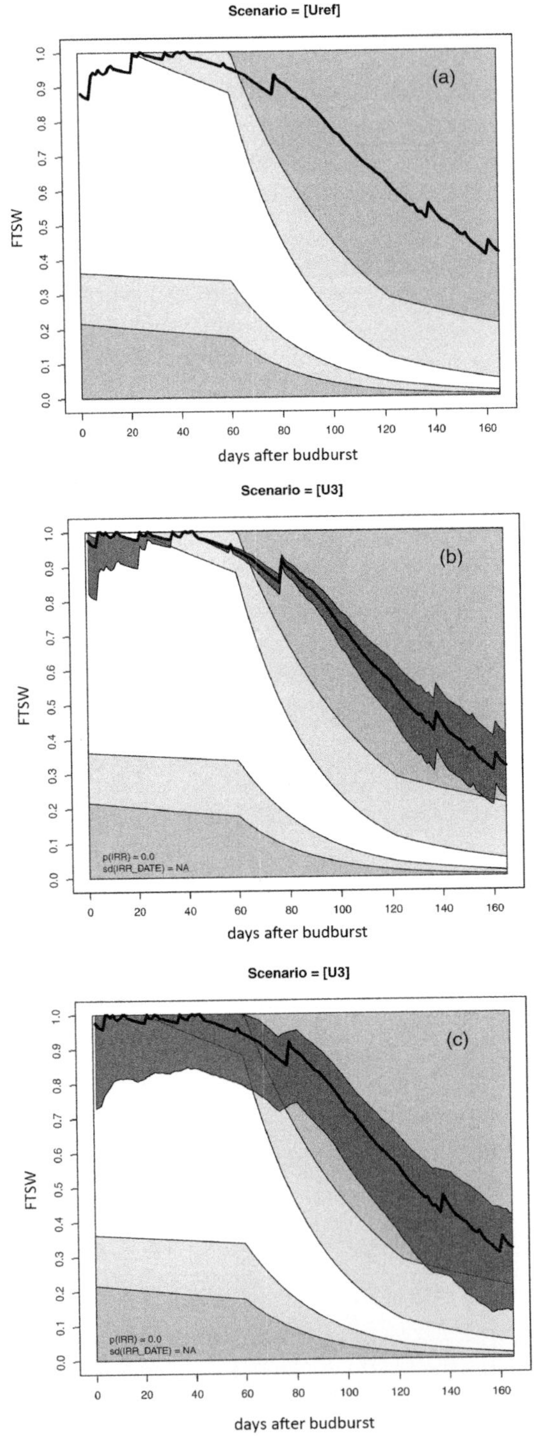

Fig. 8–7. Comparison of uncertainties due to input variables and total uncertainty for the year 2004. From left simulated FTSW(t), 90% confidence interval resulting from uncertainty in input variables for scenario U3 and 90% confidence interval resulting from total uncertainty (uncertainty in input variables and uncertainty in the model equations and parameters and residual error). The lower limit of the white area is the threshold for irrigation; irrigation is indicated if simulated FTSW(t) goes below that threshold.

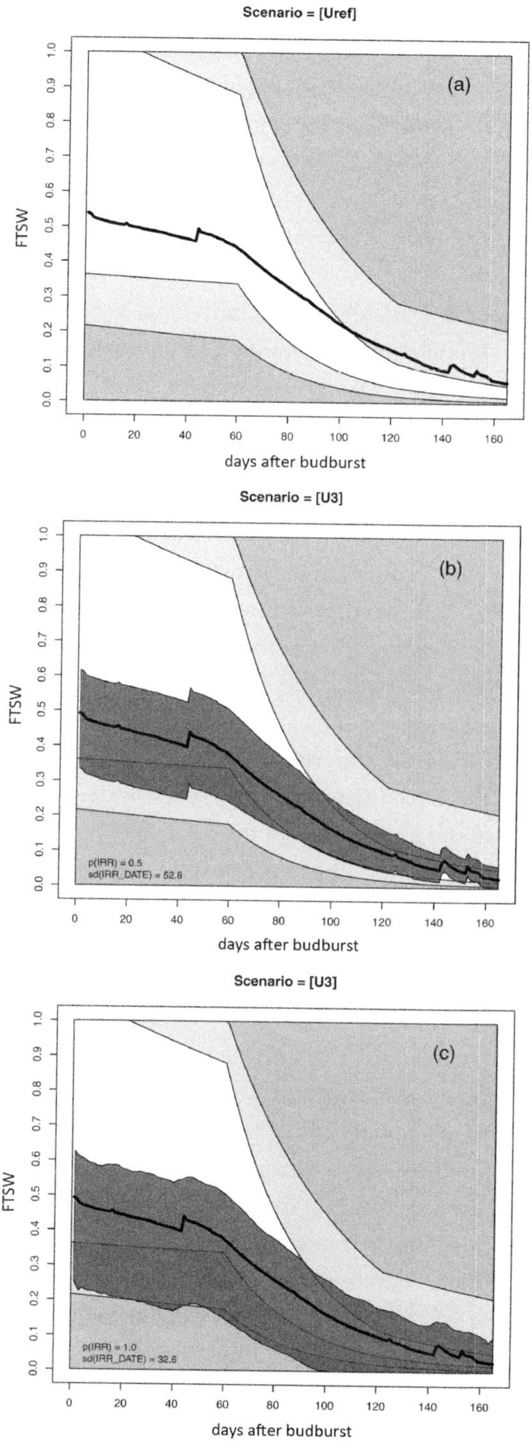

Fig. 8–8. Comparison of uncertainties due to input variables and total uncertainty for the year 2006. From left simulated FTSW(*t*), 90% confidence interval resulting from uncertainty in input variables for scenario U3 and 90% confidence interval resulting from total uncertainty (uncertainty in input variables and uncertainty in the model equations and parameters and residual error). The lower limit of the white area is the threshold for irrigation; irrigation is indicated if simulated FTSW(*t*) goes below that threshold.

simulated values that the model produces. The extension services have themselves expressed a need for uncertainty information, both for their own use and for transmission to growers. The usefulness of the model as a decision aid would be increased, according to the users, if uncertainty information were added.

The extension services, using the approach described above, could produce graphs similar to those of Fig. 8–7, in real time, based on the input variables that are approximated in their test vineyards. Such graphs would make clear the level of uncertainty in the simulated results.

One can imagine making the model available to each individual grower. In this way, each grower could obtain simulation results tailored to their particular vineyards. In this case, the growers would probably rely more on the model results than they currently do, making the uncertainty information even more valuable.

Often users replace measured values for input variables by approximations that are easier and/or cheaper to obtain, but this is at the price of increasing uncertainty in the simulated results. It is then of importance to the user to know how much uncertainty this adds in order to decide if it is worthwhile to invest in measurements of more input variables. This would require a somewhat different presentation of uncertainty, since one is evaluating the reduction in uncertainty on the average over years.

The model presented here is also used as a research tool, in particular for studying the consequences of introducing a cover crop into a vineyard (Ripoche et al., 2011). It has been shown that the introduction of grass cover between the rows of a vineyard can reduce runoff and erosion, but at the price of extra water consumption (Celette et al., 2009a) and competition for nitrogen (Celette et al., 2009b), which can lead to yield reduction. The model can be used to compare the two alternative cropping systems, that is, vineyards with and without grass cover. Here again, adding uncertainty information would be valuable, allowing one to better understand the reliability of the simulated results.

Our procedure treats separately two sources of uncertainty. The first is the uncertainty in the model assuming no error in the input variables. This includes uncertainty in the model equations, in the model parameters, and the model residual error. We propose estimating this uncertainty using the error in hindcasts. This estimates an average error. There is an important underlying assumption here, which is that the model equations and parameters are used as provided—the user does not recalibrate the model. Therefore, we are not interested in the uncertainty in the estimated parameter values. Rather, the model plus parameter values are treated as fixed, and we are just interested in the error in the model with those parameters. Note that as more information is accumulated, the estimate of error for measured inputs could be improved.

The second source of uncertainty is uncertainty in the input variables. This is very dependent on the model and on the way in which the input variables are approximated. Here we propose methods of estimating those input variables that are the most difficult to obtain. We also propose probability distributions of $\eta_i = T_i - \hat{T}_i$ for each input that is estimated. This is similar to the step in classical uncertainty analysis where one fixes the distribution for each uncertain factor. The difference here is that the uncertainties must be based on user practice if the uncertainty calculations are to be relevant. We have used historical data to test the uncertainty distributions for estimated inputs. This is good practice to ensure that these distributions are reasonable.

Given the distribution of error for measured inputs and the distribution of the uncertain inputs, one can do a Monte Carlo calculation to obtain a distribution of $\text{FTSW}_i(t) - \hat{\text{FTSW}}\left(t, \hat{T}_i, X_i; \theta\right)$. This gives the detailed information about uncertainty that one seeks.

This procedure is a practical implementation of the principles for estimating overall uncertainty proposed by Refsgaard et al. (2007).

We have proposed a graphical representation of uncertainty information. This would have to be tested with users to see if it is useful in practice. The choice of illustrating 90% confidence intervals could also be questioned. This might be too pessimistic (or optimistic). Perhaps bounds that encompass less (or more) of the probability distribution would be preferable.

Conclusions

Models can be very valuable as one of the elements of a decision support system. However, it is clear that there are substantial uncertainties attached to the simulations, and the models would be even more valuable if these uncertainties were explicitly acknowledged and quantified.

We have proposed a procedure for adding uncertainty information to a vineyard soil moisture model that is used as an irrigation decision aid by extension services. The uncertainty calculations are based on user practices. The general procedure could be applied very generally to other models used as decision aides, in cases where uncertainties in inputs are an important consideration.

Acknowledgments

This study was a part of the project Associate, a level of error in predictions of models for agronomy (network RMT modélisation et agriculture, www.modelia.org) led by ACTA (French Technical Institute for Agriculture). It was funded by a grant from the Ministry of Agriculture and Fishing of France.

References

Bellocchi, G., M. Rivington, M. Donatelli, and K. Matthews. 2010. Validation of biophysical models: Issues and methodologies. A review. Agron. Sustainable Dev. 30(1):109–130.

Bennett, N.D., B.F.W. Croke, G. Guariso, J.H.A. Guillaume, S.H. Hamilton, A.J. Jakeman, S. Marsili-Libelli, L.T.H. Newham, J.P. Norton, C. Perrin, S.A. Pierce, B. Robson, R. Seppelt, A.A. Voinov, B.D. Fath, and V. Andreassian. 2013. Characterising performance of environmental models. Environ. Modell. Software 40:1–20. doi:10.1016/j.envsoft.2012.09.011

Boughton, W.C. 1989. A review of the USDA SCS curve number method. Aust. J. Soil Res. 27(3):511–523. doi:10.1071/SR9890511

Celette, F., A. Findeling, and C. Gary. 2009a. Spatial and temporal changes to the water regime of a Mediterranean vineyard due to the adoption of cover cropping. Eur. J. Agron. 30:41–51. doi:10.1016/j.eja.2008.07.003

Celette, F., A. Findeling, and C. Gary. 2009b. Competition for nitrogen in an unfertilized intercropping system: The case of an association of grapevine and grass cover in a Mediterranean climate. Eur. J. Agron. 30(1):41–51. doi:10.1016/j.eja.2008.07.003

Celette, F., A. Ripoche, and C. Gary. 2010. WaLIS—A simple model to simulate water partitioning in a crop association: The example of an intercropped vineyard. Agric. Water Manage. 97(11):1749–1759. doi:10.1016/j.agwat.2010.06.008

Corbane, C., F. Jacob, D. Raclot, J. Albergel, and P. Andrieux. 2012. Multitemporal analysis of hydrological soil surface characteristics using aerial photos: A case study on a Mediterranean vineyard. Int. J. Appl. Earth Obs. Geoinf. 18:356–367. doi:10.1016/j.jag.2012.03.009

Delpuech, X., F. Celette, and C. Gary. 2010. Validation of the water balance model WaLIS in cover-cropped vineyards. In: 21ème Conférence Du COLUMA, Journées Internationales Sur La Lutte Contre Les Mauvaises Herbes, Dijon, France. 8–9 Dec. 2010. Association Française de Protection des Plantes, Alfortville, France. p. 414–424.

Duru, M., M. Adam, P. Cruz, G. Martin, P. Ansquer, C. Ducourtieux, C. Jouany, J.P. Theau, and J. Viegas. 2009. Modelling above-ground herbage mass for a wide range of grassland community types. Ecol. Modell. 220(2):209–225. doi:10.1016/j.ecolmodel.2008.09.015

Gary, C., J.C. Payan, K. Kansou, A. Pellegrino, and J. Wery. 2005. A model-based diagnosis tool to evaluate the water stress experienced by vine in relation with production and quality objectives. In: XIV International GESCO Viticulture Congress, Geisenheim, Germany. 23–27 Aug. 2005. Groupe d'Etude des Systemes de Conduite de la Vigne, Geisenheim, Germany. p. 449–456.

Gaudin, R., and C. Gary. 2012. Model-based evaluation of irrigation needs in Mediterranean vineyards. Irrig. Sci. 30(5):449–459. doi:10.1007/s00271-012-0349-x

Gijsman, A.J., S.S. Jagtap, and J.W. Jones. 2002. Wading through a swamp of complete confusion: How choose a to method for estimating soil water retention parameters for crop models. Eur. J. Agron. 18(1–2):77–106. doi:10.1016/S1161-0301(02)00098-9

Jakeman, A.J., R.A. Letcher, and J.P. Norton. 2006. Ten iterative steps in development and evaluation of environmental models. Environ. Modell. Software 21(5):602–614. doi:10.1016/j.envsoft.2006.01.004

Lebon, E., V. Dumas, P. Pieri, and H.R. Schultz. 2003. Modelling the seasonal dynamics of the soil water balance of vineyards. Funct. Plant Biol. 30(6):699–710. doi:10.1071/FP02222

Ojeda, H. 2007. Irrigation qualitative de précision de la vigne. Prog. Agric. Vitic. 124:133–141.

Pellegrino, A., E. Goze, E. Lebon, and J. Wery. 2006. A model-based diagnosis tool to evaluate the water stress experienced by grapevine in field sites. Eur. J. Agron. 25(1):49–59. doi:10.1016/j.eja.2006.03.003

Pellegrino, A., E. Lebon, M. Voltz, and J. Wery. 2005. Relationships between plant and soil water status in vine (*Vitis vinifera* L.). Plant Soil 266(1–2):129–142. doi:10.1007/s11104-005-0874-y

Refsgaard, J.C., J.P. van der Sluijs, A.L. Højberg, and P.A. Vanrolleghem. 2007. Uncertainty in the environmental modelling process—A framework and guidance. Environ. Modell. Software 22(11):1543–1556. doi:10.1016/j.envsoft.2007.02.004

RGA. 2000. RGA recensement agricole. Service de La Statistique et de La Prospective, Du Ministère de L'agriculture et de La Pêche, Paris.

RGA. 2010. RGA recensement agricole. Service de La Statistique et de La Prospective, Du Ministère de L'agriculture et de La Pêche, Paris.

Riou, C., C. Valancogne, and P. Pieri. 1989. A simple model for interception of solar radiation by a vineyard, comparison with field data. Agronomie 9(5):441–450.

Ripoche, A., J.-P. Rellier, R. Martin-Clouaire, N. Paré, A. Biarnès, and C. Gary. 2011. Modelling adaptive management of intercropping in vineyards to satisfy agronomic and environmental performances under Mediterranean climate. Environ. Modell. Software 26(12):1467–1480. doi:10.1016/j.envsoft.2011.08.003

Roux, S., F. Brun, and D. Wallach. 2014. Combining input uncertainty and residual error in crop model predictions: A case study on vineyards. Eur. J. Agron. 52:191–197. doi:10.1016/j.eja.2013.09.008

Soubeyroux, J-M., J-P. Vidal, J. Najac, N. Kitova, M. Blanchard, P. Dandin, E. Martin, C. Pagé, and F. Habets. 2011. Projet ClimSec: Impact du changement climatique en France sur la sécheresse et l'eau du sol. Météo-France, Toulouse, France.

USDA–NRCS. 2007. National engineering handbook, Part 630, hydrology. USDA–NRCS, Washington, DC.

Wallach, D., D. Makowski, J.W. Jones, and F. Brun. 2014. Working with dynamic crop models. 2nd ed. Academic Press, London.

Determination of Irrigation Depths Using a Numerical Model and Quantitative Weather Forecasts and Comparison with an Experiment

H. Fujimaki,* I. Tokumoto, T. Saito, M. Inoue, M. Shibata, M. Okazaki, K. Nagaz, and Fathia El-Mokh

Abstract

A scheme for determining irrigation depth using a numerical model of crop response to irrigation and quantitative weather forecasts was presented. To optimize each irrigation depth, a concept of virtual income, which is proportional to an increment in transpiration during an irrigation interval, is introduced. The optimization module has been incorporated into a numerical model that simulates water, solute, and heat movement and crop growth. Two field experiments were performed to evaluate the effectiveness of the presented scheme. Results indicated that the predicted water content and salinity agreed well with observations and that the net return was similar to treatment irrigated by an automatic irrigation system, which applies water whenever the monitored value reaches the prescribed trigger value. These results revealed that the proposed scheme can improve irrigation scheduling systems and save on costs for equipment such as soil moisture probes, dataloggers, and solenoid valves.

H. Fujimaki (fujimaki@alrc.tottori-u.ac.jp), M. Inoue (m11minoue@cpost.plala.or.jp), M. Shibata (viva-shiba@h2.dion.ne.jp), and M. Okazaki (alian_johns_wpc@yahoo.co.jp), Arid Land Research Center, Tottori University, 1390 Hamasaka, 680-0001 Tottori, Japan. *Corresponding author.

T. Saito, Faculty of Agriculture, Tottori University, 4-101 Koyamacho-Minami, Tottori 680-8550, Japan (tadaomi@muses.tottori-u.ac.jp).

I. Tokumoto, Faculty of Agriculture, Saga University, 1 Honjo-machi, Saga 840-8502, Japan (yasu@cc.saga-u.ac.jp).

K. Nagaz (nagaz.kameleddine@ira.rnrt.tn) and F. El-Mokh (F.mokh@yahoo.fr), Institut des Régions Arides, 4119 Médenine, Tunisia.

doi:10.2134/advagricsystmodel5.c9

Practical Applications of Agricultural System Models to Optimize the Use of Limited Water
Lajpat R. Ahuja, Liwang Ma, and Robert J. Lascano, Editors
Advances in Agricultural Systems Modeling, Volume 5. Lajpat R. Ahuja, Series Editor

USDA-NRCS

Soaring food prices and the intensified scarcity of water resources put a new emphasis on efficient use of water in irrigation. The determination of irrigation depth has still widely relied on the experience (i.e., fixed amount) or intuition of farmers, even in industrialized countries. Such conventional irrigation may cause yield reduction or waste water. To precisely meet crop water requirements and respond to water and/or salinity stresses quickly, automatic irrigation systems using sensors have been developed. Automatic irrigation systems generally open the solenoid valve of a pressurized pipeline for a prescribed interval whenever a monitored value such as the water content, suction, or temperature difference between the canopy and air reaches a prescribed trigger value. Such systems, however, require high initial investment and have difficulty adjusting the irrigation depth to weather forecasts. For example, it is obviously wasteful to apply full irrigation when rain is forecast for the next day. "Water stat" systems may nevertheless do it if the currently monitored value reaches its threshold value.

Expensive monitoring using sensors can be improved by a numerical simulation of water, solute, and heat transport in soil and crop response to them. Personal computers with high-speed, large memory are becoming affordable even for farmers in developing countries. Not only can a numerical model monitor the current status, it can also predict near future conditions. Numerical prediction requires quantitative weather forecasts. Today, freely accessible quantitative weather forecasts, whose accuracy is improving, have been provided on the web (e.g., Weather Underground). These progresses have enabled the optimization of irrigation depths using quantitative weather forecasts as input data for numerical models such that the net return is maximized. This chapter presents a procedure for the determination of irrigation depths using quantitative weather forecasts. In the first section, concepts, major assumptions, and the procedure are presented. An example of the numerical model and governing equations used in the scheme is presented in the second section. We also tested its effectiveness with two field experiments. The methods and results of the experiments are reported in the third and fourth sections and discussions are included in these sections.

New method to determine irrigation depth

Maximization of Virtual Net Return

Like other inputs such as labor or fertilizer, the purpose of irrigation is not necessarily to obtain the highest yields, or even water use efficiency, but to maximize the net returns. While the timing of irrigation is generally restricted to social factors such as rotation or availability of labor, irrigators have more discretion with regard to the amount of irrigation. Thus, we focus on the optimization of the amount.

If we can calculate the net return until the next irrigation, the irrigation depth can be optimized such that the net return is maximized. Although in reality income is realized when a crop is harvested and sold, we assume that a farmer can obtain virtual income, which is proportional to an increment of dry matter attained during an interval. Also, water must be priced high enough to give irrigators incentive to save water. The virtual net return, N_r (\$ ha^{-1}), during an interval is then defined as

$$N_r = P_c \varepsilon \tau_i - P_w W - C_{ot}, \qquad [1]$$

where P_c is the price of the crop (\$ kg^{-1} of dry matter), ε is the water use efficiency of the crop, τ_i is the cumulative transpiration during an interval (1 cm = 10^5 kg ha^{-1}), P_w is the price of water (\$ kg^{-1}), W is the irrigation depth (1 cm = 10^5 kg ha^{-1}), and C_{ot} is other costs (\$ ha^{-1}). The amount of the valuable part (fruit, grain, etc.) of the crop is assumed to be proportional to dry matter production, which is well known to be approximately proportional to cumulative transpiration. Under given conditions, N_r is a function of W. The problem is thus a simple one-dimensional search. Conditions such as rainfall or solar radiation may also largely affect N_r, but they are not the decision parameter to be optimized.

To estimate transpiration, τ_i, which dynamically responds to matric and osmotic potential in soil, and therefore irrigation, a sophisticated model of the crop response to irrigation is required. Numerical models, including those developed in the realm of soil physics, for example, HYDRUS (Šimůnek et al., 2006), the Soil–Water–Atmosphere–Plant (SWAP) model (Van Dam et al., 1997), and the Root Zone Water Quality Model (RZWQM) (Ahuja et al., 2000), can be used as such.

Correction of Virtual Net Return

One possible negative result of the concept of virtual income is that initial growth may not be obtained, since transpiration of the initial stage is smaller than the later stage, while initial growth may be equally as important. To correct the possible underestimation of the contribution of transpiration during the initial stage, the income correction factor, k_i, may be used:

$$N_{rc} = P_c \varepsilon \tau_i k_i - P_w W - C_{ot}, \qquad [2]$$

where N_{rc} is the corrected virtual net return. The k_i value may be larger than unity during the early stage and lower than unity during the maturity stage. Unlike the original virtual net return, the sum of N_{rc} will not correspond to the actual net return even if predicted dry matter production or yield matches the actual one. Nevertheless, the correction may enhance actual net return.

Estimation of Optimum Irrigation Depth

If transpiration, τ_i, during an interval is quickly predicted from a numerical model, a normal one-dimensional search such as the golden section method or Brents's method may be used. But sophisticated models for predicting multi-dimensional water flow often require heavy computation. Since time spent for optimization should be completed within a few hours practically, the following estimation of optimum irrigation depth may be used.

Response of τ_i to irrigation, W, may be empirically described as

$$\tau_i = \int T_r \mathrm{d}t = a_t\left[1 - \exp\left(b_t W\right)\right] + \tau_0, \qquad [3]$$

where T_r is the transpiration rate (cm s^{-1}), a_t and b_t are the fitting parameters, and τ_0 is τ at $W = 0$. Other empirical functions that are convex upward and have a constant asymptote may be used, but Eq. [3] may be the simplest and best-fit one. This empirical function can fit the relationship between W and yield as shown by Heermann et al. (1990) and Fereres and Soriano (2006) as generalized relationships. If yield is proportional to transpiration, the relationship between W and τ may also be described with Eq. [3].

The maximum net return is obtained at W, at which the derivative of Eq. [2] becomes zero.

$$\frac{\mathrm{d}N_{rc}}{\mathrm{d}W} = -P_c \in k_i a_t b_t \exp\left(b_t W\right) - P_w = 0 \qquad [4]$$

$$W = -\frac{1}{b_t}\ln\left(\frac{P_w}{P_c \in k_i a_t b_t}\right) \qquad [5]$$

Therefore, if a_t and b_t are known, the optimum irrigation depth can be estimated. To obtain the two values, an additional two points—transpiration at intermediate irrigation depth (W_1, τ_1) and at the upper limit of irrigation depth (W_2, τ_2)—must be known.

$$\tau_1 = a_t\left[1-\exp\left(b_t W_1\right)\right]+\tau_0 \quad [6]$$

$$\tau_2 = a_t\left[1-\exp\left(b_t W_2\right)\right]+\tau_0 \quad [7]$$

Rearranging Eq. [6] gives

$$a_t = \frac{\tau_1-\tau_0}{\left[1-\exp\left(b_t W_1\right)\right]}. \quad [8]$$

Eq. [6] − [7] gives

$$a_t = \frac{\tau_1-\tau_2}{\left[\exp\left(b_t W_2\right)-\exp\left(b_t W_1\right)\right]}. \quad [9]$$

Therefore,

$$\frac{\tau_1-\tau_2}{\left[\exp\left(b_t W_2\right)-\exp\left(b_t W_1\right)\right]}-\frac{\tau_1}{\left[1-\exp\left(b_t W_1\right)\right]}=0. \quad [10]$$

The value of b_t can be easily searched using the bisection method (Press et al., 1992), which repeatedly bisects an interval and then selects a subinterval in which a root must lie. Thus, by obtaining τ at three irrigation depths—zero, the upper limit, and the intermediate of the two—we may estimate the optimum irrigation depth.

Procedure

The procedure of the new method is illustrated in Fig. 9–1. First, (i) using the records of climatic condition, (ii) a numerical simulation is performed to estimate the current status. Then, (iii) quantitative weather forecasts are downloaded as input data, and (iv) we repeat simulations three times at upper limit, zero, and intermediate irrigation depths. Then irrigation is performed. In the early morning of the next irrigation day, (v) the current status is estimated by simulation using the actual records of irrigation depth and (vi) climatic condition from 00:00 of the last irrigation day until 23:59 of the last day. This cycle continues until the last irrigation.

The author has developed software (WeatherForecastDownloader) to download the HTML file and parse the HTML for Weather Underground (http://www.

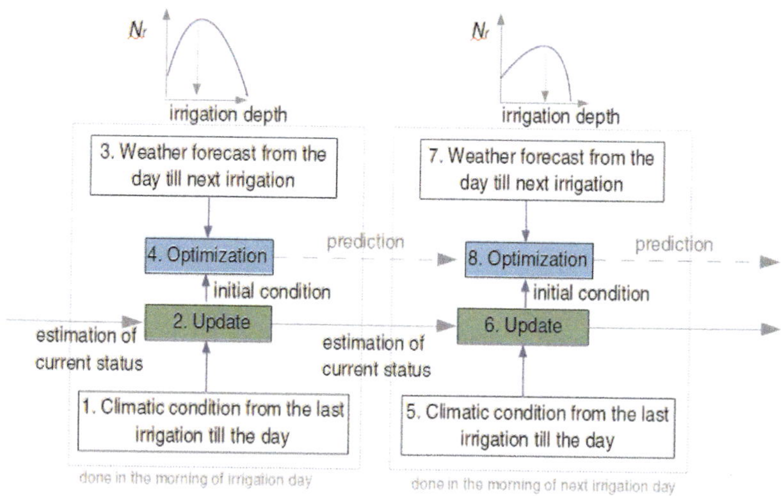

Fig. 9–1. New procedure to determine irrigation depth.

wunderground.com) and Yahoo! JAPAN (http://weather.yahoo.co.jp/weather/jp/31/6910/31302.html) for consecutive days. Usually, quantitative weather forecasts provide figures except for solar radiation. For example, Weather Underground provides cloud cover instead of solar radiation. In such a case, the empirically known relationship between cloud cover and the ratio of extraterrestrial to inner terrestrial radiation may be applied. Yahoo! JAPAN provides four qualitative descriptions: "clear," "cloudy," "rain," and "heavy rain." In this case, the empirically determined average ratio of extraterrestrial to inner terrestrial radiation for each description (e.g., "clear" = 0.82, "cloudy" = 0.63, "rain" = 0.32) may be multiplied to extraterrestrial short-wave radiation.

Numerical Model

The algorithm described above and the user interface for parameter setting have been incorporated into a numerical model, WASH_2D, which solves governing equations for two-dimensional movement of water, solute, and heat in soils with the finite difference method. The software is freely distributed with source code under the general public license from the website of the Arid Land Research Center, Tottori University (http://www.alrc.tottori-u.ac.jp/fujimaki/download/WASH_2D).

Governing Equation for Water Flow

The two-dimensional water balance equation for the combined liquid and gaseous phases is given by

$$\frac{\partial\theta}{\partial t}=-\left(\frac{\partial q_{\mathrm{lx}}}{\partial x}+\frac{\partial q_{\mathrm{lz}}}{\partial z}\right)-\left(\frac{\partial q_{\mathrm{vx}}}{\partial x}+\frac{\partial q_{\mathrm{vz}}}{\partial z}\right)-S\ , \qquad [11]$$

where θ is volumetric water content, t is time (s), q_l is liquid water flux (cm s^{-1}), q_v is water vapor flux (cm s^{-1}), x is horizontal distance, z is depth (cm), and S is sink term. Subscripts x and z refer to the direction of each flux.

The liquid water flux, q_l, is described using Darcy's law:

$$q_{\mathrm{lx}}=-K\frac{\partial\psi}{\partial x} \qquad [12a]$$

$$q_{\mathrm{lz}}=-K\left(\frac{\partial\psi}{\partial z}-1\right), \qquad [12b]$$

where K is the hydraulic conductivity (cm s^{-1}) and ψ_w is the metric potential (cm).

Water vapor flux is divided into two terms because of water potential and thermal gradient, and the latter term is multiplied by the mechanical enhancement factor η to incorporate the "liquid island" effect (Philip and De Vries, 1957):

$$q_{\mathrm{vx}}=-a\tau\rho_{\mathrm{w}}^{-1}h_{\mathrm{r}}D_{\mathrm{va}}\left\{\frac{\rho_{\mathrm{vsat}}}{R_{\mathrm{v}}T_{\mathrm{s}}}\frac{\partial\psi}{\partial x}+\eta\frac{\partial\rho_{\mathrm{vsat}}}{\partial T_{\mathrm{s}}}\frac{\partial T_{\mathrm{s}}}{\partial x}\right\} \qquad [13a]$$

$$q_{\mathrm{vz}}=-a\tau\rho_{\mathrm{w}}^{-1}h_{\mathrm{r}}D_{\mathrm{va}}\left\{\frac{\rho_{\mathrm{vsat}}}{R_{\mathrm{v}}T_{\mathrm{s}}}\frac{\partial\psi}{\partial z}+\eta\frac{\partial\rho_{\mathrm{vsat}}}{\partial T_{\mathrm{s}}}\frac{\partial T_{\mathrm{s}}}{\partial z}\right\}, \qquad [13b]$$

where a is the air-filled porosity, τ is the tortuosity for gas transport, D_{va} is the diffusion coefficient of water vapor in free air (g cm^{-2} s^{-1}), ρ_w is the density of water (0.997 g cm^{-3} at 25°C), h_r is the relative humidity, ψ is the water potential (cm), η is the enhancement factor for thermal water vapor movement, ρ_{vsat} is the saturated water vapor density (g cm^{-3}), T_s is the soil temperature (K), and R_v is the gas constant for water vapor (4697 cm K^{-1}). Assuming thermodynamic equilibrium between the liquid and gaseous phases, the relative humidity in a soil (h_r) or at the soil surface (h_{rs}) can be calculated using (Philip and De Vries, 1957)

$$h_{\mathrm{r}}=\exp\left(\frac{\psi_{\mathrm{w}}}{R_{\mathrm{v}}T_{\mathrm{s}}}\right)\ . \qquad [14]$$

Governing Equation for Solute Movement

The two-dimensional solute balance is given by

$$\frac{\partial(\theta c)}{\partial t}=-\left(\frac{\partial q_{sx}}{\partial x}+\frac{\partial q_{sz}}{\partial z}\right)-S_c \qquad s=0\cap c<c_{max}\,, \tag{15}$$

where c is the concentration of the solute (mg cm^{-3}), q_s is solute flux density (mg cm^{-2} s^{-1}), S_c is the sink term, s is the crystal content (the mass of crystal per unit volume, in mg cm^{-3}), and c_{max} is the saturated concentration (mg cm^{-3}). If precipitation and dissolution are assumed to occur instantaneously, s is calculated by

$$\begin{aligned}\frac{\partial s}{\partial t}&=-\left(\frac{\partial q_{sx}}{\partial x}+\frac{\partial q_{sz}}{\partial z}\right)-S_c-c_{max}\frac{\partial\theta}{\partial t} \qquad s>0\cap c=c_{max}\\ \frac{\partial c}{\partial t}&=0\end{aligned}\;. \tag{16}$$

Solute fluxes are calculated by the convection–dispersion equation:

$$q_{sx}=-\theta D_{xx}\frac{\partial c}{\partial x}-\theta D_{xz}\frac{\partial c}{\partial z}+q_{lx}c \tag{17a}$$

$$q_{sz}=-\theta D_{zz}\frac{\partial c}{\partial z}-\theta D_{zx}\frac{\partial c}{\partial x}+q_{lz}c\,, \tag{17b}$$

where D is the dispersion coefficient (cm^2 s^{-1}), and the first and second subscripts refer to the direction of flux and concentration gradient, respectively. The dispersion coefficient for each direction is given by

$$\theta D_{xx}=\theta D_{iw}\tau_s+\frac{\lambda_L q_{lx}^2+\lambda_T q_{lz}^2}{|q_l|} \tag{18a}$$

$$\theta D_{zz}=\theta D_{iw}\tau_s+\frac{\lambda_L q_{lz}^2+\lambda_T q_{lx}^2}{|q_l|} \tag{18b}$$

$$\theta D_{xz}=\theta D_{zx}=\frac{(\lambda_L-\lambda_T)q_{lx}q_{lz}}{|q_l|}\,, \tag{18c}$$

where D_{iw} is the ionic diffusion coefficient, τ_s is the tortuosity factor for ionic diffusion, λ_L is longitudinal dispersivity (cm), and λ_L is transversal dispersivity (cm).

The sink term S_c is calculated by

$$S_c=k_{su}cS\,, \tag{19}$$

where k_{su} is the passive uptake ratio. The current version of WASH_2D can simulate only one solute.

Governing Equation for Heat Movement

Heat conservation is described as

$$C_{hs}\frac{\partial T_s}{\partial t} = -\left(\frac{\partial q_{hx}}{\partial x}+\frac{\partial q_{hz}}{\partial z}\right) - L\rho_w\left(\frac{\partial q_{vx}}{\partial x}+\frac{\partial q_{vz}}{\partial z}\right), \quad [20]$$

where C_{hs} is the heat capacity of soil (J cm^{-3} K^{-1}), q_h is the sensible heat flux (W cm^{-2}), and L is the latent heat of water (J g^{-1}). Heat flux for each direction is given by

$$q_{hx} = -k_h\frac{\partial T_s}{\partial x} + C_{hw}T_s q_{lx} \quad [21a]$$

$$q_{hz} = -k_h\frac{\partial T_s}{\partial z} + C_{hw}T_s q_{lz}, \quad [21b]$$

where C_{hw} is the heat capacity of water (4.18 cm^{-3} K^{-1}), and k_h is the thermal conductivity of soil (W cm^{-1} K^{-1}).

Governing Equation for Evaporation Rate

Evaporation rate (E) is calculated by a bulk transfer equation:

$$E = \frac{\rho_{vss}h_{rs} - \rho_{vsa}h_{ra}}{r_a + r_{sc}}, \quad [22]$$

where ρ_{vss} is the saturated vapor concentration at the soil surface (g cm^{-3}), ρ_{vsa} is the saturated vapor concentration at reference height (g cm^{-3}), h_{rs} is the relative humidity at the soil surface (g cm^{-3}), h_{ra} is the relative humidity at reference height, r_a is the aerodynamic resistance (s cm^{-1}), and r_{sc} is the resistance due to salt crust (s cm^{-1}). For a uniform bare field, r_{a0}, r_a of the bare soil surface is calculated from wind velocity at the height of 2 m, u_2 (cm s^{-1}) (van Bavel and Hillel, 1976):

$$r_{a0} = \frac{\ln\left(\frac{200}{z_m}\right)^2}{\kappa^2 u_2}, \quad [23]$$

where z_m is the surface roughness (cm), and κ is the Karman constant (0.4). An increase in aerodynamic resistance due to plant cover is also expressed as a function of the leaf area index:

$$r_a = r_{a0}(1 + a_{ra}I), \quad [24]$$

where a_{ra} is a plant-specific parameter, and I is the leaf area index.

Heat Flux at the Soil Surface

Heat flux at the soil surface is given by the heat balance equation considering convective heat transfer by rainfall and irrigation:

$$q_{h0} = (1-\alpha_r)R_a + R_l - Lq_{v0} - C_{ha}\frac{T_{s0}-T_a}{r_a} + C_{hw}T_w q_{l0} \quad [25]$$

where R_a is the short-wave radiation flux (W cm^{-2}), R_l is the long-wave radiation flux (W cm^{-2}), α_r is the albedo, T_a is the temperature at the reference height (K), and T_w is the temperature of the infiltrating water or soil surface. The subscript 0 indicates the value at the soil surface. Since short-wave radiation flux that arrives at the soil surface is decreased because of vegetation cover, the short-wave flux is expressed as a function of the leaf area index:

$$R_s = R_{sc}\exp(-a_{Rs}I), \quad [26]$$

where R_{sc} is R_s at the canopy, and a_{Rs} is a plant-specific parameter. Campbell (1985) presented a typical value of a_{Rs} as 0.82.

Root Water Uptake

The transpiration rate, T_r (cm s^{-1}), is computed by integrating the water uptake rate, S, over the root zone:

$$T_r = L_x^{-1}\int_0^{L_x}\int_0^{L_z} S\,dx\,dz, \quad [27]$$

where L_z and L_x are the depth and width of the calculated zone, respectively. A widely used macroscopic root water uptake model (Feddes and Raats, 2004) was used in predicting S:

$$S = T_p\beta\alpha_w, \quad [28]$$

where T_p, β, and α_w are the potential transpiration rate (cm s^{-1}), normalized root density distribution, and reduction coefficient, respectively. T_p is given by multiplying the reference evapotranspiration rate (E_p, cm s^{-1}) using the Penman–Monteith equation (Allen et al., 1998) and basal crop coefficient (K_c):

$$T_{rp} = E_p K_c. \quad [29]$$

The root activity β was described as

$$\beta = 0.75(b_{rt}+1)d_{rt}^{-b-1}(d_{rt}-z)^{b} g_{rt}\left(1-x^2 g_{rt}^{-2}\right), \tag{30}$$

where b_{rt} is a fitting parameter, and d_{rt} and g_{rt} are the depth and width of the root zone (cm).

The reduction coefficient α_w is a function of matric and osmotic potential. This function is called the stress response function. WASH_2D uses a so-called additive form stress response function:

$$\alpha = \frac{1}{1+\left(\frac{\psi}{\psi_{50}}+\frac{\psi_o}{\psi_{o50}}\right)^{p}}, \tag{31}$$

where ψ_{50}, ψ_{o50}, and p are fitting parameters (van Genuchten, 1987).

Expressions for Plant Growth

The leaf area index, I, used in Eq. [26] is expressed as a function of cumulative transpiration from germination, τ, since it may primarily depend on cumulative transpiration:

$$I = a_{LAI}\left[1-\exp\left(b_{LAI}\tau\right)\right], \tag{32}$$

where a_{LAI} and b_{LAI} are fitting parameters. Since K_c depends mainly on leaf area, K_c is also expressed as a function of cumulative transpiration:

$$K_c = a_{kc}\left[1-\exp\left(b_{kc}\tau\right)\right]+c_{kc}, \tag{33}$$

where a_{kc}, b_{kc}, and c_{kc} are fitting parameters. The K_c values for major crops have conventionally been given as a function of the date after sowing. This description is valid only under standard or optimum conditions. By linking K_c to cumulative transpiration or leaf area, we can predict transpiration under drought and salinity stressed conditions more dynamically.

The depth of the root zone, d_{rt}, is also expressed as a function of cumulative transpiration:

$$d_{rt} = a_{drt}\left[1-\exp\left(b_{drt}\tau\right)\right]+c_{drt}, \tag{34}$$

where a_{drt}, b_{drt}, and c_{drt} are fitting parameters.

Plant growth also largely depends on uptake of nutrients. Growth parameters used above and water use efficiency should be functions of cumulative nutrient uptake. This topic is beyond the scope of this chapter and further efforts for parameterizing these processes and studies on appropriate parameter values should be made in the near future.

Income Correction Factor

The income correction factor, k_i, may be set based on the crop coefficient. WASH_2D uses the ratio of average K_c over the expected growth period to K_c at each day:

$$k_i = \frac{\overline{K_c}}{K_c} = \frac{\int K_c d\tau}{\tau_f K_c} = \frac{(a_{kc} + c_{kc})\tau_f - \frac{a_{kc}}{b_{kc}}\left[\exp(b_{kc}\tau_f - 1)\right]}{\tau_f K_c}, \qquad [35]$$

where τ_f is the expected final τ.

Numerical Scheme

WASH_2D solves the water flow equation including vapor movement with the alternating-direction finite difference method based on the mass-conservative iterative scheme proposed by Celia and Bouloutas (1990). Time steps, Δt, are adjusted automatically such that, at each time step, the number of iterations is approximately five and the maximum change in $\ln(\psi)$ is smaller than the prescribed value. Hysteresis is considered using a simple method by Kool and Parker (1987).

The convection–dispersion equation for solute movement is also solved with the alternating direction implicit finite difference method using the same spatial and time discretization as for the water flow analysis. Numerical dispersion, which results in an overestimation of the spreading of the solute due to discretization of the convection term, is corrected using a scheme proposed by Bresler (1973), in which the numerical dispersion coefficient is subtracted from dispersion coefficient at each time step and for each element.

The governing equation for heat movement is solved with the explicit finite difference method using the same spatial and time discretization as for the water flow analysis.

Field Experiment for Barley

Treatments

To evaluate the effectiveness of the proposed method as compared with an automatic irrigation method in terms of net return under assumed prices of water and crops, a field experiment was performed at the Institut des Régions Arides in

Medenine, Tunisia in 2011 to 2012. Two treatments were established: in Treatment A, the crop was irrigated with an automatic irrigation system, and in Treatment S our proposed simulation method was applied to the other treatment. Each treatment had three 8 by 2 m plots as replicates in a screen house. To monitor water content, five 5TE probes (Decagon Inc.) were inserted for each treatment. Two pots whose soil surfaces were covered with polystyrene foam were installed such that the soil surface was the same height as the ground to measure the potential daily transpiration rate. The weight of the pots were measured every morning, and water with low salinity (0.5 dS m^{-1}) was applied to the pots to recover to a volumetric water content of 0.2 when the average water content of the pot becomes lower than 0.15. Under such management, plants on the pots might have grown under no drought and salinity stresses. Wind velocity, air temperature, humidity, and short-wave radiation were measured in the screen house at the height of 2 m. These meteorological data were used to calculate reference ET using the Penman–Monteith equation.

Soil

The soil of the field is loamy sand whose hydraulic properties are shown in Fig. 9–2. Solute transport parameters such as dispersivity (0.36 cm) and dependence of the tortuosity factor for ionic diffusion on water content were also measured in a laboratory. Independently measured thermal properties such as the dependence of thermal conductivity and albedo on water content and were also used.

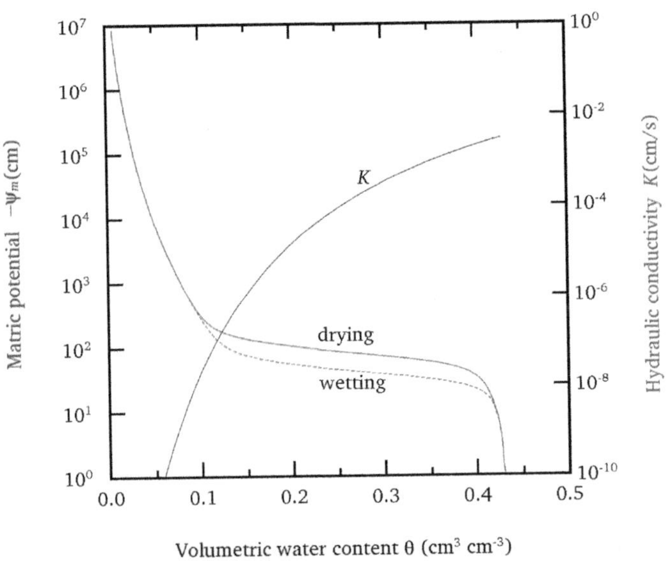

Fig. 9–2. Hydraulic properties of El-Fejaa loamy sand.

Table 9–1. Parameter values for plant stress response and growth properties.

Parameter	Barley	Corn
ψ_{50} (cm)	−2000	−1000
ψ_{o50} (cm)	−15,000	−3000
p	3	3
a_{ra}	0.2	0.2
b_{rt}	1	0.12
g_{rt}	20	30
a_{drt}	35	40
b_{drt}	−0.3	−0.4
c_{drt}	5	5
a_{kc}	2	1.3
b_{kc}	−0.4	−0.3
c_{kc}	0.05	0.1
a_{LAI}	4	2.5
b_{LAI}	−0.08	−0.3
p	3	3

Plant

A landrace of barley (*Hordeum vulgar* L. 'Ardhaui') was sown on 28 Nov. 2011 and thinned into about 20 plants m^{-1} along the lateral line. Since the stress response function of the plant has not been determined, tentatively selected parameter values listed in Table 9–1 were used. The h_{50} was set slightly lower than that of canola (Yanagawa and Fujimaki, 2013), considering high drought tolerance. The h_{o50} was set at −15,000 cm according to the salinity at which yield becomes half of the potential one (Ayers and Westcot, 1985). Since the crop coefficient for barley has not been reported as a function of cumulative transpiration, we had to set rather arbitrary parameter values at the beginning and adjust the parameter values according to deviation from the measured transpiration. As shown in Fig. 9–3, parameter values for the crop coefficient function (Eq. [34]) were updated on 21 March such that simulated transpiration matched measured values. Large values of the crop coefficient were required probably because of an "oasis effect" and difficulty in evaluating reference ET using the Penman–Monteith equation in a screen house. Parameters values for root distribution and leaf area index were chosen on the basis of data obtained through growth measurement. The water use efficiency was set at 0.007. The value of a_{ra} in Eq. [24] was set at 0.2 considering the empirically known ratio of evaporation to ET under high leaf area index.

Irrigation and Crop Management

Irrigation was applied through a drip irrigation system whose lateral distance was 40 cm and emitter distance was 20 cm. In Treatment A, 20 mm of water was

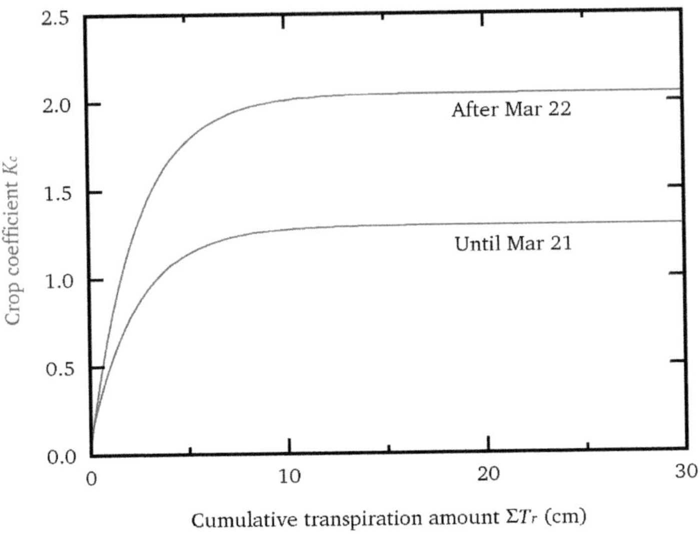

Fig. 9–3. Crop coefficient (for transpiration) as a function of cumulative transpiration for the barley.

applied manually when the average θ of two probes at the depth of 10 cm became less than 0.1. The salinity of irrigation water ranged from 8.7 to 9.5 dS m^{-1}.

The interval for Treatment S was set at 4 d. The climatic conditions were measured in the field, and quantitative weather forecasts were downloaded from the website of Weather Underground (http://wunderground.com) with the location code of 60770. Since it provides cloud cover instead of short-wave radiation, an empirical coefficient, $1 - 0.006 \times$ cloud cover (%), was multiplied by extra-terrestrial short-wave radiation. The crop and water prices were set at 0.25 and 0.000015 \$ kg^{-1}, respectively.

Organic manure was spread at 8.3 t ha^{-1} before sowing. The nutrient supply included N, P_2O_5, and K_2O at rates of 30, 30, and 20 g m^{-2}, respectively, which were adopted from the local practices. The P and K fertilizers were applied as a basal dose before sowing. Nitrogen was divided and delivered with the irrigation water during the early vegetative growth stage.

Since the salinity of irrigation water was high and the current version of WASH_2D can simulate only one solute, we neglected the movement of nutrients and approximated that all of the solutes were NaCl.

Results and Discussion

The time evolution of leaf area and biomass is shown in Fig. 9–4. Both leaf area and biomass rapidly increased around early March, and the crop coefficient function therefore had to be modified upward on 21 March. Figure 9–5 compares

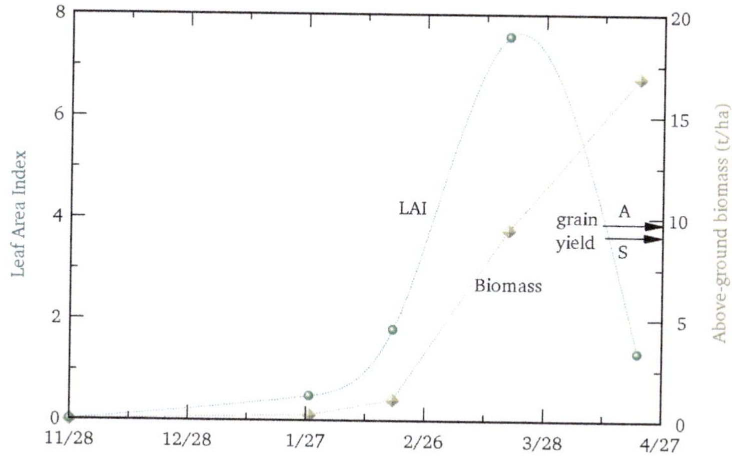

Fig. 9–4. Time evolution of leaf area and biomass for the barley (Treatment S).

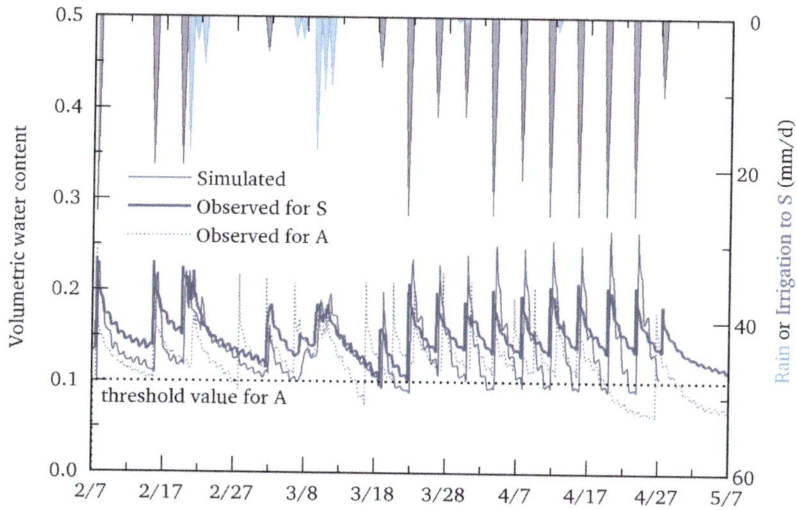

Fig. 9–5. Comparison of observed and simulated water content at the depth of 10 cm below the drip tube.

measured and simulated water content at the depth of 10 cm below the drip tube. The increase in water content after irrigation or rainfall and the subsequent fall due to internal drainage and evapotranspiration were accurately simulated, particularly for late February and mid-March. During April, WASH_2D overestimated the water content just after irrigation and underestimated it before the next irrigation. The overestimation may partly be due to the inaccuracy of hydraulic properties, while underestimation might indicate an overestimation of transpira-

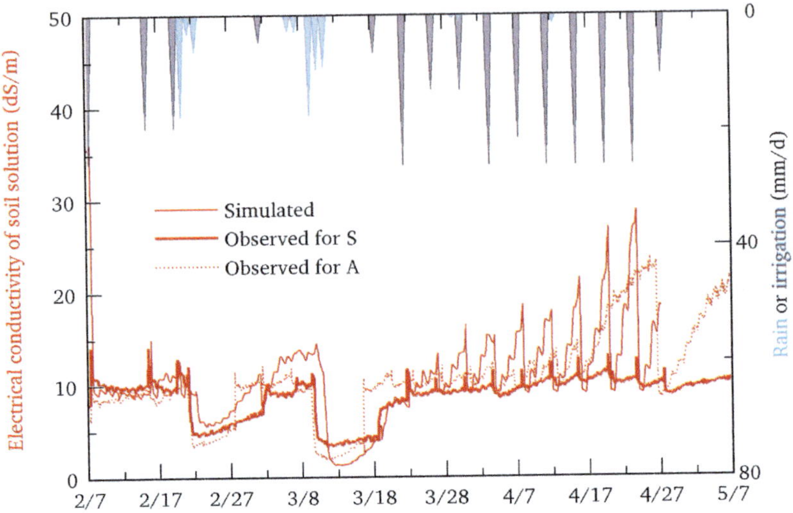

Fig. 9–6. Measured and simulated soil solution salinity for plot S.

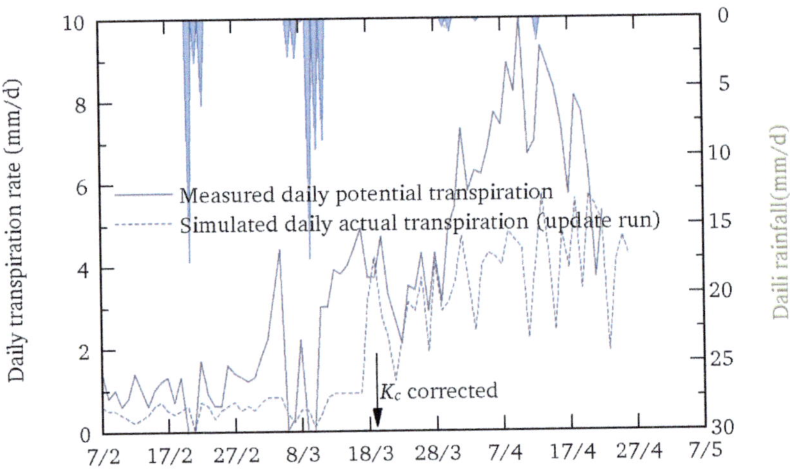

Fig. 9–7. Time evolution of daily transpiration for the barley.

tion. The measured and simulated salinity of the soil solution at the depth of 10 cm below the drip tube is shown in Fig. 9–6. The decrease in salinity after rain and the recovery by irrigation were simulated in fair agreement. The overestimation during April may be due to the underestimation of the water content. Figure 9–7 shows potential and simulated daily transpiration. In mid-March, we realized that the large underestimation could not be corrected by simulated growth, and we shifted the crop coefficient function upward. Also, we found that potential transpiration had been halved by a bug in the code. The unrealistically low crop coefficient until 21 March reflects that mistake. After that correction, the simulated transpiration rate matched the potential value, but underestimated again around mid-April. Note that the slightly lower transpiration rate is natural because of mild drought and salinity stresses. Even so, the simulation transpiration rate is largely lower than the potential value. This contradicts the slower decrease in measured water content. Plants grown in the pot were somewhat delayed in growth compared with surrounding plants. Therefore, in contrast to surrounding plants, a reduction in crop coefficient during the maturity stage, which commonly occurs among annual crops, might not have occurred for the potted plants. Also, a somewhat isolated configuration (two plants in a 20-cm diameter pot) might have caused an "oasis effect" and overestimated the potential transpiration rate.

Figure 9–8 compares "attained" net returns. The total irrigation depth for Treatment S (simulated) was 514 mm, slightly lower than the depth of 548 mm for Treatment A (automatic). The grain yield for Treatment S was 84 kg a^{-1}, also lower than the yield of 98 kg a^{-1} for Treatment A. Since the cost for water was quite small compared with income, Treatment A attained a higher net return. This result implies that at a low water price, it is difficult to realize the advantage of the presented scheme. Even so, we may state that the presented scheme can attain at least a comparable net return to automatic irrigation and we may save on costs for the automatic irrigation system. Note that for the cost of sensors, the controller and solenoid valve is not included in the comparison. If the cost for automatic irri-

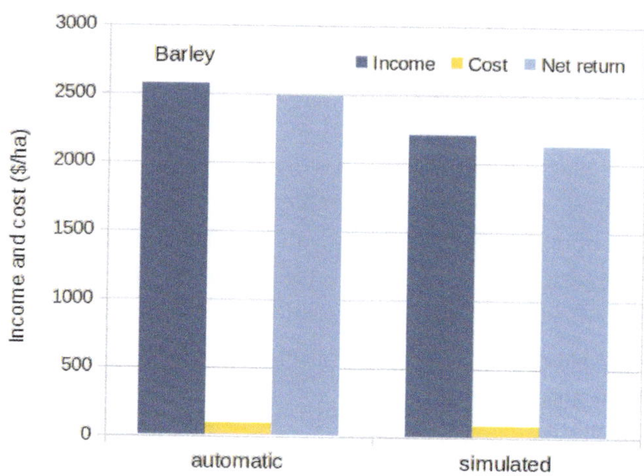

Fig. 9–8. Comparison of yield and income for the barley cropping.

gation system is 2 \$ a^{-1} yr^{-1} (= 200 \$ ha^{-1} yr^{-1}), the net income becomes nearly the same as the proposed scheme. One of the reasons for lower yield may be drought or salinity stress during the flowering stage, which is the most sensitive period. In fact, aboveground dry matter production for Treatments A and S were 102 and 98 kg a^{-1}, a smaller difference than that for grain yield. Therefore, another income correction factor may be introduced to avoid stresses during such a period. If the price of a crop is doubled during the flowering stage, for example, the software will suggest a larger irrigation depth. Another reason for lower yield might be due to simulation inaccuracy. Proper calibration of parameter values for stress response and plant growth functions as well as hydraulic, transport, and thermal properties of the soil may be a prerequisite for presented scheme.

Field Experiment for Corn

Treatments

Another field experiment was performed at the Arid Land Research Center in Tottori, Japan (35°32′09″ N, 134°12′39″ E) in 2013. Two treatments were established: in Treatment A, the crop was irrigated with an automatic irrigation system, and in Treatment S our proposed simulation method was applied to the other treatment. Each treatment had two 20 by 16 m plots as replicates. To monitor water content, 12 TDR probes were inserted for each treatment. Evapotranspiration was measured for Treatment S with a weighing lysimeter whose diameter was 150 cm.

Irrigation

Irrigation was applied through a drip irrigation system whose lateral distance was 90 cm and whose emitter distance was 20 cm. Automatic irrigation was set such that water was applied for 1 h when θ at the depth of 15 cm became less than 0.09.

The interval for Treatment S was set at 2 d. The records of climatic condition were downloaded from the website of Japan Meteorological Agency (http://www.kishou.go.jp/), and quantitative weather forecasts were downloaded from the website of Yahoo! JAPAN (http://www.yahoo.co.jp). The crop and water prices were set at 0.45 and 0.00025 \$ kg^{-1}, respectively. Liquid fertilizer (N = 10%, P_2O_5 = 4%, K_2O = 8%) and calcium chloride were mixed such that daily application rate became constant throughout growing season. Totally applied nitrogen was 15 g m^{-2}. Since the salinity of irrigation water was very low and the current version of WASH_2D can simulate only one solute, we used a "conceptual nutrient solute" whose passive uptake ratio is unity. We did not consider the effect of the nutrient uptake on growth.

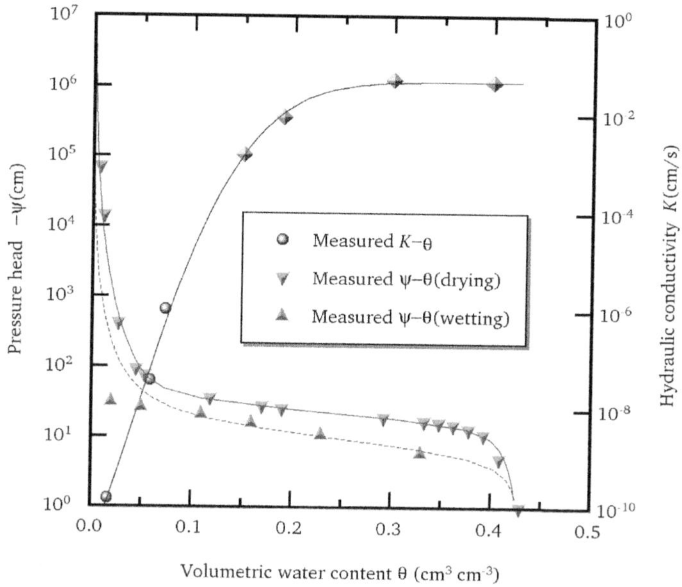

Fig. 9–9. Hydraulic properties of Tottori sand.

Soil

The soil of the field is sand, and the hydraulic properties are shown in Fig. 9–9. Solute transport parameters such as dispersivity (0.59 cm) and dependence of the tortuosity factor for ionic diffusion on water content were also measured in laboratory. Independently measured thermal properties such as dependence of the thermal conductivity and albedo on water content and were also used.

Plant

A cultivar of sweet corn (*Zea mays* L. 'Amaenbou86') was sown on 4 June and thinned into one plant per emitter. Since the stress response function of the plant has not been determined, tentatively selected parameter values listed in Table 9–1 were used. These values are comparable for canola (Yanagawa and Fujimaki, 2013). As shown in Fig. 9–10, parameter values for the crop coefficient function (Eq. [33]) were updated such that simulated evapotranspiration matched measured values. Parameter values for root distribution and leaf area index were chosen on the basis of data obtained through growth measurement. Water use efficiency was set at 0.004 on the basis of results from the preliminary experiment performed in 2012. Unfortunately, damage by corn borers (*Ostrinia nubialis* Hübner) was widespread in the field in August in spite of repeated applications of pesticide, and it negatively affected the performance of the automatic irrigation system.

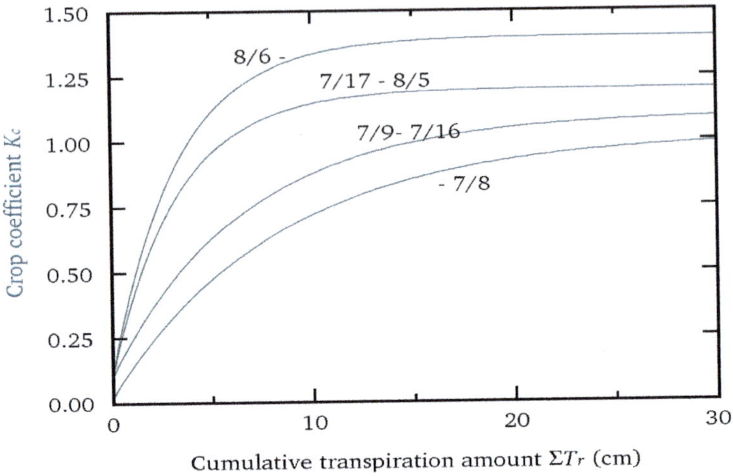

Fig. 9–10. Crop coefficient (for transpiration) as a function of cumulative transpiration for the corn.

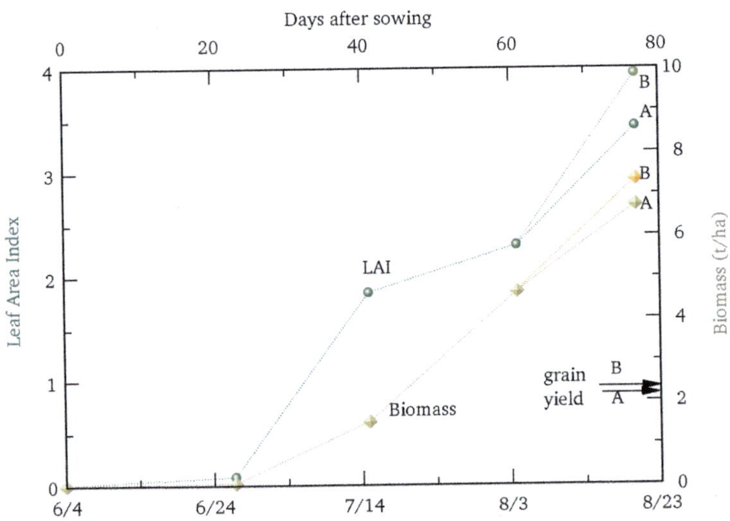

Fig. 9–11. Time evolution of leaf area and biomass for the corn.

Results and Discussion

The time evolution of leaf area and biomass is shown in Fig. 9–11. The leaf area jumped 23 times from 27 June to 15 July, and the crop coefficient function thus had to be modified upward. Once the leaf area is reported as a function of cumulative transpiration for a crop, such an adjustment may not be necessary. Figure 9–12 compares measured and simulated water content for a selected period. The value

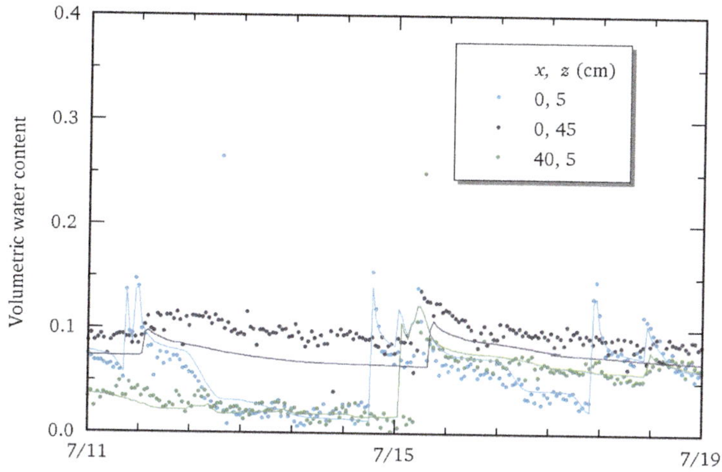

Fig. 9–12. Measured and simulated water content and concentration for plot S.

of x in the legend is the horizontal distance from the drip tube. Increases in water content below the drip tube, $(x, z) = (0, 5$ cm), after irrigation on 11 and 14 July and the subsequent fall due to internal drainage were accurately simulated. Also, water content 40 cm apart from the drip tube, $(x, z) = (40, 5$ cm), did not increase after irrigation but increased after rain (15 July). Simulated values agreed well with the measured ones at the location as well. The model slightly underestimated the water content at a deeper layer. Figure 9–13 compares measured and simulated ET and reference ET with the Penman–Monteith equation for a selected period. In July, the model tended to slightly underestimate ET because of the unexpected fast growth of the leaf area. Practically, it is not easy for a farmer to measure ET or transpiration and adjust parameter values for the crop coefficient on the basis of the deviation from measured values in the interim of the growing season, but cumulative transpiration τ can be estimated from attained dry matter production (by dividing with water use efficiency), which may be easier for a farmer. If esti-

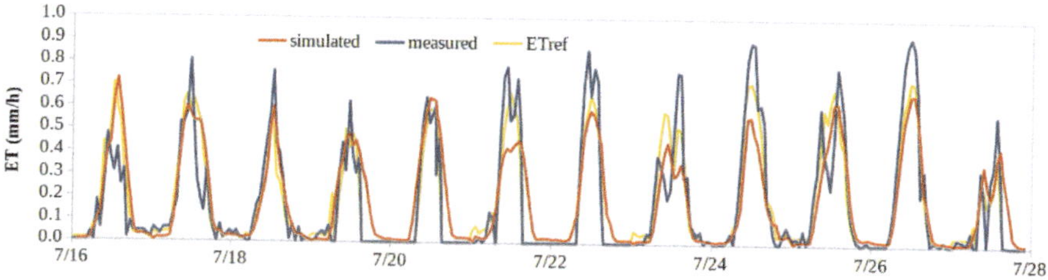

Fig. 9–13. Comparison of measured and simulated ET.

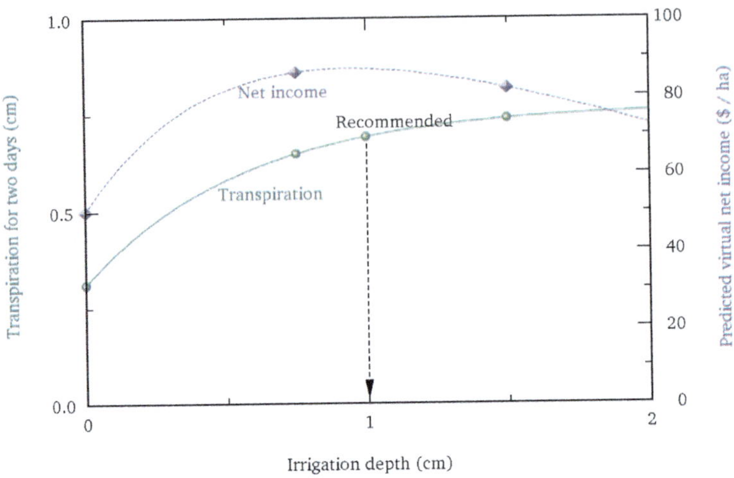

Fig. 9–14. Example of predicted transpiration and virtual net return (20 August).

mated τ is lower than the simulated value, a_{kc} and b_{kc} may be increased and initial τ in the simulation may be replaced by the estimated value.

Figure 9–14 shows an example of predicted transpiration and virtual net return (20 August). The solid curve is Eq. [3], and it is shown that transpiration at optimum irrigation depth is exactly on the curve. This demonstrates the validity of Eq. [3] in describing the relationship. The figure is also an example that optimum irrigation depth does not give maximum transpiration if the water price is high.

Finally, a comparison of "attained" net return is shown in Fig. 9–15. Although the presented scheme (simulated) gave 1.5 times more water compared with automatic irrigation, the former attained slightly higher yield, and since the cost for water was small compared with income, the former attained a higher net return. One of the reasons why automatic irrigation gave a smaller amount might be due

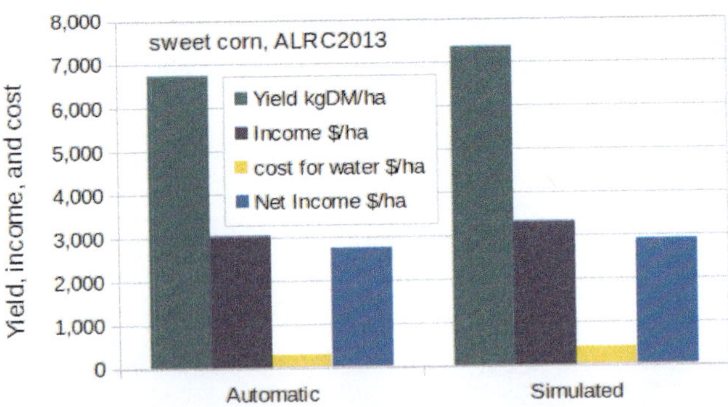

Fig. 9–15. Comparison of yield and income for the corn cropping.

to damage in one of the monitored plants by corn borers. This might be another drawback of automatic irrigation, and the number of monitored plants should be more than three. In spite of widespread damage, the model did not consider the decrease in the crop coefficient, which might have caused overestimation of ET.

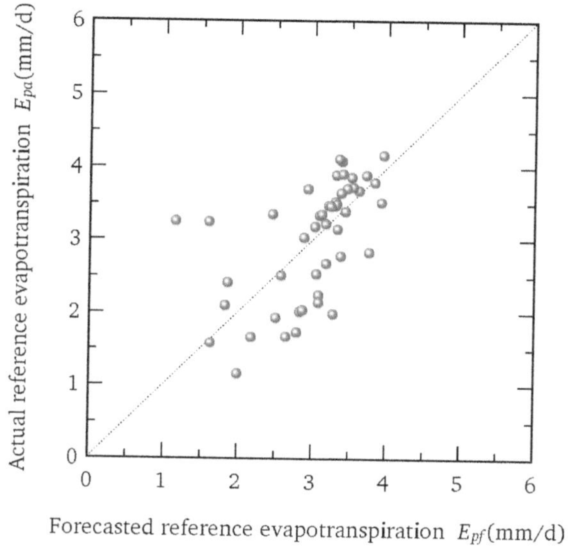

Fig. 9–16. Comparison of forecasted and actual reference evapotranspiration in Tottori, Japan from 3 July to 19 Aug. 2014.

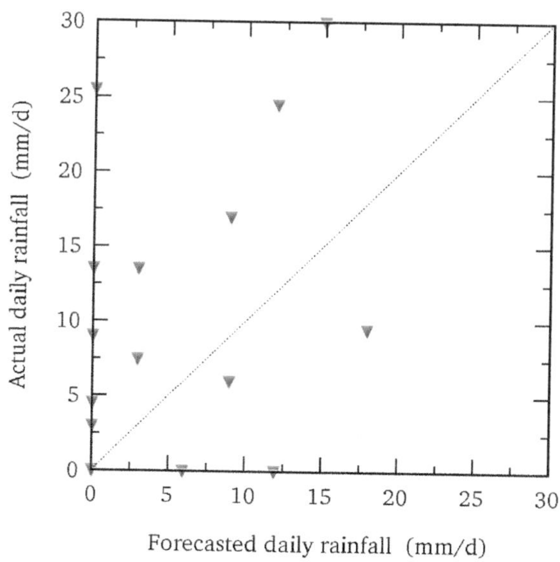

Fig. 9–17. Comparison of forecasted and actual rainfall in Tottori, Japan from 3 July to 19 Aug. 2014.

This might be another reason for larger irrigation by the presented scheme. An additional linear term ($-\ d_{kc}\tau$) may be added to Eq. [33] to describe such an effect if it is inevitable.

As shown in Fig. 9–16, forecasted reference evapotranspiration (E_p) using forecasted weather conditions relatively agreed with actual E_p at an RMSE of 0.67 mm d^{-1}. In contrast, there were large discrepancies between forecasted and actual rainfall as shown in Fig. 9–17 at an RMSE of 17.0 mm d^{-1}. Weather forecasts not only failed to predict the amount of rain, but they also often failed to predict the occurrence of a rain event. The largest error occurred on 1 August when forecasts failed to predict a rain event with 114 mm. Such a large error does not necessarily cause a large underestimation of stored water because of internal drainage to the field capacity. If we assume that the largest effective rainfall is 20 mm d^{-1}, RMSE is reduced to 6.0 mm d^{-1}. In addition to the error in prediction due to improper modeling and parameter values, the advantage of optimization may be partly offset by inaccuracy of weather forecasts.

Conclusions

In this chapter, a scheme for determining irrigation depth using a numerical model of crop response to irrigation and weather forecasts was presented. Results of field experiments indicated that (i) at a low water price, the advantage of using the presented scheme does not appear; (ii) proper calibration of parameter values for stress response and plant growth functions as well as hydraulic, transport, and thermal properties of the soil may be a prerequisite for the presented scheme; and (iii) the proposed scheme can at least alter automatic irrigation systems and save on costs for equipment such as soil moisture probes, dataloggers, and solenoid valves. Further validation studies are required using more reliable parameter values in the model.

References

Ahuja, L.R., K.W. Rojas, J.D. Hanson, M.J. Shafer, and L. Ma, editors. 2000. Root Zone Water Quality Model: Modeling management effects on water quality and crop production. Water Resources Publ., Highlands Ranch, CO.

Allen, R., L. Pereira, D. Raes, and M. Smith. 1998. Crop evapotranspiration: Guidelines for computing crop water requirements. Irrig. Drain. Pap. 56. FAO, Rome.

Ayers, R.S., and D.W. Westcot. 1985. Water quality for agriculture. Irrig. Drain. Pap. 29, Rev. 1. FAO, Rome.

Bresler, E. 1973. Simultaneous transport of solutes and water under transient unsaturated flow conditions. Water Resour. Res. 9:975–986.

Campbell, G.S. 1985. Soil physics with BASIC. Elsevier, New York.

Celia, M.A., and E.T. Bouloutas. 1990. A general mass-conservative numerical solution for the unsaturated flow equation. Water Resour. Res. 26:1483–1496. doi:10.1029/WR026i007p01483

Feddes, R.A., and P.A.C. Raats. 2004. Parameterizing the soil–water–plant root system. In: R.A. Feddes et al., editors, Unsaturated-zone modeling: Progress, challenges, and applications. Wageningen UR Frontis Ser. Vol. 6. Kluwer Acad., Dordrecht, The Netherlands. p. 95–141.

Fereres, E., and M.A. Soriano. 2006. Deficit irrigation for reducing agricultural water use. J. Exp. Bot. 58:147–159. doi:10.1093/jxb/erl165

Heermann, D.F., D.L. Martin, R.D. Jackson, and E.C. Stegman. 1990. Irrigation scheduling controls and techniques. In: B.A. Stewart and D.R. Nielson, editors, Irrigation of agricultural crops. Agron. Monogr. 30. ASA, Madison, WI. p. 509–535.

Kool, J.B., and J.C. Parker. 1987. Development and evaluation of closed-form expressions for hysteretic soil hydraulic properties. Water Resour. Res. 23:105–114. doi:10.1029/WR023i001p00105

Philip, J.R., and D.A. De Vries. 1957. Moisture movement in porous materials under temperature gradients. Eos Trans. AGU 38:222–232. doi:10.1029/TR038i002p00222

Press, W.H., B.P. Flanner, S.A. Teukolsky, and W.T. Veterling. 1992. Numerical recipes in C: The art of scientific computing. Cambridge Univ. Press, New York.

Šimůnek, J., M.Th. van Genuchten, and M. Šejna. 2006. The HYDRUS software package for simulating the two- and three-dimensional movement of water, heat, and multiple solutes in variably-saturated media. PC Progress, Prague, Czech Republic.

van Bavel, C.H.M., and D. Hillel. 1976. Calculating potential and actual evaporation from a bare soil surface by simulation of concurrent flow of water and heat. Agric. Meteorol. 17:453–476. doi:10.1016/0002-1571(76)90022-4

Van Dam, J.C., J. Huygen, J.G. Wesseling, R.A. Feddes, P. Kabat, P.E.V. Van Walsum, P. Groenendijk, and C.A. Van Diepen. 1997. Theory of SWAP, version 2.0: Simulation of water flow, solute transport and plant growth in the soil–water–atmosphere–plant environment. Rep. 71. Department of Water Resources, Wageningen Agricultural Univ., Wageningen, The Netherlands.

van Genuchten, M.Th. 1987. A numerical model for water and solute movement in and below the root zone. Research Report. U.S. Salinity Lab, Riverside CA.

Yanagawa, A., and H. Fujimaki. 2013. Tolerance of canola to drought and salinity stresses in terms of root water uptake model parameters. J. Hydrol. Hydromech. 61:73–80. doi:10.2478/johh-2013-0009

Modeling to Evaluate and Manage Climate Change Effects on Water Use in Mediterranean Olive Orchards with Respect to Cover Crops and Tillage Management

J.A. Gómez,* M.T. Rodríguez-Carretero, I.J. Lorite, and E. Fereres

Abstract

Simulation models are being used in climate change studies to aid in the development and evaluation of adaptation strategies. The Water Balance for Olive (WABOL) model used in this study describes the water balance components in a dual system—the olive tree [*Olea europaea* L.] and the area outside the tree canopy projection—integrating orchards and soil management strategies. We evaluated the water balance components and its impact on yield in a Mediterranean crop, olives, and their two most common soil management strategies: bare soil using tillage and temporary cover crop. To consider the climate and orchard variability, five locations located in the traditional olive growing areas in Andalusia (southern Spain) were considered. To reduce uncertainty, four regional models were considered and their predictions used for evaluating two climate scenarios: current (1971–2001) and future (2068–2098). The results obtained indicate that when the cover crop is properly managed (cover crop is properly implanted and its transpiration is controlled in early spring by mowing or herbicides), this soil management can be viable in a large set of weather and agronomical conditions with olive annual transpiration similar to that obtained in the tilled system. This is because for the conditions simulated, the reduction in runoff and in direct evaporation from the bare soil compensates the increase in transpiration by the cover crop. A similar result was obtained for projected climate scenarios, in which olive transpiration is reduced in general by the decrease of rainfall. In these scenarios, superficial runoff and deep percolation are significantly reduced with the use of a cover crop.

Abbreviations: WABOL, Water Balance for Olive model.

J.A. Gómez and E. Fereres, Institute for Sustainable Agriculture, CSIC, P. Box 4084, 14080 Córdoba, Spain. *Corresponding author (joseagomez@ias.csic.es).

M.T. Rodríguez-Carretero (mariat.rodriguezcarretero.ext@juntadeandalucia.es) and I.J. Lorite (ignacioj.lorite@juntadeandalucia.es), IFAPA-Alameda del Obispo, P. Box 3092. 14080 Córdoba, Spain.

doi:10.2134/advagricsystmodel5.c10

Practical Applications of Agricultural System Models to Optimize the Use of Limited Water
Lajpat R. Ahuja, Liwang Ma, and Robert J. Lascano, Editors
Advances in Agricultural Systems Modeling, Volume 5. Lajpat R. Ahuja, Series Editor

Water availability is the main limiting factor for crop yield productivity, especially under arid and semiarid conditions (Henderson, 1979). In these regions, different factors such as the human population growth or the increasing need for water in the preservation of the environment are competing with agriculture for the available water resources (Hsiao et al., 2007; Falkenmark, 2013). In addition, resource shortages associated with climate change contribute to the consideration of water management in agriculture as a key component for ensuring its medium- to long-term sustainability. Additionally, climate change contributes to the increase of uncertainty in the estimation for environmental components and crop yields, complicating the development and evaluation of agricultural practices for improvement of water management.

One approach to evaluate different agricultural alternatives for evaluating water balance and yield is using field experiments in which several alternatives are tested for a number of years and several variables are measured to be able to perform a statistical analysis among treatments (e.g., Monteiro and Lopes, 2007; Wortmann et al., 2010; Van den Putte et al., 2010). However these experiments usually require a huge effort, are time-consuming, and are often limited to a greatly reduced number of alternatives under very specific conditions.

The use of simulation models in crop production and water balance components studies, cropping systems analyses, climate change impacts, and evaluation of adaptation strategies has proven to be valuable for improving our knowledge of the functioning of agricultural systems (Fraisse et al., 2006; Thorp et al., 2008; Casadesús et al., 2012). Crop models simulate crop growth under different environmental and management conditions, taking into account various limiting factors, and thus they are also good tools for assessing the temporal and spatial variability of crop behavior over large area scales (Safir et al., 2008; Resop et al., 2012) and may reduce the need for lengthy and expensive field experimentation.

The extension of crop simulation beyond individual experiments at the field scale meets the requirements of an emerging need to support decision-making in agriculture at several scales, through either spatial or long-term analyses for policy formulation. These simulation models mathematically describe the processes of soil water balance, plant growth, and yield expression to calculate their evolution during the simulation period. There are many examples of these models in the literature, from the most simplified models as those proposed by Allen et al. (1998) to the more complex models such as The Decision Support System for Agrotechnology Transfer (DSSAT) (Jones et al., 2003) or the Soil and Water Assessment Tool (SWAT) (Jayakrishnan et al., 2005). As the complexity of the models increases, so do the data requirements of the models.

A substantial restriction in most of the models is the consideration of a single crop. Thus, a scarce number of models have been designed for intercropping conditions such as the one developed by Hulugalle and Lal (1986). These authors studied the water use efficiency of intercropped maize (*Zea mays* L.) and cowpea [*Vigna unguiculata* (L.) Walp.] by modeling the water balance of the root zone throughout the growing season in tropical Nigeria. Most of the previous models have been developed for annual crops, although there are also some examples of water balance models aimed at tree crops. One of these models was developed by Celette et al. (2010) to study water use in intercropped vineyard assuming two soil compartments: one for the cover crop and the grapevine (*Vitis vinifera* L.) and the other exclusively for the grapevine. Abid Karray et al. (2008) and Er-Raki et al. (2010) also developed simple models for partitioning the soil and plant component of evaporation in olive orchards in semiarid regions. Castro et al. (2006) also developed a highly simplified three-compartment model to evaluate the water balance in rainfed olive groves under tillage or cover crop management. In spite of the development of these new models, the need for more refined simulations of the key process in crop development and olive yield response to water deficits has been cited by some authors (Villalobos, 1997; Castro and Gómez, 2006). Detailed analysis of water extraction by the olive and cover crop is necessary for an appropriate understanding of the impact of cover crops on the soil water balance and erosion risk in the whole range of soil and climate conditions typical of Mediterranean environments (Castro and Gómez, 2006).

Olive tree is a typical Mediterranean crop since Romans times, albeit it has followed several expansion–contraction cycles during the last 2000 yr. Currently, olive orchards are one of the most important crops in the semiarid Mediterranean region. Olives covered about 4.3 million hectares in Europe in 2007, of which 50% were located in Spain (Eurostat, 2013). Traditional olive orchards in Spain are rainfed (72% of the area cultivated; Ministry of Agriculture, Food and Environment, 2010) with around 100 trees ha^{-1} and ground cover rarely exceeding 25%, and they are generally located on hillsides.

In arid and semiarid regions, rainfed agriculture has progressed by using crops and varieties better adapted to an optimum use of rainfall water and by developing techniques such as weed control or direct seeding, which are aimed at reducing runoff and water losses by direct evaporation from the soil to increase crop transpiration (Henderson, 1979). In the case of rainfed tree orchards in arid or semiarid environments, their design and management has also been focused on ensuring the survival of the plantation during dry periods within each season and also during drought years that periodically occur.

Agronomical management of olive orchards under water limiting conditions has thus evolved by using low tree density, control of tree transpiration by limiting the canopy crown size through periodic pruning, and weed control by periodic tillage and/or, in recent decades, herbicide use. A combination of low ground cover on sloping areas with Mediterranean climate has resulted in high erosion rates in many olive growing areas (e.g., Gómez et al., 2009). The high erosion rates, noted as one of the main environmental problems of this crop (Beaufoy, 2001), have spurred the development and introduction of cover crop soil management in several Mediterranean countries, among them being Spain. Since most of the olive orchards are located in arid or semiarid areas, the recommended cover crop management is based on control of the cover crop transpiration by mowing or herbicide application in late winter to prevent competition with the olive tree. Because of the risk of competition for soil water, the cover crop used in many orchards only covers a fraction of the soil, with this fraction varying among farms. Because of this large variability in possible combinations, the scarcity of experimental studies, and the heterogeneity in the olive orchards located in Andalusia, it is difficult to carry out general recommendations about the management of a cover crop and its impact on soil water balance and olive yield. For instance, the establishment of a cover crop in orchards in colder areas should require a delay of the killing of the cover crop (to ensure completion of the plant cycle for self-seeding next year and effective soil protection during summer and early fall the next season), but there is very limited information about the impact that this delay might have on the orchard water balance and ultimately olive yield.

To manage the dual system that is typical in the olive orchards located under Mediterranean conditions, the development of a water balance model seems to be a necessary step to explore the optimal strategy in a broad range of conditions. Water balance components in an olive orchard vary during the growing period and among orchards, mainly due to variation in crop canopy and density and in climatic conditions. Scientific efforts to measure evapotranspiration of olive orchards are all relatively recent (Testi et al., 2006), and direct measurement of the total water vapor flux density of an orchard has been performed by using the eddy covariance technique (Villalobos et al., 2000; Testi et al., 2006). However, these methods generally sample ET from single trees or parts of orchards. Whole orchard ET has been measured in olive by using the water balance method (e.g., Palomo et al., 2002) and estimated by using daily soil water balance–atmospheric process models such as the relatively simple FAO 56 model, which is based on the concepts of reference evapotranspiration ET_o, crop coefficient K_c, and root zone water balance (Allen et al., 1998). Er-Raki et al. (2010) have evaluated the ability of the FAO 56 dual crop coefficient model to reproduce the temporal evolution of ET and its components in olive. However, there is no simple way to calculate crop coefficients for olives, as they must integrate several factors depending on agricultural practices and orchard architecture.

This study presents the use of a simulation model specifically designed for olive orchards, WABOL (Abazi et al., 2013), which was originally intended for rainfed conditions but was modified to include drip irrigation for this study in order to evaluate the impact of two different soil management strategies (tillage or cover crop) on water balance and potential yield under a broad range of conditions, including future climate scenarios in southern Spain.

Material and Methods

Water Balance Model Description

The WABOL model is a functional model developed to simulate the effect of the different soil management alternatives available to farmers in olive orchards on runoff and soil water balance. It has been designed with the aim of providing a tool for exploring the best soil management alternatives for olive orchards under variable environmental conditions and for evaluating the effect of a specific management strategy in a particular situation, Fig. 10–1.

The model calculates daily the water balance components in an olive orchard for a complete hydrological year. It requires as input daily climate data (temperature, rainfall, irrigation, and reference evapotranspiration, ET_o) and parameters describing the olive tree and cover crop characteristics, the soil properties, and management. The olive orchard is conceptually represented in three compartments: the olive canopy area (understood as the area under the olive canopy projection), the bare lane area (the area outside the canopy projection that is not covered by a cover crop), and the cover crop lane area (the area outside the canopy projection where actually a cover crop is grown). For each of these compartments, the model updates daily the soil water content after calculating each of the compartments of the water balance, Fig. 10–2. It is worth noting that no lateral fluxes are allowed among the three soil compartments with the exception of the extraction by the olive tree roots. The soil is divided in layers of 0.2-m depth and limited to a maximum depth of 2.6 m. Figure 10–2 depicts a conceptual description of the model.

First, the model calculates the daily runoff, if there is any rainfall, according to the curve number method as calibrated for olive orchards by Romero et al. (2007), and it distributes the infiltrated water among the layer using a cascade model. Second, it calculates the soil evaporation from each of the three com-

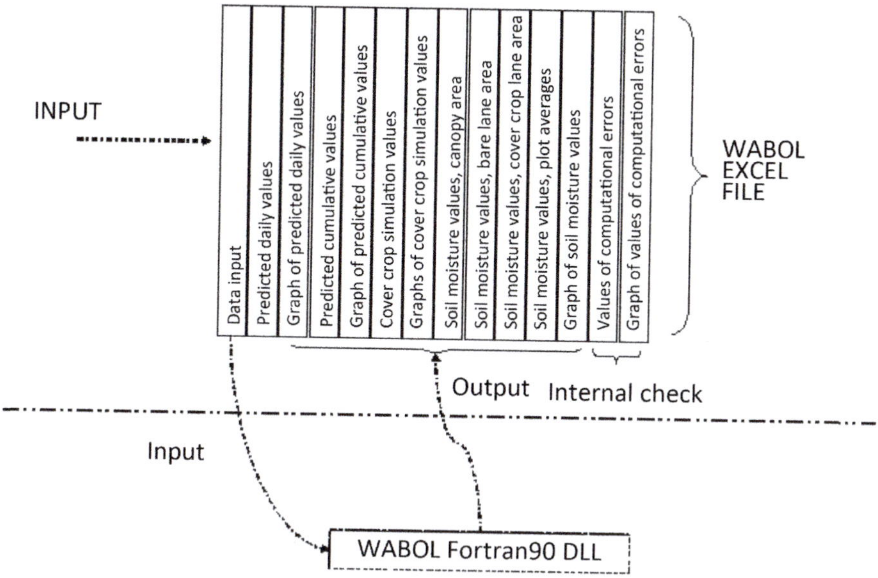

Fig. 10–1. Programming structure and interface of the WABOL model.

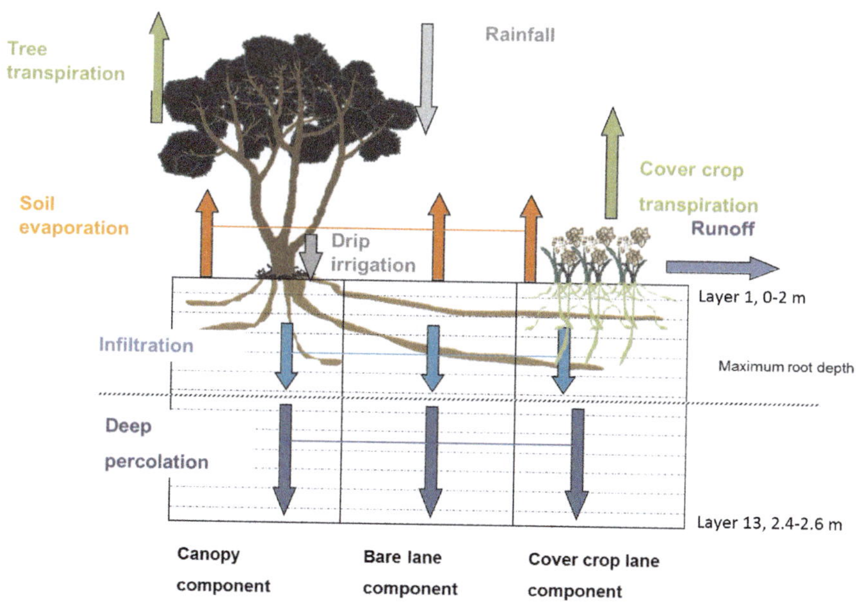

Fig. 10–2. Conceptual description of the WABOL model.

partments, extracting that water from the soil using a two-phase model. In the first phase, it considers a constant rate stage with the soil evaporation limited by the energy supply, and in the second phase it considers a falling rate stage with the evaporation controlled by the hydraulic properties of the soil (Ritchie, 1972). Afterward, it calculates the olive tree transpiration and extracts that water from the soil using the methodology proposed for olive orchards by Orgaz et al. (2006), which considers functional relationships depending on the fraction of intercepted diffuse radiation, density, and canopy volume for calculating the transpiration component. The model allows extraction of the olive tree from the three compartments into which the soil volume is divided depending on the root distribution defined by the user. If the cover crop option is considered, a subroutine simulates the development of the cover crop according to the climate data and seeding date following the procedure of the AquaCrop model (Raes et al., 2009). The development of the cover crop is used to determine the daily transpiration of the cover crop, which extracts soil water only from the cover crop compartment by using the same approach proposed in the AquaCrop model (Raes et al., 2010). A full description of the model can be found in Abazi et al. (2013). The version of WABOL used in this study included a further modification consisting of the incorporation of a drip irrigation evaluation in which the depth and wetted surface of the wetted bulb from the drippers were calculated according to Maia and Aguilar (2010), which considered the flow, time of irrigation, and soil texture, redistributing the irrigated water not evaporated during the day of irrigation into the wetted layers of the tree compartment (Fig. 10–2) the day after irrigation.

Olive Yield Assessment

Olive yield, as for the rest of crops in arid and semiarid conditions, responds nonlinearly to seasonal evapotranspiration as demonstrated, for instance, by Moriana et al. (2003) for olive trees. Because of the difficulties separating transpiration and evaporation from the soil, Moriana et al. (2003) uniquely studied the total ET through measurements of the soil water content using neutron probes, demonstrating in an experiment on an adult olive orchard in Córdoba (southern Spain) under different irrigation regimes (from fully irrigated to rainfed) that there is a polynomial correlation between seasonal (understood as January to December) olive orchard ET and olive yield, with a decreasing slope for higher ET values, indicating a diminishing increase in olive yield for higher ET values.

With the objective of determining a robust relationship between actual olive tree transpiration and olive yield, we used the WABOL model described above to separate the two components of ET in the experiments demonstrated by Moriana et al. (2003). To carry out this simulation, meteorological conditions for 1997, 1998, and 1999 and three irrigation treatments (fully irrigated, 50% of required irrigation, and rainfed) were considered. To include irrigation processes not included in the initial versions of WABOL, specific modules for drip irrigation and evaporation from the soil surface in the wet bulb were incorporated into the model. In addition to these experiments performed in Córdoba, we also studied the components of the hydrological balance for orchards under rainfed conditions in two traditional olive production areas in Andalusia, southern Spain: one of very low yields, Obejo, and another of average yields, Baena. For these two areas a comprehensive description of the orchards and average yields for three consecutive years could be found in Álvarez et al. (2007) and Soriano et al. (2012). In the Obejo and Baena cases we used the triannual average annual yield, while in the Córdoba experiments we used the biannual average annual yield, orchard evapotranspiration, and orchard transpiration. In this approximation we assume that the model performs a proper separation of olive transpiration and soil evaporation, an assumption already made in previous studies (Mesa-Jurado et al., 2010) using the same equations implemented in WABOL.

In Fig. 10–3 the relationship between the observed yield and the estimated soil transpiration for each of the different conditions simulated in this analysis, as well as the empirical equation fitted to the data, is shown. It is apparent how there is a very good fit between olive yield and orchard seasonal transpiration, $R^2 = 0.98$ and RMSE = 0.60 t yr^{-1}, with a diminishing increase for higher transpiration

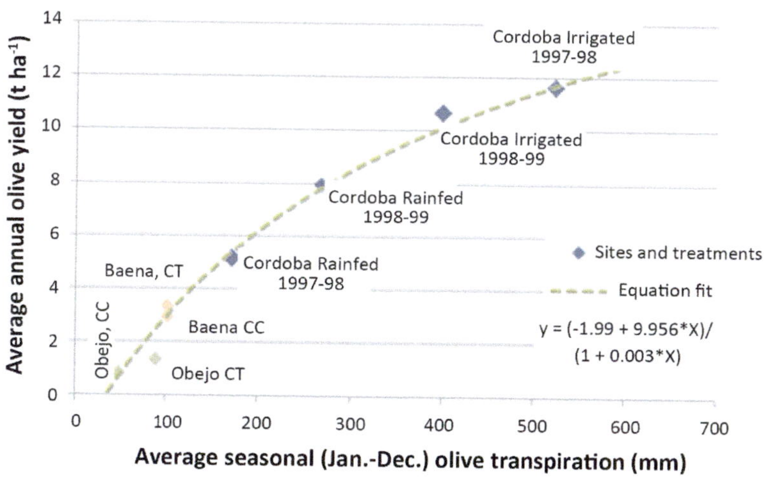

Fig. 10–3. Correlation between predicted transpiration and measured yield for the different simulated conditions (blue [Moriana et al., 2003], brown [(Soriano et al., 2012], and green [Álvarez et al., 2007] dots indicate sources of the yield data). Also shown is the fitted regression between seasonal transpiration and olive yield.

values as observed by Moriana et al. (2003) for the aggregated transpiration and ET values. That regression can be used to convert the modeled olive transpiration into expected olive yields (assuming that no other environmental or agronomical limitations appear). This tool will allow for evaluation of the impact of changing climate or management conditions into yield changes, caused by the variation of the available water for olive tree transpiration.

Climatic Models

The climatic models considered in this study were provided by the European project ENSEMBLES funded by the European Commission in the Sixth Framework Programme. The ENSEMBLES project covers almost the entire spectrum of climate change science. The climate projections generated in ENSEMBLES describe Europe experiencing tangible and measurable climate change. As the century progresses, the projected climate moves increasingly farther away from its current state, so that by 2100 the climate of Europe will be very different from today. Even under a mitigation scenario, the climate of Europe during the next few decades is still calculated to depart significantly from that of the present (van der Linden and Mitchell, 2009). ENSEMBLES results show how the impacts resulting from these climate changes, including changes in climate mean, variability, and extremes, affect all the systems and sectors studied (van der Linden and Mitchell, 2009).

All the estimations provided by the regional climatic change models show a high uncertainty due to the huge complexity and the different inputs required for the simulated processes. Quantifying this uncertainly requires the consideration of several models. The module RTB2 in ENSEMBLES focused on the generation of regional climatic change scenarios at high spatial resolution for Europe. ENSEMBLES determined numerous climatic models, and for this study four of them have been chosen (Table 10–1). Complete information about the procedure for estimation of the control and future climatic data overviewed in Table 10–1 can be obtained at http://ensemblesrt3.dmi.dk/.

The selection of these models was made to consider different driving general circulation models and different institutions generating different results. For each model daily rainfall and the maximum and minimum temperature were obtained. The spatially distributed climatic data were corrected using a statistical bias correction technique developed by Dosio and Paruolo (2011). This technique was applied to each regional climate model described in Table 10–1 and considers intermodal variability and longer natural climate variability, improving the probability distribution functions of both temperature and rainfall and increasing the capacity of reproducing extreme events (Dosio and Paruolo, 2011).

Table 10–1. Overview of the ENSEMBLES products of module RT2B (transient experiments 1951–2100 driven by global experiments) considered in this study.

Acronym	Model	Driving GCM†	Institution	Spatial resolution	Scenario
				km	
C4IRCA3	RCA3	HadCM3Q16	C4I	25	A1B
CNRM-RM4.5	ALADIN	ARPEGE	CNRM	25	A1B
ETHZ-CLM	CLM	HadCM3Q0	ETHZ	25	A1B
METO-HC_HadRM3Q0	HadRM3Q0	HadCM3Q0	HC	25	A1B

† GCM, general circulation model.

Scenarios Considered

Locations and Olive Orchard Characteristics

The olive crop in southern Spain shows a big extension in a complex geographical context. This heterogeneity implies the existence of different typologies of orchards characterized by different tree densities and canopy volumes and also of soil, slope, and climate conditions (Guzmán et al., 2009). Among these different types of olive orchards, two major types of soil management, tillage and temporary cover crop, are present. This heterogeneity has been included in this study with the selection of five different areas of study located within the Andalusia region in southern Spain. The five selected zones were Archidona in the Málaga province; Baena, Córdoba, and Obejo in the Córdoba province; and Mancha Real in the Jaén province (Fig. 10–4). In each of these five locations, two different soil management alternatives (tillage and temporary cover crop) were simulated, always considering rainfed conditions. Despite an expansion of irrigation in recent decades, the majority of olives

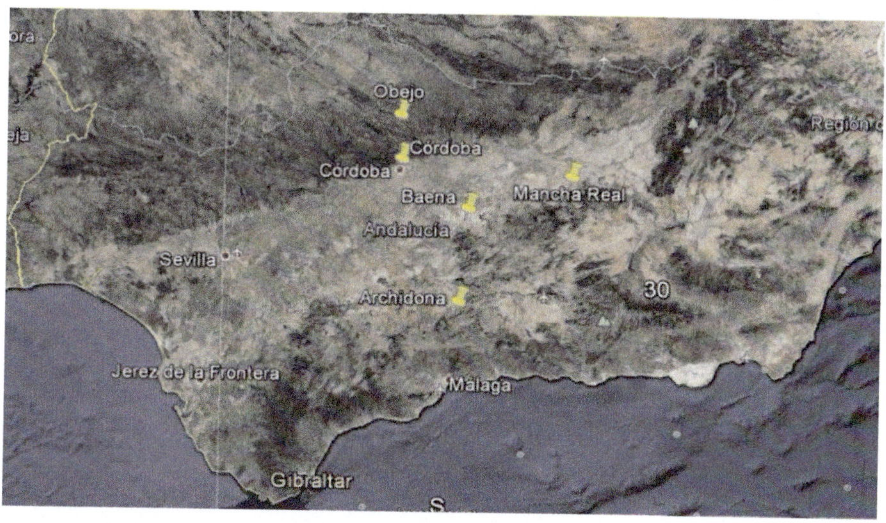

Fig. 10–4. Locations of the areas of study. Source: Google.

trees in the region have still been rainfed, and this is the most sensitive situation to water stress rather than the competition by the cover crop or a projected decrease in rainfall due to changes in climate conditions.

Each of the five locations had different weather, soil, and olive orchard characteristics. This information was obtained from several technical visits to the main olive areas in southern Spain and has been described previously (Abazi et al., 2012; Rodríguez-Carretero et al., 2013). A summary of the main characteristics are shown in Table 10–2.

All the model simulations for each individual year were run for hydrologic years (1 September to 31 August) to minimize the uncertainty in initializing soil moisture content. At this time of the year, the end of the dry season, the soil profile is close to permanent wilting point.

Regional Climate Models Estimations in Southern Spain

The selected regional climate models described in Table 10–1 provided daily weather data for two periods. The first one—the "control period"—was considered

Table 10–2. Main variables of WABOL, including weather, soil, and olive orchard characteristics for the five selected areas within southern Spain.

Main variables†	Córdoba	Archidona	Baena	Mancha Real	Obejo
Type of climate‡	Csa	Bsh	Bsh	Csa	Csa
Rainfall (mm)	502	589	467	506	526
ET_o (mm)	1407	1205	1321	1279	1276
FAO soil	Vertisol	Calcic Cambisol	Calcic Cambisol	Chromic Vertisols	Eutric Regosol
American soil	Typic Chromoxerert	Typic Xerochrepts	Typic Xerochrepts	Typic Chromoxererts	Typic Xerorthents
Texture	clay loam	clay loam	clay loam	clay	sandy loam
PWP ($cm^3\ cm^{-3}$)	0.2	0.2	0.2	0.21	0.09
FC ($cm^3\ cm^{-3}$)	0.4	0.4	0.4	0.44	0.21
WHC ($mm\ m^{-1}$)	200	200	200	230	120
Soil depth (m)	1.4	1.2	1.2	1.2	0.8
CN	76	76	76	80	72
Stone content (%)	3	24	21	4	36
Slope (%)	11	14	12	12	21
Density (trees ha^{-1})	138	69	100	100	138
Frame (m × m)	9 × 8	12 × 12	10 × 10	10 × 10	9 × 8
Diameter canopy (m)	5	5	5.5	4.5	4.5
Canopy cover (%)	29	13	25	18	23
Width cover crop (m)	3.5	5	4	4	3.5

† ET_o, reference evapotranspiration; PWP, permanent wilting point; FC, field capacity; WHC, water holding capacity; CN, curve number.

‡ Csa, temperate with dry or hot summer; BSh (hot steppe), according to the Köppen–Geiger Climate Classification for the Iberian Peninsula.

from September 1971 to August 2001, while the second period—the "Future Period"—was considered from September 2068 to August 2098. Averaged values and variations in rainfall and temperature for both periods are shown in Fig. 10–5 and Table 10–3, respectively.

With the provided daily data from the different regional climate models (Table 10–1), the number of days with rainfall, the rainfall intensity, and the reference evapotranspiration (ET_0) were determined. Considering the available climatic information (rainfall and temperature) for ET_0 assessment, the Hargreaves equation (Hargreaves and Samani, 1985) was considered. In spite of the small data requirements of this approach, the performance of the Hargreaves equation under semiarid conditions detected in southern Spain has been very satisfactory (Gavilán et al., 2006). For the ET_0 assessment, the REF-ET Software Reference Evapotranspiration Calculator by Allen (1991) was used. The ET_0 values calculated using the four regional climate models for the five locations are described in Fig. 10–5 and Table 10–4.

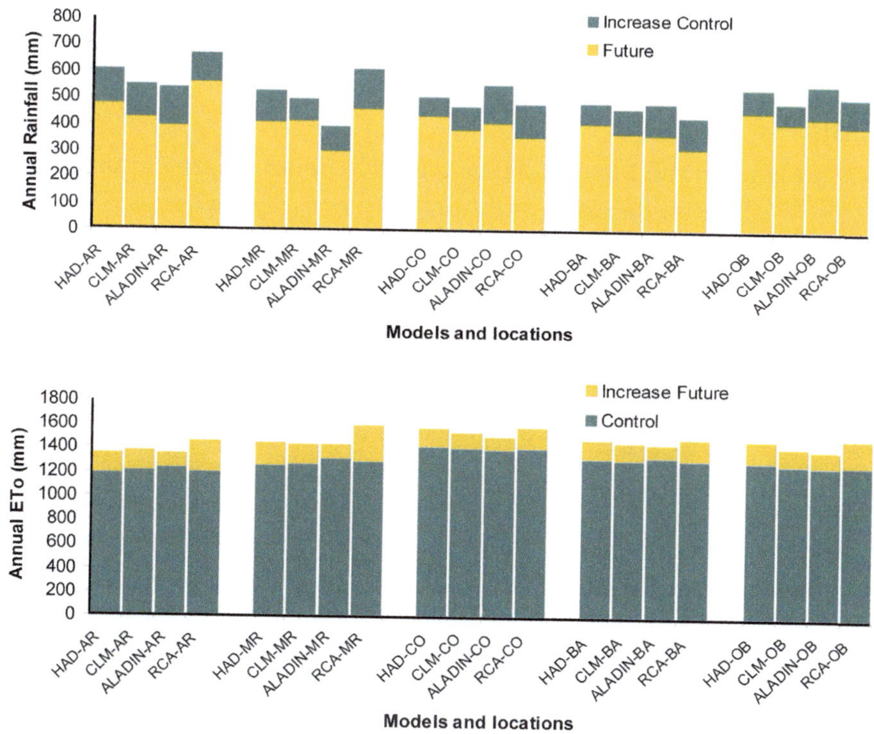

Fig. 10–5. Annual rainfall and reference evapotranspiration for each regional climate model and location.

Table 10–3. Variation of the average annual rainfall for the four considered models and the five locations included in this study.

Model	Archidona	Baena	Córdoba	Mancha Real	Obejo
	%				
C4IRCA3	−16.69	−27.43	−24.98	−24.35	−21.55
CNRM-RM4.5	−27.37	−24.13	−26.06	−24.42	−21.97
ETHZ-CLM	−23.19	−19.18	−18.97	−16.76	−15.33
METO-HC_HadRM3Q0	−21.07	−16.26	−13.89	−22.36	−16.72
Average	−22.08	−21.75	−20.98	−21.97	−18.89

Table 10–4. Variation of the average reference evapotranspiration for the four considered models and the five locations included in this study.

Model	Archidona	Baena	Córdoba	Mancha Real	Obejo
	%				
C4IRCA3	21.3	13.51	12.48	23.54	17.81
CNRM-RM4.5	9.17	8.13	7.14	8.9	10.65
ETHZ-CLM	13.21	9.93	9.24	12.77	11.52
METO-HC_HadRM3Q0	14.13	11.52	10.62	15.58	13.77
Average	14.45	10.77	9.87	15.2	13.44

Results

Components of the Water Balance under Current Climate Conditions

Figure 10–6 presents the values of the different components of the water balance for the different orchard scenarios and locations described above. For simplicity, only the results obtained using the climate input from the CNRM-RM4.5 model are depicted, although the results obtained from the other climate models are similar. It is apparent how soil evaporation and tree transpiration tend to be the major components of the water balance, although in Obejo, an area of soils with low water retention capacity and high content of rock fragments with roots concentrated in the shallower soil horizons, percolation below the olive root depth is also a relevant component. It is also apparent from results depicted in Fig. 10–6 that for the two soil management strategies simulated in this exercise, the losses by transpiration of the cover crop in the cover crop system, which the model assumed to be killed in early April as recommended in the region (e.g., Gómez and Giráldez, 2009), compared with the tilled system tend to be compensated by moderate reductions in runoff and direct evaporation from the soil surface. The result is that the predicted average tree transpiration tends to be slightly higher in the cover crop systems, albeit these differences are nonstatistically significant

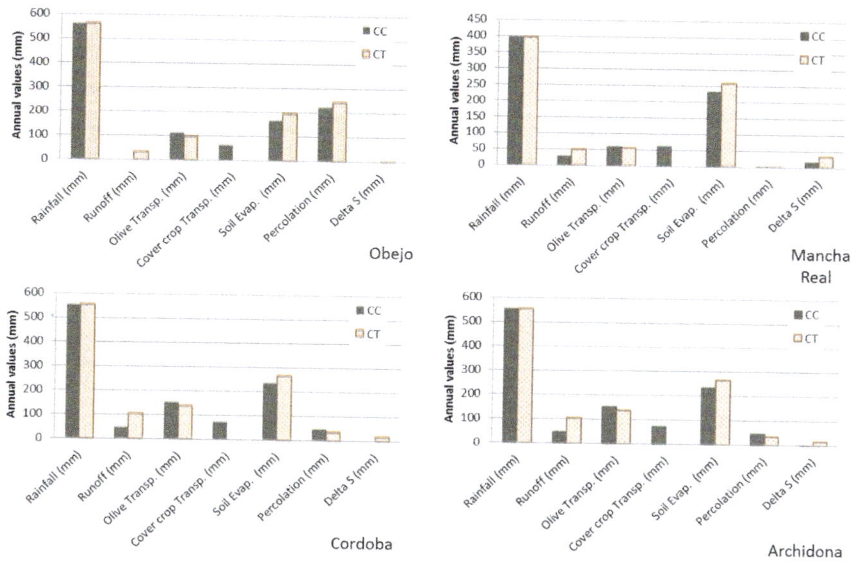

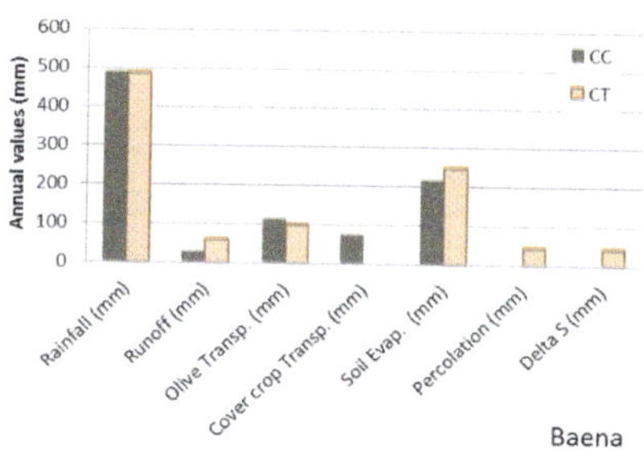

Fig. 10–6. Average values of the components of the soil water balance for the five locations during the period from 1971 to 2001 for the CNRM-RM4.5 model. Delta S is the difference in soil water content in the soil depth explored by the olive roots between the last and the first day of the simulation.

when compared using a paired *t* test (average coefficient of variation of 26% for olive tree transpiration). This result suggests that when properly managed, the cover crop system can be viable in a large set of conditions where rainfed olive trees are grown in the region. Being properly managed means that cover crops are well implanted and have a significant impact on reducing surface runoff, something that is not always achieved in farms, especially during the transition period from tillage to cover crop (Gómez and Giráldez, 2009), and transpiration of the cover crop is controlled effectively by mechanical or chemical methods in early spring. The low values of ΔS (the difference in soil water content in the soil depth explored by the olive roots between the last and the first day of the simulation) obtained indicate that the initialization of soil water content was appropriate and does not have an impact on the results of our simulations comparing soil management and locations.

As it was expected for an analysis in a Mediterranean climate area, there was a large year-to-year variability of rainfall values, which resulted in a large variability of the related components of the water balance. Figure 10–7 depicts the annual values obtained from the simulations for the Córdoba conditions with the climate inputs of the CNRM-RM4.5 model for the period from 1971 to 2001, with the other sites achieving similar year-to-year variability.

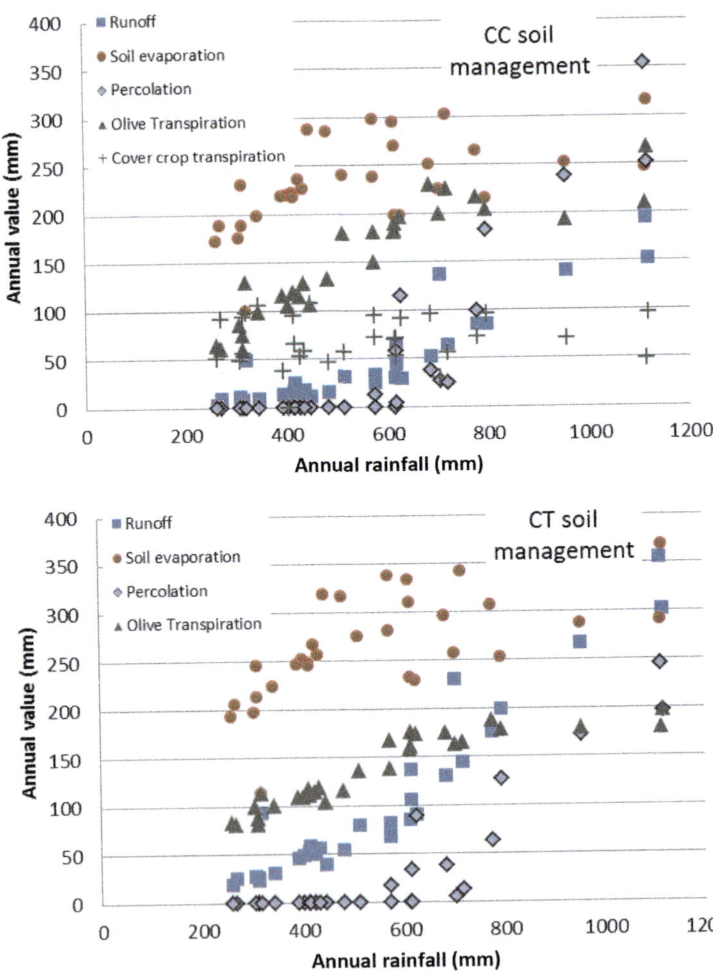

Fig. 10–7. Annual values of some of the water balance components component obtained from the simulations for the Córdoba conditions with the climate inputs of the CNRM-RM4.5 model for the period from 1971 to 2001.

Averaged Values for Four Regional Climate Models

Four regional climate models were considered in the study (C4IRCA3, CNRM-RM4.5, ETHZ-CLM, and METO-HC-HadRM3Q0; Table 10–1). The consideration of four different models provided a lower level of uncertainty to the estimations, instead providing a range of possible results to provide a single value. Analyzing the averaged results for the components of the water balance, the four models showed similar trends although differences in the absolute values were determined.

Average effects of climate change on rainfall are described in Table 10–5 and Fig. 10–8. For example, for the Córdoba location average rainfall reduction was around 21% (from the control period to future period), while an average increase

Table 10–5. Average variation of weather conditions and water balance components considering four regional climate models for the five locations considered.

Component	Córdoba	Baena	Mancha Real	Archidona	Obejo
	%				
Weather					
Rainfall	−21.0	−21.7	−22.0	−22.1	−18.9
Reference evapotranspiration	9.9	10.8	15.2	14.5	13.4
Tilled soil					
Runoff	−21.8	−22.4	−26.2	−22.7	−3.8
Olive transpiration	−24.3	−15.9	−24.5	−12.3	−6.8
Cover crop transpiration					
Soil evaporation	−16.9	−17.1	−18.2	−15.8	−16.5
Infiltration	−20.8	−21.7	−21.7	−22.0	−19.8
Percolation	−21.1	−43.1	−45.6	−43.2	−10.4
S†	−26.1	−52.2	−29.0	−25.1	−44.3
Cover crop					
Runoff	−14.1	−14.4	−21.8	−16.1	14.0
Olive transpiration	−26.0	−20.6	−26.2	−14.5	−13.3
Cover crop transpiration	−25.6	−18.3	−23.6	−9.3	−4.6
Soil evaporation	−16.8	−17.6	−18.9	−17.4	−18.6
Infiltration	−21.4	−22.1	−22.3	−22.6	−19.3
Percolation	−22.5	−40.2	−41.7	−39.8	−11.1
S	−28.8	−16.0	78.9	15.2	-2706.3
Rainfall efficiency					
Control/tilled soil	17.7	20.5	8.6	9.8	17.1
Control/cover crop	22.5	24.6	11.1	13.0	20.5
Future/tilled soil	16.9	22.0	8.3	11.1	19.7
Future/cover crop	21.1	24.9	10.5	14.3	22.0

† *S* stands for variation in soil water storage to the root depth during the hydrological year.

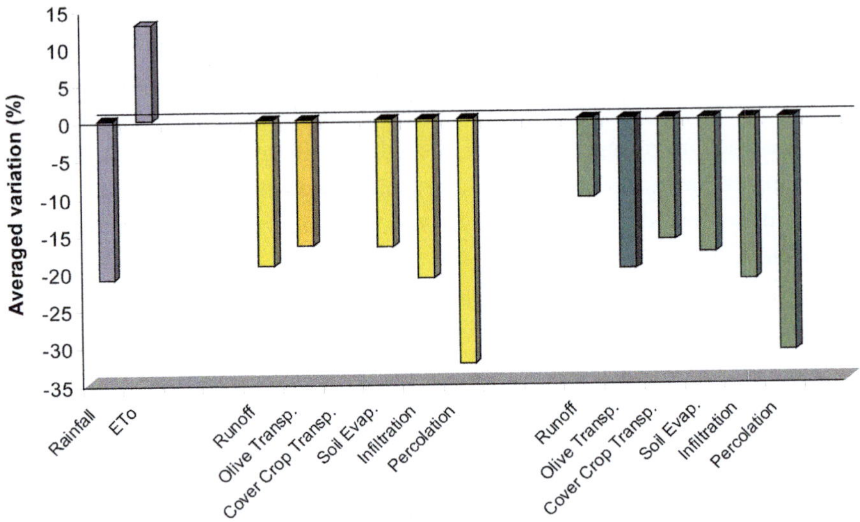

Fig. 10–8. Averaged variation of weather conditions and components of the water balance considering four regional climate models and five locations.

in ET_0 was equal to 9.9%. These values are the average for four regional climate models. Individually analyzing each model, the changes among locations ranged between reductions of 13.9 and 26.1%. This big range indicates the large variability in the different climatic models considered.

Analyzing the water balance components, two soil managements were considered. Considering the Córdoba location, for tillage management the average impact of climate change on simulated runoff was a reduction of 21.8% (Table 10–5). Analyzing the different regional climate models, the runoff ranged from a slight increase in runoff (1.03% with the METO-HC_HadRM3Q0 model) to a reduction of 35.8% for CNRM-RM4.5. The average reduction of the transpiration olive was 24.3% in the future period in comparison with the control period, ranging from −14.0 to −32.4% among models. These reductions generated an average fall in yield of 34.6%, ranging from −20.4 to −48.1% (Fig. 10–9). The average rainfall efficiency was equal to 17.7%, ranging from 17 to 19% depending on the considered model.

When soil management with cover crops was considered, also considering the Córdoba location, the average impact for runoff was −14.1% (Table 10–5), ranging from increases of 11.3% to reductions of 31.5%. The average reduction in olive transpiration was 26%, ranging from −17.2 to −34.5%, generating an average reduction in yield of 32.6% (−21.1 to −45.3%; Fig. 10–10). The average rainfall

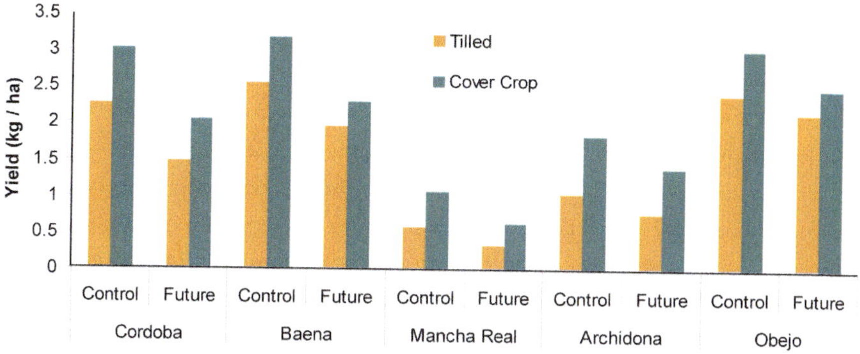

Fig. 10–9. Averaged estimated yield for each location and weather period (control and future).

efficiency was equal to 22.5%, ranging from 21.6 to 24% depending on the selected regional climate model.

Equally, the effect of changing the soil management from tilled to the consideration of a cover crop has been evaluated for four regional climate models. In general the differences between models were smaller than when the differences between control and future periods were considered. To simplify the exposition of the results, unique Córdoba location results are described. To change from tilled to cover crop under the control period generated an averaged runoff

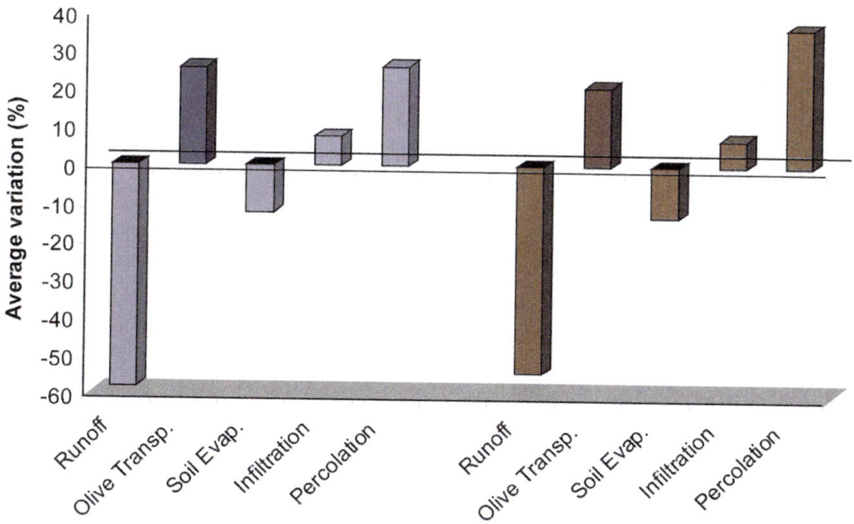

Fig. 10–10. Average variation in percentage for the water balance components from changing the soil management from tilled soil to cover crop, for control (blue charts) and future (brown charts) periods.

reduction of 56.7% (ranging from −54.2 to −59.7%). Olive transpiration increased 27.1%, ranging from 23.6 to 30.9% depending on the climatic model considered. This significant increase had as consequence an average yield increase of 34.7% (ranging from 32.6 to 39.8%). For the future period the conclusions were similar, simulating an average yield increase of 38.8%, ranging from 31.3 to 39.8%.

The elevated homogeneity in the results for different climatic models reduced the uncertainty in the selection of the optimal soil management to reduce erosion without affecting the yield. With the provided results, the optimal soil management was considering cover crop, significantly reducing the runoff (and thus water erosion). Although the differences in yield calculated from differences in olive transpiration are relatively large (around 39%), they are mostly nonstatistically significant but nevertheless indicate the sustainability of cover crop soil management under the projected reduction in rainfall for the end of the 21st century.

Specific Results for CNRM-RM4.5 Regional Climate Model

Considering the regional climate model CNRM-RM4.5 for the Córdoba location, rainfall was reduced from 551.6 mm (average for the control period of 1971–2001) to 407.8 mm (average for the future period of 2068–2098), a reduction of 26.1% (Fig. 10–5). Analyzing the number of rainfall events in the control and future periods, it was reduced by 18.2% in the future period, with the amount of rainfall per event being reduced by 11.3%. Interannual variability (variability in annual rainfall among the 30 yr considered for control and future periods) increased 6.5% for the future period in comparison with the control period (CV varied from 0.41 to 0.44).

Tilled Soil Management

With tillage soil management, runoff in the control period was 87.9 mm and in the future period was reduced to 56.5 mm (35.8% reduction; Table 10–6). The higher reduction in runoff was caused by the reduction of rainfall events without increasing the amount of rainfall per event. Soil evaporation for the control period was equal to 260.2 mm and was reduced to 219.5 mm in the future period (15.7% reduction; Table 10–6). This reduction almost coincides with the reduction of the number of rainfall events more than with the total rainfall reduction. This is because fewer events imply a reduction of the losses due to intense evaporation from the wetted soil surface during the 1 to 3 day period after a rainfall event. Percolation below the tree roots depth was reduced from 45.7 to 12.0 mm for the control period and future period, respectively, with a reduction of 73.8% (Table 10–6), and was caused by the severe reduction of rainfall that in many years was not able to fill the soil profile and generate percolation. Finally, the increase in water stored in the soil profile varied from 61.2 to 52.2 mm for the control period

Table 10–6. Weather data and water balance components for two soil management practices considering the data provided by the regional climate model CNRM-RM4.5.

Component	Tillage soil management		Cover crop management	
	Average	CV	Average	CV
	mm			
Control period				
Rainfall	551.6	0.41	551.6	0.41
Reference evapotranspiration	1398.4	0.03	1398.4	0.03
Runoff	87.9	0.97	40.3	1.19
Olive transpiration	96.6	0.48	123.1	0.51
Cover crop transpiration	–	–	172.5	0.37
Soil evaporation	260.2	0.20	227.0	0.20
Infiltration	463.7	0.32	511.3	0.36
Percolation	45.7	1.69	63.1	1.49
S †	61.2	0.26	−74.4	−0.68
Future period				
Rainfall	407.8	0.44	407.8	0.44
Reference evapotranspiration	1498.2	0.03	1498.2	0.03
Runoff	56.5	0.98	27.6	1.15
Olive transpiration	69.7	0.61	91.1	0.67
Cover crop transpiration	–	–	120.7	0.53
Soil evaporation	219.5	0.28	192.4	0.28
Infiltration	351.4	0.38	380.2	0.41
Percolation	12.0	2.89	21.3	2.47
S	50.2	0.53	−45.3	−0.97

† *S* stands for variation in soil water storage to the root depth during the hydrological year.

and future period, respectively (18% reduction; Table 10–6). These values resulted in transpiration of the olive varying from 96.6 to 69.7 mm for the control period and future period, respectively, indicating a reduction of 27.8%, similar to that detected for the rainfall.

Rainfall efficiency was defined as the ratio between the transpiration of the olive and the total rainfall. For the control period for bare soil the rainfall efficiency was equal to 17.5%. This value was reduced to 17.1% in the future period. Both values were especially low but are reasonable considering the low density of the analyzed orchard. Finally, using the functions described in Fig. 10–3, estimated yield ranged between 2.48 and 1.58 t ha^{-1} for the control period and future period, respectively, implying a reduction of 36.4%.

The small increase of variability considering the control and future periods for rainfall (6.5%) generated a negligible change in the variability of runoff that increased 1.3%. However, the variability of other components of the water balance increased significantly. Thus, soil evaporation variability was increased 38.9% and the deep percolation 71%. This fact indicates that the annual rainfall was not significantly affected in the future, but the distribution of this rainfall throughout the year (number of rainfall events and intensity) is expected to vary in a very significant way. Olive transpiration and yield also increased significantly the variability from the control to future period (27.3 and 41.2%, respectively). This fact implies a higher uncertainty for the farmer, increasing the oscillation in yield (and thus profit) and alternating yields close to zero with high-yield years.

Cover Crop Management

When the cover crop was considered, runoff in the control period was 40.3 mm and was reduced to 27.6 mm in the future period (31.5% reduction; Table 10–6). Soil evaporation in the control period was equal to 227 mm, reducing to 192.4 mm in the future period (15.2% reduction). Cover crop transpiration also reduced in the future period, moving from 172.5 to 120.7 mm with an important reduction of 30%. As with tilled soil management, deep percolation was significantly reduced from 63.1 to 21.3 mm, with a reduction of 66.3%. All these components concluded with a reduction of 26% in the transpiration of the olive orchard (from 123.1 to 91.1 mm). Rainfall efficiency showed the same values for both periods and was equal to 22.3%, but estimated yield had a lower reduction. Thus, the estimated yield for the control and future periods with cover crop was 3.32 and 2.30 t ha^{-1}, respectively, with a reduction equal to 30.8%.

Interannual variability for cover crop orchards generated similar figures than those for tilled management. Although variability for runoff was slightly reduced (3.4%), similar values than those for tilled management of variability for the different components were determined. Thus, variability in cover crop transpiration, soil evaporation, and deep percolation was significantly increased (43.5, 39.2, and 65.8%, respectively). This implied an increase in variability for olive transpiration equal to 32%, which then generated increases in yield variability equal to 43.5%.

Evaluation of Different Soil Management Practices

Tilled Management to Cover Crop for Control Period

Comparing soil management practices, in the control period the runoff was significantly reduced when cover crop is considered in comparison with tilled management. Thus, a runoff reduction of 54.2% with the CNRM-RM4.5 model was determined. Also a reduction of soil evaporation was determined (12.8%). With the opposite signal, the infiltration and the deep percolation were increased (10.3 and 38.2%, respectively). These values implied an increase of

27.4% in the olive transpiration, and the olive yield was increased by 33.6%, indicating the optimal practice of the utilization of cover crop under the conditions described previously.

Analyzing the interannual variability, in general the behavior was different for the water balance components: runoff variability was increased by 22.5%, while percolation variability was reduced by 11.9%. Olive transpiration variability was increased by 5.3%; however, the yield variability with cover crop was reduced by 9.1%.

Tilled Management to Cover Crop for Future Period

When these soil management alternatives were analyzed under the weather conditions for the future period, the results were very similar those considering the control period, showing that the satisfactory results with the use of cover crops would be similar (or perhaps slightly better) in the future period with an increase in olive transpiration of 30.6% and yield increase of 45.5%. As a consequence of this improvement, rainfall efficiency was also improved when cover crop was considered, moving from around 17.3 to 22.3%.

Analyzing variability for the future period, changing from tilled to cover crop increased variability in all components except for percolation, which was reduced by 14.6%. Olive transpiration variability increased by 9.1%, but yield variability was reduced by 7.6% when cover crop was considered. This is an additional advantage in the use of cover crops as soil management.

Comparison between Scenarios (Control with Tilled to Future with Cover Crop)

Finally, we compared the performance of an olive orchard system located in Córdoba currently (weather conditions of the control period with tillage soil management) with a simulated scenario (considering weather conditions of a future period with an adaptation strategy to climate change as the consideration of a cover crop). With this scenario, runoff will be reduced by 68.6%, significantly affecting the reduction of erosion. Soil evaporation would be reduced by 26%, while deep percolation would be reduced by 53.4%. This important reduction in deep percolation would have two opposite consequences: the possible depletion of groundwater resources and the mitigation of possible diffuse contamination associated with percolation of nitrates and pesticides.

Considering the olive crop, the transpiration would be reduced in a small percentage (5.7%), with a similar reduction in yield (7.5%). However, analyzing the variation in water stored in the soil profile, an important reduction of this term was detected, indicating a severe reduction of water stored at the end of the crop cycle in comparison with the stored water at the beginning. This fact could generate problems in the driest years, severely affecting the olive transpiration and thus yield.

Interannual variability was increased in all the water balance components, ranging from increases of 18.3% for runoff to 46.1% in deep percolation. Equally, yield variability was increased by 30.4%.

Application to Other Locations

In general, the extension of the study to different locations within the Andalusia region determined small differences in the estimated change in the weather conditions (Table 10–5). However, the variability in the olive orchard characteristics (Table 10–2) had a higher impact on the results.

Rainfall reduction was very similar for the five selected locations and ranged from −18.89 for Obejo to −22.08 for Archidona. For reference evapotranspiration (ET_0), a general increase was determined, ranging from +9.87 for Córdoba to +15.20 for Mancha Real. In spite of these small differences of weather among locations, the impact of climate change on olive yield depicted huge differences among locations caused by the specific characteristics of each selected orchard (Table 10–2). Thus, as a conclusion of the analysis of the water balance components, the soil evaporation was a critical variable that highly influenced the final yield estimation. Therefore, the percentage of soil covered with the tree canopy and by stones is directly related with soil evaporation and thus with olive transpiration.

In order to determine the impact of the specific characteristics of each orchard on the results, Baena and Córdoba locations were considered. Thus, the Baena location showed higher yield than the Córdoba location. In spite of the rainfall and ET_0 being similar (Table 10–2), the percentage of covered soil was higher for Baena than for Córdoba (the sum of cover crop plus stones was 46 vs. 32%, respectively) implying more available water for olive transpiration. Because of this fact, a lower impact of climate change on olive transpiration (reductions were equal to 15.9 and 20.6% for tilled and cover crop soil management, respectively) for the Baena location than for the Córdoba location was determined. These values implied reductions in yield of 23.9 and 27.7%, respectively, for tilled and cover crop management. Rainfall efficiency ranged between 20.5 and 24.6, respectively. The impact of changes in soil management was smaller for Baena than for Córdoba. For the Baena location, the yield increase was 24.6 and 18.1% for the control period and future period, respectively, which is significantly lower than that obtained for the Córdoba location (Fig. 10–9).

For the Mancha Real location, a significant small yield was determined for all the scenarios due to the elevated percentage of bare soil (uniquely, 22% of the soil is covered with the tree canopy or stones), generating elevated values of soil evaporation and small values of olive transpiration. When climate change scenarios are considered, the olive transpiration rates were very small compared with other locations. Analyzing Fig. 10–3, the relationship between olive transpiration and yield is not linear, increasing the slope of the yield reduction with low values of olive transpiration. Estimated yield reduction reached 48.1 and 44.8% for tilled and cover crop, respectively. Rainfall efficiency was especially low with values of 8.4 and 11% for bare and cover crop soil management, respectively. Because of the specific conditions of the olive orchard located in Mancha Real, the system is especially sensible to climate change conditions. Unfortunately, the conditions detected in Mancha Real (small size of the trees and small stones on the surface) are very common in southern Spain, indicating the huge importance of the use of some type of cover to avoid the high evaporation losses generated for the bare soil. For Mancha Real, the increases in yield when the soil management was moved from tillage to cover crop were equal to 86.5 and 110.1% for the control period and future period, respectively (Fig. 10–9), indicating the huge importance of covering the soil to reduce soil evaporation losses.

Although the Archidona location showed similar changes in weather, the average estimated rainfall for the future period was the highest of the analyzed locations (461 mm). This rainfall reaches a satisfactory level for an olive orchard, and thus the estimated olive transpiration was reduced, but by a less amount than for the Córdoba location. Thus, olive transpiration was reduced 12.3 and 14.5% for tilled and cover crop soil management, respectively, generating reductions in olive yield equal to 27.9 and 24.5%, respectively (Fig. 10–9).

Finally, the Obejo location demonstrates the huge importance of the cover of soil in the reduction of climate change impacts on the components of the water balance in an olive orchard. Although the reduction in rainfall was similar to the rest of the locations, the reduction of olive transpiration from the control period to the future period was 6.8 and 13.3% for tilled and cover crop, respectively, generating a reduction in yield equal to 10.3 and 18.0%, respectively. These excellent results are based on a reduced soil evaporation generated for the high percentage of soil surface with stones (Table 10–2). As an example of the effect of the cover crop, analyzing the effect of moving from tilled to cover crop, the impact on yield was positive but small compared with previous locations (26.3 and 15.3% for the control period and future period, respectively; Fig. 10–9).

Conclusions

The determination of the impact of climate change on the Mediterranean agricultural systems is a critical issue for ensuring the sustainability of these systems in the future. Once the impacts are evaluated, the next step is determining adaptation strategies that reduce the impact of climate change on the agricultural systems. The most adequate methodology for the determination of impacts and adaptation strategies is the use of simulation models, and in this study a model specifically designed for olive orchards (WABOL) was considered.

The WABOL evaluated the performance of a typical Mediterranean agricultural system consisting of an olive orchard and two different soil management strategies: tillage and cover crop. WABOL was applied in five locations and for four regional climate models, determining an optimal adaptation strategy for the Mediterranean olive orchards located in southern Spain. The analysis of the water balance obtained for each scenario for current climate conditions using WABOL concludes that when the cover crop is properly managed (meaning that it is properly implanted to control runoff and its transpiration is ceased in early spring by mowing or herbicides), this soil management practice can be as viable as the conventional tillage over a large set of weather and agronomical conditions.

We define viable as a situation in which the mean annual olive transpiration (and thus potential yield) is not significantly different from the one predicted for the tilled system. The evaluation of the components of the water balance indicates that for the conditions simulated, the reduction in runoff and in direct evaporation from the bare soil compensates the increase in transpiration by the cover crop. For projected climate scenarios, the results of the model simulations indicate that the implementation of a well-managed cover crop in the future compared with a current traditional tillage management reduced the runoff by around 69%, significantly affecting the generation of erosion. Equally, deep percolation was reduced by around 53%, clearly mitigating possible diffuse contamination associated with the percolation of nitrates and pesticides. Finally, olive transpiration was uniquely reduced, compared with current scenarios, around 6% because of the consideration of a correctly managed cover crop.

Finally, this study proved that simulation models are useful for the identification of optimal soil management strategies when multiple alternatives must be evaluated and analyzed.

Acknowledgments

This study has been made possible by contributions from different research projects, including P08-AGR-03925 (Andalusian Government), AGL2009-12936-C03-01 (Spanish Ministry of Science and Innovation), AGL2012-40128-C03-01, and FEDER funds.

References

Abazi, U., I.J. Lorite, B. Cárceles, A. Martínez-Raya, V.H. Durán, J.R. Francia, and J.A. Gómez. 2013. WABOL: A conceptual water balance model for analyzing rainfall water use in olive orchards under different soil and cover crop Management strategies. Comput. Electron. Agric. 91:35–48. doi:10.1016/j.compag.2012.11.010

Abazi, U., I.J. Lorite, and J.A. Gomez. 2012. Impact of cover crops on soil coservation in olive orchards under different agroecological conditions combing a conceptual soil water balance model (WABOL) and RUSLE. EGU General Assembly 2012. Geophys. Res. Abstr. 14:EGU2012-4174.

Abid Karray, J., P.J. Lhome, M.M. Masmoudi, and N. Mechlia Ben. 2008. Water balance of the olive tree-annual crop association: A modeling approach. Agric. Water Manage. 95:575–586. doi:10.1016/j.agwat.2007.12.006

Allen, R. 1991. REF-ET. Reference evapotranspiration computation software. Department of Biological and Irrigation Engineering, Utah State University, Logan.

Allen, R., L.S. Pereira, D. Raed, and M. Smith. 1998. Crop evapotranspiration—Guidelines for computing crop water requirements. Irrig. Drain. Pap. 56. FAO, Rome.

Álvarez, S., M.A. Soriano, B.B. Landa, and J.A. Gómez. 2007. Soil properties in organic olive orchards as compared to natural areas in a mountainous landscape in southern Spain. Soil Use Manage. 23:404–416. doi:10.1111/j.1475-2743.2007.00104.x

Beaufoy, G. 2001. EU policies for olive farming—unsustainable on all counts. BirdLife International-WWF, Brussels, Belgium.

Casadesús, J., M. Mata, J. Marsal, and J. Girona. 2012. A general algorithm for automated scheduling of drip irrigation in tree crops. Comput. Electron. Agric. 83:11–20. doi:10.1016/j.compag.2012.01.005

Castro, G., and J.A. Gómez. 2006. Analysis of hydrological fluxes and optimum dates for killing cover crops in the study areas. OLIVERO Proj. Commun. No. 14. CSIC, Madrid. http://hdl.handle.net/10261/7652 (accessed 20 Nov. 2013).

Castro, G., J.A. Gómez, and E. Fereres. 2006. Determinación de la fecha de siega de la cubierta protectora en olivar. Vida Rural 228:36–41.

Celette, F., A. Ripoche, and G. Christian. 2010. WaLIS—A simple model to simulate water partitioning in a crop association: The example of an intercropped vineyard. Agric. Water Manage. 97:1749–1759. doi:10.1016/j.agwat.2010.06.008

Dosio, A., and P. Paruolo. 2011. Bias correction of the ENSEMBLES high-resolution climate change projections for use by impact models: Evaluation on the present climate. J. Geophys. Res. 116:D16106. doi:10.1029/2011JD015934

Er-Raki, S., A. Chehbouni, G. Boulet, and D.G. Williams. 2010. Using the dual approach of FAO-56 for partitioning ET into soil and plant component for olive orchards in a semiarid region. Agric. Water Manage. 97:1769–1778.

Eurostat. 2013. Agricultural statistics. European Commission. http://epp.eurostat.ec.europa.eu/portal/page/portal/agriculture/introduction (accessed 15 Jan. 2014).

Falkenmark, M. 2013. Growing water scarcity in agriculture: Future challenge to global water security. Phil. Trans. R. Soc. A 371:20120410. doi:10.1098/rsta.2012.0410

Fraisse, C.W., N.E. Breuer, D. Zierden, J.G. Bellow, J. Paz, V.E. Cabrera, A. Garcia y Garcia, K.T. Ingram, U. Hatch, G. Hoogenboom, J.W. Jones, and J.J. O'Brien. 2006. AgClimate: A climate forecast information system for agricultural risk management in the southeastern USA. Comput. Electron. Agric. 53:13–27. doi:10.1016/j.compag.2006.03.002

Gavilán, P., I.J. Lorite, S. Tornero, and J. Berengena. 2006. Regional calibration of Hargreaves equation for estimating reference ET in a semiarid environment. Agric. Water Manage. 81:257–281. doi:10.1016/j.agwat.2005.05.001

Gómez, J.A., and J.V. Giráldez. 2009. Erosión y degradación de suelos. In: J.A. Gómez, editor, Sostenibilidad de la preducción de olivar en Andalucía. Consejería de Agricultura y Pesca Junta de Andalucía. Seville, Spain.

Gómez, J.A., M.G. Guzmán, J.V. Giraldez, and E. Fereres. 2009. Experimental assessment of the influence of cover crops and tillage on water and sediment yield, and on nutrient, and organic matter losses in an olive orchard on a sandy loam soil. Soil Tillage Res. 106:137–144. doi:10.1016/j.still.2009.04.008

Guzmán, J.R., J.A. Gómez, and L. Rallo. 2009. El olivar en Andalucía: Lecciones para el futuro de un cultivo milenario. In: J.A. Gómez, editor, Sostenibilidad de la producción de olivar en Andalucía. Consejería de Agricultura y Pesca Junta de Andalucía, Seville, Spain.

Hargreaves, G.H., and Z.A. Samani. 1985. Reference crop evapotranspiration from temperature. Appl. Eng. Agric. 1:96–99.

Henderson, D.W. 1979. Soil management in semi-arid environments. In: A.E. Hall, G.H. Cannell and H.W. Lawton, editors, Agriculture in semiarid environments. Chap. 9. Springer-Verlag, New York.

Hulugalle, N.R., and R. Lal. 1986. Soil water balance of intercropped maize and cowpea grown in a tropical hydromorphic soil in western Nigeria. Agron. J. 78:86–90. doi:10.2134/agronj1986.00021962007800010019x

Hsiao T.C., P. Steduto, and E. Fereres. 2007. A systematic and quantitative approach to improve water use efficiency in agriculture. Irrig. Sci. 25:209–231.

Jayakrishnan, R., R. Srinivasan, C. Santhi, and J.G. Arnold. 2005. Advances in the application of the SWAT model for water resources management. Hydrol. Processes 19:749–762. doi:10.1002/hyp.5624

Jones, J.W., G. Hoogenboom, C.H. Porter, K.J. Boote, W.D. Batchelor, L.A. Hunt, P.W. Wilkens, U. Singh, A.J. Gijsman, and J.T. Ritchie. 2003. The DSSAT cropping system model. Eur. J. Agron. 18:235–265. doi:10.1016/S1161-0301(02)00107-7

Maia, C.E., and S.L. Aguilar. 2010. Estimate of wetted bulb dimensions in surface drip irrigation using response surface model. Cienc. Rural 40:1302–1308. doi:10.1590/S0103-84782010005000099

Mesa-Jurado, M.A., J. Berbel, and F. Orgaz. 2010. Estimating marginal value of water for irrigated olive grove with the production function method. Span. J. Agric. Res. 8:197–206. doi:10.5424/sjar/201008S2-1362

Ministry of Agriculture, Food and Environment. 2010. Encuesta sobre superficies y rendimientos de cultivos 2010. MAGRAMA. http://www.magrama.gob.es/es/estadistica/temas/estadisticas-agrarias/agricultura/esyrce/ (accessed 12 Jan. 2014).

Monteiro, A., and C.M. Lopes. 2007. Influence of cover crop on water use and performance of vineyard in Mediterranean Portugal. Agric. Ecosyst. Environ. 121:336–342. doi:10.1016/j.agee.2006.11.016

Moriana, A., F. Orgaz, M. Pastor, and E. Fereres. 2003. Yield responses of a mature olive orchard to water deficits. J. Am. Soc. Hortic. Sci. 128:425–431.

Orgaz, F., L. Testi, F.J. Villalobos, and E. Fereres. 2006. Water requirements of olive orchards II: Determination of crop coefficients for irrigation scheduling. Irrig. Sci. 24:77–84. doi:10.1007/s00271-005-0012-x

Palomo, M.J., F. Moreno, J.E. Fernández, A. Dìiaz-Espejo, and I.F. Girón. 2002. Determining water consumption in olive orchards using the water balance approach. Agric. Water Manage. 55:15–35. doi:10.1016/S0378-3774(01)00182-2

Raes, D., P. Steduto, T.C. Hsiao, and E. Fereres. 2009. AquaCrop—The FAO model to simulate yield response to water: II. Main algorithms and software descriptions. Agron. J. 101:438–447. doi:10.2134/agronj2008.0140s

Raes, D., P. Steduto, T.C. Hsiao, and E. Fereres. 2010. AquaCrop version 3.1 reference manual. Chap. 3. FAO, Rome.

Resop, J.P., D.H. Feisher, Q.G. Wang, D.J. Timlim, and V.R. Reddy. 2012. Combining explanatory crop models with geospatial data for regional analysis of crop yield using field-scale models. Comput. Electron. Agric. 89:51–61. doi:10.1016/j.compag.2012.08.001

Ritchie, J. 1972. Model for predicting evaporation from a row crop with incomplete cover. Water Resour. Res. 8:1204–1213. doi:10.1029/WR008i005p01204

Rodríguez-Carretero, M.T., I.J. Lorite, M. Ruiz-Ramos, A. Dosio, and J.A. Gómez. 2013. Impact of climate change on water balance components in Mediterranean rainfed olive orchards under tillage or cover crop soil management. EGU General Assembly 2013. Geophys. Res. Abstr. 15:EGU2013-11446.

Romero, P., G. Castro, J.A. Gómez, and E. Fereres. 2007. Curve number values for olive orchards under different soil management. Soil Sci. Soc. Am. J. 71:1758–1769. doi:10.2136/sssaj2007.0034

Safir, G.R., S.H. Gage, M. Colunga-García, P. Grace, and S. Rowshan. 2008. Simulation of corn yields in the Upper Great Lakes Region of the USA using a modelling framework. Comput. Electron. Agric. 60:301–305. doi:10.1016/j.compag.2007.09.004

Soriano, M.A., S. Álvarez, B. Landa, and J.A. Gómez. 2012. Soil properties in organic olive orchards following different weed management in a rolling landscape of Andalusia, Spain. Renewable Agric. Food Syst. 29:83–91. doi:10.1017/S1742170512000361

Testi, L., F.J. Villalobos, F. Orgaz, and E. Fereres. 2006. Water requirements of olive orchards: I simulation of daily evapotranspiration for scenario analysis. Irrig. Sci. 24:69–76. doi:10.1007/s00271-005-0011-y

Thorp, K.R., K.C. DeJonge, A. Kaleita, W.D. Batchelor, and J.O. Paz. 2008. Methodology for the use of DSSAT models for precision agriculture decision support. Comput. Electron. Agric. 64:276–285. doi:10.1016/j.compag.2008.05.022

van der Linden, P., and J.F.B. Mitchell, editors. 2009: ENSEMBLES: Climate change and its impacts: Summary of research and results from the ENSEMBLES project. Met Office Hadley Centre, Exeter, UK.

Van den Putte, A., G. Govers, J. Diels, K. Gillinjns, and M. Demuzere. 2010. Assessing the effect of soil tillage on crop growth: A meta-regression analysis on European crop yields under conservation agriculture. Eur. J. Agron. 33:231–241. doi:10.1016/j.eja.2010.05.008

Villalobos, F. 1997. Modelling the carbon and water balances of olive (*Olea europaea* L.). In: J.P. Abreu et al., editors, International Workshop on Simulation in Agricultural and Environmental Systems. Sociedade de Ciências Agrárias de Portugal, Lisbon.

Villalobos, F.J., F. Orgaz, L. Testi, and E. Fereres. 2000. Measurement and modeling of evapotranspiration of olive (*Olea europaea* L.) orchards. Eur. J. Agron. 13:155–163. doi:10.1016/S1161-0301(00)00071-X

Wortmann, C.S., R.A. Drijber, and T.G. Franti. 2010. One-time tillage of no-till crop land five year post-tillage. Agron. J. 102:1302–1307. doi:10.2134/agronj2010.0051

Modeling Water Management and Food Security in India under Climate Change

Adlul Islam,* Paresh Bhaskar Shirsath, Soora Naresh Kumar, Nataraja Subash, Alok K. Sikka, and Pramod K. Aggarwal

Abstract

Climate change and variability will impact water availability and the food security of India. Trend analyses of historical data indicate an increase in temperature and changes in rainfall pattern in different parts of the country. The general circulation models (GCMs) also project increased warming and changes in precipitation patterns over India. This chapter presents examples of model applications in water management and crop yield simulation in India, focusing on climate change impact assessment. Simulation models have been successfully applied for rotational water allocation, deficit irrigation scheduling, etc. in different canal commands. Application of a universal soil loss equation in a distributed parametric modeling approach by partitioning watershed into erosion response units suggests that by treating only 14% of the watershed area, a 47% reduction in soil loss can be achieved. Simulation studies conducted using different hydrological models with different climate change projections and downscaling approaches showed varied hydrological responses of different river basins to the future climate change scenarios, depending on the hydrological model, climate change scenarios, and downscaling approaches used. Crop yield modeling showed decreases in irrigated and rainfed rice (*Oryza sativa* L.) yields under the future climate change scenarios, but the decrease is marginal for rainfed rice. Maize (*Zea mays* L.) yields in monsoon may be adversely affected by a rise in atmospheric temperature, but increased rain can partly offset those losses. Wheat (*Triticum aestivum* L.) yields are likely to be reduced by 6 to 23% and 15 to 25% during the 2050s and 2080s, respectively. A combined bottom-up participatory process and top-down integrated modeling tool could provide valuable information for locally relevant climate change adaptation planning.

Abbreviations: APSIM, Agricultural Production Systems Simulator; AWD, alternate wetting and drying; CF, continuous flooding; CMIP, Coupled Model Intercomparison Project; DO, dissolved oxygen; DSSAT, Decision Support System for Agro Technology Transfer; GCM, general circulation model; GHG, greenhouse gas; GHGES, Greenhouse Gas Emission Scenarios; HEC-RAS, Hydrologic Engineering Center–River Analysis System; IGP, Indo-Gangetic Plain; IPC, irrigation potential created; IPCC, Intergovernmental Panel on Climate Change; IPU, irrigation potential utilized; MMM, multi-model mean; NICRA, National Initiative on Climate Resilient Agriculture; PRECIS, Providing Regional Estimates for Climate Impact Studies; PRMS, Precipitation Runoff Modeling System; RCM, regional climate model; RCP, representative concentration pathway; SWAP, Soil–Water–Atmosphere–Plant; SWAT, Soil and Water Assessment Tool; UIP, ultimate irrigation potential; USLE, universal soil loss equation; VPD, vapor pressure deficit; WEPP, Water Erosion Prediction Project.

Adlul Islam (adlulislam@yahoo.com) and Alok K. Sikka (aloksikka@yahoo.co.in), NRM Division, Indian Council of Agricultural Research, KAB-II, IARI Campus, Pusa, New Delhi 110 012, India. *Corresponding author.

Paresh Bhaskar Shirsath (p.bhaskar@cgiar.org) and Pramod K. Aggarwal (P.K.Aggarwal@cgiar.org), Climate Change Agriculture and Food Security - South Asia, International Water Management Institute, NASC Complex, Pusa, New Delhi 110 012, India.

Soora Naresh Kumar, Centre for Environment Science and Climate Resilient Agriculture, Indian Agricultural Research Institute, Pusa, New Delhi 100 012, India (nareshkumar.soora@gmail.com).

Nataraja Subash, Project Directorate for Farming Systems Research, Modipuram, Meerut 250 110, Uttar Pradesh, India (nsubashpdfsr@gmail.com).

doi:10.2134/advagricsystmodel5.c11

Practical Applications of Agricultural System Models to Optimize the Use of Limited Water
Lajpat R. Ahuja, Liwang Ma, and Robert J. Lascano, Editors
Advances in Agricultural Systems Modeling, Volume 5. Lajpat R. Ahuja, Series Editor

Water is the most precious natural resource and a vital component of a socio-ecological system, providing life support for living beings and the environment. Availability of freshwater resources is of particular interest in India as the country is supporting an increasingly large population with decreasing per capita water supplies (Mall et al., 2006). With the anticipated climate change and variability, changes in air temperature and precipitation will affect hydrologic processes and water resources availability. A small perturbation in the magnitude and/or frequency of precipitation can result in significant impacts on regional water resources and the frequency of drought or flood events (Whitfield and Cannon, 2000). Any temporal and spatial variations in rainfall will have a reflective effect on water availability in both irrigated and rainfed areas, affecting the agriculture-based economy. There are preliminary reports that the recent trend of decline in grain yields of rice and wheat in the Indo-Gangetic Plain (IGP) could have been partly due to weather changes (Aggarwal et al., 2004). Changes in regional water availability due to climatic and land use changes will affect agricultural production, hydropower generation, flood control, municipal and industrial water supply, and fish and wildlife management (Ragab and Prudhomme, 2002; Gibson et al., 2005). The potential climatic impacts may have significant ramifications for decision-making about water allocation and management that are likely to be made in the coming decades. Improved understanding of climatic causes of hydrological variability is of paramount importance for the sustainable development and management of vital water resources.

The management of water resources is complex and requires a multidisciplinary approach for evaluating their effects on a social, economic, and environmental level. Water resources management requires access to a number of data sources, including meteorological, spatial such as digital elevation models, land use, and remote sensing. The integration of the data sources together with models of the underlying systems provides a way to better plan and manage water resources. A number of simulation models have been developed and applied for irrigation water management and improving water productivity (e.g., Mishra et al., 2005; Islam et al., 2008; Bhadra et al., 2009; Subash et al., 2014). Different hydrologic models have been applied for assessing water and sediment yield, deciding and prioritizing best management practices, and assessing potential effects of climate change on hydrology and water resources availability at the basin or watershed scale (e.g., Sikka and Paul, 2005; Singh et al., 2011b; Islam et al.,

2012a). Our objective is to give an overview of climate change and climate variability in India, application of simulation models in irrigation water management, and modeling climatic change impacts on water resources and crop production.

Climate Change and Climatic Variability in India

India is a large country with diverse climates, ranging from a hot and humid tropical climate to a dry desert ecosystem. The long-term average annual rain for the country as a whole is 1160 mm and is highly variable both in space and time. In India, the two monsoon seasons, namely the southwest monsoon (June to September) and the northeast monsoon (November to December) seasons, bring the rain. The term monsoon is generally used to refer to the seasonal reversal of winds and the associated rainfall. About 75% of the long-term average rainfall occurs during the southwest monsoon (rainy) season. Although the average rainfall is adequate, about 21% of the area of the country receives <750 mm rainfall and 15% area receives >1500 mm (Kumar et al., 2005). The year-to-year variability in monsoon rainfall leads to extreme hydrological events such as drought and floods in different parts of India, affecting agricultural production and the national economy. As reported by the Central Water Commission, Ministry of Water Resources, Government of India, the annual average area affected by floods is 7.56 million ha. Similarly, the drought-prone area in the country has been assessed as 51.12 million ha (Kumar et al., 2005).

The mean annual air temperature across the country varies from <10°C in the extreme north region to >28°C in southern parts of the country. On the basis of the analysis of instrumental records, several studies indicated a continuous increasing trend in mean surface air temperature over decades in most parts of the Indian subcontinent (Hingane et al., 1985; Govind Rao, 1993; Rupa Kumar et al., 1994). Arora et al. (2005) reported increases in annual mean air temperature, mean maximum air temperature, and mean minimum air temperature at the rate of 0.4, 0.9, and 0.1°C per 100 yr based on the analysis of the time series data of 125 stations distributed over India. The stations located in the southern and western regions showed a rising trend of 1.1 to 0.4°C per 100 yr, respectively, whereas the stations located at the northern plains showed a falling trend of 0.4°C per 100 yr. Dash et al. (2007) reported an increase in the annual mean and maximum air temperatures by about 0.7 and 0.8°C, respectively, over India. Maximum increase in the maximum air temperature has been predicted in the west coast (1.2°C) followed by the northeast (1.0°C), western Himalayas (0.9°C), north central (0.8°C), northwest (0.6°C), east coast (0.6°C), and interior peninsula (0.5°C) areas. Mean atmospheric surface temperature in India has increased by 1.0 and 1.1°C during the winter and post-monsoon months, respectively. This study also indicated a small increase in rainfall during the winter months of January and February, the pre-monsoon months of March to May, and the post-monsoon months of October to December. Basin-wise trend analysis (Jain and Kumar, 2012) showed an increasing trend (0.3–10.2 mm yr^{-1}) in annual rainfall in six river basins and a decreasing trend (0.5–4.9 mm yr^{-1}) in annual rainfall in 15 river basins of India. Further, most of the basins had the same direction of trend in rainfall and rainy days at the annual and seasonal scale. These studies suggest an increase in temperature and changes in rainfall pattern over India during the last century.

Our trend analyses of the India Meteorological Department's gridded (1 by 1°) rainfall (1951–2007) and temperature (1969–2005) datasets (Rajeevan et al., 2006; Srivastava et al., 2009) using a modified Mann–Kendall test (Hamed and Rao, 1998) at the 5% significance level revealed the increasing trend of winter season (January and February) rainfall in the northern and northeastern parts of the country. The IGP showed a significant increase in rainfall during the pre-monsoon (March to May) period. There is a decrease in monsoon (June to September) season rainfall in most of the northern part of the country, excluding the eastern part of Jammu and Kashmir and the eastern coast. The trend analyses of temperature datasets showed an increasing trend during pre-monsoon season in the southern region and the western coast. During monsoon season, however, temperature showed increasing trends in the north central, western coast, and northeastern regions. The post-monsoon season (October to December) witnessed an increasing trend in temperature. On an annual basis the temperature had a significant increasing trend in most parts of the country except Odisha and Chhattisgarh. These results of trend analysis (Fig. 11–1) also corroborate with the earlier findings (Govind Rao, 1993; Rupa Kumar et al., 1994; Dash et al., 2007).

Available GCMs also suggest an area-averaged annual mean warming of about 3 and 5°C in the decade of the 2050s and 2080s, respectively, over the land regions of Asia due to an increase in the atmospheric concentration of greenhouse gases (IPCC, 2001). Under the combined influence of greenhouse gas and sulfate aerosols, surface warming would be about 2.5 and 4°C during the 2050s and 2080s, respectively (IPCC, 2001). The Fourth Assessment Report (AR4) of the Intergovernmental Panel on Climate Change (IPCC) projected an area-averaged increase in the median annual mean temperature in the range of 2.5 (Southeast Asia) to 4.3°C (northern Asia) by the end of 21st century (2080–2099) under the A1b emission scenario (IPCC, 2007). The annual precipitation change is projected to vary in the range of −3 (central Asia) to +15% (northern Asia) during the same period. Analysis based on projections of the Providing Regional Climates for Impacts Studies (PRECIS) regional climate model (RCM), developed by the Hadley Centre for Climate Prediction and Research, also indicated a marked increase

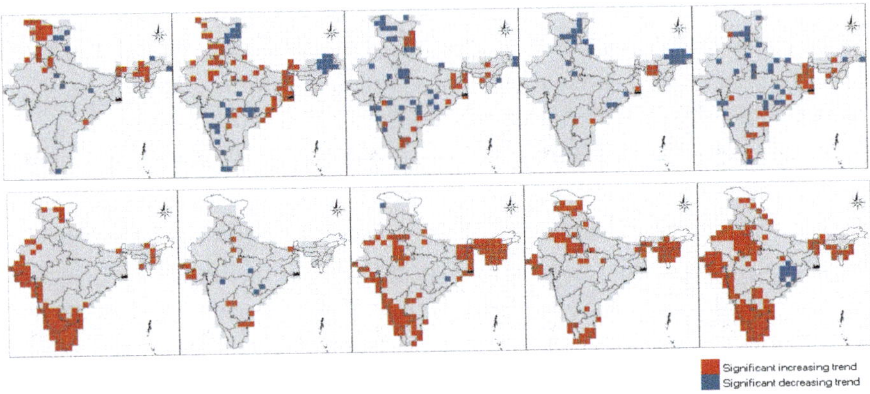

Fig. 11–1. Trends in mean rainfall (upper row) and temperature (lower row) using a modified Mann–Kendall Test at the 0.05 significance level. From left to right: winter (Jan.–Feb.), pre-monsoon (Mar.–May), monsoon (June–Sept.), post-monsoon (Oct.–Dec.), and annual.

in both rainfall and temperature toward the end of the 21st century. A substantial increase in extreme precipitation over a large area, and particularly over the west coast of India and west central India, is projected (Rupa Kumar et al., 2006).

The Coupled Model Intercomparison Project Phase 5 (CMIP5)-based model ensemble, which is based on new emission scenarios termed as representative concentration pathways (RCPs), projects a warming of 1.5, 2.4, 2.8, and 4.3°C for India under the RCP2.6, RCP4.5, RCP6.0, and RCP8.5 scenarios, respectively, for the 2080s (2071–2100) compared with the baseline period of 1961 to 1990 (Chaturvedi et al., 2012). There is a consistent warming trend over the country in short-term, mid-term, and long-term scenarios (Table 11–1). The all-India annual mean ambient temperature increases by 1.7 to 2.0°C during the 2030s under different RCP scenarios and by about 2.0 to 4.8°C during the 2080s, relative to the preindustrial base of the 1880s (1861–1900). Precipitation is projected to increase over almost all of India except for a few regions in short-term projections (2030s). The CMIP5–based ensemble projects all-India precipitation increases of 6, 10, 9, and 14% under the RCP2.6, RCP4.5, RCP6.0, and RCP8.5 scenarios, respectively, for the 2080s compared with the 1961 to 1990 baseline. All-India annual precipitation increases by 1.2 to 2.4% and 3.5 to 11.3% during the 2030s and 2080s, respectively, relative to the preindustrial base under different RCP scenarios. Analysis based on MIROC-ESMCHEM projections for RCP4.5 scenario also indicated a consistent positive trend in the frequency of extreme precipitation days (e.g., >40 mm d^{-1}) for decades 2060 and beyond (Chaturvedi et al., 2012). Higher confidence for the temperature projections and relatively low confidence for the precipitation projections from CMIP5 models

Table 11–1. CMIP5 model ensemble-based projected annual temperature and precipitation change for the 2030s, 2060s, and 2080s relative to the preindustrial baseline (1880s) (adapted from Chaturvedi et al., 2012).

Representative concentration pathway	Temperature change			Precipitation change		
	2030s	2060s	2080s	2030s	2060s	2080s
	°C			%		
RCP2.6	1.7	1.9	2.0	1.2	2.7	3.5
RCP4.5	1.9	2.5	2.9	0.8	4.3	7.0
RCP6.0	1.7	2.4	3.3	1.6	3.5	6.8
RCP8.5	2.0	3.3	4.8	2.4	6.6	11.3

were reported. The model agreement in precipitation projections increases from RCP2.6 to RCP8.5, and also the model agreement gets stronger from short- to long-term projections, thereby indicating that long-term projections are generally more reliable than their short-term counterparts.

As the performance of the Indian agrarian economy is very much dependent on monsoon season rainfall, the reliability of monsoon rainfall projections is particularly more critical for informed decision-making and to take sufficient adaptation measures. Menon et al. (2013) reported a consistent increase in seasonal mean rainfall during the summer monsoon periods. All the CMIP5 models simulate stronger seasonal mean rainfall in the future compared with the historic period under the RCP8.5 scenario. A consistent positive trend in the interannual variability of the Indian monsoon rainfall was also reported under unabated global warming. Sperber et al. (2013) suggested that because of the higher spatial resolution, CMIP5 models have a better representation of rainfall as compared with CMIP3 models, especially in the vicinity of steep topography (like the Western Ghats). The CMIP5 multi-model mean (MMM) is more skillful than CMIP3 MMM in simulating several aspects (monsoon annual cycle and the onset of the monsoon as well as the time of the monsoon peak) associated with Asian summer monsoons. The spatial extent of the monsoon and the magnitude of intraseasonal variance in the CMIP5 MMM is more realistic than in the CMIP3 MMM. However, while evaluating simulations of Asian summer monsoons from 25 CMIP5 and 22 CMIP3 simulations of the late 20th century, Ramesh and Goswami (2014) reported no improvement in the ability to project basic quantities like observed seasonal mean and trend and hence to project the regional climate system with reasonable certainty since CMIP3 in terms of reliability. Kitoh et al. (2013) also reported the limited ability of the models to reproduce the current monsoon climate and large scatter among the model projections, indicating only low confidence in the projections. Thus, an ultrahigh-resolution model considering fine-scale processes is critical for reasonable probabilistic projections of regional-scale climate change to achieve useful future projections of Indian monsoons and extremes (Rajendran et al., 2013). In the absence of dependable regional projections for the Indian summer monsoon rainfall, identification of appropriate adaptation options and contingency plans seems to be a good policy.

Water resources status

With a geographical area of 3.29 million km^2, India supports more than 18% of the world's population and has 4.2% of the world's freshwater resources. The water resources potential of the country has been estimated as 1869 km^3, which includes replenishable groundwater resources (CWC, 2013). Within the limitations of physiographic conditions, a socio-political environment, legal and constitutional constraints, and the technology available at hand, the utilizable water resources of the country have been assessed as 1121 km^3, of which surface water and groundwater contribute 690 and 431 km^3, respectively (CWC, 2013). This 690 km^3 of utilizable surface water can be harnessed only if matching storages are built. The limitations of storage and trans-basin transfer of water restrict the potential for irrigation. The ultimate irrigation potential (UIP), an estimate of the overall potential for irrigation in India, has been estimated to be 139.9 million ha without interbasin sharing of water and 175 million ha with interbasin sharing of water. The irrigation potential created (IPC), the total gross area proposed to be irrigated under different crops, has been estimated as 107.2 million ha (76.6% of UIP) up to 2009–2010. The irrigation potential utilized (IPU), the gross area actually irrigated during the reference year out of the gross proposed area to be irrigated, during the same period has been estimated to be 86.9 million ha (81% of IPC) (CWC, 2013). The ratio of utilized to created irrigation potential has come down from 1 in 1951 to 0.8 by 2009–2010, indicating increasing gap between IPC and IPU.

India is facing water stress conditions with per capita availability of water declining sharply from 5177 m^3 in 1951 to 1544 m^3 in 2011 (CWC, 2013). It is projected to reduce further to 1465 m^3 and 1235 m^3 by the year 2025 and 2050, respectively, under high population growth scenarios (Kumar et al., 2005). The water scarcity situation for various uses such as agriculture, drinking water, and domestic and industrial needs may further worsen if the anticipated impacts of climate change on hydrology and water resources are also considered. Unless timely and properly managed, water scarcity may lead to adverse situations. Solving these water resources problems will require an improved understanding of the fundamental physical, biological, economic, and social processes and a better knowledge of how all of these components operate together within the basin.

Simulation models help to understand the complex climatic and ecosystem processes and assist in "what-if" analysis to provide possible solutions.

Simulation Modeling in Water Management

Management of water resources is a complex task and may involve different activities conducted by stakeholders. Because of the complex nature of problems addressed in water management, computer simulation models are commonly used for better understanding of real-world processes and to provide quantitative information to support decision-making activities. These models provide an informed and scientifically defensible basis for proactive management strategies, enhancing our prospects to maximize the adaptive capacity of the system as a whole (Winz et al., 2009). With recent advances in computer technology and convenient access to desktop computers and software, simulation models are increasingly used to address different aspects of water management. In India, several models have been developed (e.g., Islam et al., 2008; Rao and Rajput, 2009; Bhadra et al., 2009) and applied to address the problems of irrigation water management (e.g., Mishra et al., 2005; Islam et al., 2008; Subash et al., 2014) and watershed management (e.g., Tripathi et al., 2003; Sikka and Paul, 2005; Singh et al., 2011b). Some examples of model application are presented in the following sections.

Application of Models in Canal Water Management

An important challenge facing the irrigation sector in India is the growing gap between IPC and IPU and the uneven distribution of water over the length of the canal system. The overall irrigation efficiency of the major and medium irrigation projects is estimated to be around 38%. The efficiency of surface irrigation systems can be improved from about 35 to 40% to around 50 to 60% and that of groundwater from about 65 to 70% to 72 to 75% (Planning Commission, 2009). The National Water Mission, institutionalized under the National Action Plan for Climate Change, has set the target to improve the efficiency of water use by at least 20%.

The performance of large irrigation systems can be improved through hardware (physical or structural) changes such as canal lining and installation of remotely operated control structures as well as through software (operational) improvements such as proper operation and recordkeeping, use of appropriate delivery rules, and improved communication between the water supply agency and water user. With declining investment in the irrigation sector, better operation and management of the system is one of the economically feasible options. In this direction, several hydraulic and hydrological simulation models have been developed and applied in different canal commands (e.g., Singh et al., 1997; Mishra et al., 2005; Islam et al., 2008). Singh et al. (1997) applied MIKE 11 and MIKE SHE to the Mahanadi Reservoir Irrigation Project in India for hydraulic and hydrologic simulations, respectively. Simulation results illustrated the versatility of the modeling system for planning and analysis of various aspects of an irrigation project. Mishra et al. (2005) developed an integrated optimization-simulation model and applied it to the Right Bank Main Canal system of the Kangsabati Irrigation Project in West Bengal, India to minimize the gap between irrigation water supply and crop water demand and improve the spatial distribution of supply, thereby minimizing the tail-end deprivation. Islam et al. (2008) developed a hydraulic simulation model "CanalMod," based on the solution of gradually varying one-dimensional open-channel flow equations using the four-point implicit Preissmann scheme, for steady and unsteady flow simulations in the irrigation canal network. The model was applied to the Right Bank Main Canal of the Kangsabati Irrigation Project for simulating the flow pattern in the canal system, and results were compared with that of the Hydrologic Engineering Center–River Analysis System (HEC-RAS) model. Comparison between the CanalMod and HEC-RAS showed that the stage and discharge simulated using both the models are comparable with measured flows (Fig. 11–2) and these models can be applied for canal operation scheduling.

Most of the canal irrigation schemes in India have rotational water allocation systems. Rotational systems, such as Warabandi practiced in North India, enforce some sort of equitable rationing of water to improve water distribution. Warabandi literally means "fixation of turn" for supply of water to the farmers. Under this system of management, the available water, whatever its volume, is equitably allocated to all farmers in the command irrespective of the location of their holdings. It is a predetermined schedule specifying the day, time, and duration of supply to each irrigator in proportion to landholdings in the outlet command. Water application under these schemes is generally invariable with respect to biophysical factors such as soil properties, crops, and their water needs during different phases of crop growth. Attempts have been made to develop models and tools for efficient canal water allocation under rotational systems considering water resources availability, biophysical factors, and constraints of the command area. Major models and tools included are decision support systems for canal water release (CWREDSS) (Rao and Rajput, 2009), Area Water Allocation Program (AWAP) (Gorantiwar, 1995), CanalMod (Islam et al., 2008), etc. In addition to optimal allocation through rotational scheduling, a detailed performance assessment framework of irrigation water management for heterogeneous schemes under rotational supply was also applied to semiarid regions of India (Gorantiwar and Smout, 2005b; Smout and

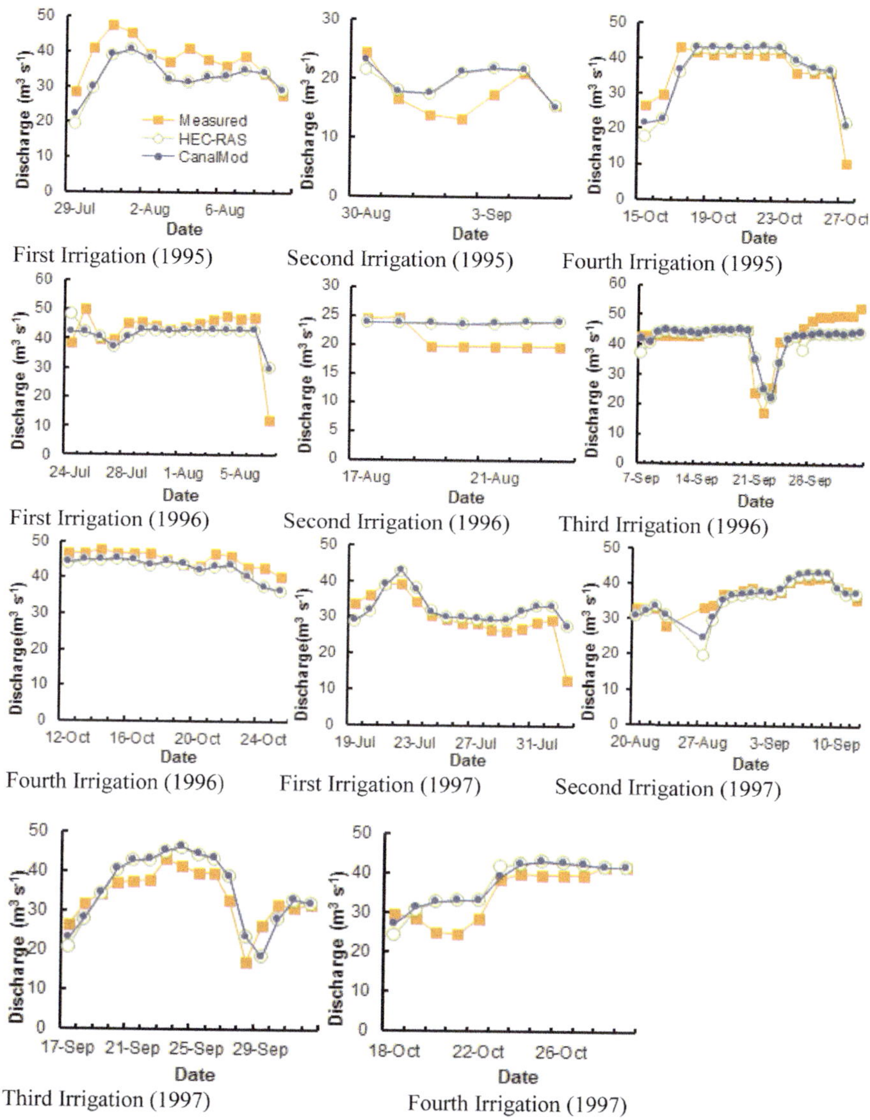

Fig. 11–2. Comparison of measured and simulated discharges for different irrigation events (adapted from Islam et al., 2008, with permission from ASCE). HEC-RAS, Hydrologic Engineering Center–River Analysis System.

Gorantiwar, 2005b). Under the water scarcity conditions, efforts have been made to optimally allocate water under deficit irrigation (Gorantiwar and Smout, 2003) as well as through demand-based water release strategies (Rao and Rajput, 2009). Deficit irrigation can be a promising strategy to maximize area and production in irrigation command areas. Through simulation experiments, Gorantiwar and Smout (2003) compared water allocation under the deficit irrigation approach for

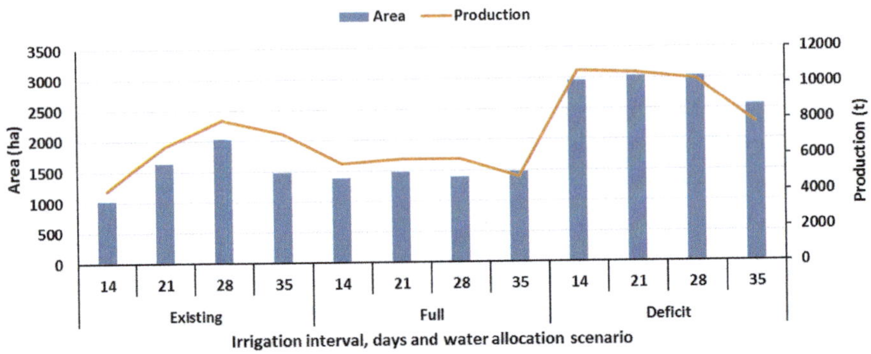

Fig. 11–3. Effect of water allocation under deficit irrigation (adapted from Gorantiwar and Smout, 2003).

wheat in an irrigation scheme in India with full irrigation (irrigation to fill the root zone to field capacity) and the existing rule of water allocation (70-mm irrigation depth and 21-d irrigation interval). Deficit irrigation leads to an increase in the irrigated area and production by 30 to 45% and 20 to 40%, respectively, over the existing rule and by 50 and 45%, respectively, over the adequate irrigation (Fig. 11–3). Further, performance-based land area and water allocation (Gorantiwar and Smout, 2006; Gorantiwar et al., 2006), a multilevel approach for optimizing land and water resources and irrigation deliveries (Gorantiwar and Smout, 2005a; Smout and Gorantiwar, 2005a), and productivity and equity of different irrigation schedules under limited water supply (Smout and Gorantiwar, 2006) were also modeled in different irrigation commands.

Application of Models in Improving Water Productivity

Scarcity of irrigation water under climate change scenarios is a great concern for rice–wheat production systems, which cover 13.5 million ha in the IGP and are vital for food security (Ladha et al., 2003). Demonstration of the impact of different irrigation regimes on rice, wheat, and rice–wheat system yields are essential to adopt suitable water saving technologies to minimize the productivity risk. Water productivity defines the relationship between crops produced and the amount of water involved in crop production and is generally expressed as crop production per unit volume of water (kg ha^{-1} mm^{-1}). Higher crop water productivity results in either the same production from less water resources or a higher production from the same water resources. Increasing agricultural water productivity is the most plausible means of mitigating problems arising from scarcity and competition for water (Kijne et al., 2003). The field-scale ecohydrological model Soil–Water–Atmosphere–Plant (SWAP) has been applied in several irrigation commands for assessing water productivity (Singh et al., 2006; Mandare et al., 2008). Singh et al. (2006) demonstrated application of SWAP in combination with field experiments, remote sensing, and GIS in a distributed manner for generating the required hydrological and biophysical variables to evaluate alternative water management scenarios at different spatial and temporal scales to improve water management and productivity in an irrigated area (Sirsa district, Haryana) in India. The results showed that reduction of seepage losses in the range of 25 to 30% of the total canal inflow and reallocation of 15% canal water inflow from the northern to the central canal commands will improve the water productivity, halt the rising or declining groundwater levels, and decrease the salinization. Mandare et al. (2008) applied the SWAP model to investigate the possibility of enhancing crop water productivity in the parts of northwest India where groundwater quality is marginal and canal water supply is severely scarce. On the basis of the simulation results, a few nomographs were prepared to provide a graphical method to compute the effect of different combinations of water quality and depth of water application on crop yield and soil salinity and to help develop some guidelines for the farming community. Water management alternatives at the field level were suggested to increase the yield and to maintain soil salinity below the threshold level. Application of the Agricultural Production Simulator (APSIM) for identification of crop management practices, especially sowing or transplanting dates and irrigation water management for improving water productivity, has been demonstrated successfully in Ludhiana, Punjab, India (Singh et al., 2011a). Subash et al. (2014) applied APSIM to simulate the effects of different irrigation regimes (IR) on yield, irrigation water requirements, and irrigation water productivity (WP_I) of selected rice, wheat, and rice–wheat systems in the Upper Gangetic Plains of India. A field experiment was conducted in Meerut (29°04′ N, 77°46′ E), a part of the Upper Gangetic region of the IGP, to generate necessary data for parameterization and validation of APSIM. A set of scenarios to reduce irrigation water use were tested using the 17 yr (1993 to 2009) weather data. Four irrigation regimes (continuous flooding [CF], 2-d alternate wetting and drying [2AWD], 5-d AWD [5AWD], and 8-d AWD [8AWD]) for rice and three phenological stage-wise irrigation schedules (crown root initiation [CRI] + jointing [2IR], CRI + tillering + jointing [3IR], and CRI + tillering + jointing + ear emergence + milking [5IR]) for wheat were imposed as scenarios for simulation studies to assess the water productivity. The treatment of 2-d AWD corresponds to current farmer practices. Irrigation water productivity for rice, wheat, and rice–wheat systems was computed for all scenario combinations. The results of the simulation studies are summarized below.

Irrigation Water Requirements and Grain Yield

Irrigation water requirements of rice ranged from 1234 to 2234 mm with 18% variability during the last 17 yr when continuous flooding IR was followed. On average the quantity of irrigation water reduced to 52% under the 8-d AWD irrigation regime compared with CF. The normal farmers' practice of 2-d AWD reduced the irrigation water requirement to 13% compared with CF. The cumulative probability curves at different irrigation regimes showed that at 60% probability, irrigation water requirements of 8-d AWD and 2-d AWD reduced to 51% and 35%, respectively, compared with CF (Fig. 11–4). As the cumulative probability increases, there is a slight broadening of irrigation requirements between different irrigation regimes.

Relationship between Water Productivity and Grain Yield

For rice, among different treatments a higher irrigation water productivity of 3.9 kg ha^{-1} mm^{-1} was noticed in 8-d AWD but at the cost of drastic yield loss (Table 11–2). However, both water productivity and grain yield are higher in CF, but at the cost of irrigation water. The lowest water productivity of 3.3 kg ha^{-1} mm^{-1} was observed in 2-d AWD. It appears that 5-d AWD may be recognized as good management option with a comparatively low yield penalty under water-limiting or water scarcity situations. In the case of wheat, a higher water productivity of 31.3 kg ha^{-1} mm^{-1} was observed in 2IR treatments, but resulted in a very high yield penalty of 42% compared with 5IR treatments, which leads to a water productivity of 21.4 kg ha^{-1} mm^{-1} (Table 11–3). As far as the rice–wheat system is concerned, the highest WP_I of 8.3 kg ha^{-1} mm^{-1} was noticed in the irrigation regime of 8AWD+5IR

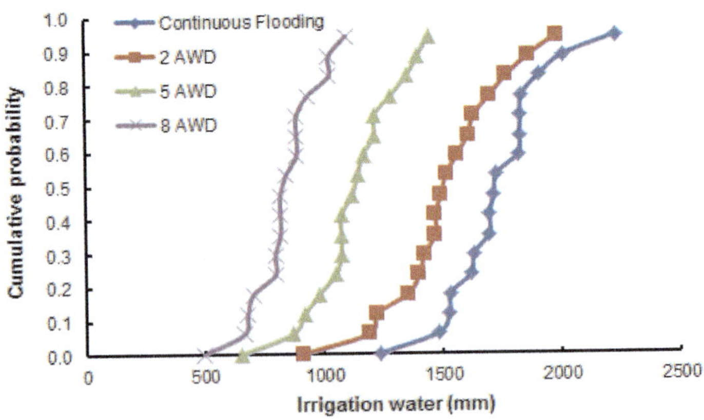

Fig. 11–4. Simulated irrigation water requirement at different cumulative probability levels under different irrigation treatments (adapted from Subash et al., 2014, with permission from Springer Science + Business Media). AWD, alternate wetting and drying.

Table 11–2. Irrigation water productivity and rice yield under different irrigation regimes (adapted from Subash et al., 2014, with permission from Springer Science + Business Media).

Irrigation treatment †	Rice yield	Water productivity
	kg ha^{-1}	kg ha^{-1} mm^{-1}
CF	6272.5	3.7
2AWD	4910.6	3.3
5AWD	3837.5	3.5
8AWD	3186.7	3.9

† CF, continuous flooding; AWD, alternate wetting and drying.

Table 11–3. Wheat grain yield, irrigation water requirements, and irrigation water productivity under different irrigation regimes (adapted from Subash et al., 2014, with permission from Springer Science + Business Media).

Treatments †	Grain yield	Irrigation water requirement	Irrigation water productivity
	kg ha^{-1}	mm	kg ha^{-1} mm^{-1}
2IR	3131	100	31.3
3IR	4339	150	28.9
5IR	5361	250	21.4

† IR, irrigation regime.

with a yield penalty of 26% (Table 11–4). The highest yield penalty of 46% was observed under the irrigation regime of 8AWD+2IR. The lowest WP_I of 5.2 kg ha^{-1} mm^{-1} was noticed in 2AWD+2IR with a higher yield penalty of 31% (Table 11–4). Relatively high irrigation water productivity was observed in the 5AWD+5IR irrigation regime under the rice–wheat system with a moderate yield penalty of 20%. Thus, a 5-d AWD regime for rice and 5 irrigations (5IR) for wheat may be a suitable irrigation management option under water-limiting situations.

Application of Models in Watershed Management

Integrated watershed management has emerged as a powerful concept for agriculture and rural development in India. Hydrological models in conjunction with spatial data manipulation in a GIS environment have been successfully applied in different watersheds in India for estimating runoff, soil erosion, and sedimentation, identifying critical areas needing treatment within the watershed for optimized investments, planning and designing of various soil and water conservation structures, and analyzing best management practices for preparing a watershed management plan. For example, Tripathi et al. (2003) applied the Soil and Water Assessment Tool (SWAT) model in a small watershed located in the eastern India for identification and prioritization of critical sub-watersheds to develop an effective management plan. The physically based

Table 11–4. Rice–wheat system yield, total irrigation water requirements, and system irrigation water productivity under different irrigation regimes (adapted from Subash et al., 2014, with permission from Springer Science + Business Media).

Treatment †	Rice–wheat system yield	Total irrigation water requirement	System irrigation water productivity	System yield deviation ‡
	kg ha^{-1}	mm	kg ha^{-1} mm^{-1}	%
CF+2IR	9,678	1,828	5.3	−20
CF+3IR	10,991	1,878	5.9	−9
CF+5IR	12,102	1,978	6.1	0
2AWD+2IR	8,316	1,608	5.2	−31
2AWD+3IR	9,629	1,658	5.8	−20
2AWD+5IR¶	10,740	1,758	6.1	−11
5AWD+2IR	7,243	1,222	5.9	−40
5AWD+3IR	8,556	1,272	6.7	−29
5AWD+5IR	9,667	1,372	7.1	−20
8AWD+2IR	6,592	935	7.1	−46
8AWD+3IR	7,905	985	8.0	−35
8AWD+5IR	9,016	1,085	8.3	−26

† CF, continuous flooding; AWD, alternate wetting and drying; IR, irrigation regime.

‡ Compared with CF+5IR irrigation regime.

¶ Represent current farmers' practices.

distributed parameter Water Erosion Prediction Project (WEPP) model has been applied successfully in a 239 ha small multi-vegetated watershed in high rainfall and high land slope conditions of the eastern Himalayan range in India to develop appropriate vegetative as well as structural measures to control sediment yield (Singh et al., 2011b). Simulations of different vegetative management scenarios indicated that replacing traditional bun agriculture and upland paddy crop with maize, soybean [*Glycine max* (L.) Merr.], and peanut (*Arachis hypogaea* L.) would reduce sediment yield by 19, 30, and 28%, respectively. Field cultivator and drill-no-tillage systems have the potential to reduce sediment yield by 13 and 22%, respectively, as compared with the existing practice of spading and country plowing. The installation of 8 check dams and 18 trash barriers in the drainage line was found to reduce (~55%) sediment yield from the watershed substantially. Simulations of combinations of management practices indicted that soybean and peanut in upland situations with a field cultivator or drill-no-tillage system and structural control in the drainage line have the potential to make agriculture sustainable in the watershed with sediment yield reduction up to 78%.

A study conducted in a watershed (KG-4–1) located in southern India, using the universal soil loss equation (USLE) in a distributed parametric modeling approach by partitioning watershed into erosion response units, suggests that by treating only 14% of the area (where soil loss is >20 t ha^{-1} yr^{-1}), a 47% reduction in soil loss can be brought about and by treating 28% of the area (where soil loss is >10 t ha^{-1} yr^{-1}), a nearly 68% reduction in soil loss can be obtained (Fig. 11–5). This study demonstrated the use of modeling to identify and prioritize critical intervention areas within a watershed for optimized investment planning (Sikka and Paul, 2005).

Modeling climate change impact on water resources

The quantitative assessment of climate change impact on water resources availability and its management requires knowledge of climate, hydrological, and water management models and particularly the relationships between each of them. Hydrologic models provide a framework to conceptualize and investigate the relationship between climate, human activities, and water resources. Hydrologic models driven by climatic data obtained either from GCM outputs through statistical downscaling techniques (e.g., Wilby and Wigley, 1997), from high-resolution RCMs (e.g., Arnell et al., 2003; Graham et al., 2007; Akhtar et al., 2009), or from hypothetical scenarios (e.g., Arnell, 1992; McCabe and Hay, 1995; Ficklin et al., 2009) are often used in the assessment of climate change impact on hydrology and water availability at the basin or watershed scale. The GCMs are the

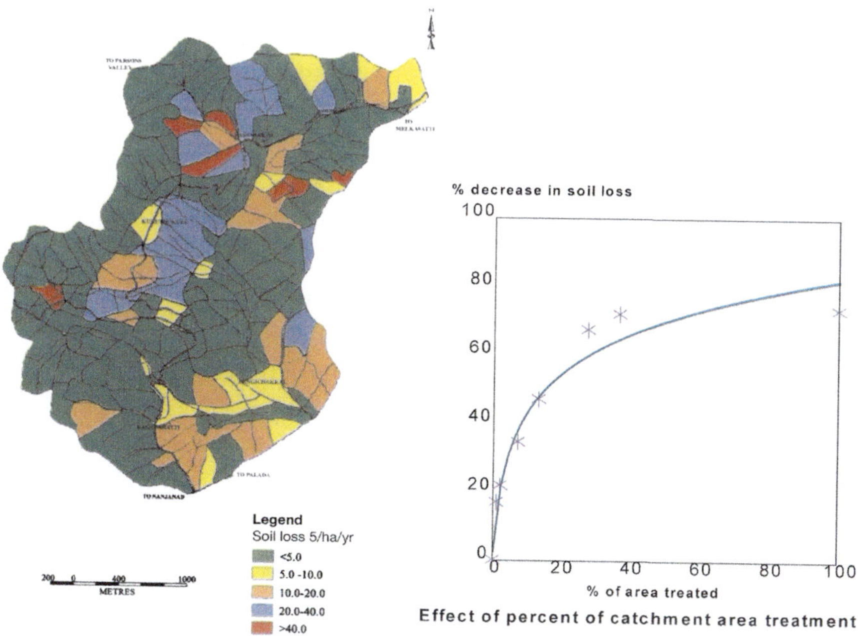

Fig. 11–5. Partial (critical) area treatment and its effect on soil loss in KG-4–1 watershed of Nilgiris (adapted from Sikka and Paul, 2005).

most credible tools designed to simulate time series of climate variables globally and are the preferred option for deriving climate scenarios. Most GCMs simulate reasonable average annual and seasonal features of present climate over large geographic areas, but are less reliable in simulating smaller spatial and temporal scale features that are relevant to regional impact assessment studies (Grotch and MacCarcken, 1991). Given the limitations of GCMs, hypothetical scenarios are preferred by several researchers (e.g., Nash and Gleick, 1991; Fontaine et al., 2001; Islam et al., 2012a). Hypothetical scenarios are constructed by simply perturbing historical records for a particular climate variable by an arbitrary amount, covering a reasonable range of possibilities.

Several authors have performed extensive studies in hydrologic impacts of climate change on Indian River basins using different hydrological models and climate change scenarios (Table 11–5). Simulation studies conducted for 12 river basins of India, using distributed hydrological model SWAT and HadRM2 RCM projected scenarios, indicated a general reduction in the quantity of the available runoff under the future (2041–2060) scenario (Gosain et al., 2006). The river basin of Luni is expected to experience acute physical water scarcity; Mahi, Pennar, Sabarmati, and Tapi are likely to experience constant water scarcity and shortage; Cauvery, Ganga, Narmda, and Krishna are likely to experience seasonal or regular water stressed conditions; and Godavari, Brahmani, and Mahanadi are expected to experience water shortage only in few locations (Table 11–6). Simulation studies also indicated the severity of droughts and the intensity of floods in various parts of India may get intensified under the future scenario. However, when PRECIS RCM A1b scenario data were used along with the SWAT hydrological model, an increase in water yield (net amount of water leaving the sub-basin and contributing to the streamflow, which includes surface runoff, lateral flow, and subsurface contribution) was reported (Table 11–6) in most of the Indian river systems during the 2030s and 2050s, except in the Pennar Basin where the decrease in water yield during the 2030s is more pronounced (Gosain et al., 2011). This may be attributed to the PRECIS RCM projected increase in the precipitation at the basin level for a majority of the river systems. The simulation results also showed an increase in flooding (varying between 10% and over 50% of the existing magnitude) and sediment yield in the majority of the river basins. Though both the studies (Gosain et al. (2006) and Gosain et al., 2011) used the same hydrological model (SWAT), there are variations in the results in some of the river basins, which may be attributed to differences in the climate change scenarios used.

An increase in streamflow from the catchment area of the Tungabhadra River has been projected (Meenu et al., 2013), with the projected rainfall and temperature scenarios of HadCM3 GCM under A2 and B2 emission scenarios

Table 11–5. Climate change impact on hydrology and water resources in India.

River basin	Model(s) used †	Climate change scenario(s)‡	Impact	Reference
The Ganges Basin	water balance model	GISS GCM	increases in mean annual runoff in all the sub-basins	Mirza (1997)
Spiti Basin	University of British Columbia watershed model	hypothetical scenarios	under a warmer climate scenario, large increase in streamflow during the pre-monsoon season followed by monsoon season	Singh and Kumar (1997)
River basins of central India	conceptual monthly rainfall–runoff model	hypothetical scenarios	the Sher basin is more sensitive to climate change followed by Kolar and Damanaganga basin	Mehrotra (1999)
Kosi Basin	lumped and grid-based deterministic water balance model	hypothetical scenarios	a 2 to 8% decrease in runoff with 4°C temperature rise and no change in precipitation	Sharma et al. (2000)
Damodar River basin	HEC-HMS	HadRM2	reduced water availability for the 2050s and 2060s scenarios	Roy et al. (2004)
Satluj River basin	a conceptual snowmelt model (SNOWMOD)	hypothetical scenarios (temperature change)	snowfed basins are more sensitive in terms of reduction in water availability	Singh and Bengtsson (2005)
River basins of India ¶	Soil and Water Assessment Tool (SWAT)	HadRM2	a general reduction in the quantity of the available runoff under the future (2041–2060) scenario	Gosain et al. (2006)
Mahanadi River basin	support vector machine and relevance vector machine downscaling	CCSR/NIES GCM B2 emission scenario	a decreasing trend in future monsoon streamflow	Ghosh and Mujumdar (2008)
Mahanadi River basin	physically based distributed hydrologic model	CGCM2	increasing intensities of flood in September and drought in April	Asokan and Dutta (2008)
Mahanadi River basin	conditional random field downscaling	CGCM2, MIROC3.2 (med. res.), GISS GCMs, A2, A1b, B1	increase in probability of extreme, severe, and moderate droughts	Raje and Mujumdar (2009)
River basins of India §	SWAT model	PRECIS RCM, A1b	an increase in water yield in most of the Indian river systems during 2030s and 2050s	Gosain et al. (2011)
Chaliyar River basin	SWAT model	PRECIS RCM, A2, B2	decrease in streamflow ($< 10\%$) during 2071–2100 under both the scenarios	Raneesh and Santosh (2011)

Table continued.

Table 11–5. Continued.

River basin	Model(s) used †	Climate change scenario(s)‡	Impact	Reference
The Brahmaputra River	Rice Irrigation System Evaluation model	PRECIS RCM, A2	increase in peak discharge and flood duration during pre-monsoon and monsoon seasons, but reduction in number of flood waves per season	Ghosh and Dutta (2012)
Tungabhadra River in India	statistical downscaling (canonical correlation analysis)	MIROC 3.2 (med. Res.) GCM, A1b	decrease in future streamflow of Tunga, Bhadra, Kumudavathi, and Haridra tributaries	Rehana and Mujumdar (2012a)
Tungabhadra River basin, India	HEC-HMS model and Statistical DownScaling Model	HadCM3 GCM, A2, B2	increase in average annual streamflow during the 2020s, 2050s, and 2080s	Meenu et al. (2013)
Brahmani River basin	PRMS	Hypothetical scenario	streamflow is more sensitive to rainfall changes than to temperature changes	Islam et al. (2012a)
Kangsabati River catchment	Water Evaluation and Planning model	REMO and HadRM3 RCM, A1b	decrease in reservoir inflows	Bhave et al. (2013)
Upper Sind River basin	SWAT model	PRECIS RCM, A1b	increase in average annual streamflow by 16.4 and 93.5% during 2021–2050 and 2071–2098, respectively	Narsimlu et al. (2013)
Ganga, Narmada, and Krishna basins	variable infiltration capacity macroscale hydrologic model	BCCR-BCM 2.0, CGCM3-T63, CSIRO-Mk3 GCMs, A2, A1b, B1	an increase in summer monsoon surface runoff, evapotranspiration, and soil moisture in Ganga and Narmada river basins, and a decrease in runoff and soil moisture for some regions in Krishna basin	Raje et al. (2014)
Brahmani River basin	PRMS model	CGCM2, CSIRO-Mk2, HADCM3, CCSR/NIES GCMs, A2 and B2	increase in annual as well as seasonal streamflow in the basin	This chapter

† HEC-HMS, Hydrologic Engineering Corps–Hydrologic Modeling Systems; SWAT, Soil and Water Assessment Tool; PRMS, Precipitation Runoff Modeling Systems.

‡ GCM, general circulation model; CCSR/NIES, Center for Climate System Research/National Institute for Environmental Studies; PRECIS RCM, Providing Regional Estimates for Climate Impact Studies regional climate model.

¶ Brahmani, Cauvery, Ganga, Godavari, Krishna, Luni, Mahanadi, Mahi, Narmada, Pennar, Sabarmati, and Tapi.

§ Baitarni, Brahmani, Brahmaputra, Cauvery, Ganga, Godavari, Indus, Krishna, Luni, Mahanadi, Mahi, Meghna, Narmada, Pennar, Sabarmati, Subernrekha, and Tapi.

Table 11–6. Changes (%) in mean annual water balance in the Indian river basins.

	HadRM2			PRECIS RCM ‡					
	2041–2060 †			2021–2050 (2030s)			2071–2098 (2080s)		
Basin	Rainfall	Runoff	Actual ET	Rainfall	Water yield	Actual ET	Rainfall	Water yield	Actual ET
					%				
Baitarni				2.5	2.8	1.9	9.9	17.6	21.6
Brahmani	18.0	24.5	11.1	8.1	12.6	1.2	8.9	24.9	35.3
Brahmaputra				−2.3	−3.5	8.1	20.7	8.7	7.7
Cauvery	2.7	−1.6	7.5	−1.7	−3.4	−1.1	10.2	−4.7	-8.5
Ganga	10.9	12.0	9.8	2.5	−0.5	7.0	23.0	27.0	16.0
Godavari	5.9	11.0	0.7	16.1	27.2	−1.4	8.0	33.6	50.0
Indus				16.6	20.0	11.9	9.3	17.5	23.9
Krishna	−5.8	−11.9	−1.6	1.5	4.3	−0.7	10.9	−4.4	-3.8
Luni	−38.5	−57.4	−34.5	13.8	51.9	10.5	2.5	7.5	59.7
Mahanadi	18.6	28.0	9.9	13.3	19.0	2.0	9.0	29.3	39.6
Mahi	−17.7	−25.3	−15.6	11.5	26.0	4.1	3.9	25.0	44.3
Meghna				25.0	33.8	3.0	11.6	37.8	48.4
Narmada	−2.4	1.7	−5.2	17.4	27.0	0.8	6.8	34.5	50.1
Pennar	−6.5	−25.8	−0.9	−3.5	−30.9	−1.9	6.2	7.4	12.8
Sabarmati	−39.3	−70.9	−34.0	13.7	38.6	−2.8	2.7	29.8	71.1
Subernrekha				1.1	1.4	0.6	8.3	17.1	22.6
Tapi	−4.8	4.4	−10.0	17.5	32.5	−0.3	5.7	32.6	54.9

† PRECIS RCM, Providing Regional Estimates for Climate Impact Studies regional climate model. Changes relative to 1981–2000 (adapted from Gosain et al., 2006).

‡ Changes relative to 1961–1990 (adapted from Gosain et al., 2011).

(Table 11–5). In this study, Hydrologic Engineering Center's Hydrologic Modeling System (HEC-HMS 3.4) and linear regression-based Statistical DownScaling Model (SDSM 4.2) were applied. The average annual streamflow was projected to increase by 4, 17, and 4% during the 2020s, 2050s, and 2080s, respectively, under the A2 scenario. Under the B2 emission scenarios, an increase of 5 and 19% in average annual streamflow was estimated during the 2050s and 2080s, respectively. However, streamflow projected using statistical downscaling with a Canonical Correlation Analysis and projections of MIROC 3.2 GCM under A1b emission scenarios showed a decreasing trend in future (2010–2040, 2040–2070, and 2070–2100) streamflow (Table 11–5) in the four tributaries (tributaries of Tunga, Bhadra, Kumudavathi, and Haridra) of the Tungabhadra river (Rehana and Mujumdar, 2012a). This resulted in impairment of water quality with the decrease in dissolved oxygen (DO) levels due to the reduced dilution capacity and low water velocities derived from the reduced streamflow and also due to the changes in reaction rates derived from the increase in water temperature. Further, simulation using the water quality model (QUAL2K) and hypothetical climate change (changes in temperature and streamflow) scenarios indicated maximum reduction in DO levels from present conditions with a 2°C increase in temperature and 20% reduction in streamflow (Rehana and Mujumdar, 2011). Simulation results also showed that water quality parameters are more sensitive to the temperature change than to streamflow change.

Islam et al. (2012a) applied the U.S. Geological Survey's Precipitation Runoff Modeling System (PRMS) (Leavesley et al., 1983) for studying the climate change impact on streamflow in the Brahmani River basin. They considered a range of climate change scenarios with ± 10 to 30% changes in rainfall and 0 to 4°C changes in air temperature from the measured baseline climate (1980–1990). Simulation results showed that the changes in temperature had a relatively lesser effect on the magnitude of annual and seasonal streamflow as compared with rainfall changes in the basin. A 10% decrease in rainfall resulted in a 23% decrease in annual streamflow, whereas a 4°C rise in temperature resulted in an 11% decrease in annual streamflow (Fig. 11–6a). A 10% decrease in rainfall resulted in 25, 12, and 21% decreases in streamflow during monsoon, pre-monsoon, and post-monsoon seasons, respectively, and a 4°C increase in temperature resulted in 12, 3, and 11% decreases in streamflow during the same seasons. The combined effect of changes in rainfall and temperature showed changes in the mean annual streamflow in the range of −33 to 62% (Fig. 11–6b). A 4°C rise in temperature coupled with a 30% increase in rainfall resulted in 62, 73, 39, 52, and 30% increases in annual, monsoon, pre-monsoon, post-monsoon, and winter season streamflow, respectively. Similarly, a temperature rise of 4°C and a 10% decrease in rainfall resulted in 33, 35, 15, 32, and 21% decreases in annual, monsoon, pre-monsoon, post-monsoon, and winter season streamflow, respectively. The greater effect of rainfall changes on streamflow as compared with that of temperature change may be due to sub-humid climatic conditions in the basin with the lower part of the basin being located in the coastal region.

Though the GCMs are the preferred option for deriving climate change scenarios, there is considerable uncertainty associated with the GCM simulations because of uncertainty in greenhouse gas (GHG) emissions, uncertainty in global GHG cycles usually simulated "offline," and uncertainty in GCM response to a particular forcing associated with model structure, parameterization, and spatial resolution (Andersson et al., 2006). Different GCMs produce significantly different outputs for some regions, posing the problem of GCM selection for impact assessment studies. Since each climate model has its own uncertainty, the use of climate change scenarios from more than one GCM is suggested for dealing with the uncertainty linked with climate projections. Several studies on hydrological

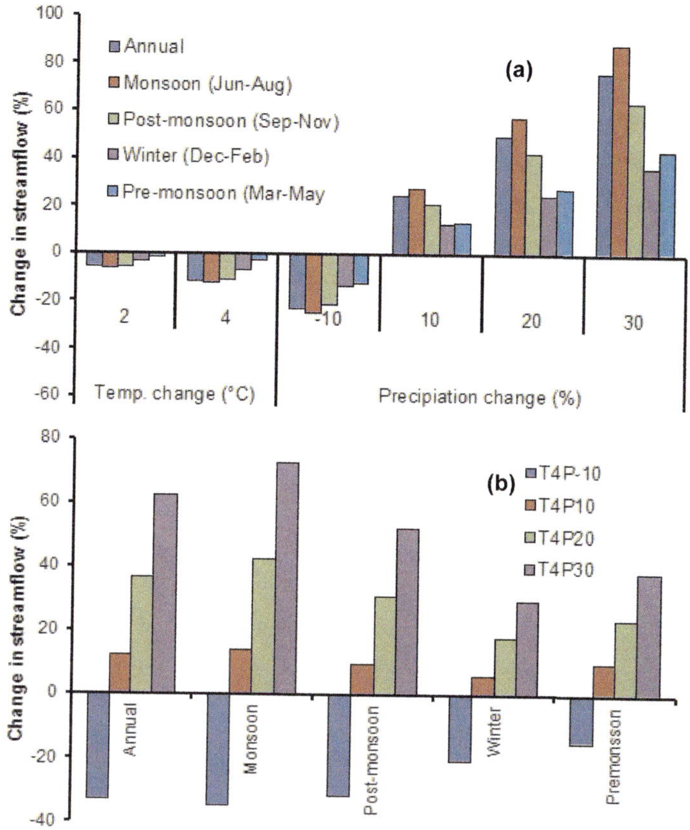

Fig. 11–6. Response of streamflow to (a) rainfall and temperature changes and (b) combined effect of rainfall and temperature changes (adapted from Islam et al., 2012a, with permission from Springer Science + Business Media).

impact assessment have been performed using more than one climate projection obtained from the combination of GCMs and greenhouse gas emission scenarios (GHGES) to address the uncertainty linked to GHGES and GCMs (Andersson et al., 2006; Maurer, 2007; Minville et al., 2008). The use of multi-model projections allows consideration of uncertainty by providing a range of possible changes of future streamflow and water resources availability.

Simulation studies conducted using the PRMS model with Third Assessment Report projections from four different GCMs [CGCM2 = Coupled Global Climate Model 2, Canadian Center for Climate Modeling and Analysis (CCCma), Canada; CSIRO-Mk2 = CSIRO Atmospheric Research Mark 2b climate model, Commonwealth Scientific and Industrial Research Organization (CSIRO), Australia; HADCM3 = Hadley Centre Coupled Model 3, Hadley Centre for Climate Prediction and Research (HCCPR), UK; CCSR/NIES = Center for Climate System

Research/National Institute for Environmental Studies, Japan] and two emission scenarios (A2 and B2) indicated an increase in annual streamflow in the Brahmani River basin for all the climate change scenarios, except with the CGCM2 model scenario (Fig. 11–7). The increase in annual streamflow varied in the range of 1 to 30%, 5 to 23%, and 7 to 75% during the 2020s, 2050s, and 2080s, respectively, depending on the GCMs and emission scenarios (Fig. 11–7). In the case of the CGCM2 model projected scenarios of rainfall and temperature change, there are 6 and 18% decreases in annual streamflow during the 2020s (A2) and 2050s (B2 emission scenario), respectively. Average (averaged over all GCMs) values indicated 6, 11, and 42% increases in annual streamflow during the 2020s, 2050s, and 2080s, respectively, under the A2 scenario. Under the B2 scenario, there are 6, 14, and 29% increases in

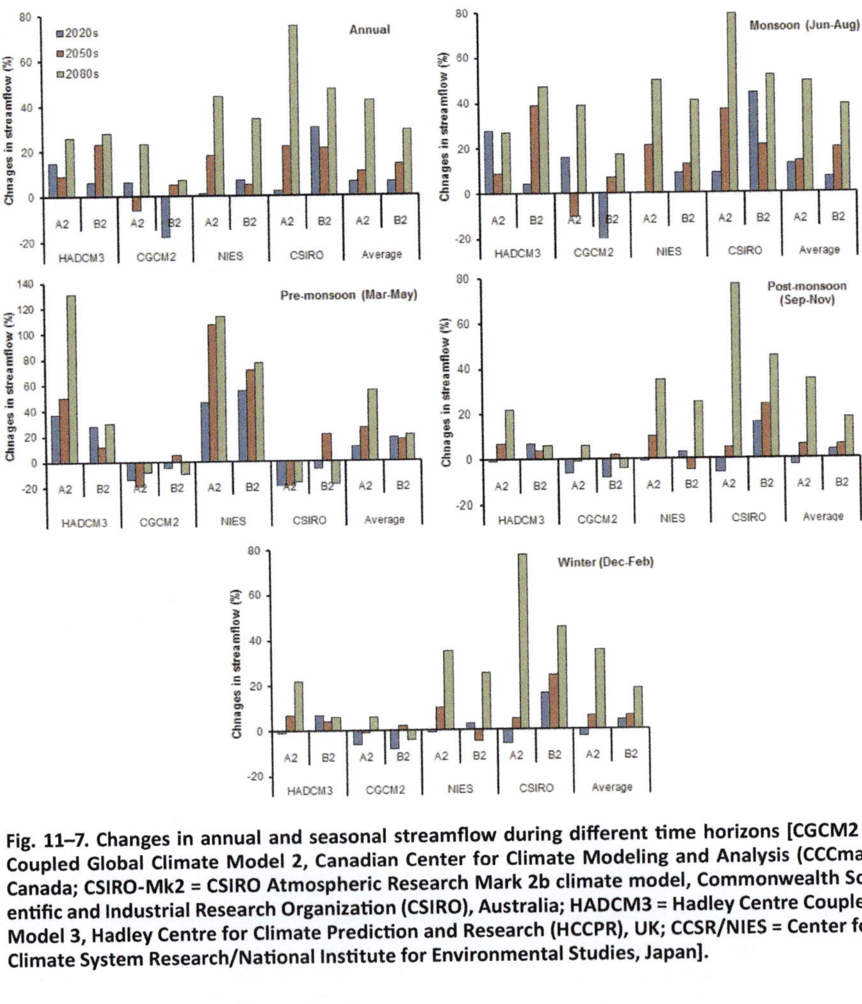

Fig. 11–7. Changes in annual and seasonal streamflow during different time horizons [CGCM2 = Coupled Global Climate Model 2, Canadian Center for Climate Modeling and Analysis (CCCma), Canada; CSIRO-Mk2 = CSIRO Atmospheric Research Mark 2b climate model, Commonwealth Scientific and Industrial Research Organization (CSIRO), Australia; HADCM3 = Hadley Centre Coupled Model 3, Hadley Centre for Climate Prediction and Research (HCCPR), UK; CCSR/NIES = Center for Climate System Research/National Institute for Environmental Studies, Japan].

annual streamflow during the same period (2020s, 2050s, and 2080s). Similar to the annual streamflow, during monsoon (rainy) season there is also an increase in streamflow for all the climate change scenarios, and it varied in the range of 13 to 49% and 7 to 39% under A2 and B2 emission scenarios, respectively. During the pre-monsoon period, climate change scenarios from CGCM2 and CSIRO-Mk2 models resulted in a decrease in streamflow in all the three time horizons under the A2 emission scenario, whereas under B2 emission scenarios there is a decrease during the 2020s and 2080s. In the post-monsoon periods, most of the climate change scenarios resulted in a decrease in streamflow during the 2020s. During the winter season, changes in streamflow remained minimal and ranged from −1 to 7% and −1 to 3% under A2 and B2 emission scenarios, respectively. The variation in seasonal streamflow could be due to changes in monthly rainfall with no definite pattern during different periods. The temporal variability in the availability of water resources in the basin under the influence of climate change indicated the need for development of different adaptation strategies. Further, there are large variations in the projected streamflow under different GCMs and emission scenarios. Thus, consideration of multiple GCMs and emission scenarios will be helpful to capture the uncertainties associated with future climate change scenarios as well as suggesting more appropriate adaptation options.

One of the major limitations in using GCMs output is their spatial and temporal resolution, which does not match the required resolution for the hydrologic models. To address this issue, a number of downscaling approaches (e.g., Wilby and Wigley, 1997; Wood et al., 2004; Fowler et al., 2007) have been developed to convert large-scale data to local-scale data. The simplest approach to downscale from a global climate model scale to a finer scale is to interpolate GCM outputs from the nearest grid boxes using some form of interpolation procedure such as Inverse Distance Interpolation (Brumbelow and Georgakakos, 2007) followed by temporal downscaling to convert monthly parameter values to the daily values using different stochastic weather generates such as WGEN (Richardson, 1981) and LARS-WG (Semenov and Barrow, 1997). Other downscaling approaches include dynamic downscaling and statistical downscaling. The dynamical downscaling approach consists of using RCMs to resolve the physical equations of atmospheric regional dynamics (Wood et al., 2004). This approach is computationally expensive, restricting their use for many applications. An RCM system known as PRECIS, developed by the Hadley Centre for Climate Predictions and Research, has been applied for India to develop high-resolution (50 by 50 km) climate change scenarios (Rupa Kumar et al., 2006). The PRECIS RCM climate change scenarios have been used in several impact assessment studies (e.g., Akhtar et al., 2009; Gosain et al., 2011; Ghosh and Dutta, 2012; Naresh Kumar et al., 2013, 2014).

The statistical downscaling approach is based on statistical relationships between large-scale GCM features (or predictors) and local-scale climatic variables (or predictands) and is quite flexible and computationally less expensive. Different statistical downscaling methods using different approaches such as a relevance vector machine (Ghosh and Mujumdar, 2008; Mujumdar and Ghosh, 2008), a support vector machine (Tripathi et al., 2006; Anandhi et al., 2008, 2009), fuzzy clustering (Ghosh and Mujumdar, 2007; Mujumdar and Ghosh, 2008), conditional random field distribution (Raje and Mujumdar, 2009), and the cumulative distribution function transform (Vigaud et al., 2013) have been applied in different studies (Table 11–5) in India. Though the different downscaling methods may perform similarly during training periods, they may give different projections when applied to GCM outputs for future scenarios. These statistical downscaling approaches are based on the assumption that predictor–predictand relationships are valid under future climatic conditions and the predictor variables and their changes are well characterized by GCMs. Further, direct downscaling of GCM predictors to streamflow simplifies the naturally complicated hydrologic cycle to a great extent and neglects the influences of land use, soil cover, and groundwater storage on streamflow.

Modeling Climate Change Impact on Irrigation Water Demand

Irrigation water demand is likely to be affected under the projected climate change scenarios because of changes in rainfall, temperature, and the evaporative demand. Rehana and Mujumdar (2012b) simulated irrigation water demand of paddy, sugarcane, permanent garden, and semidry crops using the MIROC 3.2 GCM projections for the A1b emission scenario. The short grass reference evapotranspiration (ET_o) calculated using the FAO-56 Penman–Monteith method (Allen et al., 1998) with the projected maximum (T_{max}) and minimum air temperature (T_{min}), relative humidity (RH), and wind speed (U_2) data showed an increase in ET_o for all the months. The annual irrigation water requirements for paddy, sugarcane, permanent garden, and semidry crops were projected to increase in the Bhadra reservoir command area, suggesting that the effect of the projected increase in rainfall on the irrigation demands is offset by the effect due to projected changes in other meteorological variables.

Several studies have showed that the effect of rising temperature on ET is moderated by elevated CO_2 concentrations (Rosenberg et al., 1989; Ficklin et al., 2009; Islam et al., 2012b, 2012c), and thereby ET_o demand may not rise significantly. A simulation study conducted using meteorological data of Varanasi (25.26° N, 82.99° E) also demonstrated similar results (Priya et al., 2014). They also used the FAO-56 Penman–Monteith method to simulate ET_o, and the combined effect of temperature and elevated CO_2 levels was studied by varying the temperature from 1 to 5°C and the CO_2 level from 330 to 660 μmol mol^{-1}. Comparison of the relative effect of changes in different climatic variables (mean temperature [T_{mean}], wind speed, vapor pressure deficit [VPD], and solar irradiance [R_g]) on annual ET_o showed that the annual ET_o is most sensitive to changes in mean temperature and least sensitive to changes in wind speed (Fig. 11–8). The change in annual ET_o varied in the range of −17 to +18% with changes in the mean temperature from −25 to +25%. For similar changes in the wind speed (−25 to +25%), the changes in annual ET_o varied in the range of −6 to +5%. Among the VPD and the solar irradiance, the ET_o is more sensitive to changes in R_g than to the changes in VPD. A 25% change in the VPD and R_g resulted in about an 8 and 14% change in annual ET_o, respectively. Thus, the relative effect of different climatic variables in annual ET_o is in the order of $T_{mean} > R_g > VPD > U_2$.

Effect of Temperature and CO_2 Concentration Changes on ET_o

Every 1°C rise in mean temperature resulted in an increase in reference ET_o by about 0.1 mm d^{-1} (2.3% increase with respect to baseline) annually (Priya et al., 2014). Among different months, December and January recorded the minimum increase (~0.06 mm d^{-1} increase with 1°C temperature rise), while April and May

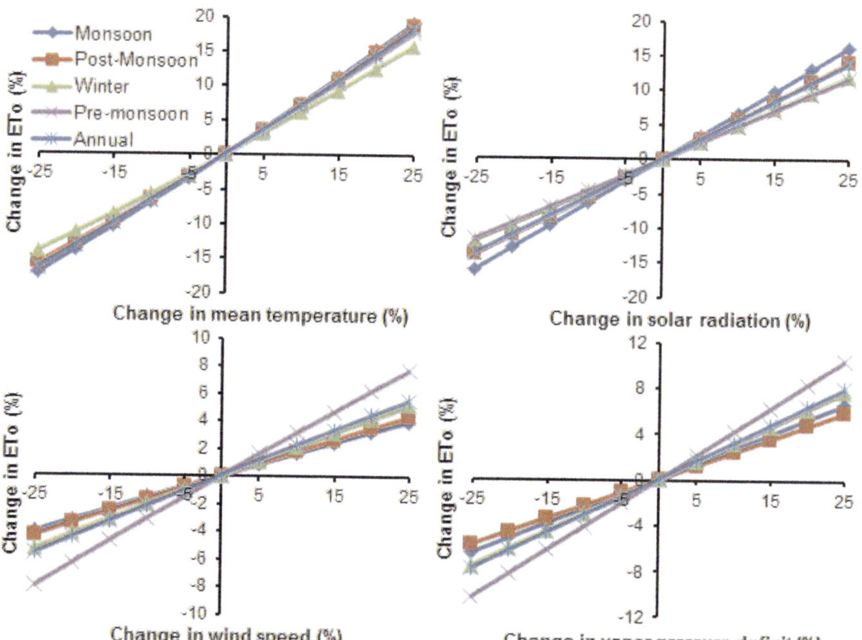

Fig. 11–8. Effect of changes in different climatic variables on seasonal reference evapotranspiration, ET_o, in Varanasi (adapted from Priya et al., 2014, with permission from Association of Agrometeorologists).

recorded the maximum increase (~0.13 mm d^{-1} increase with 1°C temperature rise). There is a decrease in ET_o with increases in CO_2 concentration (Fig. 11–9), due to a decrease in stomatal conductance and increase in stomatal resistance with an increase in CO_2 concentration. There is about a 6% decrease in annual ET_o demand with doubling (660 μmol mol^{-1}) of CO_2 concentration and ambient temperature remaining constant. Simulating the combined effect of temperature and elevated CO_2 showed that the effect of about a 1.0°C rise in temperature is offset by an increase in CO_2 levels up to 495 μmol mol^{-1}, and a 2.5°C rise in temperature is offset by an increase in CO_2 levels up to 660 μmol mol^{-1} (Fig. 11–9). There is a decrease in ET_o in most of the months with a 2.0°C rise in temperature coupled with an increase in CO_2 concentrations to 660 μmol mol^{-1}. However, with a 4.0°C increase in temperature there is an increase in monthly ET_o even with an increase in CO_2 concentrations to 660 μmol mol^{-1}, but the increase in ET_o remained <10% in most of the months (Table 11–7). Thus, crop water demand may not rise significantly under the climate change scenarios because of the moderating effect of rising CO_2 concentrations on evapotranspiration demand. These simulation results provide valuable information for planning future irrigation water management strategies considering the possible impacts of climate change on evapotranspiration demand.

Simulation results presented above show large temporal and spatial variability in the projected streamflow, water yield, and evapotranspiration demand depending on the hydrological model, climate change scenarios, and downscaling approaches used. Even with the same hydrological model, there is variation in simulated streamflow with changes in regional climate change scenarios

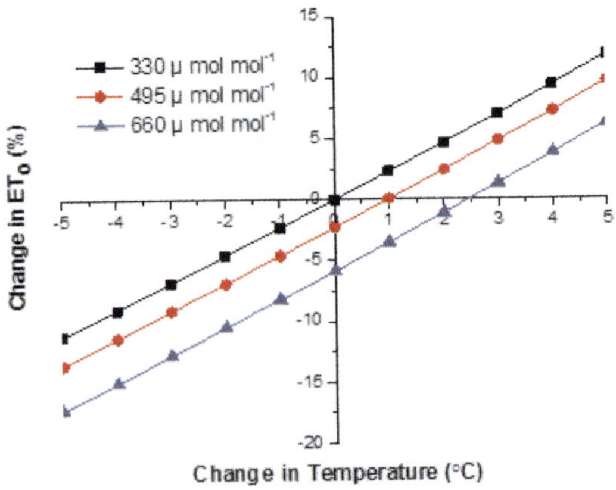

Fig. 11–9. Effect of changes in temperature and CO_2 concentration on reference evapotranspiration, ET_o (adapted from Priya et al., 2014, with permission from Association of Agrometeorologists).

Table 11–7. Effect of temperature and different levels of CO_2 on ET_o (adapted from Priya et al., 2014, with permission from Association of Agrometeorologists).†

	Change in ET_o (%) with temperature and CO_2 change			
	Temperature increase = 2°C		Temperature increase = 4°C	
Month	CO_2 = 495 μmol mol^{-1}	CO_2 = 660 μmol mol^{-1}	CO_2 = 495 μmol mol^{-1}	CO_2 = 660 μmol mol^{-1}
	%			
Jan.	3.8	0.2	10.2	6.6
Feb.	3.1	−0.8	9.0	5.1
Mar.	2.2	−1.6	6.9	3.2
Apr.	1.3	−2.4	5.1	1.5
May	1.4	−2.4	5.3	1.6
June	1.5	−2.4	5.7	1.9
July	1.9	−2.2	6.6	2.5
Aug.	2.4	−1.3	7.5	3.8
Sept.	3.1	0.0	8.4	5.4
Oct.	3.9	1.8	9.5	7.4
Nov.	4.2	2.1	10.1	8.0
Dec.	4.2	1.2	10.5	7.5
Annual	2.4	−1.1	7.2	3.8

† ET_o, reference evapotranspiration.

(Table 11–5). The spatial variability of streamflow responses is largely conditioned by climatic and physiographical characteristics of the basins. Impact studies today lack in high spatial and temporal resolution of the input data. The uncertainty in hydrologic impacts due to the spatial scale of climate scenarios has been one of the most studied areas. Use of high-resolution climate change scenarios for regional impact studies significantly affects estimations of changes in water yield under climate change conditions when compared with results from coarse resolution GCMs (Stone et al., 2003). The explicit spatial downscaling methods, which explicitly consider spatial differences of climate scenarios and variability, are more appropriate for assessing the site-specific impacts of climate change on natural resources, especially for those heavily influenced by local conditions such as soil erosion and crop production (Zhang, 2007). Different downscaling methods often lead to statistically different realizations of daily weather series and therefore result in an different impact on runoff predictions for a given climate change scenario. Murphy (1999) showed that while dynamical and statistical downscaling approaches yield similar reproductions of current climate, they differ significantly in their projections of future climate conditions. Several studies suggest bias correction of climate model output to assure meaningful results in applications like hydrologic and water resources assessments (Murphy, 1999; Wilby et al., 2000). Unbias corrected RCM inputs are unable to reproduce observation-driven simulations and after bias-correction coupling GCMs with RCMs has little to no benefit for hydrological impacts assessment (Wood et al., 2004). Proper consideration of local climate and environmental and land use conditions is essential for assessment of climate change impact on hydrological regimes.

Several studies also suggest that the rising levels of CO_2 concentration and climate change may have significant effects on the hydrology and water resources availability (Fontaine et al., 2001; Jha et al., 2006; Wu et al., 2012). There are other studies suggesting that an increase in leaf area with rising CO_2 concentration may potentially offset the effect of reduction of stomatal conductance to some extent (Betts et al., 1997; Kergoat et al., 2002). Under a doubling of CO_2 concentration, global mean runoff has been projected to increase by ~5% as a result of reduced ET due to CO_2 enrichment alone (Leipprand and Gerten, 2006; Betts et al., 2007). Without considering the effects of CO_2, the drought predicted would be overestimated and the risk of flooding would be underestimated, and future watershed management decisions are likely to be misleading. Land use change is another major factor affecting the hydrological processes, and some studies suggest that the consequences may outweigh those from climate change (Vorosmarty et al., 2000; DeFries and Eshleman, 2004). This needs to be integrated in hydrological impact assessment studies for reliable assessment of climate change impact on hydrology and water resources. The integration of land use information and disaggregation of the landscape into area characterized by significantly different hydrological process behavior is essential for meso- and macro-scale hydrological modeling. Another major limitation of most of the available hydrological models is that they lack extreme event modeling capability, resulting in underprediction of high flows. To enhance our ability to predict future climate change impact at a regional scale, hydrological models with extreme event modeling capabilities are needed.

Modeling climate change impact on crop production

Several simulation studies have been performed for quantifying the impacts of climate change on Indian agriculture (Table 11–8) and are documented either for entire country (NATCOM, 2004, 2012; Aggarwal, 2008; Naresh Kumar, 2011) or for specific regions (Aggarwal et al., 2010; Naresh Kumar et al., 2011). The adaptation analysis has been done by quantifying the response of different varieties, sowing time, nutrient management, water management, introduction of new crops, shift in cropping sequences, altered resource management, and introduction of new technologies, etc. under different climate change scenarios so as to derive the best suitable technology package for reducing negative impacts of climate change at a

Table 11–8. Impact of climate change on crop production in India.

Crop	Model(s)	Study area	Climate change scenario(s)	Impact	Reference
Wheat	WTGROWS	72 locations spread across wheat growing region	hypothetical scenarios (temperature)	a 2°C temperature rise and 425 μmol mol^{-1} CO_2 level reduced potential yields in most of the wheat growing regions, but smaller effect on irrigated yields in northern India, and more pronounced effect in central India	Aggarwal and Sinha (1993)
Rice and wheat	CERES rice and wheat	northwest India	hypothetical scenarios (temperature and CO_2)	a 2 and 3°C temperature rise offsets the positive effect of elevated CO_2 on rice and wheat yields, respectively	Lal et al. (1998)
Rice	CERES-Rice	Kerala, India	hypothetical and ECHAM3 model scenarios	increase in rice yields under rainfed conditions	Saseendran et al. (2000)
Rice	CERES-Rice ORYZA 1N	different agroclimatic regions (four sites)	hypothetical scenarios (temperature and CO_2)	increase in rice yields in the range of 1.0–16.8% and 3.5–33.8% for pessimistic and optimistic climate change scenarios, respectively	Aggarwal and Mall (2002)
Rice	ORYZA1 and INFOCROP	eastern India	hypothetical as well as GFDL, GISS, and UKMO scenarios	for every 1°C increase in temperature, ORYZA1 and INFOCROP rice models predicted 7.2 and 6.7% decrease in yield, respectively, at CO_2 level of 380 μmol mol^{-1}; for future climate change (doubled CO_2) scenarios, ORYZA1 and INFOCROP predicted yield decrease of 7.6–15.9% and 9.0–21.4%, respectively, depending on the GCMs used	Krishnan et al. (2007)
Maize	InfoCrop	IGP	HadCM3 A2	decrease in winter maize grain yield with an increase in temperature in mid IGP and Southern Plateau, and increase in yield up to a 2.7°C temperature rise in in the upper IGP	Byjesh et al. (2010)
Rice and wheat	DSSAT	Ganga Basin (India)	REMO and HadRM3 RCMs, A1b	in the upper Ganga Basin, decrease in rice and wheat yield for both the scenarios; however, in lower Ganga basin, HadRM3 projections showed increase while REMO RCM projections showed decrease in rice and wheat yields	Mishra et al. (2013)
Rice	InfoCrop	India	MIRO3.2.HI GCM, A1b, B1 and PRECIS RCM A1b, A2, B2	a decrease of 4% (2020s) to 10% (2080s) in irrigated rice yield and a 6% decrease in rainfed rice yield during the 2020s, but decrease is marginal during the 2050s and 2080s	Naresh Kumar et al. (2013)
Wheat	InfoCrop	India	MIRO3.2.HI GCM A1b, B1 and PRECIS RCM A1b, A2, B2	decrease in wheat yields in the range of 6–23% and 15–25% by the 2050s and 2080s, respectively; yield is projected to decrease in areas with mean seasonal maximum and minimum temperature >27 and >13°C, respectively	Naresh Kumar et al. (2014)

† IGP, Indo-Gangetic Plain.

‡ GCM, general circulation model; PRECIS RCM, Providing Regional Estimates for Climate Impact Studies regional climate model.

regional level, then up-scaling to the state and national level. The following sections describe the results of simulations studies conducted for the major crops under different climate change scenarios.

Impact of Climate Change on Rice in India

For analyzing the impacts of climate change on irrigated and rainfed rice in India, a simulation analysis was done using the InfoCrop-RICE model (Naresh Kumar et al., 2013) to quantify impacts and adaptation gains and to identify the vulnerable regions for irrigated and rainfed rice cultivation in future climates in India. For this, a global climate model (GCM-MIRO3.2.HI), a regional climate model (RCM-PRECIS), and four emission scenarios (A1b, A2, B1, and B2) were considered. The analysis indicated that in general, the direction of impact is similar in MIRO3.2.HI and PRECIS based assessments, even though the magnitude of impact is different. For India, the climate change is projected to reduce irrigated rice yields by about 4, 7, and 10% during the 2020s (2010–2039), 2050s (2040–2069), and 2080s (2070–2099), respectively. Conversely, rainfed rice yields in India are likely to be reduced by ~6% during the 2020s scenario, but during the 2050s and 2080s they are projected to decrease marginally (<2.5%). The magnitude of impact has significant spatial variations (Fig. 11–10). Some regions are likely to be affected more than others, while some regions may gain because of climate

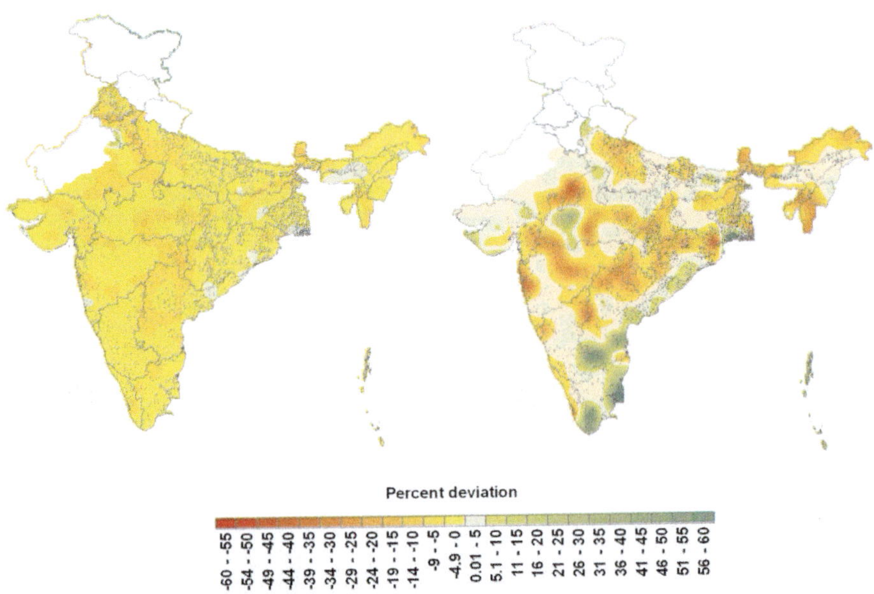

Fig. 11–10. Impact of climate change on irrigated and rainfed rice in the 2020s time period (adapted from Naresh Kumar et al., 2013, with permission from Springer Science + Business Media).

change. Basically, three types of influences that are likely to be caused by climate change are (i) regions that are adversely affected by climate change and can gain in net productivity with adaptation; (ii) regions that are adversely affected and still remain vulnerable despite adaptation gains; and (iii) rainfed regions (with low rainfall) that are likely to benefit because of an increase in rainfall and can further benefit by adaptation.

Several low-cost and easy to adopt adaptation strategies were also simulated to quantify the adaptation gains. Adaptation strategies comprising agronomical management can offset negative impacts in the near future, particularly in the rainfed conditions; however, in the longer run, developing suitable varieties coupled with improved and efficient crop husbandry becomes essential (Fig. 11–11). The analysis indicated that the regions falling in the vulnerable category even after suggested adaptation to climate change require more intensive, specific, and innovative adaptation options. Analysis indicated that current varieties may perform better under increased input and efficiency regimes, offsetting yield loss during the 2020s scenario. However, during the 2050s and 2080s, irrigated rice yield may be vulnerable with yield losses of 6 and 8%, respectively. Adopting improved varieties, efficient application, and rescheduling of inputs to suit crop phenology can reduce yield loss in future climate. Further, application of higher dose of fertilizers (about 25% higher than amount currently applied by farmers) can increase all India irrigated rice yields by about 15, 12, and 8% over current values during the 2020s, 2050s, and 2080s scenarios, respectively.

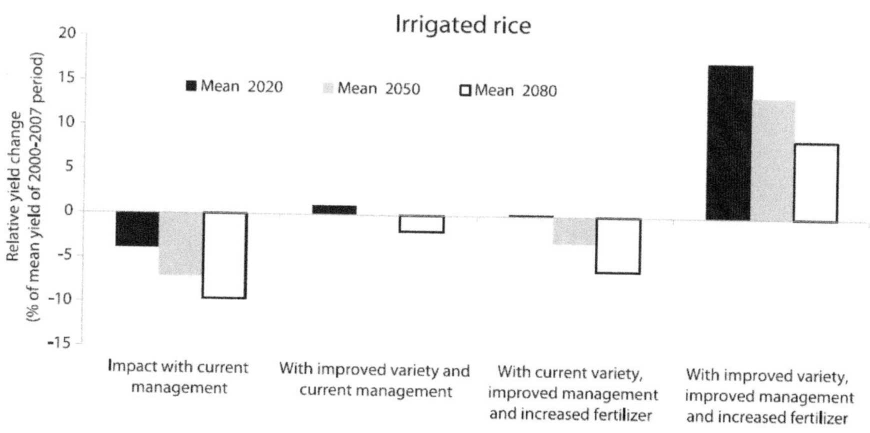

Fig. 11–11. Adaptation gains in irrigated rice in the 2020s, 2050s, and 2080s climate scenarios with different adaptation options (adapted from Naresh Kumar et al., 2013, with permission from Springer Science + Business Media).

Impact of Climate Change on Maize in India

Using the InfoCrop-MAIZE model, the impacts of the increase in temperature and the change in rainfall on yield of maize was studied. Simulation was also done using the HadCM3 A2 scenario for the 2020s, 2050s, and 2080s (Byjesh et al., 2010). Simulation results indicated that maize yields in monsoons may be adversely affected because of the rise in atmospheric temperature, but increased rain can partly offset those losses. During winter, maize grain yield is projected to be reduced with an increase in temperature in the mid IGP and Southern Plateau. However, in the upper IGP, yield is projected to increase up to a 2.7°C rise in temperatures, where relatively low temperatures prevail during winter (Fig. 11–12). In this region, during monsoon season a 20% increase in rainfall is projected to offset the yield loss due to 1°C rise in temperature. Similarly, a 30% increase in rainfall offset the adverse impact of 2°C rise in temperature. The spatiotemporal variations in projected changes in temperature and rainfall are likely to lead to differential impacts in the different regions. Analysis indicated that timely sowing of improved cultivars can increase the monsoon maize yields by ~17% in the upper IGP, but in other two regions, climate change negative impacts can be offset (Byjesh et al., 2010).

Simulation with HadCM3 projected climate change scenarios indicated a gain in the winter maize crop in the upper IGP region (Fig. 11–13), and these positive impacts of climate change can be further improved by adaptation strategies by about 6 and 11% during the 2020s and 2050s, respectively, relative to current

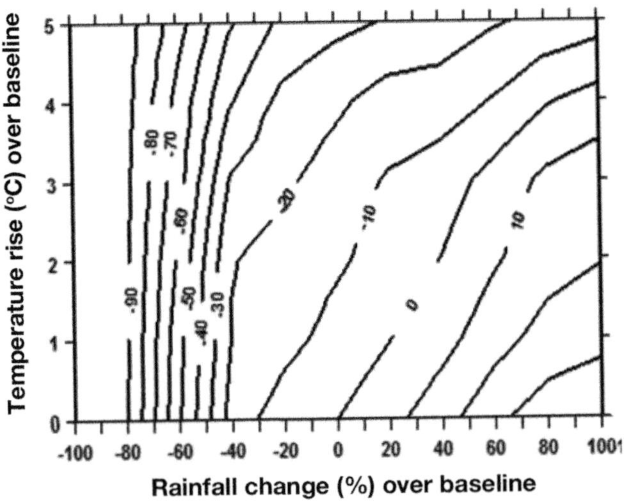

Fig. 11–12. Impact of change in temperature and rainfall on the maize yield during monsoon in the upper Indo-Gangetic Plain (adapted from Byjesh et al., 2010, with permission from Springer Science + Business Media).

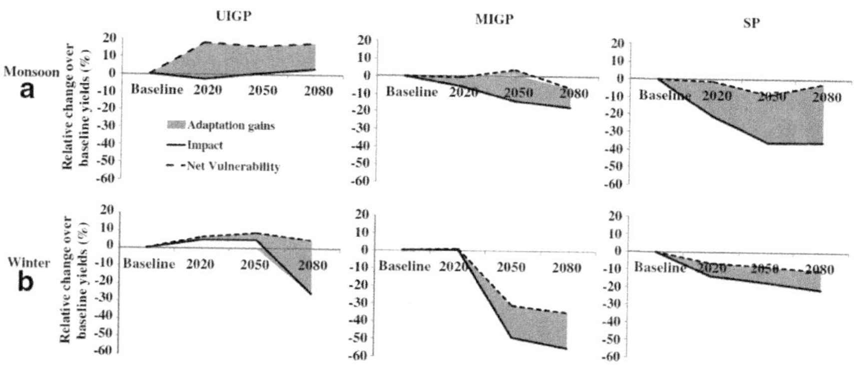

Fig. 11–13. Impact of climate change on maize with and without adaptation during the (a) monsoon and (b) winter seasons in the 2020s, 2050s, and 2080s climate scenarios of the HadCM3 model (UIGP, upper Indo-Gangetic Plain; MIGP, middle and lower Indo-Gangetic Plain; SP, Southern Plateau) (adapted from Byjesh et al., 2010, with permission from Springer Science + Business Media).

yields. However, maize in the mid IGP region is projected to be very sensitive to climate change. Even with adaptation, maize crop productivity may reduce by 28 and 30% during the 2050s and 2080s, respectively. Similarly in the Southern Plateau region, adaptation can reduce the impacts to 6, 8, and 10% during the 2020s, 2050s, and 2080s, respectively (Byjesh et al., 2010).

Impact of Climate Change on Wheat in India

Using the InfoCrop-WHEAT model, the regional vulnerability of wheat production to climate change in India was assessed (Naresh Kumar et al., 2014). The study projected that climate change will reduce the wheat yield in India in the range of 6 to 23% by the 2050s and 15 to 25% by the 2080s under the projected climate scenarios of a global (MIROC3.2.HI) and a regional (PRECIS) climate model. A considerable spatial variation in impacts is projected. Warmer central and south central regions of India may be more affected. However, simple and low-cost adaptation options, such as improvement in sowing times and increased and efficient use of inputs, could not only offset yield reduction but could also improve yields until the middle of the century (Fig. 11–14). Converting late-sown areas into timely sown regions can significantly improve yield even with the existing varieties in the near future (Naresh Kumar et al., 2014).

The above results are based on simulation studies conducted using a single crop model (InfoCrop), with selected GCM and RCM projections. Though the direction of change (increase or decrease) in yield is similar for different models (Table 11–8), magnitudes of projected changes in yield are quite different. For example, Decision Support System for Agro Technology Transfer (DSSAT) model

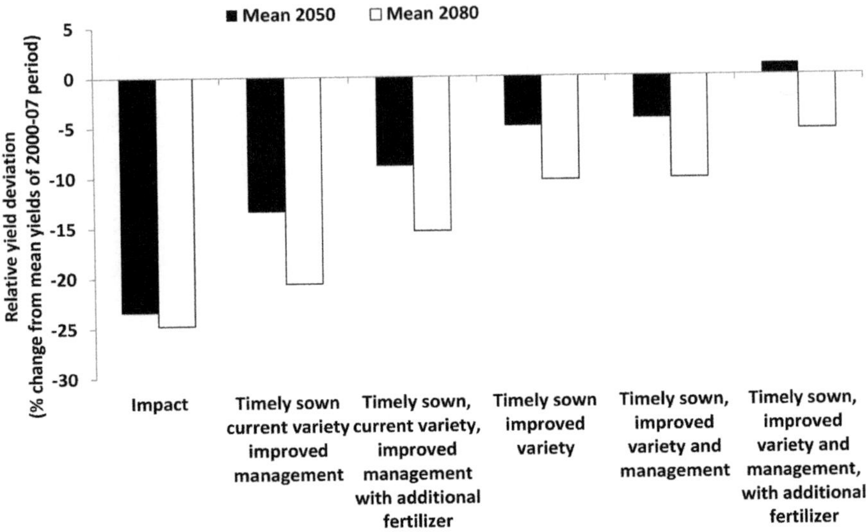

Fig. 11–14. Impact of climate change on wheat yield in 2050s and 2080s climate scenarios with different adaptation options (adapted from Naresh Kumar et al., 2014, with permission from Inter-Research).

simulation showed changes in the rice yield in the range of −4.7 (lower Indo-Gangetic Basin [IGB]) to −23.8% (upper IGB) and 1.2 (lower IGB) to −5.9% (upper IGB) for REMO and HadRM3 projected scenarios, respectively, under the commonly adopted management level in the IGB (Mishra et al., 2013). Similarly, they reported a change in wheat yield in the range of −1.7 (lower IGB) to −12.9% (upper IGB) and 5.4 (lower IGB) to −6.1% (upper IGB) under REMO and HadRM3 projected scenarios, respectively. As the simulated climate change impact varies across models because of differences in model structure and its parameterization (Asseng et al., 2013), multi-model ensemble analysis may be helpful for better quantification of climate change impact on crop yield and preparing adaptation plans.

Differences in treating climate natural variability, temporal structuring of daily climate variables, and large-scale forcing of local climate variables between the change factor and sophisticated spatial downscaling methods are some of the factors responsible for differences in impact assessments (Diaz-Nieto and Wilby, 2005). Dynamical downscaling may not add significant value to agricultural impacts assessments, but yield estimates made with both GCMs and RCMs require some form of bias correction to climate inputs (Glotter et al., 2014). Once a bias correction is applied, GCM and RCM-driven yields are essentially indistinguishable (Glotter et al., 2014). Further, changes in climate variability may have a more profound effect on yield than changes in mean climate (e.g., Mearns et al., 1996; Porter and Semenov, 2005). Therefore, policy analysis should not rely on scenarios of future climate involving only changes in means (Katz and Brown, 1992; Semenov and Barrow, 1997), but changes in mean climate and climate variability including impact of extreme events (e.g., Easterling and Apps, 2005; Moriondo et al., 2011) should also be included in crop modeling approaches to avoid the risk of incorrect crop yield estimation leading to incorrect policies for coping with climate change. For impact studies, focus should also be on proper consideration of local environmental conditions as well as management practices, as the uncertainties associated with these components may be of larger importance than the variation due to projected climate change.

Adaptation to Climate Change

Climate change is expected to produce water stresses in several parts of the country and thereby affect crop production. Substantial adaptation efforts will be needed to ensure adequate supply and efficient utilization of available water resources. The supply and demand management measures either meant at conserving, enhancing, or improving the water supply may be capable of alleviating such impacts to a limited extent in the water and food sector. The greatest potential for short-term adaptation is in demand control and more efficient and integrated management of surface and groundwater supplies. Adoption of drip or micro irrigation, rainwater harvesting, recharging groundwater, encouraging water-saving techniques such as water recycling, and reducing losses in canal systems could form some of the adaptation options. Participatory integrated watershed management approaches covering biophysical and socio-economic interventions have shown rich dividends for ensuring resilience to climate change. Managed aquifer storage is one of the adaptation options to improve water availability and address region-specific changes in precipitation distribution and increased evaporation rates (Shah, 2009). In the agriculture sector, the adaptive measures to counter detrimental impacts of climate variations include changing planting or sowing dates, changing crops, and having crop varieties that are more tolerant to climatic variations.

Under the National Initiative on Climate Resilient Agriculture (NICRA), a network project of the Indian Council of Agricultural Research, different interventions like in situ moisture conservation, biomass mulching, residue incorporation instead of burning, brown and green manuring, water harvesting and recycling for supplemental irrigation, improved drainage in flood prone areas, conservation tillage where appropriate, artificial ground water recharge, and water-saving irrigation methods have been implemented under the technology demonstration component of NICRA in the vulnerable districts of India to enhance the resilience of Indian agriculture to climate change and climate variability (Venkateswarlu et al., 2012). The National Water Mission under the National Action Plan on Climate Change focuses on ensuring integrated water resources management to conserve water; minimizing wastage and ensuring equitable distribution across and within states; developing a framework to optimize water use; recycling wastewater and ensuring equitable and efficient management structures; and optimizing efficiency of existing irrigation systems, incentives for water neutral or water-positive technologies, underground recharge, etc. Though many adaptation options are generic, activities with an explicit focus on adaptation and climate change will also be required. Options to address increasing water scarcity through better comanagement of water at the watershed, aquifer, and river basin level are needed in many water-stressed areas. Use of seasonal climate forecast in crop planning and management could possibly reduce the losses due to weather variability and provide opportunities for diversification. The major challenge remains in proper planning and effective implementations of different developmental programs and schemes at the local level. Identification of vulnerable regions and vulnerable sectors is essential for prioritization and implementation of adaptation plans.

Interbasin water transfer is also identified as a potential adaptation tool to improve both demand and supply management. Water transfer can promote the efficient allocation of water and has been attempted worldwide as a water management alternative in both developed and developing countries (Gupta, 2001). Though the river water transfer can help to address the issues of imbalances in water supply and demand, it has generated several controversies and cautions throughout the world (Gupta, 2001; Gichuki and McCornick, 2008; Gupta and van der Zaag, 2008). The interlinking of rivers projects in India aims to redistribute water from "surplus" to "deficit" zones within India and to counter groundwater depletion in western and southern India (Kumar et al., 2005). This is expected to provide a solution to the paradox of floods and drought in India and will also provide water for irrigation and power generation. However, the large centralized water development schemes, large dams, and diversions for water resources planning and management have also been criticized for its impacts on the natural environments and societies. Amarasinghe et al. (2008) claimed that from an overall economic perspective, groundwater development is a cheaper option mainly because most of the suitable sites for large dams in India have already been utilized. Alternative options such as increased water use efficiency, importance of decentralized and traditional water systems, improved maintenance of existing water infrastructure, reallocation of water from water-inefficient crops to more efficient crop types or varieties, increased awareness of water saving and water use efficiency, increased stakeholder participation, and increased alliances between and across governmental and nongovernmental stakeholders at local, regional, and national levels will not only address the issues of water scarcity under changing climate scenarios but will also accommodate the concerns and issues of different stakeholders.

Integrated Modeling Approaches

Climate change represents a complex strategic risk, and thus robust adaptation options are required that will provide benefits under various future climate scenarios (Willows and Connell, 2003). Integration of climate risk information in development planning is essential to prepare for climate change impacts across different sectors (Wilby et al., 2009). It is important that adaptation options are founded on sound scientific knowledge and also consider future socio-economic changes driving the water demand. Further, to increase resilience, adaptation plans should be made location specific. With the experiential understanding of local biophysical and socioeconomic systems, stakeholders are valuable knowledge bearers of potential adaptation options (Bhave et al., 2013). An integrated approach, involving top-down and bottom-up approaches, provides a roadmap to identify, prioritize, and assess adaptation options. No-regret options that would be justified under all plausible future scenarios, including the absence of human-induced climate change (Willows and Connell, 2003), also need consideration to facilitate mainstreaming of climate change adaptation.

Not many models and tools are available for such purposes, especially considering the complexities of the agricultural production process. Decision support tools are therefore required to assist policymakers in prioritizing investment to make agriculture climate-smart. These tools will also help in identifying the priorities within wider planning for climate change adaptation. Quantitative decision support tools will bring analytical rigor to this planning process, particularly supporting the developing countries in their preparation of National Adaptation Programmes of Action and Nationally Appropriate Mitigation Actions under the United Nations Framework Convention on Climate Change (UNFCCC) framework (UNFCCC, 2002).

Attempts have been made to bring analytical rigor in the planning process through development of a prioritization toolkit by the Consultative Group on International Agricultural Research program on Climate Change Agriculture and Food Security (Tyagi et al., 2014). In its current formulation, the toolkit is flexible in its capability to model agricultural production at a wide range of spatial (federal, state, district, farm-level) and temporal (seasonal, annual, decadal) scales. Application of this toolkit needs an integrated approach, often involving several models and tools like crop simulation models and irrigation and emission calculators. This toolkit provides a simple and quantitative framework for prioritization of climate-smart interventions using detailed bottom-up biophysical and socioeconomic datasets. The toolkit allows the user to identify robust decisions under a set of uncertain circumstances. It is also possible to carry out tradeoff analysis of alternative climate-smart agriculture development pathways. On the basis of the dynamic pathways, the toolkit can support decisions on what crops to cultivate; which climate-smart agricultural technologies and practices to invest in; where to target that investment; and when those investments should be made. The toolkit prioritizes investment decisions for the short term, while locating those priorities within a medium- and long-term agricultural growth pathway. The toolkit generates information for policymakers, which will substantially increase their ability to make better decisions, even under existing and future uncertainties.

Summary

The declining per capita land and freshwater availability coupled with global climate change have made sustainable agriculture a great challenge in India. The temperature had a significant increasing trend in most parts of the country and rainfall in general showed a declining trend in northern India. With increasing water demand from other competing sectors, the agricultural water use in India will face stiff competition in the future. Simulation models provide the opportunity to test various scenarios and management options to balance and optimize resource use, assess the impacts, and prepare and prioritize adaptation strategies. Model applications in water resources management bring the analytical rigor in the planning process and highlight their potential for deriving more informed decisions. This chapter describes the application of models in canal operation scheduling, improvement of water productivity, watershed management intervention prioritization, climate change impact assessment, etc.

Canal irrigation covers approximately one-third of the areas under irrigation in India, and its overall irrigation efficiency is around 38%. Several models have been used for the efficient management and water allocation in canal commands. Major improvements in the overall performance of the irrigation project can be achieved by just changing the delivery rules. Scarcity of irrigation water under climate change scenarios is a great concern for rice–wheat production systems, which cover 13.5 million ha in the IGP and are vital for food security. Increasing agricultural water productivity is the most plausible means of mitigating problems arising from scarcity and competition for water. Field-scale ecohydrological models like SWAP and APSIM have been used for identification of crop management practices and effects of irrigation regimes on yields. Moreover, these simulation studies help quantify the impacts of alternate management practices, which lead to more useful and informed answers. Likewise, several process-based as well as empirical models have been used in integrated watershed management, which include USLE, WEPP, and SWAT. These models helped target the interventions where gains can be maximized.

The quantitative assessment of the impact of climate change on water resources availability and its management requires knowledge of climate, hydrological, and water management models and particularly the relationships between each of them. Application of the PRMS model for studying the impact of climate change on streamflow in the Brahmani River basin showed that the changes in temperature had a relatively lesser effect on the magnitude of annual and seasonal streamflow as compared with rainfall changes. A 10% decrease in rainfall resulted in a 23% decrease in annual streamflow, whereas a 4°C rise in temperature resulted in an 11% decrease in annual streamflow. Further, simulation studies conducted using the future climate projections of different GCMs and emission scenarios indicated an increase in the annual streamflow in the Brahmani River basin in the range of 1 to 30%, 5 to 23%, and 7 to 75% during the 2020s, 2050s, and 2080s, respectively, depending on the GCMs and emission scenarios used. The altered climatic and hydrological conditions affect the crop production with varying degree of impacts. Crop simulation models like DSSAT, InfoCrop, and APSIM have been used to quantify the climate change impacts as well as different adaptation options. The climate change impact on rice indicated a decline in irrigated and rainfed rice yields. Maize yields in monsoons may be adversely affected because of a rise in atmospheric temperature, but increased rain can partly offset those losses. During winter, maize yield is projected to be reduced with an increase in ambient temperature in the mid IGP and Southern Plateau. However, in the upper IGP, yield is projected to increase up to a 2.7°C rise in temperature, where relatively low temperature prevails during winter. Wheat yield may be reduced by 6 to 23% and 15 to 25% during the 2050s and 2080s, respectively, under the GCM and RCM projected climate scenarios. Simulation results also showed that simple agronomic adaptation strategies can offset the negative impacts in the near future, but in the longer run alternate adaptation technologies are essential.

Simulation results revealed large spatial and temporal variability in the projected streamflow, water yield, and evapotranspiration demand across different river basins and also within the basin. The spatial variability of streamflow responses is largely due to climatic and physiographical characteristics of the basins. Impact studies today lack high spatial and temporal resolution of the input data. Proper consideration of local climate, environmental and land use conditions, and bias corrections of GCM and RCM outputs is essential for the assessment of climate change impact on hydrological regimes. Hydrological models with extreme event modeling capabilities are needed to enhance our ability to predict future climate change impacts at a regional scale. Similarly, crop models need to be further improved to model the extreme climatic events like heat stress, frosts, and freeze damages. High-resolution climate models considering fine-scale processes are also required for achieving reliable and useful future projections including climatic extremes. Simulated climate change impact varies across models, and hence multi-model analysis using multi-model ensemble climate change scenarios may be helpful for better quantification of climate change impact on water resources availability and crop yield and preparing adaptation plans. As climate change represents a complex and strategic risk, combined bottom-up participatory process and top-down modeling tools could be helpful to identify, prioritize, and assess the adaptation options for providing locally relevant climate change adaptation policy.

References

Aggarwal, P.K. 2008. Global climate change and Indian agriculture: Impacts, adaptation and mitigation. Indian J. Agric. Sci. 78:911–919.

Aggarwal, P.K., P.K. Joshi, J.S.I. Ingram, and R.K. Gupta. 2004. Adapting food systems of the Indo-Gangetic plains to global environmental change: Key information needs to improve policy formulation. Environ. Sci. Policy 7:487–498. doi:10.1016/j.envsci.2004.07.006

Aggarwal, P.K., and R.K. Mall. 2002. Climate change and rice yields in diverse agro-environments of India. II. Effect of uncertainties in scenarios and crop models on impact assessment. Clim. Change 52:331–343. doi:10.1023/A:1013714506779

Aggarwal, P.K., S. Naresh Kumar, and H. Pathak. 2010. Impacts of climate change on growth and yield of rice and wheat in the Upper Ganga Basin. WWF-India Report. New Delhi. http://assets.wwfindia.org/downloads/impacts_of_climate_change_on_growth_and_yield_of_rice_and_wheat_in_the_upper_ganga_basin.pdf (accessed 10 Apr. 2014).

Aggarwal, P.K., and S.K. Sinha. 1993. Effect of probable increase in carbon dioxide and temperature on wheat yields in India. J. Agric. Meteorol. 48:811–814. doi:10.2480/agrmet.48.811

Akhtar, M., N. Ahmad, and M.J. Booij. 2009. Use of regional climate model simulations as input for hydrological models for the Hindukush-Karakorum-Himalaya region. Hydrol. Earth Syst. Sci. 13:1075–1089. doi:10.5194/hess-13-1075-2009

Allen, R.G., L.S. Pereira, D. Raes, and M. Smith. 1998. Crop evapotranspiration: Guidelines for computing crop water requirements. Irrig. Drain. Pap. 56. FAO, Rome.

Amarasinghe, U., T. Shah, and P. McCornick. 2008. Seeking calm waters: Exploring policy options for India's water future. Nat. Resour. Forum 32:305–315. doi:10.1111/j.1477-8947.2008.00203.x

Anandhi, A., V.V. Srinivas, D. Nagesh Kumar, and R.S. Nanjundiah. 2009. Role of predictors in downscaling surface temperature to river basin in India for IPCC SRES scenarios using support vector machine. Int. J. Climatol. 29:583–603. doi:10.1002/joc.1719

Anandhi, A., V.V. Srinivas, R.S. Nanjundiah, and D. Nagesh Kumar. 2008. Downscaling precipitation to river basin in India for IPCC SRES scenarios using support vector machine. Int. J. Climatol. 28:401–420. doi:10.1002/joc.1529

Andersson, L., J. Wilk, M.C. Todd, D.A. Hughes, A. Earle, D. Kniveton, R. Layberry, and H.H.G. Savenije. 2006. Impact of climate change and development scenarios on flow pattern in the Okavango River. J. Hydrol. 331:43–57. doi:10.1016/j.jhydrol.2006.04.039

Arnell, N.W. 1992. Factors controlling the effects of climate change on river flow regimes in a humid temperate environment. J. Hydrol. 132:321–342. doi:10.1016/0022-1694(92)90184-W

Arnell, N.W., D.A. Hudson, and R.G. Jones. 2003. Climate change scenarios from a regional climate model: Estimating change in runoff in southern Africa. J. Geophys. Res. 108(D16):4519. doi:10.1029/2002JD002782

Arora, M., N.K. Goel, and P. Singh. 2005. Evaluation of temperature trend over India. Hydrol. Sci. J. 50:81–93. doi:10.1623/hysj.50.1.81.56330

Asokan, S.M., and D. Dutta. 2008. Analysis of water resources in the Mahanadi River Basin, India under projected climate conditions. Hydrol. Processes 22:3589–3603. doi:10.1002/hyp.6962

Asseng, S., F. Ewert, C. Rosenzweig, J.W. Jones, J.L. Hatfield, A.C. Ruane, K.J. Boote, P.J. Thorburn, R.P. Rötter, D. Cammarano, N. Brisson, B. Basso, P. Martre, P.K. Aggarwal, C. Angulo, P. Bertuzzi, C. Biernath, A.J. Challinor, J. Doltra, S. Gayler, R. Goldberg, R. Grant, L. Heng, J. Hooker, L.A. Hunt, J. Ingwersen, R.C. Izaurralde, K.C. Kersebaum, C. Müller, S. Naresh Kumar, C. Nendel, G. O'Leary, J.E. Olesen, T.M. Osborne, T. Palosuo, E. Priesack, D. Ripoche, M.A. Semenov, I. Shcherbak, P. Steduto, C. Stöckle, P. Stratonovitch, T. Streck, I. Supit, F. Tao, M. Travasso, K. Waha, D. Wallach, J.W. White, J.R. Williams, and J. Wolf. 2013. Uncertainty in simulating wheat yields under climate change. Nat. Clim. Change 3:827–832. doi:10.1038/nclimate1916

Betts, R.A., O. Boucher, M. Collins, P.M. Cox, P.D. Falloon, N. Gedney, D.L. Hemming, C. Huntingford, C.D. Jones, D.M.H. Sexton, and M.J. Webb. 2007. Projected increase in continental runoff due to plant responses to increasing carbon dioxide. Nature 448:1037–1041. doi:10.1038/nature06045

Betts, R.A., P.M. Cox, S.E. Lee, and F.I. Woodward. 1997. Contrasting physiological and structural vegetation feedbacks in climate change simulations. Nature 387(6635):796–799. doi:10.1038/42924

Bhadra, A., A. Bandyopadhyay, R. Singh, and N.S. Raghuwanshi. 2009. Integrated reservoir-based canal irrigation model (IRCIM). I: Description. J. Irrig. Drain. Eng. 135:149–157. doi:10.1061/(ASCE)0733-9437(2009)135:2(149)

Bhave, A.G., A. Mishra, and N.S. Raghuwanshi. 2013. A combined bottom-up and top-down approach for assessment of climate change adaptation options. J. Hydrol. doi:10.1016/j.jhydrol.2013.08.039.

Brumbelow, K., and A. Georgakakos. 2007. Consideration of climate variability and change in agricultural water resources planning. J. Water Resour. Plan. Manage. 133:275–285. doi:10.1061/(ASCE)0733-9496(2007)133:3(275)

Byjesh, K., S. Naresh Kumar, and P.K. Aggarwal. 2010. Simulating impacts, potential adaptation and vulnerability of maize to climate change in India. Mitig. Adapt. Strategies Glob. Change 15:413–431. doi:10.1007/s11027-010-9224-3

Chaturvedi, R.K., J. Joshi, M. Jayaraman, G. Bala, and N.H. Ravindranath. 2012. Multi-model climate change projections for India under representative concentration pathways. Curr. Sci. 103:791–802.

CWC. 2013. Water and related statistics. Water Resources Information System Directorate, Information System Organization, Water Planning & Project Wing, Central Water Commission, New Delhi. http://www.cwc.nic.in/main/downloads/Water%20and%20Related%20Statistics-2013.pdf (accessed 28 July 2013).

Dash, S.K., R.K. Jenamani, S.R. Kalsi, and S.K. Panda. 2007. Some evidence of climate change in twentieth-century India. Clim. Change 85:299–321. doi:10.1007/s10584-007-9305-9

DeFries, R., and K.N. Eshleman. 2004. Land-use change and hydrologic processes: A major focus for the future. Hydrol. Processes 18:2183–2186. doi:10.1002/hyp.5584

Diaz-Nieto, J., and R.L. Wilby. 2005. A comparison of statistical downscaling and climate change factor methods: Implications on low flows in the River Thames, United Kingdom. Clim. Change 69:245–268. doi:10.1007/s10584-005-1157-6

Easterling, W., and M. Apps. 2005. Assessing the consequences of climate change for food and forest resources: A view from the IPCC. Clim. Change 70:165–189. doi:10.1007/s10584-005-5941-0

Ficklin, D.L., Y. Luo, E. Luedeling, and M. Zhang. 2009. Climate change sensitivity assessment of a highly agricultural watershed using SWAT. J. Hydrol. 374:16–29. doi:10.1016/j.jhydrol.2009.05.016

Fontaine, T.A., J.F. Klassen, T.S. Cruickshank, and R.H. Hotchkiss. 2001. Hydrological response to climate change in the black hills of South Dakota, USA. Hydrol. Sci. J. 46:27–40. doi:10.1080/02626660109492798

Fowler, H.J., S. Blenkinsop, and C. Tebaldi. 2007. Linking climate change modeling to impacts studies: Recent advances in downscaling techniques for hydrological modeling. Int. J. Climatol. 27:1547–1578. doi:10.1002/joc.1556

Ghosh, S., and S. Dutta. 2012. Impact of climate change on flood characteristics in Brahmaputra basin using a macro-scale distributed hydrological model. J. Earth Syst. Sci. 121:637–657. doi:10.1007/s12040-012-0181-y

Ghosh, S., and P.P. Mujumdar. 2007. Nonparametric methods for modeling GCM and scenario uncertainty in drought assessment. Water Resour. Res. 43:W07405. doi:10.1029/2006WR005351

Ghosh, S., and P.P. Mujumdar. 2008. Statistical downscaling of GCM simulations to streamflow using relevance vector machine. Adv. Water Resour. 31:132–146. doi:10.1016/j.advwatres.2007.07.005

Gibson, C.A., J.L. Meyer, N.L. Poff, L. Hay, and A. Georgakakos. 2005. Flow regime alteration under changing climate in two river basins: Implications for freshwater ecosystems. River Res. Appl. 21:849–864. doi:10.1002/rra.855

Gichuki, F., and P.G. McCornick. 2008. International experiences of water transfers: Relevance to India. In: U.A. Amarasinghe and B.R. Sharma, editors, Strategic analyses of the National River Linking Project (NRLP) of India, Series 2; Proceedings of the Workshop on Analyses of Hydrological, Social and Ecological Issues of the NRLP. International Water Management Institute, Colombo, Sri Lanka, p. 345–371. http://publications.iwmi.org/pdf/H041812.pdf (accessed 8 July 2014).

Glotter, M., J. Elliotty, D. McInerney, N. Besty, I. Fostery, and E. Moyer. 2014. Evaluating the utility of dynamical downscaling in agricultural impacts projections. Proc. Natl. Acad. Sci. USA doi:10.1073/pnas.1314787111.

Gorantiwar, S.D. 1995. A model for the planning and operation of heterogeneous irrigation schemes in semi-arid regions under rotational water supply. Ph.D. diss., Loughborough Univ., Loughborough, UK. https://dspace.lboro.ac.uk/2134/11706 (accessed 5 Mar. 2014).

Gorantiwar, S.D., and I.K. Smout. 2003. Allocation of scarce water resources using deficit irrigation in rotational systems. J. Irrig. Drain. Eng. 129:155–163. doi:10.1061/(ASCE)0733-9437(2003)129:3(155)

Gorantiwar, S.D., and I.K. Smout. 2005a. Multilevel approach for optimizing land and water resources and irrigation deliveries for tertiary units in large irrigation schemes. II: Application. J. Irrig. Drain. Eng. 131:264–272. doi:10.1061/(ASCE)0733-9437(2005)131:3(264)

Gorantiwar, S.D., and I.K. Smout. 2005b. Performance assessment of irrigation water management of heterogeneous irrigation schemes: 1. A framework for evaluation. Irrig. Drain. Syst. 19:1–36. doi:10.1007/s10795-005-2970-9

Gorantiwar, S.D., and I.K. Smout. 2006. Model for performance based land area and water allocation within irrigation schemes. Irrig. Drain. Syst. 20:345–360. doi:10.1007/s10795-006-9012-0

Gorantiwar, S.D., I.K. Smout, and K. Vairavamoorthy. 2006. Performance-based optimization of land and water resources within irrigation schemes. I: Method. J. Irrig. Drain. Eng. 132:332–340. doi:10.1061/(ASCE)0733-9437(2006)132:4(332)

Gosain, A.K., S. Sandhya Rao, and A. Arora. 2011. Climate change impact assessment of water resources of India. Curr. Sci. 101:356–371.

Gosain, A.K., S. Sandhya Rao, and D. Basuray. 2006. Climate change impact assessment on hydrology of Indian river basins. Curr. Sci. 90:346–353.

Govind Rao, P. 1993. Climate changes and trend over a major river basin in India. Clim. Res. 2:215–233. doi:10.3354/cr002215

Graham, L.P., S. Hagemann, and S. Jaun. 2007. On interpreting hydrological change from regional climate models. Clim. Change 81:97–122. doi:10.1007/s10584-006-9217-0

Grotch, S.L., and M.C. MacCarcken. 1991. The use of general circulation models to predict regional climate change. J. Clim. 4:286–303. doi:10.1175/1520-0442(1991)004<0286:TUOGCM>2.0.CO;2

Gupta, J., and P. van der Zaag. 2008. Interbasin water transfers and integrated water resources management: Where engineering, science and politics interlock. Phys. Chem. Earth 33:28–40. doi:10.1016/j.pce.2007.04.003

Gupta, R.K. 2001. Human rights dimensions of regional water transfer: Experience of Sardar Sarovar Project. Water Resour. Dev. 17:125–147. doi:10.1080/713672565

Hamed, K.H., and A.R. Rao. 1998. A modified Mann–Kendall trend test for autocorrelated data. J. Hydrol. 204:182–196. doi:10.1016/S0022-1694(97)00125-X

Hingane, L.S., K. Rupa Kumar, and B.V. Ramanamurty. 1985. Long-term trends of surface air temperature in India. Int. J. Climatol. 5:521–528. doi:10.1002/joc.3370050505

IPCC. 2001. Climate change 2001: The scientific basis. Contribution of Working Group I to the Third Assessment Report of the Intergovernmental Panel on Climate Change, edited by J.T. Houghton, Y. Ding, D.J. Griggs, M. Noguer, P.J. van der Linden, X. Dai, K. Maskell, and C.A. Johnson. Cambridge Univ. Press, Cambridge, UK.

IPCC. 2007. Climate change 2007: The physical science basis. Contribution of Working Group I to the Fourth Assessment Report of the Intergovernmental Panel on Climate Change, edited by S. Solomon, D. Qin, M. Manning, Z. Chen, M. Marquis, K.B. Averyt, M. Tignor, and H.L. Miller. Cambridge Univ. Press, Cambridge, UK.

Islam, A., L.R. Ahuja, L.A. Garcia, L. Ma, and A.S. Saseendran. 2012b. Modeling the effect of elevated CO_2 and climate change on reference evapotranspiration in the semi-arid Central Great Plains. Trans. ASABE 55:2135–2146. doi:10.13031/2013.42505

Islam, A., L.R. Ahuja, L.A. Garcia, L. Ma, A.S. Saseendran, and T.J. Trout. 2012c. Modeling the impact of climate change on irrigated corn production in the Central Great Plains. Agric. Water Manage. 110:94–108. doi:10.1016/j.agwat.2012.04.004

Islam, A., N.S. Raghuwanshi, and R. Singh. 2008. Development and application of hydraulic simulation model for irrigation canal network. J. Irrig. Drain. Eng. 134:49–59. doi:10.1061/(ASCE)0733-9437(2008)134:1(49)

Islam, A., A.K. Sikka, B. Saha, and A. Singh. 2012a. Streamflow response to climate change in the Brahmani River basin, India. Water Resour. Manage. 26:1409–1424. doi:10.1007/s11269-011-9965-0

Jain, S.K., and V. Kumar. 2012. Trend analysis of rainfall and temperature data for India. Curr. Sci. 102:37–49.

Jha, M., J.G. Arnold, P.W. Gassman, F. Giorgi, and R.R. Gu. 2006. Climate change sensitivity assessment on upper Mississippi River basin streamflows using SWAT. J. Am. Water Resour. Assoc. 42:997–1015. doi:10.1111/j.1752-1688.2006.tb04510.x

Katz, R.W., and B.G. Brown. 1992. Extreme events in a changing climate: Variability is more important than averages. Clim. Change 21:289–302. doi:10.1007/BF00139728

Kergoat, L., S. Lafont, H. Douville, B. Berthelot, G. Dedieu, S. Planton, and J.F. Royer. 2002. Impact of doubled CO_2 on global-scale leaf area index and evapotranspiration: Conflicting stomatal conductance and LAI responses. J. Geophys. Res. 107(D24):4808. doi:10.1029/2001JD001245

Kijne, J.W., R. Barker, and D. Molden, editors. 2003. Water productivity in agriculture: Limits and opportunities for improvement. Comprehensive Assessment of Water Management in Agriculture Series 1. CABI Publ., Wallingford, UK.

Kitoh, A., H. Endo, K. Krishna Kumar, I.F.A. Cavalcanti, P. Goswami, and T. Zhou. 2013. Monsoons in a changing world: A regional perspective in a global context. J. Geophys. Res. 118:3053–3065. doi:10.1002/jgrd.50258

Krishnan, P., D.K. Swain, B.C. Bhaskar, S.K. Nayak, and R.N. Dash. 2007. Impact of elevated CO_2 and temperature on rice yield and methods of adaptation as evaluated by crop simulation studies. Agric. Ecosyst. Environ. 122:233–242. doi:10.1016/j.agee.2007.01.019

Kumar, R., R.D. Singh, and K.D. Sharma. 2005. Water resources of India. Curr. Sci. 89:794–811.

Ladha, J.K., D. Dawe, H. Pathak, A.T. Padre, R.L. Yadav, B. Singh, Y. Singh, Y. Singh, P. Singh, A.L. Kundu, R. Sakal, N. Ram, A.P. Regmi, S.K. Gami, A.L. Bhandari, R. Amin, C.R. Yadav, E.M. Bhattarai, R.K. Gupta, and P.R. Hobbs. 2003. How extensive are yield declines in long-term rice-wheat experiments in Asia? Field Crops Res. 81:159–180. doi:10.1016/S0378-4290(02)00219-8

Lal, M., K.K. Singh, L.S. Rathore, G. Srinivasan, and S.A. Saseendran. 1998. Vulnerability of rice and wheat yields in NW India to future changes in climate. Agric. For. Meteorol. 89:101–114. doi:10.1016/S0168-1923(97)00064-6

Leavesley, G.H., R.W. Lichty, B.M. Troutman, and L.G. Saindon. 1983. Precipitation-Runoff Modeling System: User's manual. Invest. Rep. 83-4238. U.S. Geological Survey, Denver, CO. http://pubs.usgs.gov/wri/1983/4238/report.pdf (accessed 5 Jan. 2014).

Leipprand, A., and D. Gerten. 2006. Global effects of doubled atmospheric CO_2 content on evapotranspiration, soil moisture and runoff under potential natural vegetation. Hydrol. Sci. J. 51:171–185. doi:10.1623/hysj.51.1.171

Mall, R.K., A. Gupta, R. Singh, R.S. Singh, and L.S. Rathore. 2006. Water resources and climate change: An Indian perspective. Curr. Sci. 90:1610–1625.

Mandare, A.B., S.K. Ambast, N.K. Tyagi, and J. Singh. 2008. On-farm water management in saline groundwater area under scarce canal water supply condition in the Northwest India. Agric. Water Manage. 95:516–526. doi:10.1016/j.agwat.2007.12.010

Maurer, E.P. 2007. Uncertainty in hydrologic impacts of climate change in the Sierra Nevada, California, under two emissions scenarios. Clim. Change 82:309–325. doi:10.1007/s10584-006-9180-9

McCabe, G.J., and L.E. Hay. 1995. Hydrological effects of hypothetical climate change in the East River basin, Colorado, USA. Hydrol. Sci. J. 40:303–318. doi:10.1080/02626669509491417

Mearns, L.O., C. Rosenzweig, and R. Goldberg. 1996. The effect of changes in daily and interannual climatic variability on CERES wheat: A sensitivity study. Clim. Change 32:257–292. doi:10.1007/BF00142465

Meenu, R., S. Rehana, and P.P. Mujumdar. 2013. Assessment of hydrologic impacts of climate change in Tunga–Bhadra river basin, India with HEC-HMS and SDSM. Hydrol. Processes 27(11):1572–1589. doi:10.1002/hyp.9220

Mehrotra, R. 1999. Sensitivity of runoff, soil moisture and reservoir design to climate change in central Indian River basins. Clim. Change 42:725–757. doi:10.1023/A:1005496932343

Menon, A., A. Levermann, J. Schewe, J. Lehmann, and K. Frieler. 2013. Consistent increase in Indian monsoon rainfall and its variability across CMIP-5 models. Earth Syst. Dyn. 4:287–300. doi:10.5194/esd-4-287-2013

Minville, M., F. Brissette, and R. Leconte. 2008. Uncertainty of the impact of climate change on the hydrology of a nordic watershed. J. Hydrol. 358:70–83. doi:10.1016/j.jhydrol.2008.05.033

Mirza, M.M.Q. 1997. The runoff sensitivity of the Ganges river basin to climate change and its implications. J. Environ. Hydrol. 5:1–13.

Mishra, A., R. Singh, and N.S. Raghuwanshi. 2005. Development and application of an integrated optimization-simulation model for major irrigation projects. J. Irrig. Drain. Eng. 131:504–513. doi:10.1061/(ASCE)0733-9437(2005)131:6(504)

Mishra, A., R. Singh, N.S. Raghuwanshi, C. Chatterjee, and J. Froebrich. 2013. Spatial variability of climate change impacts on yield of rice and wheat in the Indian Ganga Basin. Sci. Total Environ. 468–469:S132–S138. doi:10.1016/j.scitotenv.2013.05.080

Moriondo, M., C. Giannakopoulos, and M. Bindi. 2011. Climate change impact assessment: The role of climate extremes in crop yield simulation. Clim. Change 104:679–701. doi:10.1007/s10584-010-9871-0

Mujumdar, P.P., and S. Ghosh. 2008. Modeling GCM and scenario uncertainty using a possibilistic approach: Application to the Mahanadi River, India. Water Resour. Res. 44:W06407. doi:10.1029/2007WR006137

Murphy, J. 1999. An evaluation of statistical and dynamical techniques for downscaling local climate. J. Clim. 12:2256–2284. doi:10.1175/1520-0442(1999)012<2256:AEOSAD>2.0.CO;2

Naresh Kumar, S. 2011. Climate change and Indian agriculture: Current understanding on impacts, adaptation, vulnerability and mitigation. J. Plant Biol. 37:1–16.

Naresh Kumar, S., P.K. Aggarwal, S. Rani, S. Jain, R. Saxena, and N. Chauhan. 2011. Impact of climate change on crop productivity in Western Ghats, coastal and northeastern regions of India. Curr. Sci. 101:33–42.

Naresh Kumar, S., P.K. Aggarwal, R. Saxena, S. Rani, S. Jain, and N. Chauhan. 2013. An assessment of regional vulnerability of rice to climate change in India. Clim. Change 118:683–699. doi:10.1007/s10584-013-0698-3

Naresh Kumar, S., P.K. Aggarwal, D.N. Swaroopa Rani, R. Saxena, N. Chauhan, and S. Jain. 2014. Vulnerability of wheat production to climate change in India. Clim. Res. 59:173–187. doi:10.3354/cr01212

Narsimlu, B., A.K. Gosain, and B.R. Chahar. 2013. Assessment of future climate change impacts on water resources of upper Sind River basin, India using SWAT model. Water Resour. Manage. 27:3647–3662. doi:10.1007/s11269-013-0371-7

Nash, I., and P. Gleick. 1991. Sensitivity of streamflow in Colorado basin to climate changes. J. Hydrol. 125:119–146.

NATCOM. 2004. India's initial national communication to the United Nations Framework Convention on Climate Change. Ministry of Environment and Forests, Government of India, New Delhi. http://unfccc.int/resource/docs/natc/indnc1.pdf (accessed 10 Mar. 2014).

NATCOM. 2012. India's second national communication to the United Nations Framework Convention on Climate Change. Ministry of Environment and Forests, Government of India, New Delhi. http://unfccc.int/resource/docs/natc/indnc2.pdf (accessed 10 Mar. 2014).

Planning Commission. 2009. Report of the task force on irrigation. Planning Commission, Government of India, New Delhi. http://planningcommission.gov.in/reports/genrep/rep_irr2112.pdf (accessed 7 Apr. 2014)

Porter, J.R., and M.A. Semenov. 2005. Crop responses to climatic variation. Philos. Trans. R. Soc. Lond. B Biol. Sci. 360:2021–2035. doi:10.1098/rstb.2005.1752

Priya, A., A.K. Nema, and A. Islam. 2014. Effect of climate change and elevated CO_2 on reference evapotranspiration in Varanasi, India—A case study. J. Agrometeorology. 16:44–51.

Ragab, R., and C. Prudhomme. 2002. Climate change and water resources management in arid and semi-arid regions: Prospective and challenges for 21st century. Biosystems Eng. 81:3–34. doi:10.1006/bioe.2001.0013

Raje, D., and P.P. Mujumdar. 2009. A conditional random field based downscaling method for assessment of climate change impact on multisite daily precipitation in the Mahanadi basin. Water Resour. Res. 5:W10404. doi:10.1029/2008WR007487

Raje, D., P. Priya, and R. Krishnan. 2014. Macroscale hydrological modelling approach for study of large scale hydrologic impacts under climate change in Indian river basins. Hydrol. Processes 28:1874–1889. doi:10.1002/hyp.9731

Rajeevan, M., J. Bhate, J.D. Kale, and B. Lal. 2006. A high resolution daily gridded rainfall for the Indian region: Analysis of break and active monsoon spells. Curr. Sci. 91:296–306.

Rajendran, K., S. Sajani, C.B. Jayasankar, and A. Kitoh. 2013. How dependent is climate change projection of Indian summer monsoon rainfall and extreme events on model resolution? Curr. Sci. 104:1409–1418.

Ramesh, K.V., and P. Goswami. 2014. Assessing reliability of regional climate projections: The case of Indian monsoon. Sci. Rep. 4:1071. doi:10.1038/srep04071

Raneesh, K.Y., and G.T. Santosh. 2011. A study on the impact of climate change on streamflow at the watershed scale in the humid tropics. Hydrol. Sci. J. 56:946–965. doi:10.1080/02626667.2011.595371

Rao, B.K., and T.B.S. Rajput. 2009. Decision support system for efficient water management in canal command areas. Curr. Sci. 97:90–98.

Rehana, S., and P.P. Mujumdar. 2011. River water quality response under hypothetical climate change scenarios in Tunga-Bhadra river, India. Hydrol. Processes 25(22):3373–3386 10.1002/hyp.8057. doi:10.1002/hyp.8057

Rehana, S., and P.P. Mujumdar. 2012a. Climate change induced risk in water quality control problems. J. Hydrol. 444–445:63–77. doi:10.1016/j.jhydrol.2012.03.042

Rehana, S., and P.P. Mujumdar. 2012b. Regional impacts of climate change on irrigation water demands. Hydrol. Processes 27(20):2918–2933 10.1002/hyp.9379.

Richardson, C.W. 1981. Stochastic simulation of daily precipitation, temperature and solar radiation. Water Resour. Res. 17:182–190. doi:10.1029/WR017i001p00182

Rosenberg, N.J., M.S. McKenney, and P. Martin. 1989. Evapotranspiration in a greenhouse-warmed world: A review and a simulation. Agric. For. Meteorol. 47:303–320. doi:10.1016/0168-1923(89)90102-0

Roy, P.K., D. Roy, and A. Mazumdar. 2004. An impact assessment of climate change and water resources availability of Damodar river basin. Hydrol. J. 27:53–70.

Rupa Kumar, K., K. Krishnakumar, and G.B. Pant. 1994. Diurnal asymmetry of surface temperature trends over India. Geophys. Res. Lett. 21:677–680. doi:10.1029/94GL00007

Rupa Kumar, K., A.K. Sahai, K.K. Kumar, S.K. Patwardhan, P.K. Mishra, J.V. Revadekar, K. Kamala, and G.B. Pant. 2006. High-resolution climate change scenarios for India for the 21st century. Curr. Sci. 90:334–344.

Saseendran, S.A., K.K. Singh, L.S. Rathore, S.V. Singh, and S.K. Sinha. 2000. Effects of climate change on rice production in the tropical humid climate of Kerala, India. Clim. Change 44:495–514. doi:10.1023/A:1005542414134

Semenov, M.A., and E.A. Barrow. 1997. Use of stochastic weather generator in the development of climate change scenarios. Clim. Change 35:397–414. doi:10.1023/A:1005342632279

Shah, T. 2009. Climate change and groundwater: India's opportunities for mitigation and adaptation. Environ. Res. Lett. 4:035005. doi:10.1088/1748-9326/4/3/035005

Sharma, K.P., C.J. Vörösmarty, and B. Moore. 2000. Sensitivity of the Himalayan hydrology to land-use and climatic changes. Clim. Change 47:117–139. doi:10.1023/A:1005668724203

Sikka, A.K., and D.K. Paul. 2005. Upper catchment management: Opportunities for developing linkages, proposals and partnerships, In: B.R. Sharma, J.S. Samra, C.A. Scott, and S.P. Wani, editors, Watershed management challenges: Improving productivity, resources and livelihoods. International Water Management Institute, Colombo, Sri Lanka. p. 283–292. http://oar.icrisat.org/2467/1/Watershed_management_challenges_improving_productivity__resources_and_livelihoods.pdf (accessed 5 Jan. 2014).

Singh, B., D.S. Gaydon, E. Humphreys, and P.L. Eberbach. 2011a. The effects of mulch and irrigation management on wheat in Punjab, India—Evaluation of the APSIM model. Field Crops Res. 124:1–13. doi:10.1016/j.fcr.2011.04.016

Singh, P., and L. Bengtsson. 2005. Impact of warmer climate on melt and evaporation for the rainfed, snowfed and glacierfed basins in the Himalayan region. J. Hydrol. 300:140–154. doi:10.1016/j.jhydrol.2004.06.005

Singh, P., and N. Kumar. 1997. Impact assessment of climate change on the hydrological response of a snow and glacier melt runoff dominated Himalayan river. J. Hydrol. 193:316–350. doi:10.1016/S0022-1694(96)03142-3

Singh, R., R.K. Jhorar, J.C. van Dam, and R.A. Feddes. 2006. Distributed ecohydrological modeling to evaluate irrigation system performance in Sirsa district, India II: Impact of viable water management scenarios. Agric. Water Manage. 82:253–278. doi:10.1016/j.agwat.2005.07.027

Singh, R.K., R.K. Panda, K.K. Satapathy, and S.V. Ngachan. 2011b. Simulation of runoff and sediment yield from a hilly watershed in the eastern Himalaya, India using the WEPP model. J. Hydrol. 405:261–276. doi:10.1016/j.jhydrol.2011.05.022

Singh, R., J.C. Refssgaard, L. Yde, G.H.J. Ørgensen, and M. Thorsen. 1997. Hydraulic–hydrological simulations of canal-command for irrigation water management. Irrig. Drain. Syst. 11:185–213. doi:10.1023/A:1005775729271

Smout, I.K., and S.D. Gorantiwar. 2005a. Multilevel approach for optimizing land and water resources and irrigation deliveries for tertiary units in large irrigation schemes. I: Method. J. Irrig. Drain. Eng. 131:254–263. doi:10.1061/(ASCE)0733-9437(2005)131:3(254)

Smout, I.K., and S.D. Gorantiwar. 2005b. Performance assessment of irrigation water management of heterogeneous irrigation schemes: 2. A case study. Irrig. Drain. Syst. 19:37–60. doi:10.1007/s10795-005-2971-8

Smout, I.K., and S.D. Gorantiwar. 2006. Productivity and equity of different irrigation schedules under limited water supply. J. Irrig. Drain. Eng. 132:349–358. doi:10.1061/(ASCE)0733-9437(2006)132:4(349)

Sperber, K.R., H. Annamalai, I.-S. Kang, A. Kitoh, A. Moise, A. Turner, B. Wang, and T. Zhou. 2013. The Asian summer monsoon: An intercomparison of CMIP5 vs. CMIP3 simulations of the late 20th century. Clim. Dyn. 41:2711–2744. doi:10.1007/s00382-012-1607-6

Srivastava, A.K., M. Rajeevan, and S.R. Kshirsagar. 2009. Development of a high resolution daily gridded temperature data set (1969–2005) for the Indian Region. Atmos. Sci. Lett. 10:249–254.

Stone, M.C., R.H. Hotchkiss, and L.O. Mearns. 2003. Water yield responses to high and low spatial resolution climate change scenarios in the Missouri River Basin. Geophys. Res. Lett. 30(4):1186. doi:10.1029/2002GL016122

Subash, N., M. Shamim, V.K. Singh, B. Gangwar, B. Singh, D.S. Gaydon, C.H. Roth, P.L. Poulton, and A.K. Sikka. 2014. Applicability of APSIM to capture the effectiveness of irrigation management decisions in rice-based cropping sequence in the Upper-Gangetic Plains of India. Paddy Water Environ. doi:10.1007/s10333-014-0443-1 (in press).

Tripathi, M.P., R.K. Panda, and N.S. Raghuwanshi. 2003. Identification and prioritisation of critical sub-watersheds for soil conservation management using the SWAT model. Biosystems Eng. 85:365–379. doi:10.1016/S1537-5110(03)00066-7

Tripathi, S., V.V. Srinivas, and R.S. Nanjundiah. 2006. Downscaling of precipitation for climate change scenarios: A support vector machine approach. J. Hydrol. 330:621–640. doi:10.1016/j.jhydrol.2006.04.030

Tyagi, N.K., P.K. Joshi, P.K. Aggarwal, and A. Khatri-Chhetri. 2014. Institutions and policies for scaling out climate smart agriculture. CCAFS Workshop Report. CGIAR Research Program on Climate Change, Agriculture and Food Security (CCAFS). Copenhagen, Denmark. https://cgspace.cgiar.org/bitstream/handle/10568/41813/Workshop%20Report%20Colombo%202013.pdf?sequence = 1 (accessed 12 Mar. 2014).

UNFCCC. 2002. Annotated guidelines for the preparation of national adaptation programmes of action. Least Developed Countries Expert Group, Bonn, Germany. http://unfccc.int/resource/docs/publications/annguid_e.pdf (accessed 12 Mar. 2014).

Venkateswarlu, B., S. Kumar, S. Dixit, C.S. Rao, K.D. Kokate, and A.K. Singh. 2012. Demonstration of climate resilient technologies on farmers' fields action plan for 100 vulnerable districts. Central Research Institute for Dryland Agriculture, Hyderabad, India. http://www.nicra-icar.in/nicrarevised/images/Books/Action%20Plan%20100%20Vulnerable%20Districts%20.pdf (accessed 12 July 2014).

Vigaud, N., M. Vrac, and Y. Caballero. 2013. Probabilistic downscaling of GCM scenarios over southern India. Int. J. Climatol. 33:1248–1263. doi:10.1002/joc.3509

Vorosmarty, C.J., P. Green, J. Salisbury, and R. Lammers. 2000. Global water resources: Vulnerability from climate change and population growth. Science 289:284–288. doi:10.1126/science.289.5477.284

Whitfield, P.H., and A.J. Cannon. 2000. Recent variation in climate and hydrology in Canada. Can. Water Resour. J. 25:19–65. doi:10.4296/cwrj2501019

Wilby, R.L., L.E. Hay, W.J. Gutowski, R.W. Arritt, E.S. Takle, Z. Pan, G.H. Leavesley, and M.P. Clark. 2000. Hydrologic responses to dynamically and statistically downscaled climate model output. Geophys. Res. Lett. 27:1199–1202. doi:10.1029/1999GL006078

Wilby, R.L., J. Troni, Y. Biot, L. Tedd, B.C. Hewitson, D.M. Smith, and R.T. Sutton. 2009. A review of climate risk information for adaptation and development planning. Int. J. Climatol. 29:1193–1215. doi:10.1002/joc.1839

Wilby, R.L., and T.M.L. Wigley. 1997. Downscaling general circulation model output: A review of methods and limitations. Prog. Phys. Geogr. 21:530–548. doi:10.1177/030913339702100403

Willows, R.I., and R.K. Connell, editors. 2003. Climate adaptation: Risk, uncertainty and decision making. UKCIP Technical Report. UK Climate Impact Programme (UKCIP), Oxford, UK. http://www.ukcip.org.uk/wordpress/wp-content/PDFs/UKCIP-Risk-framework.pdf (accessed 18 July 2014).

Winz, I., G. Brierley, and S. Trowsdale. 2009. The use of system dynamics simulation in water resources management. Water Resour. Manage. 23:1301–1323. doi:10.1007/s11269-008-9328-7

Wood, A.W., L.R. Leung, V. Sridhar, and D.P. Lettenmaier. 2004. Hydrologic implications of dynamical and statistical approaches to downscaling climate model outputs. Clim. Change 62:189–216. doi:10.1023/B:CLIM.0000013685.99609.9e

Wu, Y., S. Liu, and O.I. Abdul-Aziz. 2012. Hydrological effects of the increased CO_2 and climate change in the Upper Mississippi River Basin using a modified SWAT. Clim. Change 110:977–1003. doi:10.1007/s10584-011-0087-8

Zhang, X.-C. 2007. A comparison of explicit and implicit spatial downscaling of GCM output for soil erosion and crop production assessments. Clim. Change 84:337–363. doi:10.1007/s10584-007-9256-1

Irrigation and Water Management Decisions Involving Sugarcane in Southern Africa

Neil Louis Lecler

Abstract

The advent of computers has opened opportunities for the development and application of representative and integrated algorithms to model natural processes. In this chapter the focus is on the development and application of modeling algorithms related to the irrigation of sugarcane (*Saccharum* spp.) from both a field- and catchment-scale perspective. Following a description of what is a unique integration of modeling algorithms for simulating the water balance and sugarcane yields in relation to different irrigation system types and levels of performance, a number of case study applications are described. These have shown the following: how models can provide valuable information to inform economic or financial analyses of various irrigation and water management options; that long-standing practices or even standards such as simply increasing water applications to account for low irrigation uniformities (uneven application) is hardly justified; how changing irrigation systems and strategies can have some unexpected but valid implications for water availability to downstream users; and how industries dependent on irrigation have benefitted from the use of models to take stock of how their irrigation systems and management strategies are performing, to receive early warning of the likelihood of serious problems developing, and to make proactive improvements. In certain cases the magnitude and timing of irrigation return flows from less "efficient" irrigation systems and strategies can actually work to the benefit of downstream users in terms of their water availability.

Abbreviations: ACRU, Agricultural Catchments Research Unit model; AED, atmospheric evaporative demand; CU, Christiansen's coefficient of uniformity; ERC, estimated recoverable crystal; IEPI, irrigation engineering performance indices; MIPU, mobile irrigation performance unit; NRH, net return per hectare; RNR, relative net return; TAWC, total available water content; ZSAES, Zimbabwe Sugar Association Experiment Station.

N.L. Lecler, Zimbabwe Sugar Association Experiment Station, P/Bag 7006, Trinagle Ngundu Road, Chiredzi, Zimbabwe; also, Bioresources Engineering, University of KwaZulu-Natal, P/Bag X01, Scottsville 3209, South Africa (nlecler@zsaes.org.zw).

Some material used in this chapter was originally presented at conferences hosted by either the South African Sugarcane Technologists' Association or the International Society of Sugarcane Technologists. Where this is the case, the material has been used with permission of both these organizations.

doi:10.2134/advagricsystmodel5.c12

Practical Applications of Agricultural System Models to Optimize the Use of Limited Water
Lajpat R. Ahuja, Liwang Ma, and Robert J. Lascano, Editors
Advances in Agricultural Systems Modeling, Volume 5. Lajpat R. Ahuja, Series Editor

USDA-NRCS

Models are useful for assessing the performance of irrigation systems and management strategies because they can provide insights into how different options are likely to affect crop yields and the beneficial and nonbeneficial components of the field water balance. Models can also help stakeholders understand interactions within a catchment water balance, for example, the impact of irrigation return flows on downstream water availability or how irrigation water availability from a reservoir (dam) or stream meets crop water demands. Knowing these, different ways of applying water with different types of irrigation systems on different areas and crops can be compared. Such comparisons are useful for selecting the best irrigation system or strategy for specific contexts, for example, soil types or climatic regimes or water availability and financial constraints.

Models rely on well-designed experiments for the development and verification of appropriate equations or algorithms to describe natural processes. Once verified, however, an advantage of models over experiments is that models can be used to evaluate a wide range of scenarios relatively efficiently and cost effectively, for example, impacts of wet and dry seasons and/or deep or shallow soils, and to determine the likelihood or probability of certain occurrences, for example, a dam running dry when supplying water to a certain irrigation scheme.

Models can be relatively simple or more complex. For example, evapotranspiration (ET) can be "modelled" in a relatively simple way by multiplying a monthly "crop factor" by a monthly reference evaporation value. Computers enable the use of more complex and representative models of, among others, ET, and the integration of models of different processes and systems, for example, the simulation of hydrological processes in a catchment that impact on the water balance of a dam, which affects and is affected by the withdrawal of irrigation water in relation to the ET of different crops under different irrigation systems and watering strategies. The relatively complex and interdependent calculations would be too tedious and/or time-consuming to do by hand, especially when repeated on a daily basis for, say, a 50-yr simulation period. Thus, the advent and accessibility of computers have facilitated a vast array of relatively complex and integrated modeling algorithms and applications providing unique insights and value to water and agricultural managers and policymakers.

Most sugarcane models that have been used in southern Africa have their roots in the Thompson (1976), Agricultural Catchments Research Unit (ACRU)

(Schulze, 1995b), CANEGRO (Inman-Bamber, 1991), or ZIMsched 2.0 (Lecler, 2000, 2003; 2004) models. Thompson (1976) developed robust relationships between actual ET and tons cane and actual ET and tons sucrose. He derived these relationships using data from Hawaii, Australia, Mauritius, and several locations in South Africa. The CANEGRO, ACRU, and ZIMsched 2.0 models have algorithms to simulate the components of the field water balance in addition to crop yields.

The ACRU model is essentially an agrohydrological model that can be used to simulate the hydrology of an entire catchment. Among other things, the simulation of streamflow, irrigation demand and supply, crop yields, and reservoir (dam) yields are integrated. Sugarcane yield estimates in ACRU are based on simulations of ET and the robust relationships between sugarcane yields and ET developed by Thompson (1976). The CANEGRO model and its derivatives have an emphasis on crop growth and development processes, while with the ZIMsched 2.0 model and its derivatives, the emphasis is more on irrigation aspects, the water balance, and associated impacts on water resources. The simulation of the water balance in the CANEGRO model was derived from concepts used in the Crop Environment Resource Synthesis model (Jones and Kiniry, 1986), while the water balance used in ZIMsched 2.0 was derived from a novel synthesis of concepts described in Allen et al. (1998) and some algorithms used in the ACRU model (Schulze, 1995b).

In southern Africa, these models and their derivatives have seen wide application, including in the following: irrigation scheduling (McGlinchey, 1999; Lecler, 2000; Singels and Smith, 2006), the analyses of various irrigation systems and strategies (Singels et al., 1999, Armitage et al., 2008; Jumman and Lecler, 2010), forecasting crop yields (Lumsden et al., 2000; Bezuidenhout, 2004), benchmarking performance on irrigation schemes (Greaves et al., 2008), assessing sugarcane as a "streamflow reduction activity" (Schulze et al., 2000), and economic analysis of burnt versus green cane harvesting options (Lecler et al., 2009).

Following a brief description of the various algorithms used in the ZIMSched 2.0 model and the associated rationale for them, some case studies involving ZIMsched 2.0 and its derivatives are described in this chapter to show:

- how models can be used to help farmers develop appropriate adaptation strategies if water licensing results in water supply constraints being imposed on them;
- that long-standing practices or even standards such as simply increasing water applications to account for low irrigation uniformities (uneven application) are hardly justified;

- that when there are new irrigation developments and/or opportunities to make changes, models can provide valuable information to inform economic or financial analyses of various options;
- that in the often "charged" environment of integrated water resources management, models can provide information and unique insights for policymakers, catchment planners, and other stakeholders so that they can better understand the impact of irrigation and any proposed changes on water resources availability; and
- how industries dependent on irrigation have benefitted from the use of models to take stock of how their irrigation systems and management strategies are performing, to receive early warning of the likelihood of serious problems developing, and to make proactive improvements.

Example of a Planning Model: ZIMsched 2.0

ZIMsched 2.0 was developed to investigate how water management, different irrigation systems, and in-field measures of irrigation system performance, such as distribution uniformity, impact crop yields, irrigation water requirements, and water balance components (Lecler, 2004). Since ZIMsched 2.0 was used in many of the case study applications described in this chapter and was also a precursor to the ACRUCane model (Moult et al., 2005), a brief description of the algorithms and associated rationale used in ZIMsched 2.0 to simulate components of the water balance and sugarcane yields is provided here. ACRUCane is integrated with a version of the ACRU agrohydrological model (Schulze, 1995b) that has the facility to simulate catchment hydrology. Thus, with ACRUCane a user can investigate impacts of irrigation water supply constraints, for example, as may occur when water is withdrawn from a certain river or dam.

Evapotranspiration

In ZIMsched 2.0, evaporation from the soil and crop (transpiration) are determined separately on the basis of the algorithms developed and described by Allen et al. (1998). Separating these two processes is important because before the development of significant canopy cover, water is predominately lost by evaporation from the soil surface (Allen et al., 1998).

In addition to the amount of canopy cover or shading, evaporation from the soil surface is dependent on the type of irrigation system being used and the scheduling or management strategy. Different types of irrigation systems have different typical wetting frequencies and wet different fractions of the soil surface. Thus, ET for subsurface drip irrigation can be very different to that from overhead or center pivot irrigation, largely due to differences in evaporation from the soil surface. A simplified model of ET such as a single crop coefficient multiplied by a reference evaporation estimate will not distinguish the difference between the ET under drip and the ET under a pivot, which before a full crop canopy are likely to be very different. In terms of modeling evaporation from the soil surface, Lecler (2004) made two changes to the Allen et al. (1998) methodology, namely to allow transpiration to occur from soil moisture stored in the upper soil layer and to account for runoff.

The ZIMsched 2.0 method of determining crop transpiration for sugarcane differs from the Allen et al. (1998) methodology in that crop canopy development and hence transpiration is a function of accumulated thermal time rather than calendar days. Lecler (2004) motivated this change because the rate of canopy development in sugarcane is dependent on temperature and the crop is harvested and regrown (ratooned) throughout a milling season, which covers a wide range of temperature conditions. Crop transpiration is then determined by multiplying the basal crop coefficient (Allen et al., 1998) with the atmospheric evaporative demand (AED). In ZIMsched 2.0, AED was initially represented by U.S. Weather Bureau Class A pan measurements. It has since been refined such that AED can also be represented by a Penman–Monteith short grass reference ET, as defined by Allen et al. (1998).

Transpiration is reduced below its maximum value if soils are too dry relative to AED, according to a relationship derived by Slabbers (1980) whereby if the AED is relatively high, plant stress will start to occur at relatively high soil water contents, whereas when it is cooler and the AED is lower, plant stressing will only start to occur at relatively lower soil water contents. Slabbers's (1980) relationship is very similar to comparable relationships described by Allen et al. (1998) and is discussed in detail in Lecler and Schulze (1995). Algorithms from Allen et al. (1998) are also used in ZIMsched 2.0 to further adjust the threshold soil water content at which stress starts according to soil texture (Lecler, 2004).

When simulated soil water contents are above field capacity or the drained upper limit, transpiration is reduced below potential because of poor root aeration. This is accounted for in ZIMsched 2.0 using an equation from the ACRU model (Schulze, 1995b), which was based on research reported by Dijkhuis and Berliner (1988).

Surface Runoff

Stormflow, or surface runoff, is an important component of the water balance simulated with ZIMsched 2.0 because it impacts on "infiltration" and "effective rainfall" estimates. Runoff is estimated using the Soil Conservation Service (USDA-SCS, 1985) stormflow equation as modified by Schulze (1995a).

For stormflow calculations, the soil water deficit is taken as the difference between water retention at porosity and the simulated actual soil water content just before the rainfall event. This approach inherently accounts for different growth stages of the crop, and being more dynamic, represents a substantial refinement to the more static curve number approach to account for, among other things, antecedent soil water conditions. The approach is discussed in detail in Schulze (1995a).

Deep Percolation

In ZIMsched 2.0, deep percolation of water below the root zone is only initiated if the soil water content is higher than the drained upper limit for the soil. Deep percolation can take place over a number of days during which the plant can extract water, but plant extraction (transpiration) is at a reduced rate because of poor aeration. The amount of drainage and the duration of drainage are dynamic and are dependent on soil texture and drainage characteristics, antecedent soil water, and the magnitude of the rainfall or irrigation event resulting in excessive soil water, as described in detail by Lecler (2004). Thus, when compared with many other water budgeting algorithms, which assume a fixed drainage time, often of only 1 d, the time for the soil to drain to its drained upper limit (i.e., field capacity) in ZIMsched 2.0 is highly variable. This is a very important aspect of the model, because the tendency to oversimplify drainage assumptions and assume drainage to field capacity within a fixed time period, which is often too short, can result in grossly inaccurate water budgets and lead to a snowballing cycle of over-irrigation and poor root aeration, with large differences between the theoretical budget and actual field conditions.

Crop Yield

An estimation of crop yield is important to establish the performance of a certain type of irrigation system and associated management strategy. The algorithm used in ZIMsched 2.0 to estimate yields of sugar as estimated recoverable crystal (ERC) is based on simulating a reference potential ERC yield and then adjusting this reference potential yield according to the timing and magnitude of soil water stress. The reference potential yield estimate is based on a robust relationship between actual ET and tons sucrose that was derived by Thompson (1976). In ZIMsched 2.0, rather than using ET, which could lead to spurious results when simulating irrigation strategies that may have resulted in high levels of evaporation from the soil surface, potential transpiration was used in a modified form of Thompson's (1976) sucrose versus ET relationship developed by Lecler (2004). The reference potential sucrose estimate in ZIMsched 2.0 is adjusted according to the timing and magnitude of any water stress that may have occurred during the main growing season and/or ripening period using procedures based on research by de Jager (1994) and Doorenbos and Kassam (1979). Details of the procedures are given by Lecler (2004).

Irrigation Systems Performance

ZIMsched 2.0 incorporates irrigation system performance parameters (i.e., either the statistical uniformity, SU, Christiansen's coefficient of uniformity, CU, or the low quarter distribution uniformity, DU_{lq}, index) to estimate the impact of irrigation system performance on crop yield, that is, the impact of nonuniform irrigation applications (Lecler, 2004). This is achieved by simulating three different water budgets on three equally sized portions of the irrigated field and assuming the variations in irrigation applications are normally distributed. At each irrigation event, each of the three areas receives a different depth of irrigation dependent on the specified performance of the simulated irrigation system. One-third of the irrigated field receives the mean application, one-third receives the mean application less a certain deviation percentage, and the remainder of the field receives the mean application plus a certain deviation percentage. This approach and options to simulate a greater number of field portions and different distributions of applied water other than the normal distribution were evaluated and reported to be sound by Ascough and Lecler (2004).

Thus, in addition to having a comprehensive and representative water budget, differences in soil water evaporation between different types of irrigation systems are represented and the actual performance of those irrigation systems in terms of application uniformity is also represented by utilizing a multiple water balance approach. ZIMsched 2.0 is therefore capable of simulating the following processes:

- evaporation from the soil surface and transpiration in relation to
  - atmospheric evaporative demand,
  - available soil water, including excess and/or deficient conditions,
  - crop and rooting characteristics, the development of which are related to temperature and thermal time,
  - irrigation system type, for example, subsurface drip irrigation versus overhead sprinkler irrigation,
  - uniformity of irrigation water applications;
- stormflow (surface runoff); and
- deep percolation.

All of these processes relate to rainfall effectiveness and sucrose yields.

Model Verification

To verify ZIMsched 2.0, Lecler (2004) used data from two irrigation trials undertaken by researchers at the Zimbabwe Sugar Association Experiment Station (ZSAES). These trials, which were undertaken between 1966 and 1972 and 1986 and 1991, provided an ideal range of soil, climate, and irrigation watering or scheduling regimes with wide variations in corresponding crop yields.

A scatter plot of observed and simulated yields of ERC, shown in relative terms (i.e., as fractions of the yield obtained by the non-water-stressed treatment for each of the two trials, namely Trial 4200/1 and Trial 4200/12) is presented in Fig. 12–1. The associated statistics of model performance are given in Table 12–1.

On the basis of the statistics of model performance shown in Table 12–1, the scatter plot shown in Fig. 12–1, and the widely accepted algorithms on which it is based, for example, from Allen et al. (1998), Lecler (2004) reported that the ZIMsched 2.0 model was considered sufficiently accurate for the evaluation of irrigation and water management systems, especially the comparison of one system to another in relative terms.

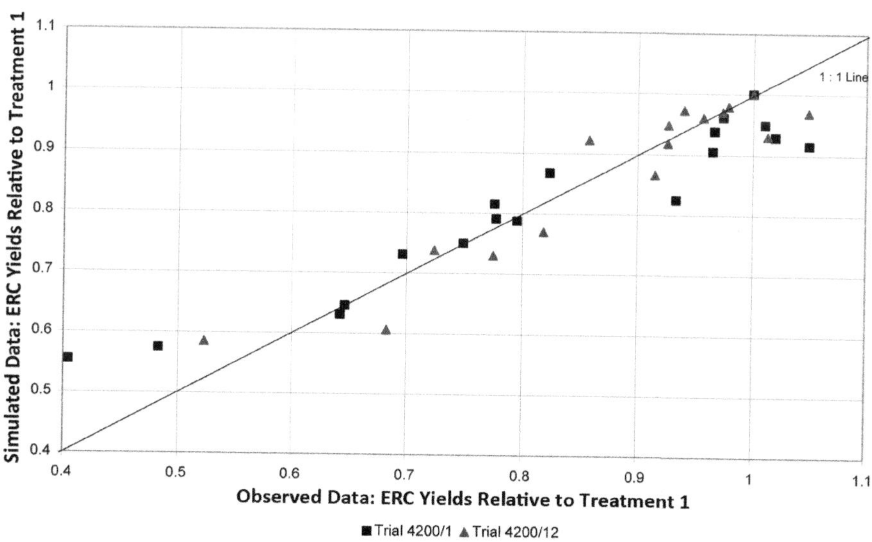

Fig. 12–1. Scatter plot of observed and simulated estimated recoverable crystal (ERC) yields shown in relative terms, that is, as a fraction of the yield obtained by the non-water-stressed treatment, Treatment 1, for each of two irrigation trials, where different crop factors were used to determine irrigation intervals, namely Trial 4200/1 (undertaken between 1966 and 1972) and Trial 4200/12 (undertaken between 1986 and 1991), conducted at the Zimbabwe Sugar Association Experiment Station on sugarcane variety NCo376 (after Lecler, 2004).

Table 12–1. Quantitative performance measures for comparing simulated estimated recoverable crystal (ERC) yields and observed ERC yields in relative terms, that is, as fractions of the yield obtained by the non-water-stressed treatment, for Trials 4200/1 and 4200/12 combined (Lecler, 2004).†

O_{mean}	S_{mean}	N	a	b	RMSE	$RMSE_u$	$RMSE_s$	d	r
0.85	0.84	30	0.16	0.80	0.056	0.043	0.036	0.96	0.94

† Terms *N*, *b*, *d*, and *r* are dimensionless, while remaining terms represent relative yields in terms of a fraction. O_{mean}, mean of observed relative yields of ERC; S_{mean}, mean of simulated relative yields of ERC; RMSE, root mean squared error; $RMSE_u$ and $RMSE_s$, root mean squared errors, unsystematic and systematic, respectively; *d*, index of agreement, in which 1.0 indicates perfect agreement and 0.0 indicates no agreement (Willmott, 1981); *r*, correlation coefficient (Pearson's *r*); *N*, number of data points; *a* (intercept) and *b* (slope) of a least squares regression between predicted relative yields as the dependent variable and observed relative yields as the independent variable.

WATER MANAGEMENT CASE STUDIES

To illustrate the potential of models to inform water management decisions, a number of example case studies are discussed here. These include the development of crop production functions for drip and sprinkler irrigated sugarcane, the application of models to inform deficit irrigation design decisions, and the use of models to assess what impacts different irrigation systems and management would have on irrigation return flows and water availability to downstream users in a catchment context.

Irrigated Sugarcane Production Functions and Water Use Licenses

With the ongoing implementation of the 1998 National Water Act in South Africa (NWA, 1998) and the associated initiatives to potentially reallocate water, the irrigation water requirements of sugarcane and other crops were coming under increasing scrutiny. To provide some perspective on the issue, irrigated sugarcane production functions, which relate crop yield response to the gross amount of irrigation water exiting sprinkler nozzles or drip emitters, were developed by Lecler and Jumman (2009) using ZIMsched 2.0. The production functions show the response of sugarcane to various levels of irrigation water applied and were for either well or poorly maintained drip or sprinkler irrigation systems, represented by different irrigation uniformities (i.e., CU) and for relatively deep and shallow soils. To develop such information using experiments would have been relatively challenging, time-consuming, and expensive. The results of this and similar exercises were presented to farmers and other stakeholders at Catchment Management Forums and/or meetings and conferences.

Methodology

Fourteen years (1985–1999) of daily climate data from the Komatipoort (Tenbosh) weather station (25°22′ S, 31°55′ E) were used for the study. A sandy clay loam soil with a depth of 0.6 m and total available water content (TAWC) of 57 mm was assumed for the shallow soil scenario. A 1.2-m deep profile with a TAWC of 114 mm was assumed for the deep soil scenario.

The assumptions or "treatments" used to simulate production functions associated with sprinkler irrigation were as follows:

- 42 mm of water applied per irrigation application (gross);
- 10% nonbeneficial spray evaporation and wind drift loss (after McNaughton [1981], Tolk et al. [1995], and Thompson et al. [1997]);
- start irrigation application when 20, 30, 50, and 65% of the available soil water (TAWC) has been depleted;
- only apply irrigation if the minimum cycle time had elapsed—5, 10, 15, 20, and 25 d were used for the minimum cycle times; and
- different irrigation uniformities:
  - CU = 60% to represent poor design and maintenance, and
  - CU = 80% to represent good design and maintenance.

The assumptions or "treatments" used to simulate production functions associated with drip irrigation were as follows:

- systems capable of applying 2, 3, 4, 6, 8, and 10 mm per day;
- no nonbeneficial spray evaporation and wind drift loss;
- start application when 40% of the available soil water (TAWC) has been depleted;
- apply only if minimum cycle time elapsed—1 d was used for the minimum cycle time; and
- different irrigation uniformities:
  - CU = 70% to represent poor design and maintenance, and
  - CU = 90% to represent good design and maintenance.

Results

The results reported reflect the sugarcane response to the gross amounts of irrigation water exiting the sprinkler nozzles or drip emitters. Any conveyance losses within a field or between the field and the irrigation water source are not accounted for. The simulated relationships between seasonal water applications (excluding rain) and predicted sucrose yields, expressed as a percentage of the maximum potential sucrose yield, are shown in Fig. 12–2 for sprinkler irrigation systems on deep and shallow soils and in Fig. 12–3 for drip irrigation systems on deep and shallow soils.

There were substantial differences in sucrose yields simulated for different irrigation uniformities, soil depths, and types of irrigation systems for the same water application amount. For instance, in terms of soil depth, a seasonal water application of 1100 mm on the shallow soil resulted in 90 and 80% of the maxi-

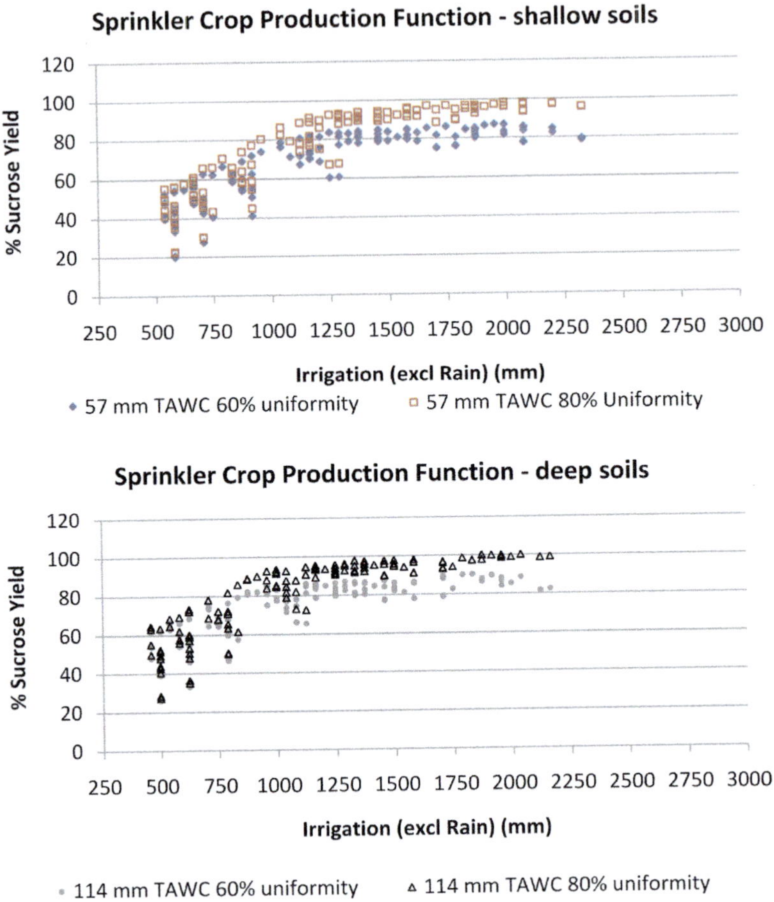

Fig. 12–2. Simulated sugarcane crop production functions for Komatipoort. The information shown is for sprinkler irrigation on soils with a total available water content (TAWC) of 57 and 114 mm and for irrigation uniformities represented by a Christiansen's coefficient of uniformity of 60 and 80%.

mum sucrose yield simulated for well and poorly maintained sprinkler irrigation systems, respectively. On a deep soil, the corresponding sucrose yield was 98 and 85% of the maximum yield simulated, that is, the deep soil was less sensitive to negative impacts of poor irrigation uniformities. In both soils, and for both types of irrigation system, a well-maintained system, represented by a more uniform application of water (i.e., a CU of 80 or 90%), resulted in substantial gains in crop yield relative to the system with a CU of 60 or 70%.

Increasing the amount of irrigation water applied to compensate for nonuniform water application is a common practice or even recommendation (Burt et al., 1997). The results shown in Fig. 12–2 and Fig. 12–3 illustrate that this practice

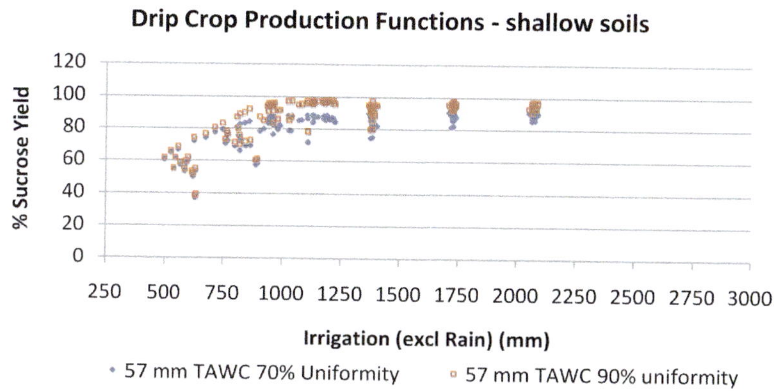

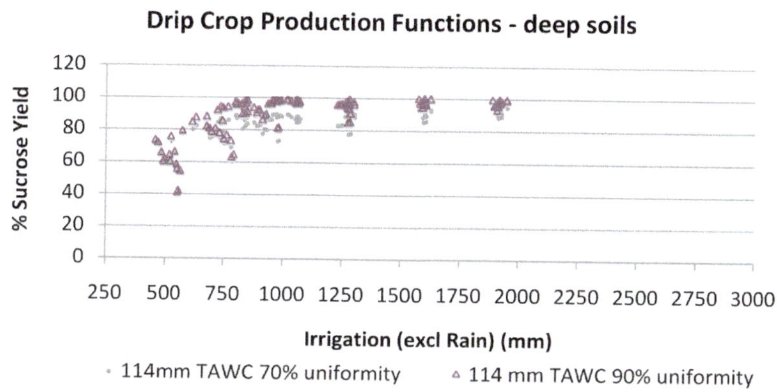

Fig. 12–3. Simulated sugarcane crop production functions for Komatipoort. The information shown is for drip irrigation on soils with a total available water content (TAWC) of 57 and 114 mm and for irrigation uniformities represented by a Christiansen's coefficient of uniformity of 70 and a 90%.

resulted in only small increments of yield gain for a large amount of additional irrigation water. Furthermore, no matter how much additional water was applied, the yield potential of the irrigation systems with the higher CUs (i.e., more uniform water applications) was never attained by the systems with the lower CUs, no matter how much water was applied. It is likely that with poor uniformities, portions of a field that are receiving relatively low amounts of water and that benefit from increasing water applications are offset by yield losses on other parts of the field that then receive excessive water and suffer because of increasingly anaerobic soil conditions. Although not simulated in this study, the areas receiving excess water would also likely suffer from nutrient losses.

Simulations showed that the optimum amount of water to apply for the different systems and soils varied as follows:

- shallow soil, sprinkler irrigation—1250 mm;
- deep soil, sprinkler irrigation—1000 mm;
- shallow soil, drip irrigation—1000 mm;
- deep soil, drip irrigation—800 mm.

For the shallow soil scenarios with sprinkler irrigation, sucrose yields declined rapidly when irrigation water applications were reduced to below approximately 1150 mm. For the deep soil scenarios, the corresponding irrigation threshold was approximately 900 mm. This illustrates that crop response to water applications cannot be generalized and differences in crop response to specific environments should be considered in assessing potential economic impacts of water allocations. Corresponding thresholds for drip irrigation were close to 900 mm for a shallow soil and 750 mm for a deep soil.

The variation in the sucrose yield response to applied water increased when irrigation water applications were reduced. This highlights the increasing risk of economic losses with lower water applications when drought conditions occur.

Strategic Design and Water Management Decisions: A Case Study on Center Pivots

The maximum water delivery capacity of an irrigation system depends on limitations of hardware, which can have considerable cost, flexibility, and efficiency implications. For example, to provide a high water delivery capacity, larger pumps, motors, pump houses, mainlines, submains, laterals, and/or canals are required. Therefore, designing or specifying irrigation systems with excess capacity can result in substantial cost implications. On the other hand, if the peak irrigation system capacity is too small, excessive crop yield losses may occur as the irrigation system may not have the capacity to supply sufficient water to compensate for ET losses and alleviate potential crop water stress. The development of a methodology to enable the selection of more optimal peak irrigation system capacities and associated operating strategies, for example, an appropriate deficit irrigation strategy, would be valuable. The use of the ZIMsched 2.0 irrigation systems simulation model (Lecler, 2003, 2004) for such an application was described by Lecler et al. (2005).

Methodology

Crop yields and associated seasonal irrigation water applications associated with different peak irrigation system capacities were simulated using daily climate data recorded at the ZSAES during the period from 1975 to 1992. The simulations were based on the following assumptions:

- irrigation system capacity limitations equivalent to 4.3, 5, 6, 7.5, and 10 mm/d;
- total available water (TAWC) equivalent to 95 mm;
- gross irrigation water applications of 30 mm were scheduled to be applied once 38 mm of soil water had been depleted, provided system capacity constraints were not exceeded;
- a CU of 90%; and
- spray evaporation and wind drift losses of 10%.

The economic implications of the different irrigation strategies were estimated by calculating a net return per hectare (NRH) and relative net return (RNR) using the simulated crop yield and water use information together with reasonable production cost and revenue assumptions. Equations [1] and [2] were used respectively to calculate the NRH and the RNR. The RNR reflects the opportunity cost of water by multiplying the NRH by a relative production area that could be achieved using a certain fixed volume of water, for example, if Strategy 1 used 1000 mm of water and Strategy 2 used 500 mm of water, 2 ha could be irrigated using Strategy 2 for every 1 ha irrigated using Strategy 1.

$$\mathrm{NRH} = (\mathrm{ERC_y} \times \mathrm{ERC_p}) - \mathrm{BPC} - (I_a \times \mathrm{W_c}) - (I_a \times \mathrm{E_c}) - (\mathrm{ERC_y} \times 100/12 \times \mathrm{HnH_c}) - \mathrm{FII_c} - \mathrm{FID_c} \qquad [1]$$

$$\mathrm{RNR} = \mathrm{NRH} \times \frac{\mathrm{MW_{ass}}}{\mathrm{WU_{gss}}} \qquad [2]$$

where NRH is net return per hectare ($/ha); RNR is relative net return ($) (i.e., return of one system relative to another assuming water is limited); $\mathrm{ERC_y}$ is estimated recoverable crystal yield (t/ha)—typically 12% of total sugarcane yield; $\mathrm{ERC_p}$ is price of estimated recoverable crystal ($); BPC is base production costs ($/ha); I_a is irrigation water applied (mm/ha); W_c is cost of irrigation water ($/mm); E_c is cost of energy for the irrigation water ($/mm); $\mathrm{HnH_c}$ is harvest and haulage costs ($/t sugarcane); $\mathrm{FII_c}$ is fixed irrigation interest costs ($/ha); $\mathrm{FID_c}$ is fixed irrigation depreciation costs ($/ha); $\mathrm{MW_{ass}}$ is maximum amount of water used considering all systems and seasons (mm); and $\mathrm{WU_{gss}}$ is water used for the given system and season (mm).

The estimated interest and depreciation charges for center pivot irrigation systems with different peak capacity limitations included variations due to mainline pipe sizes, pump houses, and pumps, but assumed in-field costs of the center pivot infrastructure were the same. Other cost and revenue information for water, electricity, harvesting and haulage, base production costs, and the price of ERC were assumed at representative rates consistent for all systems.

Results

Simulated crop yields and irrigation water applications for center pivots with different peak water design capacities are shown in Fig. 12–4.

Analysis of the results showed that

- the peak irrigation system capacity had a considerable effect on the total amount of irrigation water applied in a season;
- the potential yield that could be obtained was dependent on the system capacity limitations;
- the marginal yield benefits decreased as the system capacities increased;
- increasing the system capacity beyond a certain limit resulted in minimal further crop yield benefits, illustrated by a comparison of the water used and yields attained for systems with capacity limitations of 10 and 7.5 mm/d, which were very similar (see Fig. 12–4);
- correct irrigation scheduling is important, as indicated in the high range of seasonal irrigation water applications simulated for a given system capacity, in the different seasons (see Fig. 12–4); and
- correct irrigation scheduling becomes more and more important with increasing system capacity as the potential range in water applied becomes larger and the potential for wastage greater (see Fig. 12–4).

Performance of Irrigation Systems and Water Use of Sugarcane: A Field- to Catchment-Scale Perspective

In catchment water resources studies, there is often a mistaken perception that much water can be released for competing "downstream" water users by improving the efficiency of irrigation or upgrading to "more efficient" irrigation systems. However, one of the issues with irrigation efficiency is that it is not well understood, especially in relation to the availability of water in the context of catchment-scale water resources assessments. Unlike more traditional definitions of efficiency, for example, in thermodynamics, water that is not used efficiently in irrigation is not necessarily lost to the system or "consumed," but often becomes available for downstream users, provided it is of adequate quality. This contributes to a situation where there is considerable misunderstanding and misconception regarding the performance of irrigation systems, the meaning of irrigation efficiency, and associated implications for water availability to competing users. Part of this lack of understanding could be addressed if the fates of water applied at the field, farm, and watershed or catchment scales were better determined and understood.

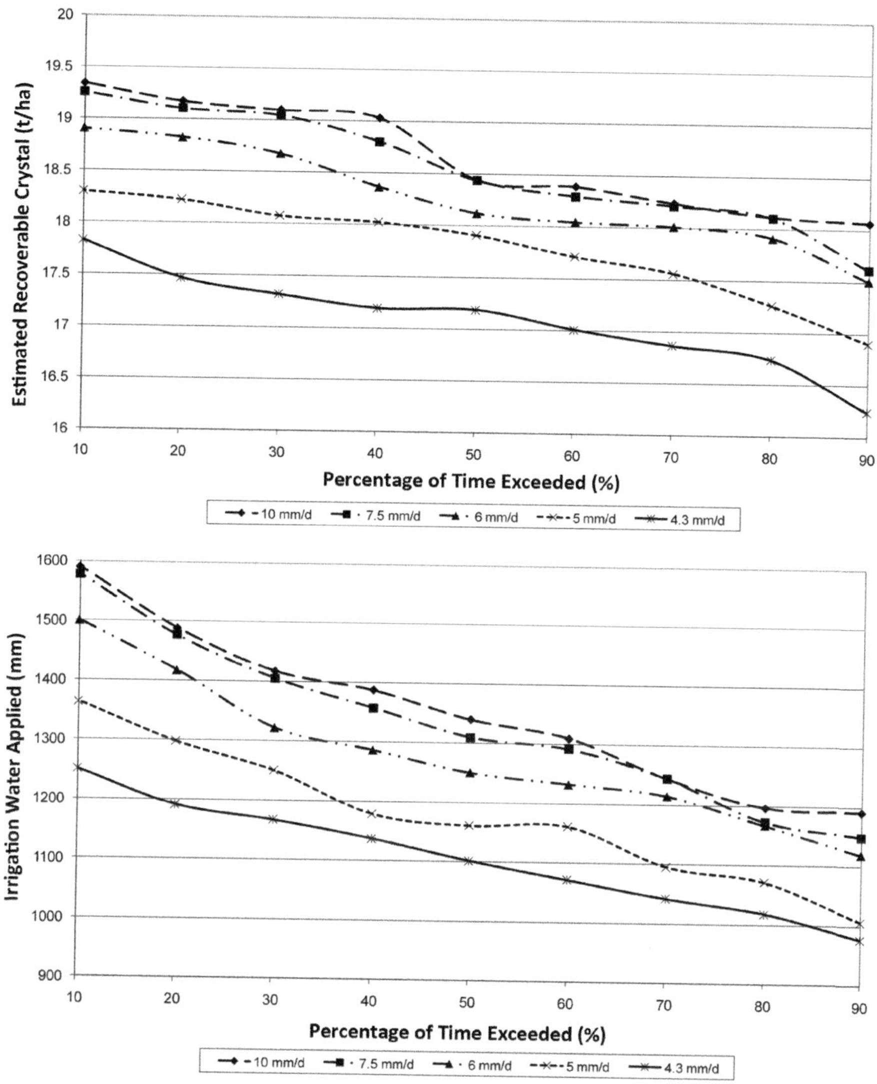

Fig. 12–4. Simulated crop yields of estimated recoverable crystal and applied irrigation water for center pivots with different peak design system capacities.

This is one of the reasons ACRUCane, a specialist sugarcane crop yield and irrigation systems simulation module, was developed (Moult et al., 2005, Moult, 2006). ACRUCane was derived from ZIMsched 2.0 and operates as a module within the ACRU agrohydrological systems model (Schulze, 1995b). It has the capacity to account for the impact of different soils, seasonal weather, irrigation systems, and water management strategies on the water balance and associated crop yields of irrigated sugarcane, in an integrated catchment context (Moult et al., 2005).

In a case study reported by Lecler et al. (2007), the results from the application of ACRUCane to predict water balances and sugarcane yields associated with poor and top performing furrow and center pivot irrigation systems were discussed.

In all the scenarios simulated in the case study, the associated impacts on water availability to competing downstream users were predicted. Field water balances were reported rather than specific irrigation performance indices such as a defined "irrigation efficiency" to highlight the value of showing components of the water balance rather than reporting on various and often misinterpreted irrigation performance indices such as "irrigation efficiency."

Methodology

The impacts associated with water management were represented through the simulation of various irrigation scheduling options and irrigation performance indices in ACRUCane. In simulations the irrigation requirement was determined for a specified scheduling option, but the amount of water that was actually applied to the crop was limited by water availability from the water supply source, as simulated with the ACRU model.

A variety of water supply options can be simulated by ACRUCane through the ACRU model. These options are illustrated in Fig. 12–5. The user can thus quantify the impact of different water supply options and constraints on the water balance and ultimately the yield of an irrigated sugarcane crop.

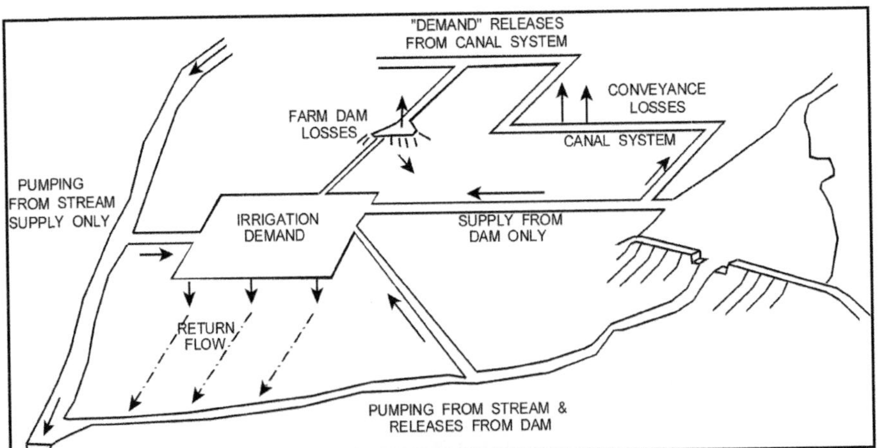

Fig. 12–5. Schematic diagram of supply options available in the ACRU model (Lecler et al., 1995).

The ACRUCane model was configured to represent a catchment with runoff feeding into a dam that supplied water for a sugarcane irrigation scheme such that return flows from the irrigation scheme would contribute inflow to the water balance of the dam. The dam was also required to release a fixed amount of water to competing water users.

The different irrigation scenarios were simulated by making adjustments to model input parameters and variables, such as irrigation system uniformity, irrigation system type, and scheduling. The key differences between the various irrigation scenarios are summarized in Table 12–2. Characteristics of the different irrigation systems, such as the uniformity of applied water, wetted area, and application amounts were derived from in-field evaluations undertaken by a mobile irrigation evaluation laboratory in Zimbabwe (Griffiths and Lecler, 2001; Lecler, 2004).

For the furrow irrigation systems, conveyance losses of 20% were assumed, and for the center pivot irrigation systems spray evaporation and wind drift losses were assumed to be 10% following work reported by McNaughton (1981), Tolk et al. (1995), and Thompson et al. (1997).

Results

Lecler et al. (2007) showed results from a frequency analysis of the water balance components for the irrigated area, the associated dam levels, and the water releases from the dam for other competing users. In the frequency analysis, the

Table 12–2. Irrigation hardware and management characteristics assumed for various scenarios (Lecler et al., 2007, reprinted with permission of ISSCT).

Irrigation system	DU_{lq}†	Irrigation application amount	Summer cycle	Winter cycle	Rain threshold to halt cycle
			——— d	———	mm
Typical furrow	49	70% of total available water content (i.e., equal to 66 mm)	10	20	15 stop for (rain amount/6) days, or maximum 1 cycle period
Top performing furrow	75	apply 45 mm when soil water depleted to 45 mm below DUL	–	–	–
Typical pivot	86	12 mm	2	4	15 stop for (rain amount/6) days, or maximum 1 cycle period
Top performing pivot	86	apply 25 mm when soil water depleted to 40 mm below DUL	–	–	–

† $DU_{lq} = (D_{zq}/D_z) \times 100$, where DU_{lq} is low quarter distribution uniformity, D_{zq} is average depth of water infiltrated in the quarter of the area that received the least infiltrated depth (mm), D_z is average depth of water infiltrated (mm); TAWC, total available water content (mm); DUL, drained upper limit (i.e., after drainage due to gravity has effectively ceased) (mm).

P-th percentile divides the values such that there is a percentage *P* of the numbers below the *P*-th percentile and a percentage 100-*P* above the *P*-th percentile (McPherson, 1990, p. 21–25).

As shown in Fig. 12–6, both the typical furrow and typical pivot scenarios showed relatively high deep percolation as a result of excessive amounts of water applied per irrigation application relative to the soil's TAWC, the low irrigation uniformity, and the fixed summer and winter irrigation cycle times (see Table 12–2). The reduction in deep percolation resulting from improving the irrigation scheduling and uniformity of both types of irrigation system indicated that deep percolation was more sensitive to the design and management of the irrigation system rather than the type of system (Lecler et al., 2007).

In all scenarios simulated runoff was less than simulated deep percolation, and most runoff came from the top performing pivot scenario (see Fig. 12–7). Pivots characteristically have high application rates, especially toward the outer towers, which cover a relatively large field area relatively quickly. Thus, when the amount of water applied per application is higher (i.e., 25 mm for the top performing pivot versus 12 mm for the typical pivot), chances of higher amounts of runoff increase (Lecler et al., 2007).

Evaporation losses from the soil surface were higher for the pivot scenarios compared with the furrow scenarios (see Fig. 12–8). With furrow irrigation, only

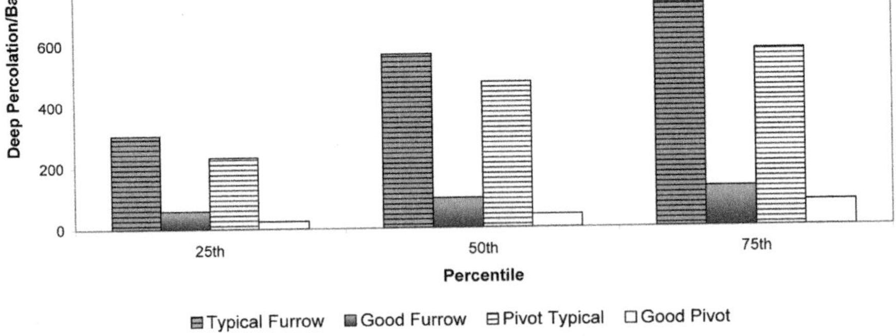

Fig. 12–6. Frequency analysis of deep percolation (leading to baseflow) simulated for the four irrigation scenarios. The *P*-th percentile divides the values such that there is a percentage *P* of the numbers below the *P*-th percentile and a percentage 100-*P* above the *P*-th percentile (after Lecler et al., 2007, reprinted with permission of ISSCT).

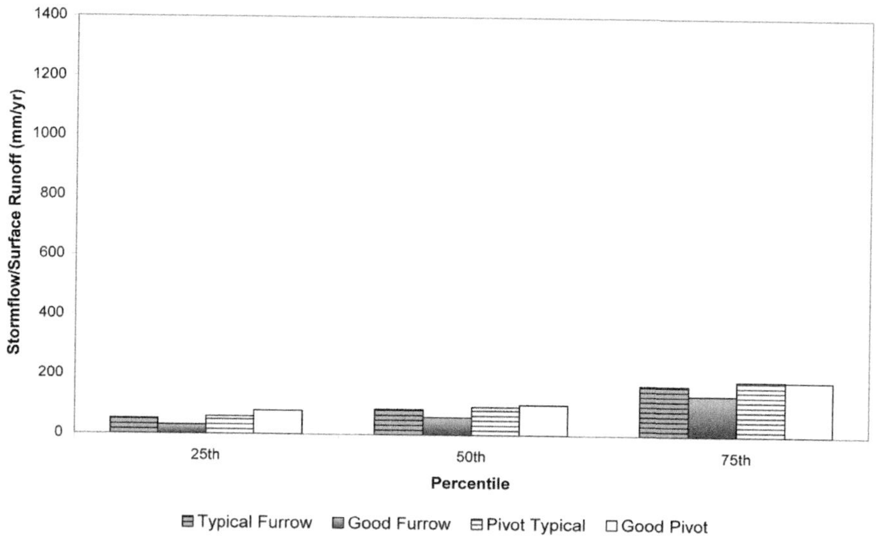

Fig. 12–7. Frequency analysis of stormflow (runoff) for the four irrigation scenarios. The *P*-th percentile divides the values such that there is a percentage *P* of the numbers below the *P*-th percentile and a percentage 100-*P* above the P-th percentile (after Lecler et al., 2007, reprinted with permission of ISSCT).

approximately 60% of the soil surface is wetted during irrigation and watering is less frequent than with pivots. Increasing the pivot application from 12 to 25 mm (see Table 12–2) resulted in reduced evaporation from the soil surface, but further increases in application depth with pivots is likely to result in excessive runoff losses.

Results for simulated transpiration and yield reflect the better uniformity of pivots and the more frequent watering, which reduced potential soil water stress (see Fig. 12–9 and Fig. 12–10). With furrow irrigation, a greater proportion of the field was under-watered or overwatered (poor uniformity) and there were greater water shortages in the dry years relative to the other irrigation systems, which resulted in relatively greater crop yield reductions.

The typical pivot scenario with the highest soil water evaporation losses and frequent water applications used the most water, likely as a result of less effective use of rainfall due to the relatively frequent watering, which also resulted in large amounts of deep percolation and stormflow (see Fig. 12–6 and 12–7). Thus, changing a typical furrow irrigation system to a pivot may not result in lower in-field water usage, unless appropriate management adjustments are also made. The effect of water shortages was greatest for the typical furrow and typical pivot scenarios shown by the relatively large differences in the 25th, 50th, and 75th per-

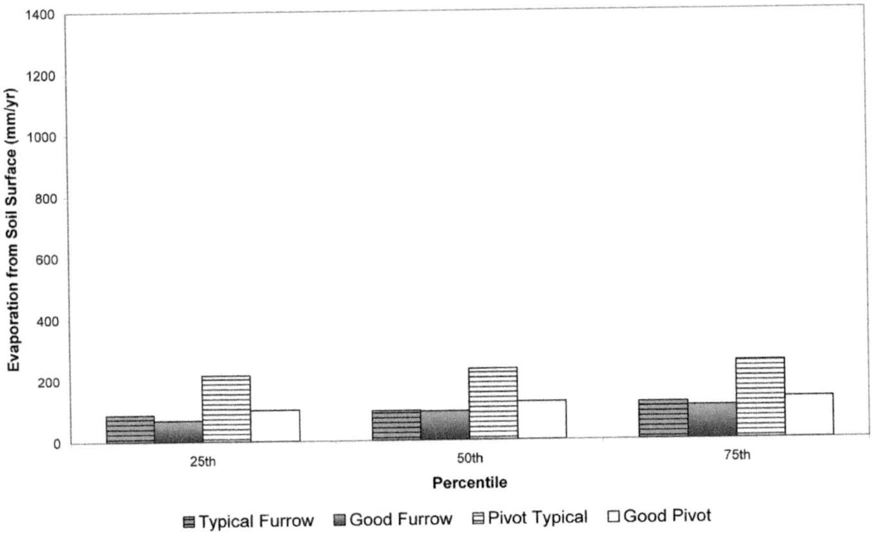

Fig. 12–8. Frequency analysis of evaporation from the soil surface for the four different irrigation scenarios. The *P*-th percentile divides the values such that there is a percentage *P* of the numbers below the *P*-th percentile and a percentage 100-*P* above the *P*-th percentile (after Lecler et al., 2007, reprinted with permission of ISSCT).

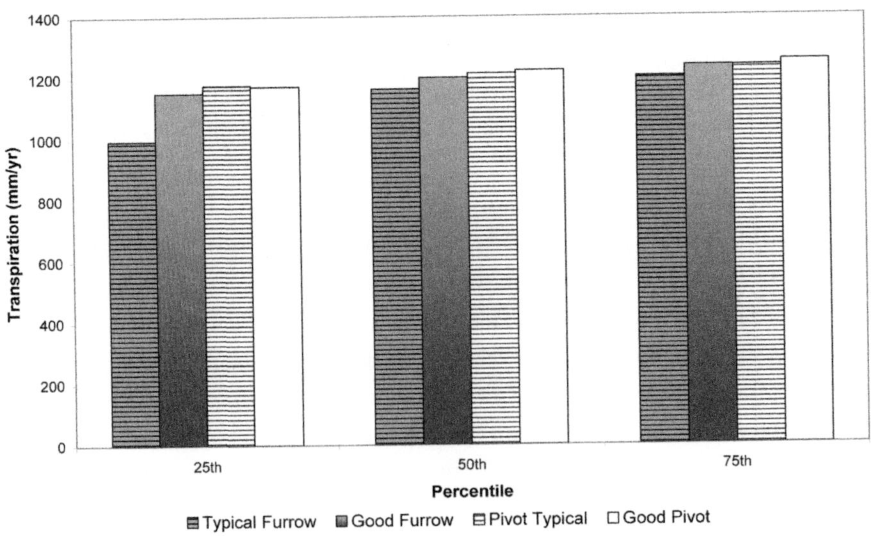

Fig. 12–9. Frequency analysis of transpiration for the four irrigation scenarios. The *P*-th percentile divides the values such that there is a percentage *P* of the numbers below the *P*-th percentile and a percentage 100-*P* above the *P*-th percentile (after Lecler et al., 2007).

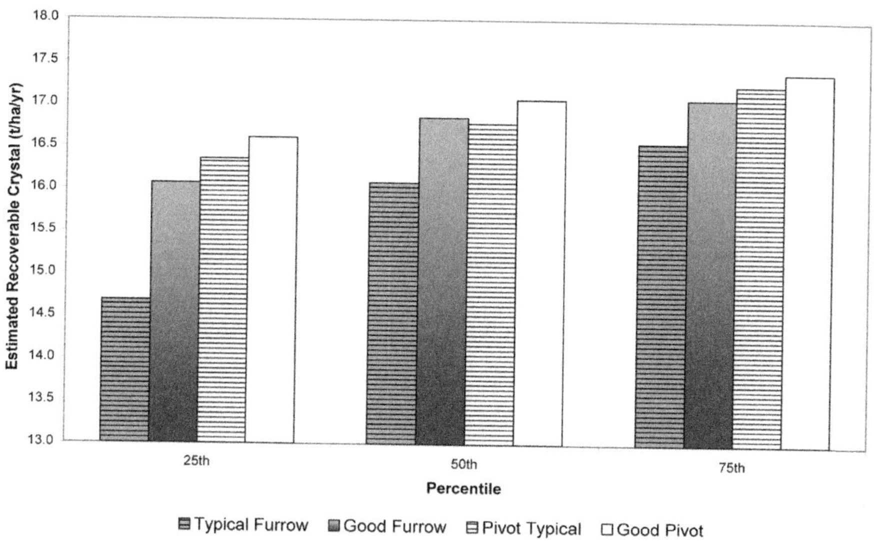

Fig. 12–10. Frequency analysis of estimated recoverable crystal for the four irrigation scenarios. The *P*-th percentile divides the values such that there is a percentage *P* of the numbers below the *P*-th percentile and a percentage 100-*P* above the *P*-th percentile (after Lecler et al., 2007, reprinted with permission of ISSCT).

centile applied water values, indicating that these systems were most impacted by water shortages in the dam (see Fig. 12–11).

Lecler et al. (2007) explained the following:

> Changing from typical furrow irrigation to pivot irrigation did not allow greater water releases to competing downstream users from the dam, even though dam levels were slightly higher with the pivot irrigation systems [see Fig. 12–12 and 12–13]. The reason for this is that with the typical furrow irrigation scenario there were higher return flows in the dryer part of the season and green water flows (i.e., evaporative losses) were less than those from the center pivot systems. Thus, although more water was withdrawn from the dam for the typical furrow system, much of this water returned to the dam later in the season as return flow. These return flows entered the dam during the drier part of the season and enabled water releases to be made. This highlights the importance of assessing performance of different scenarios in a systems context where the various components of the water balance/hydrological cycle are represented.

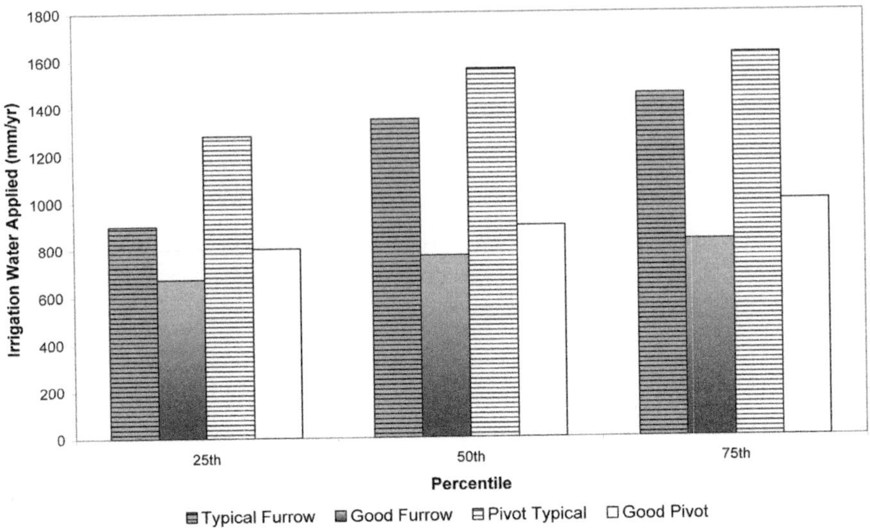

Fig. 12–11. Frequency analysis of applied irrigation water for the four irrigation scenarios. The *P*-th percentile divides the values such that there is a percentage *P* of the numbers below the *P*-th percentile and a percentage 100-*P* above the *P*-th percentile (after Lecler et al., 2007, reprinted with permission of ISSCT).

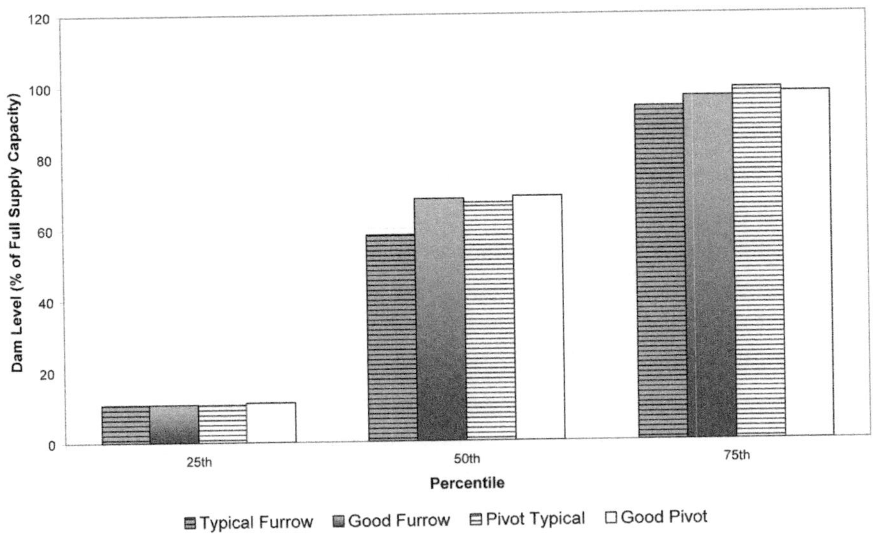

Fig. 12–12. Frequency analysis of dam levels for the four irrigation scenarios. The *P*-th percentile divides the values such that there is a percentage *P* of the numbers below the *P*-th percentile and a percentage 100-*P* above the *P*-th percentile (after Lecler et al., 2007).

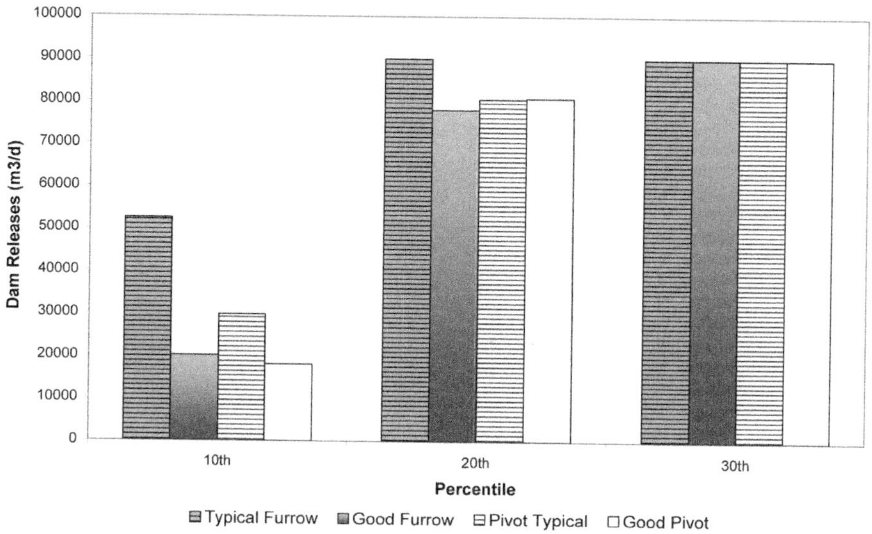

Fig. 12–13. Frequency analysis of dam release to other users for the four irrigation scenarios. The target release was 90,000 m^3/d, and no differences between systems were observed above the 30th percentile. The *P*-th percentile divides the values such that there is a percentage *P* of the numbers below the *P*-th percentile and a percentage 100-*P* above the *P*-th percentile (after Lecler et al., 2007, reprinted with permission of ISSCT).

Evaluation of Irrigation and Water Management Systems in the Lowveld of Zimbabwe

Growers of sugarcane in the Lowveld of Zimbabwe were interested in knowing how the performance of the various irrigation and water management systems compared and which systems were best suited to particular circumstances. The sugar industry had already nearly collapsed as a result of crippling water shortages following the 1991 and 1992 drought (Kaseke, 1998). Therefore, adopting mitigation strategies based on increased insight into the performance of the various irrigation systems and water management approaches was needed to help avert future crises.

Methodology

Lecler and Griffiths (2003) and Lecler (2004) reported on the use of the ZIMsched 2.0 simulation model to translate field-derived "irrigation engineering performance indices" (IEPI), such as the distribution uniformity of water applications, to potential impacts on crop yields and water budgets taking into consideration various watering approaches or scheduling methods prevailing in the Lowveld sugar industry. The field-derived IEPIs were derived using a mobile irrigation performance unit (MIPU) that was initiated during the project.

Adopting an experimental approach to cover a representative range of irrigation systems performing at different levels, over many seasons, and for different soils and water management strategies would have been too time-consuming, costly, and disruptive to be practically feasible (Lecler, 2004).

Results

Lecler (2004, p. 126–129) reported the following.

> Questions regarding the actual and potential performance of the various types of irrigation systems were addressed. In this regard, it was found that most of the floppy, centre pivot and overhead sprinkler irrigation systems were performing at levels close to those expected, had they been designed and installed to appropriate engineering standards. However, most of the drip irrigation systems were performing below potential. Simulations showed that potential crop yields were limited by an average of approximately 12% for one-third of all the drip systems evaluated. This was largely due to in-field variations in water applied, caused by substandard system design, installation and/or maintenance. The crop yield potentials and irrigation efficiencies of the furrow irrigation systems were limited by large variations in water applied to individual furrows, and water applications which were, on average, excessively high. Although the simulations showed that the large water applications did compensate, to a degree, for the variations in applied water between and along the furrows, they also compromised efficiencies and could lead to further development of other problems, including raised water tables and increased soil salinity levels. Also, adequate drainage was assumed for the simulations. If drainage is impaired, yield reductions due to poor soil aeration are likely to occur. Most furrow designs and layouts were not operator friendly. The high variation in water applied to individual furrows showed that it was difficult for operators to control water applications and application variability using siphons. This is an area which could be improved through better, albeit more complex and expensive, designs and installations. For example, well designed supply furrows discharging through pipe spiles could be used to control flows into the furrows more evenly and would be much more operator friendly.
>
> In terms of the potential performance of the various irrigation systems, it was shown that, provided drip irrigation systems are designed, installed and commissioned to an appropriate standard, and then

operated and maintained correctly; subsurface drip has a slight edge over the other irrigation systems which were evaluated. Average water savings ranged from approximately 2.2 to 1.5 ML/ha relative to sprinkler type systems, and 3.5 to 2.3 ML/ha relative to typical furrow irrigation systems, depending on how water applications were scheduled. With subsurface drip, water savings occur as a result of reduced evaporation from the bare soil surface, no spray evaporation or wind drift losses, and the inherent flexibility in applying water which makes it possible to control runoff and deep percolation relatively easily compared with, say, furrow irrigation. However, this study revealed that, in practice, most drip systems in the Lowveld were not performing at potential optimum levels. There needs to be a greater level of professionalism in the design and installation of drip systems and growers need to have appropriate drip irrigation management training.

Furthermore, it was shown that on deep soils with high TAWCs and on fairly level topography, furrow irrigation could, theoretically, be almost as efficient as subsurface drip irrigation. The reason for this is that with furrow irrigation, evaporation from the bare soil surface is limited because only a portion of the soil surface is wetted, there are no spray evaporation or wind drift losses and with proper design and layout, irrigation water applications can be applied uniformly. However, this study revealed that with furrow irrigation, there is a big gap between potential theoretical performance and that generally achieved in practice in the Lowveld conditions, particularly on the sandier soils with low water holding capabilities. With many furrow irrigation systems, feeder water losses constitute an additional, but very variable loss. Water losses in unlined feeders of up to 30% have been measured on an estate in the Lowveld.

Simulations with the ZIMsched 2.0 irrigation system model showed that drainage under furrow irrigation, in particular, was excessive. Unless remedial actions are initiated, for example, through improvements to irrigation scheduling and better control of water applications, the development of high water tables and associated salinity problems is likely to get increasingly worse. In many cases, soils and topography may dictate that changes to another type of irrigation system would improve profitability. In this regard, a unique potential application of ZIMsched 2.0 is to interpret in-field irrigation systems evaluations performed by a Mobile Irrigation Performance Evaluation Unit, MIPU, and then to compare these results, through an economic analysis, by using the simulated crop yields and water budgets predicted when other types of irrigation systems, or an upgrade to the existing system, are assumed.

The centre pivots and overhead hand-moved sprinkler systems which were evaluated by the MIPU were found to apply water with very good uniformities. It was also found that operators were able to control the amount of water applied with these systems relatively easily, compared with furrow irrigation. As a result, simulated efficiencies where higher than with furrow irrigation, generally leading to an annual saving in water of at least 1 ML/ha. Nevertheless, a potential limitation regarding the efficiency of floppy, sprinkler, and centre pivot irrigation systems was highlighted, namely, that of water losses attributable to evaporation from the wet soil surface. The reason for this is that with floppy, sprinkler and centre pivot irrigation systems the entire soil surface is wetted and if water is applied frequently, the evaporation losses from the bare soil surface can be excessively high before the development of full crop canopies. Early season water management approaches should, therefore, take cognisance of the potential for such losses and aim at applying relatively larger water amounts less frequently, i.e., initiating irrigation applications after at least 35 mm of water has been depleted.

Applying such large water applications with overhead sprinkler irrigation systems, particular with centre pivots, can, however, present a challenge in the sense that surface runoff is likely to increase. Therefore, appropriate surface mulching, for example, using the trash and tops from sugarcane should also be considered. When the soil surface is protected from the impacts of water droplets, it is less prone to crusting and higher infiltration rates can be maintained. If the soil surface is mulched, the water budget/schedule should also be adjusted accordingly, otherwise excessive water is likely to be applied in the early season with increased potential for leaching, denitrification and lowering of crop yields. The reason for this is that with surface mulching the water budget is impacted in a positive way through reduced evaporation from the soil surface. Sugarcane varieties which ratoon well under trash conditions should also be targeted, otherwise the potential for delays in germination and canopy development resulting from the trash blanket may offset any potential gains from water savings.

Discussion and Conclusions

The typical practice of simply increasing the amount of irrigation water applied to compensate for irrigation systems with a low CU was shown in the study of production functions to be largely ineffective. Thus, when converting net irrigation water requirements to gross irrigation water requirements, no adjustment for low irrigation uniformities should be made. Low irrigation uniformity should be specifically addressed through better design, evaluation, and maintenance practices. Improving the uniformity of irrigation water applications was shown to have substantial crop yield benefits, particularly on the relatively shallow soils assumed in simulations.

Crop yield response to irrigation cannot be generalized into one production function, even for a specific location. Production functions were shown to be dependent on soils, the type of irrigation system, and the uniformity of a given irrigation system, which reflects design and maintenance standards. Thus, policymakers should be made aware that an allocation of water that may be suitable for a particular farm and soil may result in substantial yield penalties for another farm with a different soil, even where both farms are in the same climatic region.

The application of appropriately representative crop yield and irrigation water balance simulation models such as ZIMsched 2.0 can be extended to investigate crop responses to various types of irrigation systems and operating strategies. The results of such simulations provide the information necessary to determine optimum strategies for making the most effective and efficient use of water and can be used in reallocation scenarios to inform the water licensing process. Water users who have their allocations cut can use the model to investigate alternative systems and strategies to deal with the reduced water.

For example, in a case study on center pivots, the system with the highest capacity of 10 mm/d gave the best yields of ERC; however, from an economic perspective this was not the best system. If availability of land limited production (relative to available water), then the system with a capacity equivalent to 7.5 mm/d was estimated to give the best overall returns. However, if water was the factor that limited production (relative to available land), then the system with a capacity of only 5 mm/d gave the best overall profitability.

The pivot case study highlighted the potential benefits of deficit irrigation, especially under water-limited production conditions, and also the value of using simulation models to inform the design process and the selection of best management practices. Whether land limited production relative to available water or vice versa was a major factor in the selection of an optimal system.

In another case study, some key determinants of the performance of irrigation systems (i.e., a water balance showing the various fates of applied irrigation water) were simulated. This was done to contribute to a better understanding of irrigation performance fundamentals and associated interpretations of irrigation efficiency. As was demonstrated in this study, the amount of water actually consumed in irrigation (i.e., the evaporated component) remains relatively unchanged at various levels of so-called "irrigation efficiency" or can even increase when changing to irrigation systems that are often perceived to be relatively more efficient such as pivots.

With reference to specific components of the water balance, the following conclusions were made by Lecler et al. (2007).

> The uniformity with which water is applied can have a substantial effect on the performance and water balance of irrigation systems. Uniformity is related to the type of irrigation system, but it is also very dependent on the standard to which a given system has been designed, and is operated and maintained. Center pivots typically have higher uniformities than furrow irrigation systems and as a result the crop yield under pivots is expected to be greater than under furrow systems. To compensate for lower uniformities and partly as a result of designs, the amount of water applied per application with most furrow irrigation systems is excessive, probably contributing to the general perception that furrow irrigation is not very efficient. In a systems context much of the excess water applied using typical furrow systems results in increased runoff and deep percolation, which can, however, contribute to downstream water requirements. At the field level the excess water can lead to raised water tables and associated salinity issues. Nevertheless, properly designed and operated furrow systems were shown to use less water than properly designed and operated center pivots, but resulted in slightly lower crop yields.
>
> The amount of irrigation water applied in relation to the available soil water storage capacity is largely determined by management, but it can also be constrained by the type and design of the irrigation system, particularly in the case of furrow irrigation. Management, through the selection and implementation of irrigation watering schedules, or scheduling methodologies, can have greater impacts on overall system performance than changing the type of irrigation system used, i.e., it is the managers who decide when and for how long the pumps should run. This is especially so if the irrigation hardware is well designed and

maintained. If the irrigation hardware is poorly designed and has a low inherent irrigation uniformity, overall performance is likely to be poor, even with appropriate watering schedules and good water management.

The most significant losses from a field are due to evaporation from the exposed soil surface, runoff from stormflow and deep percolation. These losses depend on both the type of irrigation system and its management. Runoff and deep percolation both return to the system as return flows and can potentially be re-used. However, the impact of the return flows on water quality needs to be carefully monitored. In a systems context where the impact of irrigation on competing users is important, changing from furrow irrigation to center pivot irrigation which is often perceived to be more efficient, did not contribute to greater water availability to other users owing to the timing and magnitude of associated return flows. Thus, assessing scenarios in a representative and integrated systems context is essential and can sometimes yield results which are unexpected, but conceptually sound.

When striving to improve irrigation systems performance, deriving representative values for a seasonal water balance can be more useful than reporting on various indices such as 'irrigation efficiency'. It is difficult to misinterpret a water balance, and, furthermore, nonbeneficial water use components which are relatively high can be specifically targeted for reduction. Performance indices such as 'irrigation efficiency' do not show where the water is going and thus lack important information needed for making improvements or changes. For example, in this study it was shown that growers using center pivots should target strategies such as surface mulching using trash blankets to improve performance by reducing surface runoff and evaporation from the soil surface. Potential water quality impacts of return flows were not simulated in this study but are of concern and should be given attention.

The rational quantification of the water balance in relation to the uniformity of applied water, the environment and the water management approach should form the basis of assessing the performance of irrigation and water management systems and should be promoted in place of the often misinterpreted performance indices.

While no model of a natural system is perfect because of the inherent complexity of nature, the development and application of robust sugarcane irrigation models as described here has provided numerous and valuable insights to irrigation and water management challenges. In summary, case study applications have shown the following:

- how models can be used to provide information that could guide farmers in developing appropriate adaptation strategies if water licensing results in water supply constraints being imposed on them, that is, models can provide valuable information to inform economic or financial analyses of various options, including when there are new irrigation developments and/or opportunities to make changes;
- that long-standing practices or even standards such as simply increasing water applications to account for low irrigation uniformities (uneven application) is hardly justified;
- that in the often "charged" environment of integrated water resources management, models can provide information and unique insights for policymakers, catchment planners, and other stakeholders so that they can better understand the impact of irrigation and any proposed changes on water resources availability; and
- how industries dependent on irrigation have benefitted from the use of models to take stock of how their irrigation systems and management strategies are performing, to receive early warning of the likelihood of serious problems developing, and to make proactive improvements.

References

Allen, R.G., L.S. Pereira, D. Raes, and M. Smith. 1998. Crop evapotranspiration: Guidelines for computing crop water requirements. Irrig. Drain. Pap. 56. FAO, Rome.

Armitage, R.M., N.L. Lecler, A. Jumman, and K. Dowe. 2008. Implementation of the Irriecon V2 decision support tool to assess net returns to irrigation systems. Proc. S. Afr. Sugar Technol. Assoc. 81:454–465.

Ascough, G.W., and N.L. Lecler. 2004. Accounting for uniformity of applied irrigation water using multiple water balances in ZIMsched 2.0. Paper presented at: The South African National Committee on Irrigation and Drainage Conference, South Africa. 16–19 Nov. 2004.

Bezuidenhout, C.N. 2004. The predictability of cane production in the South African sugar industry using seasonal climate outlooks and the CANESIM yield forecasting system. Proc. S. Afr. Sugar Technol. Assoc. 78:65–68.

Burt, C., A. Clemmens, T. Strelkoff, K. Solomon, R. Bliesner, L. Hardy, T. Howell, and D. Eisenhauer. 1997. Irrigation performance measures: Efficiency and uniformity. J. Irrig. Drain. Eng. 123(6):423–443. doi:10.1061/(ASCE)0733-9437(1997)123:6(423)

de Jager, J.M. 1994. Accuracy of vegetation evaporation ratio formulae for estimating final wheat yield. Water S.A. 20(4):307–314.

Dijkhuis, F.J., and P.R. Berliner. 1988. A simple model for the prediction of dryland wheat yield. Department of Agriculture and Water Supply, Soil and Irrigation Research Institute, Pretoria, South Africa.

Doorenbos, J., and A.H. Kassam. 1979. Yield response to water. Irrig. Drain. Pap. 33. FAO, Rome.

Greaves, K.R., N.L. Lecler, and J.C. Smithers. 2008. Benchmarking irrigation water use and sugarcane yield in an irrigation scheme. Proc. S. Afr. Sugar Technol. Assoc. 81:466–469.

Griffiths, B.A.K., and N.L. Lecler. 2001. Irrigation system evaluation. Proc. S. Afr. Sugar Technol. Assoc. 75:58–67.

Inman-Bamber, N.G. 1991. A growth model for sugarcane based on a simple carbon balance and the CERES-Maize water balance. S. Afr. J. Plant Soil 8(2):93–99.

Jones, C.A., and J.R. Kiniry. 1986. CERES-Maize: A simulation model of maize growth and development. Texas A&M Univ. Press, College Station.

Jumman, A., and N.L. Lecler. 2010. Deficit irrigation: A strategy to counteract escalating electricity tariffs and water shortages. Proc. S. Afr. Sugar Technol. Assoc. 83:160–173.

Kaseke, E. 1998. Development of drought management strategies for the Runde catchment in Zimbabwe. Unpublished Ph.D. thesis, School of Engineering, James Cook Univ., Townsville, Australia.

Lecler, N.L. 2000. ZIMsched: An irrigation management and yield forecasting tool. Proc. S. Afr. Sugar Technol. Assoc. 74:124–130.

Lecler, N.L. 2003. A model for the evaluation of irrigation and water management systems in the Lowveld of Zimbabwe. I: Model development and verification. Proc. S. Afr. Sugar Technol. Assoc. 77:322–333.

Lecler, N.L. 2004. Performance of irrigation and water management systems in the Lowveld of Zimbabwe. Unpublished Ph.D. thesis, School of Bioresources Engineering and Environmental Hydrology, University of KwaZulu-Natal, Pietermaritzburg, South Africa.

Lecler, N.L., D.J. Clark, R.E. Schulze, B.A.K. Griffiths, and J.C. Smithers. 2007. Performance of irrigation systems and water use of sugarcane: A field to catchment scale perspective. In: M. Hogarth, editor, Proceedings of the 26th Congress of the International Society of Sugar Cane Technologists, Durban, South Africa. 29 July to 2 Aug. 2007. ISSCT, Durban, South Africa. p. 89–102.

Lecler, N.L., and B.A.K. Griffiths. 2003. A model for the evaluation of irrigation and water management systems in the Lowveld of Zimbabwe. II: Model application. Proc. S. Afr. Sugar Technol. Assoc. 77:334–346.

Lecler, N.L., and A. Jumman. 2009. Irrigated sugarcane production functions. Short communication. Proc. S. Afr. Sugar Technol. Assoc. 82:604–607.

Lecler, N.L., F.C. Olivier, and M. Jones. 2009. Economic analysis of harvesting burnt versus green cane under irrigated conditions. Sugarcane Int. 27(6):248–249.

Lecler, N.L., and R.E. Schulze. 1995. Irrigation crop water demand. In: R.E. Schulze, editor, Hydrology and agrohydrology: A text to accompany the ACRU 3.00 Agrohydrology Modelling System. WRC Report TT69/95. Water Research Commission, Pretoria, South Africa. p. 17–1–17–16.

Lecler, N.L., R.E. Schulze, and W.J. George. 1995. Irrigation water supply and return flows. In: R.E. Schulze, editor, Hydrology and agrohydrology: A text to accompany the ACRU 3.00 Agrohydrology Modelling System. WRC Report TT69/95. Water Research Commission, Pretoria, South Africa. p. 18–1–18–8.

Lecler, N.L., M. Zartmann, and R.E. Schulze. 2005. Strategic irrigation design and water management decisions: A case study on centre pivots. Proc. S. Afr. Sugar Technol. Assoc. 79:447–451.

Lumsden, T.G., R.E. Schulze, N.L. Lecler, and E.J. Schmidt. 2000. Assessing the potential for improved sugarcane yield forecasting using seasonal rainfall forecasts and crop yield models. Proc. S. Afr. Sugar Technol. Assoc. 72:131–139.

McGlinchey, M.G. 1999. Computer crop model applications: Developments in Swaziland. Proc. S. Afr. Sugar Technol. Assoc. 73:35–38.

McNaughton, K.G. 1981. Net interception losses during sprinkler irrigation. Agric. Meteorol. 24:11–17. doi:10.1016/0002-1571(81)90030-3

McPherson, G. 1990. Statistics in scientific investigation. Springer-Verlag, New York.

Moult, N.G., N.L. Lecler, J.C. Smithers, C.N. Bezuidenhout, and D.J. Clark. 2005. A catchment scale irrigation systems model for sugarcane. Proc. S. Afr. Sugar Technol. Assoc. 79:452– 455.

Moult, N.G. 2006. The development of a catchment scale irrigation systems model for sugarcane. M.Sc(Eng) diss., School of Bioresources Engineering and Environmental Hydrology, Univ.of KwaZulu-Natal, Pietermaritzburg, South Africa.

NWA. 1998. National Water Act of 1998. RSA Government Gazette No. 36 of 1998: 26 August 1998, No. 19182. Cape Town, South Africa.

Schulze, R.E. 1995a. Streamflow. In: R.E. Schulze, editor, Hydrology and agrohydrology: A text to accompany the ACRU 3.00 Agrohydrology Modelling System. WRC Report TT69/95. Water Research Commission, Pretoria, South Africa. p. 10–1–10–6.

Schulze, R.E., editor. 1995b. Hydrology and agrohydrology: A text to accompany the ACRU 3.00 Agrohydrological Modelling System. WRC Report TT69/65. Water Research Commission, Pretoria, South Africa.

Schulze, R.E., M.J.C. Horan, and E.J. Schmidt. 2000. Hydrological complexities in assigning rainfed sugarcane a "stream flow reduction activity". Proc. S. Afr. Sugar Technol. Assoc. 74:140–150.

Singels, A., C.N. Bezuidenhout, and E.J. Schmidt. 1999. Evaluating strategies for supplementing irrigation of sugarcane in South Africa. Proc. Conf. Aust. Soc. Sugar Cane Technol. 21:1–8.

Singels, A., and M.T. Smith. 2006. Provision of irrigation scheduling advice to small-scale sugarcane farmers using a web based crop model and cellular technology: A South African case study. Irrig. Drain 55:363–372. doi:10.1002/ird.231

Slabbers, P.J. 1980. Practical prediction of actual evapotranspiration. Irrig. Sci. 1:185–196. doi:10.1007/BF00270883

Thompson, A.L., J.M. Martin, J.A. Norman, T.A. Tolk, J.R. Howell, J.R. Gilley, and A.D. Schneider. 1997. Testing of a water loss distribution model for moving sprinkler systems. Trans. ASABE 40(1):81–88. doi:10.13031/2013.21251

Thompson, G.D. 1976. Water use by sugarcane. S. Afr. Sugar J. 60:593–635.

Tolk, J.A., T.A. Howell, J.L. Steiner, D.R. Krieg, and A.D. Schneider. 1995. Role of transpiration suppression by evaporation of intercepted water in improving irrigation efficiency. Irrig. Sci. 16:89–95. doi:10.1007/BF00189165

USDA–SCS. 1985. National engineering handbook, section 4: Hydrology. U.S. Gov. Print. Office, Washington, DC.

Willmott, C.J. 1981. On the validation of models. Phys. Geogr. 2:184–194.

A Modeling Approach to Explore Water Management Strategies for Late-Sown Maize and Double-Cropped Wheat–Maize in the Rainfed Pampas Region of Argentina

Jorge L. Mercau and Maria E. Otegui*

Abstract

Most of Argentine maize (*Zea mays* L.) production takes place under rainfed farming in the Pampas region, where water availability is the main limiting resource. Most crops are still sown early in spring to have the critical period of silking before the frequent mid-summer water deficit. Early sowing dates (September) have been also linked to the need for escaping the increased incidence of insect pests associated with late sowings, a restriction that has been drastically reduced with the introduction of Bt hybrids. Therefore, very late sowing dates (December) became an actual alternative that must be evaluated. We estimated the combined effect of late sowings, variable levels of plant available soil water (PASW) at sowing in single- and double-cropped crops, and variable levels of soil N supply on the interannual variation of maize yield in four sites of the region. For this purpose we used 41 yr of daily, measured weather data and the Crop Environment Resource Synthesis-Maize (CERES-Maize) simulation model. Rainfed yield was very stable across December sowings ($\geq$7.5 Mg ha^{-1}) for latitudes $\leq$36° S, provided PASW (1.8 m) at sowing was 100% and initial N availability reached approximately 130 kg ha^{-1} N. This strategy had almost no yield penalty as compared with potential yield. The response rate to mentioned N availability turned negative for PASW at sowing $\leq$35%. Because of the apparent yield stability across December sowings, farmers should be recommended to survey PASW rather than tie late sowing of maize to scattered rainfall events.

Abbreviations: AAPRESID, Argentine Association of No-Till Farmers; CERES-Maize, Crop Environment Resource Synthesis-Maize; DCM, double-cropped wheat–maize; ENSO, El Niño–Southern Oscillation; LSM, late-sown, single-cropped maize; PASW, plant available soil water; PET, potential evapotranspiration; RM, relative maturity; WUE, water use efficiency.

J.L. Mercau, Grupo de Estudios Ambientales del Instituto de Matemática Aplicada, Universidad Nacional de San Luis (GEA-IMASL) and Consejo Nacional de Investigaciones Científicas y Tecnológicas (CONICET), San Luis, Argentina. jorgemercau@gmail.com

M.E. Otegui, Instituto de Investigaciones Fisiológicas y Ecológicas Vinculadas a la Agricultura (IFEVA-CONICET) and Departamento de Producción Vegetal, Facultad de Agronomía, Universidad de Buenos Aires, Av. San Martin 4453 (C1417DSE), Buenos Aires, Argentina. *Corresponding author (otegui@agro.uba.ar).

doi:10.2134/advagricsystmodel5.c13

Practical Applications of Agricultural System Models to Optimize the Use of Limited Water
Lajpat R. Ahuja, Liwang Ma, and Robert J. Lascano, Editors
Advances in Agricultural Systems Modeling, Volume 5. Lajpat R. Ahuja, Series Editor

USDA-ARS

Since the advent of the second industrial era in the 1880s, agricultural production has been challenged by the steady increase of demand (Evans, 1997). Until the 1960s this increase was satisfied by the expansion of crops into new land, but since then we have depended predominantly on increasing yield (production per unit land area). At the end of the 20th century, however, some studies highlighted the need for intensifying cropping systems to ensure production keeps pace with demand, focusing on the annual yield per unit land area cropped to several crops rather than on the yield of individual crops (Evans, 1997, 1999). International concern about these issues grew during recent years, when gloomy reports provided warnings about the apparent incapacity of global yield gains to fulfill predicted demand trends (Fischer and Edmeades, 2010; Hall and Richards, 2013; Ray et al., 2013). Forecasts on yield gains were less critical for maize than for wheat (*Triticum aestivum* L.) and rice (*Oryza sativa* L.) (Ray et al., 2013), and the Pampas region of Argentina (ca. 30–39° S and 56–65° W) was identified as one of the world areas where maize had a steady and rapid yield increase (Ray et al., 2012). Commercial maize production in Argentina rose from 8.5 ± 1.6 million megagrams in the 1970s to 17.5 ± 3.8 million megagrams in the 2000s, with almost no change in the area cropped to this species (ca. 3 million hectares). The increase was due exclusively to enhanced yields, and gains for this trait were estimated at 120 kg ha^{-1} yr^{-1} (2.8% yr^{-1}) for the 1970 to 2010 period (based on public statistics; SAGPYA, 2013). This progress can be attributed to both genetics and management practices and has been reached without irrigation.

Most of Argentine maize production takes place under rainfed farming in the Pampas region of the country (Fig. 13–1), for which rainfall has a large interannual variability (Messina et al., 1999). Consequently, water availability is usually the most limiting resource to yield of this crop, and a strong association has been detected between maize yield and the occurrence of the extreme phases of the El Niño–Southern Oscillation (ENSO) phenomenon during the 20th century (Podestá et al., 1999). This trend held for the period with steady increase in maize yield (1969–1970 to 2010–2011). Years with below-expected values (30% of the series with a decrease of ≥10% of expected value, based on residuals analysis) usually corresponded (54%) to the cool La Niña phase, which brings below-normal spring and summer rainfall to Argentina. On the contrary, the warm El Niño phase (wetter than normal for Argentina) had a larger incidence (62%) among years with above-expected values (30% of the series with an increase ≥10% of

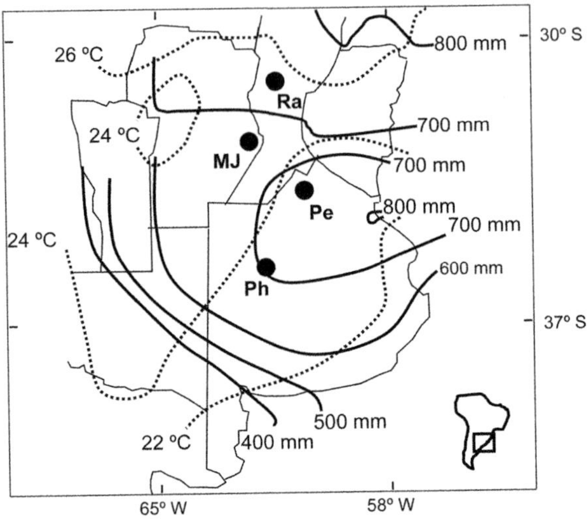

Fig. 13–1. The enlarged area represents the Pampas region of Argentina (ca. 30–39° S and 56–65° W), indicated with a square within South America in the inset. Solid and dotted lines indicate isohyets and mean temperature isotherms, respectively, during the warm season (21 September to 21 March). Solid circles correspond to the evaluated sites of Pergamino (Pe), Pehuajó (Ph), Marcos Juárez (MJ), and Rafaela (Ra).

expected value). These trends had no link to a bias of the phases along the studied period, which registered an even distribution of La Niña events in 33% of the years and of El Niño events in 35% of the years.

In spite of the abovementioned evidence, the combined effect of management practices (soil water at sowing, sowing date, stand density, fertilization) that affect crop water use did not receive adequate attention until the end of the 20th century. This situation was due to some permanent restrictions, such as the increased incidence of important insect pests (e.g., European corn borer [*Ostrinia nubilalis* Hübner], fall armyworm [*Spodoptera frugiperda* J.E. Smith], corn earworm [*Helicoverpa zea* Boddie]) with delayed sowing, which made late-sown maize and double-cropped winter crop–maize (i.e., maize sown immediately after the harvest of any winter crop) nonvalid alternatives in areas where double-cropped wheat–soybean is widely spread since the 1980s (Hall et al., 1992; Messina et al., 1999). This constraint, together with the strong paradigm of the advantage of early sowing for maximizing maize yield in temperate regions (Cirilo and Andrade, 1994; Otegui et al., 1995), has limited late sowing of this species (December–January) to the subtropical, low-yielding non-Pampas areas of the northwest and northeast parts of the country, linked to their pronounced monsoonal regime and heat-shock risk during the summer. The insect problem began to decrease in 1998 with the appearance of the first transgenic maize in the Argentine market (Bt corn; Williams et al., 1997) and has been currently minimized with the inclusion of additional transgenic events ("stacked events") for enhanced crop protection against a wide insect spectrum. However, the lack of a thorough approach to the analysis of the interaction among management practices persisted for a few more years, because the necessary information for the correct use of integrative tools such as crop models was not available. Calibration of CERES-Maize (Jones and Kiniry, 1986) to the conditions of the Pampas region of Argentina commenced in the 1990s (Otegui et al., 1996; Guevara et al., 1999), but it was its adoption by AACREA (Argentine Association of Regional Agricultural Experimentation Consortia) that granted the rapid assembly of all needed information on soils and climate (Mercau et al., 2001) and boosted its use as a decision-making tool in the last decade (Bert et al., 2006, 2007).

In this new technological scenario, it is possible to revisit some recurrent questions raised by farmers along the extended sowing period for summer crops in the Pampas region, but now including its broad variation in soil types and interannual rainfall regimes (Hall et al., 1992). Examples of these types of questions include the following:

1. When early sowing (September–October) of maize is delayed and the time to start soybean sowing is reached (November), is it better to finish sowing maize or to postpone this activity until December, after all soybean has been sown?
2. When soil water availability during spring is usually scarce and early sowing promotes a large interannual variation in maize yield, as in many western areas of the Pampas region (Ferreyra et al., 2001), or a winter cover or forage crop does not allow to sow early enough to avoid silking to take place before the critical summer water deficit, what yield level could be expected of a delayed sowing?
3. When sowing of soybean in the double-cropped wheat–soybean sequence reaches mid-December and a significant yield decline is expected for each day delay in its sowing (Peltzer, 2007), is it better to continue with soybean or to start with maize?

In this chapter we will present several elements for assisting decision-making on these issues and will compare them with the few available data on late-sown, single-cropped maize (LSM) crops from on-farm research organized in recent years by the Argentine Association of No-Till Farmers (AAPRESID). Our results are based on a previous study (Otegui et al., 2002) developed for a technological meeting on maize management. This study was synthesized in an assistant tool (Mercau et al., 2002) that is periodically updated. Different production levels will be analyzed (van Ittersum and Rabbinge, 1997), from conditions with no stress (i.e., potential yields) to water-limited and water- + N-limited crops (i.e., attainable yields). All results (observed and simulated) correspond to crops grown free of biotic constraints (insects, diseases, and weeds) and not influenced by the soil water table (Nosetto et al., 2009), because their effects are not included in crop simulation models used in this study.

Late-Sown Maize and Double-Cropped Wheat–Maize in the Central Area of the Pampas Region

The few records of potential on-farm maize yields (i.e., harvested in areas >1 ha) in the Pampas region indicate that they range between 13.9 and 17.8 Mg ha^{-1} (dry weight). They have been obtained with transgenic, single-cross maize hybrids of relative maturity (RM) 116 to 125 (FAO 500–650; Otegui et al., 1995, 1996), sown early in September and grown under irrigation and high N fertilizer rates near Teodelina (34° S, 61° W), Santa Fe (Otegui et al., 2007). These yields were close to values registered during the previous decade (1990s) in experimental plots cropped in the same area under September sowings (Otegui et al., 1995) and 80,000 plants ha^{-1}, but with nontransgenic, single-cross hybrids of RM 110 (12.8–15.4 Mg ha^{-1}) and three-way cross hybrids of RM 125 (15.7–17.7 Mg ha^{-1}). On the basis of these studies and a series of 24 yr of weather records (1970–1971 to 1993–1994) of the region, it was estimated that potential yield of maize hybrids (RM 125) grown at the mentioned stand density in the area ranged between 15.3 ± 3.4 and 15.2 ± 1.0 Mg ha^{-1} for September and October sowings, respectively, and declined to 14.2 ± 1.1 Mg ha^{-1} for November sowings (Otegui et al., 1996). These trends have been recently confirmed by another model-based study (Maddonni, 2012), which expanded the analysis to other sites of the Pampas region (31°10′–35°48′ S, 61°30′–64°21′ W) but did not include sowing dates beyond 1 December. Evidently, the sowing delay along November does not represent a strong penalty to potential yield for latitudes ≤36° S. The trend, however, cannot be extrapolated to later sowings, because several aspects may lie beyond the limits of the models used for estimation (e.g., reduced air temperature that causes a reduction in radiation use efficiency; Otegui et al., 1996), yielding wrong predictions (Jones and Kiniry, 1986). Moreover, there is evidence of a strong yield decline for sowing dates ≥November at latitudes ≥37°45′ S (Cirilo and Andrade, 1994), but no evidence of the ability of models such as CERES-Maize for predicting them in the environment of interest. For these reasons, the first step was the evaluation of errors when a model is used for comparison with observed data from very late sowings.

Model Goodness of Fit for Predicting Crop Performance in Very Late Sowing Dates

The evaluated model was a modified version of MZCER980, included in the Decision Support System for Agrotechnology Transfer (DSSAT)-CERES-Maize V3.5 (Jones et al., 1998). Four changes were introduced to the original version: (i) the optimum temperature that affects maximum grain-filling rate was set at 26°C as in a previous version (Jones and Kiniry, 1986), (ii) the use efficiency of carbon remobilized to grain filling was set at 0.3 after Kiniry et al. (1997), (iii) the response of the kernel number per plant to plant growth during the critical period of silking + 170°C day was modified to account for positive effects of low temperatures on the duration of the period (Otegui and Bonhomme, 1998), which may affect the total assimilate production through extended light interception as well as through daily plant growth rate (Ritchie and Alagarswamy, 2003), and (iv) the effect of the soil root growth factor on root depth capacity, introduced in MZGECER980, was removed and limited to root density, as in a previous version (Jones and Kiniry, 1986). This modified model has been used satisfactorily for describing the performance of rainfed maize crops grown under different N rates and conventional sowing dates in the area of interest (Mercau et al., 2001). Water and N routines, therefore, were assumed to work properly for the Pampas region environment. Data for model evaluation in late sowing dates were provided by A. Cirilo (personal communication) and corresponded to 48 experimental plots, grown free of abiotic and biotic stress between 1996–1997 and 2000–2001 at the Pergamino experimental unit (33°56′ S, 60°33′ W; Buenos Aires province) of the National Institute of Agricultural Technology. Plots were cropped on soils of silty clay loam type (Typic Argiudoll; USDA taxonomy), and sowing dates ranged between 31 October and 25 January. Five hybrids of RM between 113 and 125 were used (DK639, DK664, DK752, DK757, DK765, and C280), for which the necessary genetic coefficients were available. Stand densities ranged between 50,000 and 80,000 plants ha^{-1}. Observed yields varied between 1.9 and 9.9 Mg ha^{-1} (all the yield values are presented on a dry basis) and were adequately predicted by the model with an RMSE of 18% of observed values (Fig. 13–2). This trend did not vary among yield components (RMSE of 18% for grain number m^{-2} and RMSE of 14% for individual grain weight), suggesting the model weighs all factors that affect crop growth along the cycle evenly. No trend in yield was detected across stand densities for late sowings under no water or N stress. Accordingly, simulations indicated an almost null effect of evaluated stand densities on yield, and there was a good agreement between observed and predicted results for this trait (RMSE for yield of 16, 20, and 17% at low, medium, and high stand densities).

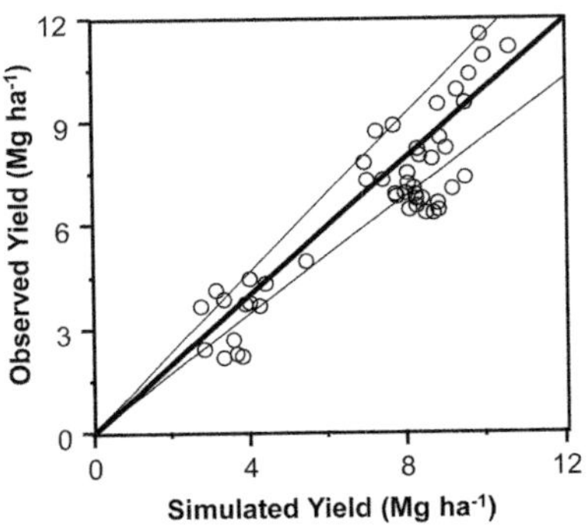

Fig. 13–2. Relationship between observed and predicted maize yields. Data (*n* = 48) correspond to late-sown (between 31 October and 25 January) experimental plots cropped at the Pergamino site (Fig. 13–1). The thick solid line represents the 1:1 relationship, and the thin lines represent a model prediction of 1.15 (bottommost) and 0.85 (uppermost) of observed yields.

Water-Limited Yields with No Nitrogen Limitation

The scenario of analysis included 41 growing seasons (1971–1972 to 2011–2012), with sowing dates evaluated at 10-d intervals between 10 December and 10 January. Simulated conditions included standard soils of the region (Typic Argiudols with 2.5% organic matter and 256 mm of available soil water up to 1.8-m depth), two single-cross maize hybrids (DK 752 of RM 125, and DK 664 of RM 116), and a stand density of 65,000 plants ha^{-1}. Three levels of soil water at sowing were tested: one representative of LSM and the other two representative of the most expected conditions at physiological maturity of a wheat crop, immediately before sowing of maize in a double-cropped wheat–maize sequence (DCM). The whole soil profile was set at field capacity for the LSM. For the DCM, one condition was defined as dry (DCM_{DRY}) and the other as wet (DCM_{WET}). For DCM_{DRY}, soil water content was (i) set at 30% PASW up to 1.2-m depth and (ii) increased to 40% of PASW for the 1.2- to 1.5-m layer and to 50% PASW for the 1.5- to 1.8-m layer (mean PASW of 35% for the whole profile). For DCM_{WET}, soil water was set at field capacity up to 0.6-m depth, and the rest of the layers were at the same PASW of the DCM_{DRY} condition (mean PASW of 58% for the whole profile). Soil water in the DCM_{DRY} condition is representative of years when wheat crops end with low water content and farmers start sowing the summer species in the double-cropped sequence with any rainfall event of small intensity (i.e., low recharge). Soil water in the DCM_{WET} condition is representative of wheat crops that received a rainfall recharge after physiological maturity and/or farmers that wait for it to reduce the risk of crop failure due to poor seedling establishment. To ensure crop germination, a rainfall of 8 mm was simulated by adding irrigation of that amount immediately after each simulated sowing date. The nitrogen routine was turned off to ensure no N limitation to crop growth.

Results for the described scenarios are summarized in Table 13–1. Estimated yields with no N limitation were highest for fully irrigated crops at the earliest evaluated sowing date (10 December) and lowest for the driest option (DCM_{DRY}) and most delayed sowing (10 January). As expected, yield declined with reduced soil water content at sowing (Table 13–1). Nevertheless, while rainfed late-sown maize crops (LSM_{RAIN}) had yield levels slightly lower than the potential expected with no water limitation along the cycle (LSM_{POT}), both double-cropped strategies (DCM_{WET} and DCM_{DRY}) had higher risk of yield reduction. This reduction was chiefly linked to reductions in kernel number m^{-2} (Fig. 13–3a), which were associated with a low maximum light interception efficiency at silking (Fig. 13–3b). Differences between DCM_{WET} and DCM_{DRY} were linked to water-stress conditions that affect leaf expansion, which led to reduced (<85%) or markedly reduced (<65%) ground cover in 39 and 10% of the years, respectively, among DCM_{DRY} crops. These negative conditions were negligible among DCM_{WET} crops (Fig. 13–3b).

Potential yield of a single-cropped maize declined with a delayed sowing date from 10 December to 10 January, in agreement with previous reports for early sown maize in the same area (Otegui et al., 1995, 1996). Interestingly, this

Table 13–1. Sowing date effects on estimated yield of maize crops grown under contrasting water regimes and no N limitation at the Pergamino site of the Pampas region.

		Sowing date			
Water regime†	Quantile	10 December	20 December	30 December	10 January
	%	kg ha^{-1}			
LSM_{POT}	25	10,442	10,024	7,858	6,212
	50	11,154	11,121	9,169	7,510
	75	12,301	11,859	10,622	8,294
LSM_{RAIN}	25	9,548	9,285	7,849	6,212
	50	10,526	10,800	8,872	7,492
	75	11,924	11,831	10,502	8,294
DCM_{WET}	25	7,204	8,429	6,200	5,343
	50	9,842	9,805	8,181	7,021
	75	11,135	11,143	9,492	8,086
DCM_{DRY}	25	5,378	7,102	5,900	4,720
	50	8,320	8,558	7,602	6,486
	75	10,056	9,496	8,700	7,210

† Water regimes included (i) single-cropped, late-sown maize crops (LSM) with no water restriction along the cycle (LSM_{POT}) or rainfed farming with the soil at field capacity at sowing (LSM_{RAIN}) and (ii) rainfed farming of double-cropped maize crops (DCM). For the latter, levels of plant available soil water (PASW) at sowing were as follows: (i) high, at field capacity in the uppermost 60-cm soil layer and 30% PASW between 60 and 120 cm (DCM_{WET}), and (ii) low, at 30% PASW in the uppermost 120-cm soil layer (DCM_{DRY}). The rest of the profile was between 40% (120–150 cm) and 50% (150–180 cm) PASW for DCM crops.

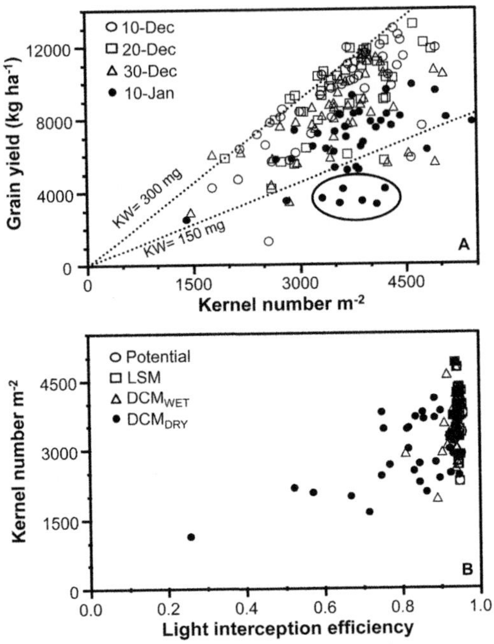

Fig. 13–3. Estimated response of (a) yield to kernel number m^{-2} and (b) kernel number m^{-2} to maximum light interception efficiency. Data in Fig. 13–3a correspond to DCM_{WET} crops grown at four sowing dates at the Pergamino site. Dotted lines represent isoquants of constant individual kernel weight (KW), and encircled symbols correspond to crops that experienced early termination of grain filling due to low temperatures. Data in Fig. 13–3b correspond to different cropping strategies (LSM_{POT}, with no water restriction; LSM_{RAIN}, rainfed farming with 100% soil water at sowing up to 1.8-m depth; DCM_{WET}, double-cropped wheat–maize with high soil water at maize sowing in the uppermost 60-cm soil layer and low water availability in deeper layers; DCM_{DRY}, same as DCM_{WET} but with low soil water at maize sowing in the uppermost soil layer) for a 20 December sowing date at the Pergamino site. All data in Fig. 13–3a and 13–3b are with no N restriction.

trend was slightly different under rainfed farming (LSM_{RAIN} and DCM), and the highest yields were obtained with sowing dates on 10 December as well as 20 December (Table 13–1, Fig. 13–4). A delayed sowing date pushed the estimated silking date from 13 February ± 4 d for the 10 December sowing to 18 March ± 4 d for the 10 January sowing. For the central area of the Pampas region, this delay in silking time is accompanied by a decline in air temperature and solar radiation and an increase in rainfall (Hall et al., 1992; Maddonni, 2012), which causes (i) a reduction in evaporative demand and an improved soil water condition and (ii) a less favorable photothermal quotient (Cantagallo et al., 1997). Consequently, the crop water balance tends to improve in the transition from summer to fall, whereas the opposite occurs in the photothermal environment (Coll et al., 2011). The relative importance of these trends affected estimated final yield

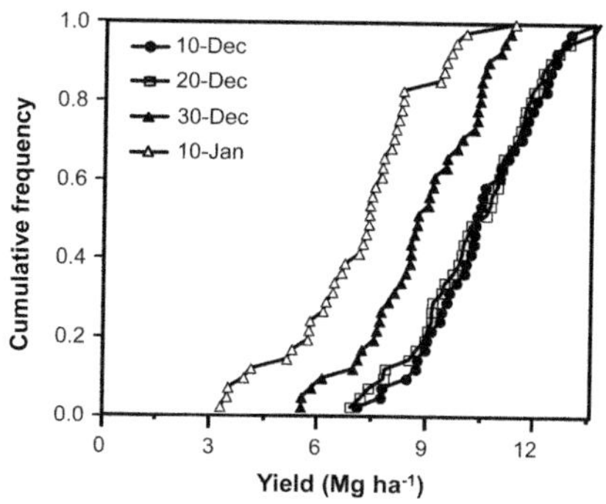

Fig. 13–4. Estimated frequency of attainable yield (LSM_{RAIN}, rainfed farming and 100% plant available soil water at sowing) across four sowing dates with no N limitation at the Pergamino site.

and its components, kernel number m^{-2} and individual kernel weight. Multiple regression analysis indicated that final yield was always significantly ($P < 0.001$) affected by both yield components, but its variations were mainly driven by variations in kernel number m^{-2} for December sowings and by variations in kernel weight afterward. These results highlight the importance given by the model to the grain-filling period and source–sink relations for late-sown maize crops, in agreement with physiological studies on this topic that included a wide range of sowing dates (Cirilo and Andrade, 1996; Borrás et al., 2004). For the area under study, it was estimated that early frosts had no terminal effect among December sowings; but this constraint, together with long periods of low temperatures, caused the anticipated termination of grain filling in 25% of the years among January sowings (Fig. 13–3a).

On the basis of the negative effect of low soil water at sowing due to poor rainfall recharge on yield, as well as on the apparent yield stability across December sowings, farmers should be recommended to survey soil water content rather than to tie maize sowing dates to the first rainfall events after the harvest of winter crops. Estimated trends among maize crops in a DCM sequence set an important difference with soybean crops in the double-cropped wheat–soybean sequence. Similar analyses made for the latter with CROPGRO-soybean (Hoogenboom et al., 1999) suggested a marked decline in yield with delayed sowing after 15 December for the nearby site of Paraná, located at 32° S and 60° W (Peltzer, 2007; Coll et al., 2011).

Water-Limited Yields: Response to Nitrogen

We simulated a range of N availability at sowing (soil N-NO_3 + soil N-NH_4 + N-fertilizer) between 40 and 190 kg ha^{-1} N. This was performed by adding different amounts of N-fertilizer to a fixed amount of N-soil (40 kg ha^{-1} N in the uppermost 0.6-m soil layer and 15 kg ha^{-1} N in the rest of the profile). The response of estimated maize yield to N availability at sowing increased with improved soil water availability ($LSM_{RAIN} > DCM_{WET} > DCM_{DRY}$) and decreased with sowing date delay after 20 December (Table 13–2). Considering 70 kg ha^{-1} N as a common soil N supply for maize crops in this area (Alvarez and Steinbach, 2009), the response rate to N fertilization in the 20 December sowing rose from 9 kg grain ha^{-1} kg N^{-1} for DCM_{DRY} to 24 kg grain ha^{-1} kg N^{-1} for DCM_{WET} and 27 kg grain ha^{-1} kg N^{-1} for LSM_{RAIN} when N availability increased from 70 to 100 units N ha^{-1}. This response declined to (i) 2.1, 16.3, and 21.5 kg grain ha^{-1} kg N^{-1}, respectively, when N availability increased from 100 to 130 kg ha^{-1} N and (ii) −0.23, 5.6, and 12.4 kg grain ha^{-1} kg N^{-1}, respectively, when N availability increased from 130 to 160 kg ha^{-1} N. These figures are in agreement with recent studies on site-specific N management that set a limit at 130 kg ha^{-1} N for Typic Haplustol soils in the western area of the Pampas region (Gregoret et al., 2011). However, the estimated response of yield to N availability among the bottommost 25% of the years (discussed in a next section) indicated that it (i) began to decrease at relatively low N levels for the

Table 13–2. Yield response to sowing date and N availability at sowing (soil N-NO_3 + soil N-NH_4 + N-fertilizer) for rainfed crops at the Pergamino site.†

		LSM_{RAIN}		DCM_{WET}		DCM_{DRY}	
Sowing date	N	Mean	SD	Mean	SD	Mean	SD
		kg ha^{-1}					
10 December	70	8,137	± 815	7,268	±1,377	6,129	±1,937
	100	8,981	± 916	7,927	±1,609	6,449	±2,238
	130	9,653	±1,041	8,345	±1,995	6,536	±2,512
	160	10,008	±1,299	8,549	±2,232	6,451	±2,675
20 December	70	8,118	± 924	7,470	±1,248	6,399	±1,630
	100	8,925	±1,056	8,193	±1,433	6,670	±1,908
	130	9,570	±1,202	8,682	±1,655	6,732	±2,108
	160	9,941	±1,355	8,849	±1,830	6,725	±2,229
30 December	70	7,260	±1,104	6,674	±1,361	5,917	±1,523
	100	7,924	±1,252	7,274	±1,528	6,164	±1,813
	130	8,455	±1,374	7,655	±1,706	6,194	±1,962
	160	8,707	±1,468	7,714	±1,905	6,063	±1,911
10 January	70	6,070	±1,367	5,666	±1,428	5,045	±1,369
	100	6,589	±1,534	6,136	±1,564	5,251	±1,563
	130	6,939	±1,676	6,386	±1,659	5,210	±1,641
	160	7,075	±1,761	6,381	±1,689	5,066	±1,593

† Water regimes included (i) single-cropped, late-sown maize crops (LSM) with no water restriction along the cycle (LSM_{POT}) or rainfed farming with the soil at field capacity at sowing (LSM_{RAIN}) and (ii) rainfed farming of double-cropped maize crops (DCM). For the latter, levels of plant available soil water (PASW) at sowing were as follows: (i) high, at field capacity in the uppermost 60-cm soil layer and 30% PASW between 60 and 120 cm (DCM_{WET}), and (ii) low, at 30% PASW in the uppermost 120-cm soil layer (DCM_{DRY}). The rest of the profile was between 40% (120–150 cm) and 50% (150–180 cm) PASW for DCM crops.

LSM_{RAIN} and DCM_{WET} strategies or (ii) remained very low and almost invariable for the DCM_{DRY} strategy. This variation highlights the importance of long-term weather forecasts together with the survey of PASW at sowing for minimizing the economic risk linked to management decisions (Messina et al., 1999).

Late-Sown Maize and Double-Cropped Wheat–Maize across the Pampas Region

Currently, most of maize production comes from an area located between 30 and 35° S and 60 and 65° W of the Pampas region (Fig. 13–1). Within this area, three additional sites were selected for the analysis of the strategies (LSM_{RAIN}, DCM_{WET}, and DCM_{DRY}) previously evaluated for the Pergamino site. The new environments were considered representative of the regional variation in soil and climate conditions and with a long enough frost-free period for the development of late-sown maize. The new sites (Fig. 13–1) were Rafaela (31°12′ S, 61°30′ W; Santa Fe province), Marcos Juárez (32°36′ S, 62°33′ W; Córdoba province), and Pehuajó (35°54′ S, 61°54′ W; Buenos Aires province). The most remarkable differences among them have been described by Maddonni (2012) and are summarized in Table 13–3 for all evaluated combinations of sites and sowing dates. We highlight (i) the strong differences in rainfall among sites (Rafaela > Pehuajó > Pergamino > Marcos Juárez) as compared with the slight differences in potential evapotranspiration (PET), but the clear improvement in water balance (rainfall–PET) with delayed sowing for all sites, (ii) a latitude gradient in mean maximum and mean minimum temperatures (Rafaela > Marcos Juárez > Pergamino > Pehuajó), (iii) differences in incident solar radiation (Pergamino > Pehuajó > Marcos Juárez > Rafaela), but particularly an abrupt decline during late summer and early autumn for the site at highest latitude (Pehuajó), and (iv) Typic Hapludol soils in Pehuajó (no argillic horizon) as compared with Typic Argiudol soils (with an argillic horizon) in all the other sites.

Water-Limited Yields with No Nitrogen Limitation

Mentioned differences in temperature caused a gradient in the length of phenological phases across the different combinations of sites and sowing dates (Table 13–4), which was not affected across strategies related to PASW at sowing (data not shown). The dates of silking and physiological maturity were delayed with increased latitude (Pehuajó > Pergamino > Marcos Juárez > Rafaela) and sowing date. These differences, however, were much smaller for the pre-silking stage than for the post-silking one (Table 13–4), because the former matched the period of high and stable temperatures of early and mid-summer whereas the latter took place during the period of highest rate of decline in this variable (Hall et al., 1992; Borrás et al., 2004). On the contrary, low temperatures during grain filling are

Table 13–3. Detail of climatic characteristics during the cycle for maize crops sown at different sites and dates.

Site	Sowing date	Temperature†		Cumulative values		
		Maximum	Minimum	Incident PAR‡	Rainfall	PET§
		°C		MJ	mm	
Pergamino	10 December	27.5 ± 1.3	14.7 ± 1.1	1175 ± 73	405 ± 11	651 ± 33
	20 December	26.3 ± 1.3	13.7 ± 1.1	1160 ± 66	432 ± 7	630 ± 32
	30 December	25.2 ± 1.2	12.8 ± 1.1	1115 ± 73	458 ± 7	596 ± 38
	10 January	24.5 ± 1.1	12.2 ± 1.0	1023 ± 91	483 ± 6	545 ± 45
Pehuajó	10 December	26.2 ± 1.7	13.3 ± 1.3	1189 ± 81	511 ± 11	643 ± 33
	20 December	25.2 ± 1.7	12.7 ± 1.2	1136 ± 75	545 ± 8	607 ± 30
	30 December	24.5 ± 1.6	12.1 ± 1.2	1053 ± 76	568 ± 8	559 ± 31
	10 January	24.0 ± 1.4	11.8 ± 1.2	949 ± 78	612 ± 14	506 ± 35
Marcos Juárez	10 December	28.6 ± 1.3	16.4 ± 0.9	1098 ± 72	229 ± 39	624 ± 39
	20 December	27.7 ± 1.3	15.6 ± 1.0	1085 ± 61	301 ± 16	609 ± 33
	30 December	26.3 ± 1.5	14.3 ± 1.1	1081 ± 54	348 ± 10	592 ± 29
	10 January	25.2 ± 1.5	13.4 ± 1.2	1022 ± 62	383 ± 6	554 ± 32
Rafaela	10 December	29.5 ± 1.2	16.9 ± 0.8	1085 ± 47	655 ± 13	627 ± 27
	20 December	28.8 ± 1.3	16.4 ± 0.9	1063 ± 52	707 ± 19	611 ± 28
	30 December	27.5 ± 1.3	15.4 ± 1.0	1057 ± 50	783 ± 20	595 ± 24
	10 January	26.0 ± 1.4	14.1 ± 1.0	1033 ± 52	932 ± 101	568 ± 26

† Mean ± SD of daily records.

‡ Photosynthetically active radiation.

§ Potential evapotranspiration.

expected to cause the early termination of the cycle and consequently to reduce the extension of the post-silking period for sowings beyond December at the site of highest latitude (Pehuajó) evaluated in this research (Table 13–4).

As described for the Pergamino site, estimated potential yields (LSM_{POT}) were generally highest ($\geq$9.9 ± 1.03 Mg ha^{-1}) for the earliest sowing date (10 December) and declined with delayed sowing (Table 13–5). There was, however, a sowing date × site interaction response (Table 13–5). The negative effects of low temperature with delayed sowing were more pronounced at the highest latitude site (Pehuajó) than at the intermediate latitudes of Pergamino (Table 13–1) and Marcos Juárez (Table 13–5), and no trend was detected in potential yield for the site located at the lowest evaluated latitude (Rafaela, Table 13–5). The response to sowing date was very similar for attainable yields with LSM crops (Fig. 13–4). Interestingly, estimated yield losses (i.e., attainable as compared with potential for each evaluated condition) were very low (0–6%) for LSM_{RAIN} crops in all evaluated sites (Table

Table 13–4. Estimated occurrence of phenological events for maize crops sown at different sites and dates.

Site	Sowing date	Date†		Period	
		Silking	Physiological maturity	Sowing to silking	Silking to physiological maturity
				d	
Pergamino	10 December	6–22 Feb.	22 Mar. to 2 June	65 ± 4	80 ± 12
	20 December	15 Feb. to 3 Mar.	30 Apr. to 18 June	65 ± 3	89 ± 11
	30 December	26 Feb. to 15 Mar.	2 May to 15 July	66 ± 3	90 ± 16
	10 January	12–25 Mar.	4 May to 15 Aug.	67 ± 4	85 ± 18
Pehuajó	10 December	9 Feb. to 1 Mar.	11 Mar. to 3 June	69 ± 4	81 ± 26
	20 December	18 Feb. to 7 Mar.	23 Mar. to 16 June	69 ± 4	84 ± 27
	30 December	2–21 Mar.	1 Apr. to 20 June	70 ± 4	82 ± 25
	10 January	21 Mar. to 2 Apr.	12 Apr. to 20 June	72 ± 4	72 ± 22
Marcos Juárez	10 December	3–20 Feb.	13 Mar. to 4 May	64 ± 4	66 ± 13
	20 December	15 Feb. to 1 Mar.	11 Mar. to 24 May	65 ± 4	73 ± 16
	30 December	25 Feb. to 13 Mar.	28 Mar. to 12 June	65 ± 4	83 ± 20
	10 January	8 Mar. to 20 Apr.	7 Apr. to 3 July	66 ± 3	86 ± 20
Rafaela	10 December	4–17 Feb.	26 Mar. to 9 May	62 ± 4	65 ± 7
	20 December	13–25 Feb.	28 Mar. to 27 May	62 ± 3	70 ± 10
	30 December	22 Feb. to 9 Mar.	13 Apr. to 3 July	62 ± 3	80 ± 10
	10 January	6–20 Mar.	27 Apr. to 18 July	63 ± 3	89 ± 14

† Maximum range.

Table 13–5. Estimated potential yields of late-sown, single-cropped maize crops grown at different sites and sowing dates with no N restriction.

Site	Quantile	Sowing date			
		10 December	20 December	30 December	10 January
	%		kg ha^{-1}		
Pehuajó	25	10,756	9,039	6,453	4,268
	50	11,540	10,037	7,474	5,309
	75	12,093	10,914	9,009	6,422
Marcos Juarez	25	9,237	9,166	8,790	7,533
	50	10,411	10,374	10,204	8,743
	75	11,083	10,914	10,713	9,869
Rafaela	25	9,118	8,658	8,615	8,404
	50	9,831	9,507	9,921	9,853
	75	10,868	10,515	10,601	10,361

13–6). This result contrasts markedly with estimated losses from early sown (15–20 September) maize crops in the region, which are expected to reach ~32% when the computation is based on mean values of potential (Otegui et al., 1996) and attainable (rainfed farming with no N restriction and 100% PASW at sowing) yields (Mercau, 2002). However, an important penalty to yield is expected among late-sown maize crops with reduced water availability at sowing (Fig. 13–5 and Table 13–6). This penalty reached 6.5 to 22% for DCM_{WET} and increased pronouncedly for DCM_{DRY} (14–47%). These trends in yield had a concomitant effect on water use efficiency (WUE, in kilograms of grain per millimeter of crop evapotranspiration; Table 13–7), which declined with reduced water availability at sowing (LSM_{RAIN} > DCM_{WET} > DCM_{DRY}). A delayed sowing date led to a decline in yield losses, as compared with potential yield among DCMs (Table 13–6), because the critical periods of silking and grain filling (Table 13–4) are expected to take place after the occurrence of maximum evaporative demand that takes place during January (Totis de Zeljkovich and Rebella, 1980; Hall et al., 1992). For crops at high N availability, however, these effects of sowing date did not have a uniform effect on WUE (Table 13–7). This trait tended to increase with delayed sowing for the Marcos Juárez and Rafaela sites, whereas it tended to decline with delayed sowing at Pehuajó (steadily)

Table 13–6. Estimated grain yield loss of rainfed maize as single-cropped late-sown crops and in the double-cropped wheat–maize sequence.†

Site	Strategy‡	Sowing date 10 December	20 December	30 December	10 January
		%			
Pergamino	LSM_{RAIN}	6	3	3	0
	DCM_{WET}	12	12	11	7
	DCM_{DRY}	25	23	17	14
Pehuajó	LSM_{RAIN}	5	5	1	0
	DCM_{WET}	16	14	9	9
	DCM_{DRY}	30	22	17	15
Marcos Juarez	LSM_{RAIN}	2	1	5	2
	DCM_{WET}	17	12	17	11
	DCM_{DRY}	31	26	31	25
Rafaela	LSM_{RAIN}	4	2	4	0
	DCM_{WET}	21	18	22	15
	DCM_{DRY}	47	38	34	32

† For each site × sowing date, loss (%) = (1 − attainable GY/potential GY) × 100. Median values were used for computation. GY, grain yield.

‡ Water regimes included (i) single-cropped, late-sown maize crops (LSM) with no water restriction along the cycle (LSM_{POT}) or rainfed farming with the soil at field capacity at sowing (LSM_{RAIN}) and (ii) rainfed farming of double-cropped maize crops (DCM). For the latter, levels of plant available soil water (PASW) at sowing were as follows: (i) high, at field capacity in the uppermost 60-cm soil layer and 30% PASW between 60 and 120 cm (DCM_{WET}), and (ii) low, at 30% PASW in the uppermost 120-cm soil layer (DCM_{DRY}). The rest of the profile was between 40% (120–150 cm) and 50% (150–180 cm) PASW for DCM crops.

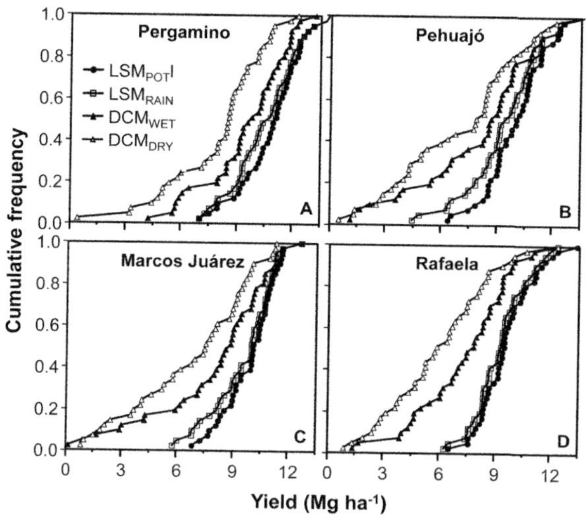

Fig. 13–5. Estimated frequency of maize yield for different cropping strategies (described in Fig. 13–3) in a 20 December sowing date and with no N restriction at four sites of the Pampas region.

and Pergamino (from 20 December onward). This difference among sites is chiefly linked to the effect of this management practice on yield (Tables 13–1, 13–5, and 13–6), because crop evapotranspiration follows PET (Table 13–3) and always declined with delayed sowing.

Water-Limited Yield: Response to Nitrogen

Nitrogen use efficiency (in kilograms of grain per kilogram of N absorbed by the crop at maturity) usually declined with reduced PASW at sowing, delayed sowing date, and enhanced N availability (Table 13–7). There was, however, a clear site × PASW at sowing interaction in yield response to N availability (Fig. 13–6). For all evaluated conditions in a 20 December sowing date, the response rate (first derivate of the quadratic model adjusted to 41 simulated years along 6 levels of N supply at sowing) was always high (19–34 kg grain ha^{-1} kg N^{-1}) when soil N (soil N-NO_3 + soil N-NH_4 + N-fertilizer) was low (≤40 kg ha^{-1} N) and soil water was high (LSM_{RAIN} and DCM_{WET}), and it declined with reduced water availability and/or enhanced soil N. The expected response for an initial soil N level of 70 kg ha^{-1} N would be always ≥25 kg grain ha^{-1} kg N^{-1} (Table 13–8) among late-sown maize crops that start growing with the whole profile at field capacity (LSM_{RAIN} strategy) in a 20 December sowing. On the contrary, the response declined to almost null or even negative values for the worst water-limited scenario (DCM_{DRY}).

Estimated mean yield across all sowing date × PASW × soil N combinations ranged between 5.1 and 10 Mg ha^{-1} for Pergamino, 3.9 and 10.4 Mg ha^{-1} for

Table 13–7. Mean values of water and N use efficiencies for different sites, cropping strategies and sowing dates.†

Site	Strategy	N level	Sowing date							
			10 December		20 December		30 December		10 January	
			WUE	NUE	WUE	NUE	WUE	NUE	WUE	NUE
		kg ha^{-1}	kg mm^{-1}	kg kg^{-1}	kg mm^{-1}	kg kg^{-1}	kg mm^{-1}	kg kg^{-1}	kg mm^{-1}	kg kg^{-1}
Pergamino	LSM_{RAIN}	70	15	74	16	71	15	61	14	52
		130	18	60	19	57	17	49	16	41
	DCM_{WET}	70	14	68	15	68	14	59	13	50
		130	16	53	18	53	16	46	15	39
	DCM_{DRY}	70	13	60	14	60	14	54	13	47
		130	14	47	15	47	15	43	13	38
Pehuajó	LSM_{RAIN}	70	17	67	16	59	15	49	12	37
		130	19	54	18	48	16	40	13	31
	DCM_{WET}	70	15	60	15	55	14	46	11	35
		130	17	49	17	45	15	38	12	29
	DCM_{DRY}	70	14	56	15	52	14	45	11	33
		130	16	46	16	44	14	38	11	29
Marcos Juárez	LSM_{RAIN}	70	14	70	15	67	16	65	16	59
		130	17	56	17	54	18	52	18	48
	DCM_{WET}	70	13	63	14	62	14	61	15	56
		130	15	50	16	49	17	48	17	45
	DCM_{DRY}	70	13	58	13	58	14	57	14	52
		130	13	45	14	46	15	45	15	42
Rafaela	LSM_{RAIN}	70	17	70	15	67	15	65	16	62
		130	13	56	17	54	18	52	19	51
	DCM_{WET}	70	14	59	13	57	14	59	15	58
		130	11	46	14	45	16	45	17	46
	DCM_{DRY}	70	11	50	12	51	13	51	14	52
		130	15	41	12	42	13	43	14	44

† Water regimes included (i) single-cropped, late-sown maize crops (LSM) with no water restriction along the cycle (LSM_{POT}) or rainfed farming with the soil at field capacity at sowing (LSM_{RAIN}) and (ii) rainfed farming of double-cropped maize crops (DCM). For the latter, levels of plant available soil water (PASW) at sowing were as follows: (i) high, at field capacity in the uppermost 60-cm soil layer and 30% PASW between 60 and 120 cm (DCM_{WET}), and (ii) low, at 30% PASW in the uppermost 120-cm soil layer (DCM_{DRY}). The rest of the profile was between 40% (120–150 cm) and 50% (150–180 cm) PASW for DCM crops. WUE, water use efficiency (kilograms of grain per millimeter of crop evapotranspiration); NUE, N use efficiency (kilograms of grain per kilogram of N absorbed by the crop at maturity).

Pehuajó, 5.4 and 9.3 Mg ha^{-1} for Marcos Juárez, and 4.7 and 9.2 Mg ha^{-1} for Rafaela. These estimates are within the range of observed results obtained by AAPRESID farmers with only 2 yr of evaluation in sites close to those included in our analysis (Gambin and Borrás, 2013). They obtained approximately 8.2 to 10.8 Mg ha^{-1} for

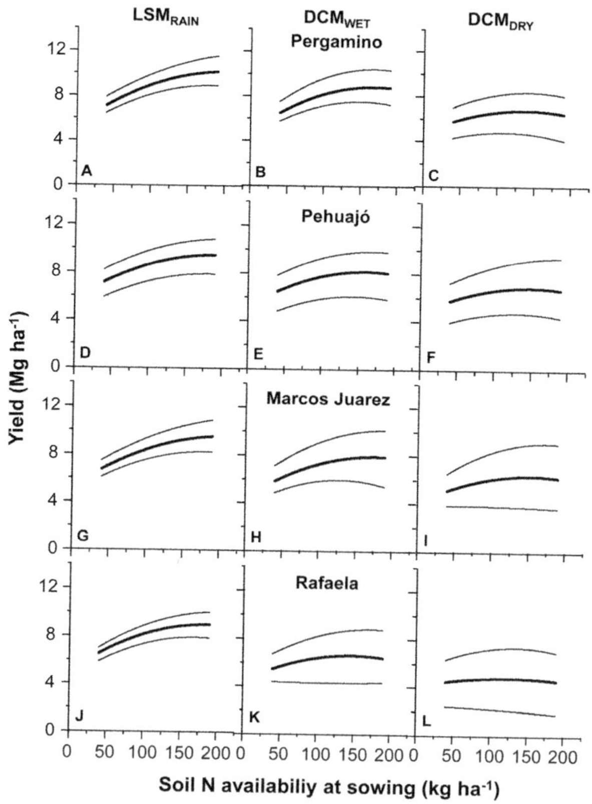

Fig. 13–6. Estimated response of yield to soil N supply at sowing (soil N-NO_3 + soil N-NH_4 + N-fertilizer) for three cropping strategies (described in Fig. 13–3) and four sites of the Pampas region. The thick solid line represents the quadratic model adjusted to 41 simulated years along 6 levels of N supply at sowing for median yield values, and thin lines represent the models adjusted to 20% (bottommost) and 80% (uppermost) quantiles.

the dry 2011–2012 growing season (classified as La Niña) and approximately 4.3 to 11.8 kg ha^{-1} for the typical 2012–2013 growing season (classified as neutral). They highlighted the sowing date and N fertilization as important management practices behind observed variations in yield. For all combinations of site × sowing date × PASW, our analysis indicated that the response rate to N supply varied in agreement with expected yield under no N limitation (Table 13–8); that is, the response rate is expected to decline with reduced PASW at sowing (LSM_{RAIN} > DCM_{WET} > DCM_{DRY}) and delayed sowing (Table 13–8). Considering that 8 to 15 kg of grain are necessary to pay for 1 kg of N applied as fertilizer (Mercau et al., 2002), the fertilizer rate to be recommended to farmers will vary considerably among evaluated sites and N levels. The continuous functions fitted in this research are expected to assist agronomists in this task.

Table 13–8. Estimated marginal yield response to two levels of nitrogen availability at sowing (soil N-NO_3+ soil N-NH_4 + N-fertilizer) evaluated for different sites, cropping strategies, and sowing dates. Data correspond to median values as those described in Fig. 13–6 for the 20 December sowing.†

Site	Strategy	N level	Sowing date: 10 December	20 December	30 December	10 January
		kg ha^{-1}	kg grain ha^{-1} kg N^{-1}			
Pergamino	LSM_{RAIN}	70	32	31	26	20
		130	17	17	13	8
	DCM_{WET}	70	26	28	23	18
		130	10	11	7	5
	DCM_{DRY}	70	16	15	13	10
		130	1	2	1	-1
Pehuajó	LSM_{RAIN}	70	29	25	18	11
		130	16	12	7	4
	DCM_{WET}	70	23	21	16	9
		130	8	7	4	2
	DCM_{DRY}	70	17	16	11	6
		130	4	3	2	1
Marcos Juárez	LSM_{RAIN}	70	29	28	27	25
		130	16	15	15	11
	DCM_{WET}	70	23	24	24	21
		130	9	10	9	7
	DCM_{DRY}	70	15	16	17	14
		130	3	4	3	3
Rafaela	LSM_{RAIN}	70	29	27	27	26
		130	14	13	13	13
	DCM_{WET}	70	18	16	17	18
		130	4	3	3	5
	DCM_{DRY}	70	6	5	5	8
		130	–1	–1	0	2

† The estimated marginal yield response was obtained from the first derivative of the mean quadratic function fitted in Fig. 13–6. Water regimes included (i) single-cropped, late-sown maize crops (LSM) with no water restriction along the cycle (LSM_{POT}) or rainfed farming with the soil at field capacity at sowing (LSM_{RAIN}) and (ii) rainfed farming of double-cropped maize crops (DCM). For the latter, levels of plant available soil water (PASW) at sowing were as follows: (i) high, at field capacity in the uppermost 60-cm soil layer and 30% PASW between 60 and 120 cm (DCM_{WET}), and (ii) low, at 30% PASW in the uppermost 120-cm soil layer (DCM_{DRY}). The rest of the profile was between 40% (120–150 cm) and 50% (150–180 cm) PASW for DCM crops.

Conclusions

Late sowing can be an avenue to increase maize production in the Pampas region. Yield potential for this alternative is lower than for early sowings (Otegui et al., 1995, 1996), but interannual variation among attainable yields (i.e., rainfed farming) is expected to be substantially reduced for LSM_{RAIN} and DCM_{WET} as compared with early sowings, particularly in areas where soil water recharge is not expected to take place until late in spring. Late sowing, therefore, would allow an increase in the land cropped to maize when soil water at sowing usually affects adequate crop establishment in early sowings. It would also be a good alternative when soil temperature is low early in spring, which is a common restriction for latitudes >36° S (not evaluated in this study). Results from our modeling approach suggest that the survey of soil water at sowing is critical for the correct planning of a late-sown maize crop, particularly those scheduled as double crops (DCM) after any winter crop (wheat, barley [*Hordeum vulgare* L.], pea [*Pisum sativum* L.], rapeseed [*Brassica napus* L.]). Adequate soil water recharge becomes more critical than the sowing date for reducing yield losses among late sown crops, particularly to the north of the region (i.e., lower latitudes). The analysis of complementary irrigation at sowing will be necessary in areas where this recharge may limit early crop establishment. Occurrence of a shallow water table requires its monitoring because it may introduce a large variation in expected maize yield in the region (Nosetto et al., 2009) and is not considered by models used in this study. Finally, our approach produced a tool for assisting farmers and their advisors on N fertilization decisions, taking into account the simultaneous modification of management practices (e.g., sowing date) and growing conditions (e.g., soil water and soil N content at sowing) in contrasting environments. This tool leads to improved fertilizer use efficiency, a key issue for reducing the negative impact of cropping systems in the context of forecasted climate change scenarios (Intergovernmental Panel on Climate Change, 2007).

References

Alvarez, R., and H.S. Steinbach. 2009. A review of the effects of tillage systems on some soil physical properties, water content, nitrate availability and crops yield in the Argentine Pampa. Soil Tillage Res. 104:1–15. doi:10.1016/j.still.2009.02.005

Bert, F.E., C.E. Laciana, G.P. Podestá, E.H. Satorre, and A.N. Menéndez. 2007. Sensitivity of CERES-Maize simulated yields to uncertainty in soil properties and daily solar radiation. Agric. Syst. 94:141–150. doi:10.1016/j.agsy.2006.08.003

Bert, F.E., E.H. Satorre, F.R. Toranzo, and G.P. Podestá. 2006. Climatic information and decision-making in maize crop production systems of the Argentinean Pampas. Agric. Syst. 88:180–204. doi:10.1016/j.agsy.2005.03.007

Borrás, L., G.A. Slafer, and M.E. Otegui. 2004. Seed dry weight response to source-sink manipulations in wheat, maize and soybean. A quantitative reappraisal. Field Crops Res. 86:131–146. doi:10.1016/j.fcr.2003.08.002

Cantagallo, J.E., C.A. Chimenti, and A.J. Hall. 1997. Crop physiology and metabolism: Number of seeds per unit area in sunflower correlates well with a photothermal quotient. Crop Sci. 37:1780–1786. doi:10.2135/cropsci1997.0011183X003700060020x

Cirilo, A.G., and F.H. Andrade. 1994. Sowing date and maize productivity: II. Kernel number determination. Crop Sci. 34:1044–1046. doi:10.2135/cropsci1994.0011183X003400040038x

Cirilo, A.G., and F.H. Andrade. 1996. Sowing date and kernel weight in maize. Crop Sci. 36:325–331. doi:10.2135/cropsci1996.0011183X003600020019x

Coll, L., O.P. Caviglia, and N.P. van Opstal. 2011. Rendimiento de soja de segunda sobre diferentes antecesores invernales. In: O. Paparotti, editor, Actualización técnica soja. Ser. Ext. 62. INTA EEA Paraná, Buenos Aires. p. 75–82.

Evans, L.T. 1997. Adapting and improving crops: The endless task. Philos. Trans. R. Soc. London, Ser. B 352:901–906. doi:10.1098/rstb.1997.0069

Evans, L.T. 1999. Steps towards feeding the ten billion: A crop physiologists view. Plant Prod. Sci. 2:3–9. doi:10.1626/pps.2.3

Ferreyra, R.A., G.P. Podestá, C.D. Messina, D. Letson, J. Dardanelli, E. Guevara, and S. Meira. 2001. A linked-modeling framework to estimate maize production risk associated with ENSO-related climate variability in Argentina. Agric. For. Meteorol. 107:177–192. doi:10.1016/S0168-1923(00)00240-9

Fischer, R.A., and G.O. Edmeades. 2010. Breeding and cereal yield progress. Crop Sci. 50:S-85–S-98. doi:10.2135/cropsci2009.10.0564

Gambin, B., and L. Borrás. 2013. Resultados de la red de ensayos de maíces tardíos. In: Red de evaluación de hibridos en maíz en siembras tardías. AAPRESID, Rosario, Argentina. p. 9–19. https://sites.google.com/a/agro.uba.ar/granero/ (accessed 23 Dec. 2013).

Gregoret, M.C., M. Díaz Zorita, J. Dardanelli, and R.G. Bongiovanni. 2011. Regional model for nitrogen fertilization of site-specific rainfed corn in haplustolls of the central Pampas, Argentina. Precis. Agric. 12:831–849. doi:10.1007/s11119-011-9224-7

Guevara, E., S. Meira, M. Maturano, and G. Coco. 1999. Maize simulation for different environments in Argentina. In: M. Donatelli et al., editors, Proceedings of the International Symposium: Modelling Cropping Systems, Lleida, Spain. 21–23 June. University of Lleida, Catalonia, Spain, p. 193–194.

Hall, A.J., C.M. Rebella, C.M. Ghersa, and J.Ph. Culot. 1992. Field-crop systems of the Pampas. In: C.J. Pearson, editor, Ecosystems of the world: Field crop ecosystems. Elsevier, Amsterdam. p. 413–449.

Hall, A.J., and R.A. Richards. 2013. Prognosis for genetic improvement of yield potential and water-limited yield of major grain crops. Field Crops Res. 143:18–33. doi:10.1016/j.fcr.2012.05.014

Hoogenboom, G., P.W. Wilkens, P.K. Thornton, J.W. Jones, L.A. Hunt, and D.T. Imamura. 1999. Decision support system for agrotechnology transfer V3.5. In: G. Hoogenboom, P.W. Wilkens, and G.Y. Tsuji, editors, DSSAT Version 3, Volume 4. Univ. of Hawaii, Honolulu. p. 1–36.

Intergovernmental Panel on Climate Change. 2007. Climate change 2007: The physical science basis. Fourth assessment report of the intergovernmental panel on climate change. Cambridge Univ. Press, Cambridge, UK.

Jones, C.A., and J.R. Kiniry. 1986. CERES-Maize: A simulation model of maize growth and development. Texas A&M Univ. Press, College Station.

Jones, J.W., G.Y. Tsuji, G. Hoogenboom, L.A. Hunt, P.K. Thornton, P.W. Wilkens, D.T. Imamura, W.T. Bowen, and U. Singh. 1998. Decision Support System for Agrotechnology Transfer: DSSAT. In: G.Y. Tsuji, G. Hoogenboom, and P.K. Thornton, editors, Understanding options for agricultural production. Kluwer Acad., Dordrecht, The Netherlands. p. 157–177.

Kiniry, J.R., J.R. Williams, R.L. Vanderlip, J.D. Atwood, D.C. Reitocosky, J.A. Mulliken, W.J. Cox, H.J. Mascagni, Jr., S.E. Hollinger, and W.J. Wiebold. 1997. Evaluation of two maize models for nine U.S. locations. Agron. J. 89:421–426. doi:10.2134/agronj1997.00021962008900030009x

Maddonni, G.A. 2012. Analysis of climatic constraints to maize production in the current agricultural region of Argentina—A probabilistic approach. Theor. Appl. Climatol. 107:325–345. doi:10.1007/s00704-011-0478-9

Mercau, J.L. 2002. Modelos de simulación y aplicación de herramientas informáticas para el manejo del cultivo. In: E.H. Satorre, editor, Guía Dekalb del cultivo de maíz. Servicios y Marketing Agropecuario, Buenos Aires. p. 203–217. https://sites.google.com/a/agro.uba.ar/granero/ (accessed 23 Dec. 2013).

Mercau, J.L., M.E. Otegui, and F.J. Menendez. 2002. Maíz tardío y maíz de segunda. Faculty of Agronomy, Univ. of Buenos Aires, Buenos Aires. https://sites.google.com/a/agro.uba.ar/granero/ (accessed 23 Dec. 2013).

Mercau, J.L., E.H. Satorre, M.E. Otegui, G.A. Maddonni, J. Cárcova, R. Ruiz, M. Uribelarrea, and F.J. Menendez. 2001. Evaluación a campo del comportamiento del modelo Ceres en cultivos de maíz del norte de la provincia de Buenos Aires. VII Congreso Nacional de Maíz. Buenos Aires. https://sites.google.com/a/agro.uba.ar/granero/ (accessed 23 Dec. 2013).

Messina, C.D., J.W. Hansen, and A. Hall. 1999. Land allocation conditioned on ENSO phases in the Argentine Pampas. Agric. Syst. 60:197–212. doi:10.1016/S0308-521X(99)00032-3

Nosetto, M.D., E.G. Jobbágy, R.B. Jackson, and G.A. Sznaider. 2009. Reciprocal influence of crops and shallow ground water in sandy landscapes of the Inland Pampas. Field Crops Res. 113:138–148. doi:10.1016/j.fcr.2009.04.016

Otegui, M.E., and R. Bonhomme. 1998. Grain yield components in maize: I. Ear growth and kernel set. Field Crops Res. 56:247–256. doi:10.1016/S0378-4290(97)00093-2

Otegui, M.E., A.G. Cirilo, and G.A. Maddonni. 2007. Fundamentos, posibilidades y limitaciones para la intensificación del cultivo de maíz: Híbridos, nutrición, densidad. In: Congreso Mundo Agro 2007. Servicios y Marketing Agropecuarios, Buenos Aires. p. 67–73. https://sites.google.com/a/agro.uba.ar/granero/ (accessed 23 Dec. 2013).

Otegui, M.E., J.L. Mercau, and F.J. Menendez. 2002. Estrategias de manejo para la producción de maíz tardío y de segunda. In: E.H. Satorre, editor, Guía Dekalb del cultivo de maíz. Servicios y Marketing Agropecuario, Buenos Aires. p. 171–184. https://sites.google.com/a/agro.uba.ar/granero/ (accessed 23 Dec. 2013).

Otegui, M.E., M.G. Nicolini, R.A. Ruiz, and P. Dodds. 1995. Sowing date effects on grain yield components for different maize genotypes. Agron. J. 87:29–33. doi:10.2134/agronj1995.00021962008700010006x

Otegui, M.E., R.A. Ruiz, and D. Petruzzi. 1996. Modeling hybrid and sowing date effects on potential grain yield of maize in a humid temperate region. Field Crops Res. 47:167–174. doi:10.1016/0378-4290(96)00031-7

Peltzer, H.F. 2007. Efecto de la fecha de siembra sobre el rendimiento potencial de soja en Entre Rios. In: O. Paparotti, editor, Actualización técnica soja. Ser. Ext. 45. INTA EEA Paraná, Buenos Aires. p. 16–18.

Podestá, G.P., C.D. Messina, M.O. Grondona, and G.O. Magrin. 1999. Associations between grain crop yield in central-eastern Argentina and El Niño–Southern Oscillation. J. Appl. Meteorol. 38:1488–1498. doi:10.1175/1520-0450(1999)038<1488:ABGCYI>2.0.CO;2

Ray, D.K., N.D. Mueller, P.C. West, and J.A. Foley. 2013. Yield trends are insufficient to double global crop production by 2050. PLoS ONE 8(6):e66428. doi:10.1371/journal.pone.0066428

Ray, D.K., N. Ramankutty, N.D. Mueller, P.C. West, and J.A. Foley. 2012. Recent patterns of crop yield growth and stagnation. Nat. Commun. 3:1293. doi:10.1038/ncomms2296

Ritchie, J.T., and G. Alagarswamy. 2003. Model concepts to express genetic differences in maize yield components. Agron. J. 95:4–9. doi:10.2134/agronj2003.0004

SAGPYA. 2013. Secretaría de agricultura, ganadería, pesca y alimentos. Sistema integrado de información agropecuaria, Buenos Aires. http://www.siia.gov.ar/ (accessed 23 Dec. 2013).

Totis de Zeljkovich, L., and C.M. Rebella. 1980. Necesidades de agua de un cultivo de maíz en la región de Pergamino. In: Actas del II Congreso Nacional de Maíz. AIANBA, Buenos Aires. p. 211–219.

van Ittersum, M.K., and R. Rabbinge. 1997. Concepts in production ecology for analysis and quantification of agricultural input-output combinations. Field Crops Res. 52:197–208. doi:10.1016/S0378-4290(97)00037-3

Williams, W.P., J.B. Sagers, J.A. Hanten, F.M. Davis, and P.M. Buckley. 1997. Transgenic corn evaluated for resistance to fall armyworm and Southwestern corn borer. Crop Sci. 37:957–962. doi:10.2135/cropsci1997.0011183X003700030042x

Model Applications for Sustainable Intensification of Maize-Based Smallholder Cropping in a Changing World

Mary Ollenburger and Sieglinde Snapp*

Abstract

The challenge of sustainable intensification (SI) of smallholder production is a growing one, as climate change is acerbating an already vulnerable situation. Rainfed cereals are the mainstay of food security across much of sub-Saharan Africa, a system that is at risk from variable rainfall. Crop production response to fertility is known to interact strongly with water availability, in complex ways. There are few examples of model applications in a smallholder context to evaluate SI technologies under climate change scenarios. We present a case study from Malawi on alternate approaches to maize (*Zea mays* L.) intensification, evaluating the response at two sites with variable rainfall intensity for a range of climate scenarios and fertility levels. The SI technologies evaluated included intercrops and rotations of maize with pigeonpea [*Cajanus cajan* (L.) Millsp.], a bushy, food-producing legume crop, and comparisons with a continuous sole maize crop. This allowed testing of biological N fixation response, water use, and complementarity of resource partitioning over space and time, as well as consequences for maize and farming system performance. Competition and synergy were assessed through modeling growth, N stress, and drought stress. Interestingly, conditioning effects of soil fertility and site rainfall patterns were observed, with complex interactions and consequences for plant species and climate scenario performance. Higher fertility soils in some cases posed a risk of crop failure under poor rainfall distribution scenarios, as plant growth outstripped water availability. In summary, crop simulation modeling elucidated how tradeoffs among N and water limitations influenced yield and crop failure risk among sole crops, rotations, and intercrop systems.

Abbreviations: APSIM, Agricultural Production Systems Simulator; HF, high fertility; LF, low fertility; SI, sustainable intensification.

M. Ollenburger, Plant Production Systems Group, P.O. Box 430, 6700 AK Wageningen,Wageningen University, Wageningen, the Netherlands (mollenburger@gmail.com).

S. Snapp, Department of Plant Soil and Microbial Sciences, Michigan State University, 1066 Bogue St., East Lansing, MI 48824. *Corresponding author (snapp@msu.edu).

doi:10.2134/advagricsystmodel5.c14

Practical Applications of Agricultural System Models to Optimize the Use of Limited Water
Lajpat R. Ahuja, Liwang Ma, and Robert J. Lascano, Editors
Advances in Agricultural Systems Modeling, Volume 5. Lajpat R. Ahuja, Series Editor

USDA-ARS

Smallholder farming systems are crucial to food security and development prospects in sub-Saharan Africa and are particularly at risk of climatic variability such as poorly timed dry spells, extended drought, and excess rainfall (Funk et al., 2008). In this complex and changing environment, technical solutions have frequently failed and rarely been brought to scale (Snapp and Heong, 2003). Participatory action research and integrated use of participatory approaches linked to crop simulation modeling have provided notable exceptions, promoting the adoption of integrated agricultural management that is tailored to local requirements (Snapp et al., 2010; Tittonell et al., 2012). A systematic systems analysis and co-learning approach with local communities has supported the identification and rapid adoption of best bet options in some cases (Giller et al., 2006). Models can play a key role in this process, as in a climatically risky environment it is difficult to derive lessons from field experimentation because of the extreme variation in response from site to site and year to year (Robertson et al., 2005). Exploring the effects of climate change adds complexity to this process, particularly when incorporating complex climate models.

A promising approach explored here is to link easily understood climate change scenarios and crop simulation modeling with action research as a means to understand performance in a variable, and changing, climate (Chikowo et al., 2008; McCown et al., 2009); however, the challenge of linking approaches remains, and examples are rare (van Wijk et al., 2012).

Understanding farming systems and applying cropping system models to further that understanding is fundamental to building local capacity. This approach has been proposed as a means to support the innovation, experimentation, and information sharing that leads to adoption and more resilient farming techniques (O'Brien, 1998). Farmer preferences often value traits such as stability of production, meeting a diversity of requirements, labor and land saving traits, and locally preferred taste and storage traits above metrics of productivity, such as high yield (Mugwe et al., 2009). "Plausible bet" options and SI technologies have been identified that show promise as means to improve soil moisture holding capacity and drought resilience, but these options need to be tested and scaled out over space and time in an often risky climate and marginal production environment (Snapp and Heong, 2003).

Questions remain on how to use a range of crop and soil simulation tools to understand tradeoffs over time, where soil nitrogen and carbon buildups can be understood both in terms of positive feedbacks with aggradation of the soil resources and the risk of occasional crop failure in a variable climate. Many technologies promoted to enhance overall farming system productivity involve resource partitioning and competition for resources within and between species, and competition can play out over space and time (Defoer, 2002). For example, agroforestry systems generally involve maize grown in mixtures with trees or in temporal sequence, where maize is grown subsequent to a tree fallow. Similarly, maize–grain legume intercrops often involve tradeoffs in terms of maize yield, where competition for water and light may be compensated for by an enhanced nitrogen resource base from the presence of a nitrogen-fixing legume species. Models can help assess risk of crop failure in such complex systems and contribute to recommendations that are challenging to develop on the basis of highly variable results from farming systems research (Robertson et al., 2002; Sirrine et al., 2010).

Clearly a case can be made that assessment of climatic risk can be greatly strengthened through application of crop simulation models. This is frequently appreciated, but examples of integration of modeling fully with on-farm experimentation remain rare, particularly in terms of smallholder farming systems in developing countries. In northern Malawi we have shown that science-based interventions in maize-based farming systems, including integrated use of new legume crop genetics and targeted fertilizers, can drive agricultural intensification in a manner that supports local resilience to variable rainfall (Mhango et al., 2013; Snapp et al., 2010). Soil organic matter builds up, which has positive consequences for soil fertility over time, but competition for moisture may also occur, suppressing maize if water uptake is excessive from bushy legume species grown either in mixed intercrops or rotational sequences (Beedy et al., 2010; Ollenburger, 2012). It is difficult to assess which resources are most limiting under different rainfall patterns and cropping sequences, as the dynamics of soil nitrogen and water availability are highly complex and the environment is heterogeneous.

Malawi Case Study Modeling Maize–Legume Diversification

The model application presented here is based on over a decade of research focused on the introduction of diversity to support SI of maize production and soil fertility gains (Snapp et al., 2002, 2010). Grain legumes were selected by researchers in partnership with farmer groups in Malawi, using several key criteria. The legumes needed to be vigorous producers of nitrogen-enriched roots and leaves and should produce some edible grain, so as not to compromise household food security. They also needed to live longer than the 4 mo typical of annual crops. Mucuna [*Mucuna pruriens* (L.) DC.] and pigeonpea are the two chief examples, which the research team termed "semi-perennial" legumes (Kerr et al., 2007). Pigeonpea is already grown as a shrubby intercrop in some parts of East Africa, including southern Malawi, where farmers appreciate its slow growth early in the season and deep rooting, which ensure minimal competition when grown as an intercrop with maize (Snapp and Silim, 2002). It has a high nitrogen-fixing capacity and is appreciated by farmers for its drought tolerance and ability to produce food on degraded sites—at the least producing modest amounts of grain (Mhango et al., 2013).

There is convincing evidence from evaluations conducted across Malawi and other southern and East African countries that maize grain yields are consistently improved when moderate amounts of fertilizer are applied, in doses such as 30 kg ha^{-1} N that farmers can afford or access under government subsidy programs (Vanlauwe et al., 2014). There is less certainty about the effect of incorporation of legumes into the system. In some countrywide trials in Malawi, fertilizer use efficiency has been shown to rise sharply in maize–pigeonpea and maize–mucuna diversified fields, with concomitant marked gains in yield stability for maize grain. Legume rotation systems could produce equivalent quantities of grain with half the amount of fertilizer and on a more stable basis (yield variability reduced from 22 to 13%) compared with monoculture, with more variable results from maize–legume intercrops (Snapp et al., 2010). Both maize and legume performance suffer in years with variable rainfall and long dry spells (Mhango et al., 2013), and this raises questions about how widely applicable results from field experimentation are and what the implications are of increasingly variable rainfall under climate change scenarios. Modeling is one way that legume diversification could be assessed more broadly as a means for smallholders to cope with a variable climate in a sustainable manner (Robertson et al., 2005).

In this case study we apply the crop and soil simulation model Agricultural Production Systems Simulator (APSIM) to compare performance of sole maize and legume-diversified maize under current and plausible future climate scenarios. Northern Malawi sites were used, and sole-cropped versus maize–pigeonpea rotations were compared at two nitrogen fertility levels. The APSIM model was used on the basis of numerous studies conducted in eastern and southern Africa that have involved extensive calibration for simulating crop growth in a smallholder field environment (Gowing et al., 2003; Mwale et al., 2011; Ncube et al., 2009; Robertson et al., 2001). Climate scenarios were generated empirically from historical records, producing easily understood examples of plausible future conditions.

Our objective was to simulate crop response to current weather variability and climate scenarios in order to evaluate performance and risk of crop failure associated with maize and pigeonpea grown under continuous and rotated management at different fertility levels. Climate change scenarios were used to develop model-assisted understanding of response to temperature increases and rainfall changes for assessment of risk associated with a crop failure, and tradeoffs were elucidated among yield, nitrogen fertility, and water use.

Materials and Methods

Generation of Climate Scenarios from Current Weather Data

Both the total amount of rainfall and the distribution of rainfall—the number of rainy days, the number and distribution of dry spells, and the length of the rainy season—is critical to crop production (Barron et al., 2003). Therefore, changes in both amount and distribution should be considered when examining the sensitivity of a cropping system to changes in rainfall. To do this we developed a set of functions for use with the R statistical software package (R Development Core Team, 2011), which add, remove, or redistribute rainfall in various ways.

Changing Rainfall Distribution by Removing Rainy Days

Changing the rainfall distribution is accomplished by a set of functions that remove a percentage of existing rainfall from a user-selected fraction of rainy days or rainy spells. Rainy days are defined by a minimum threshold rainfall amount, with a default of 0.85 mm. The existing rainfall record provides, in each year, a stochastic rainfall pattern, so the removal is deterministic: a count is kept of rainy days, and the *n*th day is removed each time. It is possible to vary the days selected by varying the start of the count in order to check for artificially inflated or reduced rainfall totals based on single large events. Rainy spells can be of length one or greater. To apply rainfall removal functions to spells, the count value is increased by one on the first day with rain, where the previous day is dry, and thus remains the same throughout the spell.

A secondary set of functions redistributes removed rainfall by dividing it evenly among a specified fraction of remaining rainy days in order to reach a specified goal percentage of original rainfall. This percentage can be either greater than or less than the original total. By restricting these functions to a date-defined fraction of the full data set, rainfall changes can be constrained to calendar months or seasons.

Changing Season Length

Adjusting growing season length requires defining and identifying the start and end of the rainy season. This is done according to criteria that can be set by the user. Stern and Cooper (2011) define the first period of 3 d with greater than

20 mm total rainfall as the beginning of the rainy season. They define the end of the rainy season as the first date with less than 10 mm total rainfall in the preceding 10 d and no rainfall in the following 20 d. We use these definitions. The search domain for these dates will depend on the seasonal rainfall patterns in the area: for Malawi we use November through January for the start date and April through June for the end date—defining search domains reduces computing time. Once start and end dates are defined, rainfall removal and redistribution functions can be applied around these dates to shorten or extend the rainy season. This can be done either by counting rainy days from start or end dates or by creating a new data structure by extracting a period of days to be used as input to the global rainfall redistribution functions. Packages previously developed for R such as plyr (Wickham, 2011) allow functions to be applied individually to years within a longer data set or simultaneously to records from multiple stations, streamlining the creation of model-usable weather data files (source code available from M. Ollenburger, mollenburger@gmail.com).

Weather Data Used

The two sites used for this study were from northern Malawi, Zombwe (11.33°S, 33.82°E, altitude 143 masl), and north-central Malawi, Kasungu (13.03°S, 22.45°E, altitude 1036 masl). The weather data available from Malawi government extension stations included daily rainfall record, from 1945 to 2011 in Zombwe and from 1927 to 2010 in Kasungu. Temperature records in Kasungu were available from 1987, and in Zombwe only long-term monthly mean temperatures were available. Weatherman in the DSSAT 4.0 application suite (Pickering et al., 1994) was used to fill in missing data and to generate solar radiation estimates. The annual average temperature at Zombwe was 20.5°C and at Kasungu was 21.7°C; the mean annual rainfall was 783 mm yr^{-1} at Zombwe and 739 mm yr^{-1} at Kasungu.

Scenarios Used in Simulations

The rainfall scenarios used here were chosen to represent the range seen in climate model projections for the area. While some models project large percentage changes in rainfall during the dry season (June through September), rainfall amounts during these months are very low (Ollenburger, 2012). Overall, the projected changes in rainfall amounts are within 25% of current rainfall, so scenarios were developed within this range. A variety of distributions were tested for each rainfall amount to examine the sensitivity of yield to differing changes in distribution. Rainfall records were modified in several ways. First, a set of moderate rainfall reduction scenarios with rainfall levels of about 90% of current rainfall were generated (Group R90). Second, a group of increased-rainfall scenarios were generated with rainfall 110 to 125% of current values (Group R+). Third, a set of worst-case rainfall reduction scenarios were generated. These included reductions in rainfall across the growing season, as well as reductions targeted to critical early-season periods, all leading to approximately 75% of current rainfall (Group R75). Rainfall removal patterns in each scenario are described in Table 14–1. Each rainfall scenario was simulated for temperature increases of 0 to 4°C in increments of 1°. The APSIM baseline CO_2 concentration is 350 ppm and was used for all results reported here. We explored higher CO_2 concentration levels, but these had almost no detectable effect on maize yield and overpredicted pigeonpea yield relative to the limited data available (Saha et al., 2012). Clearly, the model requires further adjustments to accurately simulate CO_2 effects on pigeonpea (Ollenburger, 2012). Temperature increases of approximately 2°C are predicted by 2050, and an increase of between 3 and 4°C is predicted by the end of the century (Christensen et al., 2007).

Simulation Setup

The treatments simulated were continuous sole maize with the addition of fertilizer at two levels, moderate and recommended (24 and 92 kg ha^{-1} N), and a maize–pigeonpea rotation with 24 kg ha^{-1} N applied to maize, simulated with both maize and pigeonpea entry points to provide a maize yield value in each year. Fertilizer was applied as urea at 4 WAP in the treatments with 24 kg ha^{-1} N and in the treatments with 92 kg ha^{-1} N as a split application, with 24 kg ha^{-1} N at planting and 68 kg ha^{-1} N at 4 WAP. Simulations were run for the length of the rainfall record available at each site: 66 yr at Zombwe and 83 yr at Kasungu.

Soil carbon and nitrogen in these simulations were reset on 15 July, which occurred yearly in continuous maize treatments but only before planting pigeonpea in the case of the rotation system, to allow simulation of carryover effects from pigeonpea to maize. Soil water was allowed to carry over between seasons. Planting was set to occur on the first instance between 15 November and 30 January where there is greater than 20 mm of rainfall over 3 d, with a minimum soil moisture of 15 mm. This is a common planting condition for the area (Stern and Cooper, 2011). Eighty-five percent (85%) of maize residues were removed, following farmer practice and to avoid immobilization of soil N by residues. Pigeonpea residues were incorporated. Each treatment was simulated for two soil types that are representative of smallholder conditions (Mhango et al., 2013). The high fertility (HF) soil has relatively higher fertility and finer texture, with 63% sand, 27% clay, and 0.78% organic carbon (OC) in the topsoil layer, while the low fertility (LF) soil has lower fertility and coarser texture, with 79% sand, 14% clay, and 0.57% OC in the topmost layer.

Table 14–1. Rainfall scenarios used in climate change simulations. As described in Materials and Methods, rainfall distribution groups describe the level of rainfall relative to baseline. Group R90 has 90% of baseline rainfall, Group R75 has 75% of baseline rainfall, and Group R+ has additional rainfall at 110 and 125%. Individual letters correspond to the methods used to change rainfall distribution. Scenarios using the same letter use the same pattern of rainfall removal or addition at different intensities.

Rainfall removal or addition pattern		Mean annual rainfall at Zombwe	Mean annual rainfall at Kasungu	Mean season length at Zombwe	Mean season length at Kasungu
		mm		d	
Group R90 (90% of current rainfall)					
A	reduce rainfall 10% on each rainy day	707	667	145	129
B	remove all rainfall from 1 in 10 rainy days	716	669	148	130
C	remove 1 in 5 rainy days, redistribute rainfall on 1/5 of remaining days to reach 90% of initial rainfall amount	704	663	148	132
D	remove all rainfall from 1 in 10 rainy spells	711	683	145	130
Group R75 (75% of current rainfall)					
AA	reduce rainfall 25% on each rainy day	590	556	139	126
BB	remove all rainfall from 1 in 4 rainy days	594	566	139	126
CC	remove 1 in 3 rainy days, redistribute rainfall on 1/5 of remaining days to reach 75% of initial rainfall amount	586	552	139	123
E	remove 50% of rainfall from all rainy days for 60 d after start of rainy season	579	541	137	123
F	remove all rainfall from 1/2 of rainy days for 60 d after start of rainy season	587	539	136	118
Group R+ (110–125% of current rainfall)					
A+	increase rainfall 10% on each rainy day	865	815	154	135
AA+	increase rainfall 25% on each rainy day	983	926	157	135
B+	distribute additional rainfall to 1 in 5 rainy days to reach 110% of initial total rainfall	862	812	155	137
BB+	distribute additional rainfall to 1 in 5 rainy days to reach 125% of initial total rainfall	976	918	159	143

The model was found to simulate crop yields in on-farm research participatory trials conducted at Zombwe to within one standard deviation of the mean, and details of the model parameterization and evaluation were reported previously (Ollenburger, 2012).

Results

Rainfall in Climate Change Scenarios

The mean total rainfall in the climate change scenarios tested ranged from 537 to 926 mm at Kasungu and from 578 to 984 mm at Zombwe, compared with averages of 740 and 787 mm, respectively, in the base rainfall scenarios. Season start and end dates varied by scenario: even for scenarios where early-season rainfall was not targeted, start dates tended to be later in reduced rainfall scenarios, and end dates tended to be earlier (Table 14–1). These effects were small (less than 1 wk). Season start dates are similar at the two sites, near 1 December in base rainfall scenarios. End dates differed by nearly 3 wk between sites, from 7 April at Kasungu to 25 April at Zombwe.

Maize Yield

Overall, continuous maize at low N (24 kg ha^{-1} N fertilizer) produced moderate yields (1.5 to 2.3 t ha^{-1}) at both sites and soil types. Rotated maize at low N produced approximately twofold higher yields than continuous low N maize for all climate scenarios and sites and was only modestly affected by climate change scenarios. Continuous maize at high N (92 kg ha^{-1} N fertilizer) produced the highest maize yields (3.5 to 5 t ha^{-1}). Under climate change scenarios that imposed high temperature and moisture stress, yield of continuous maize at high N was reduced to levels equivalent to rotated maize at low N.

An examination of maize yield response to climate change scenarios shows that temperature effects varied with farming practice (Fig. 14–1). The moderate yields associated with low N continuous maize did not respond to changes in temperature. In contrast, substantial responses to temperature were observed for the higher yielding treatments. In continuous high N maize, yields at the Kasungu site were nearly constant with temperature increases up to 2°C, but above this yields fell to 15% below baseline yields. In Zombwe, all temperature increases resulted in substantial yield decreases in high N maize, with a 39% decrease at +4°C in the HF soil and a 17% decrease in the LF soil. Interestingly, a different pattern was seen for maize yields produced in a pigeonpea rotation. At both sites, maize yields in rotation increased slightly with temperature increases of 1 to 2°C and decreased slightly at higher temperatures (Fig. 14–1). The substantial decrease observed for yield of continuous maize in Zombwe at a 4°C increase in temperature was not observed for yield of rotated maize.

Overall, maize yield response to altered rainfall amount and distribution was modest for the 10% rainfall reduction scenarios (Fig. 14–1). However, for the 25% reduction scenarios, maize yield differences were observable at the more arid of the two sites (Kasungu). The least responsive farming practice was the low N continuous maize, which had moderate yields that did not vary much for any scenario. Yields were reduced in the rotation system under a 25% rainfall reduction, notably so at Kasungu in the sandy and low-fertility soil type.

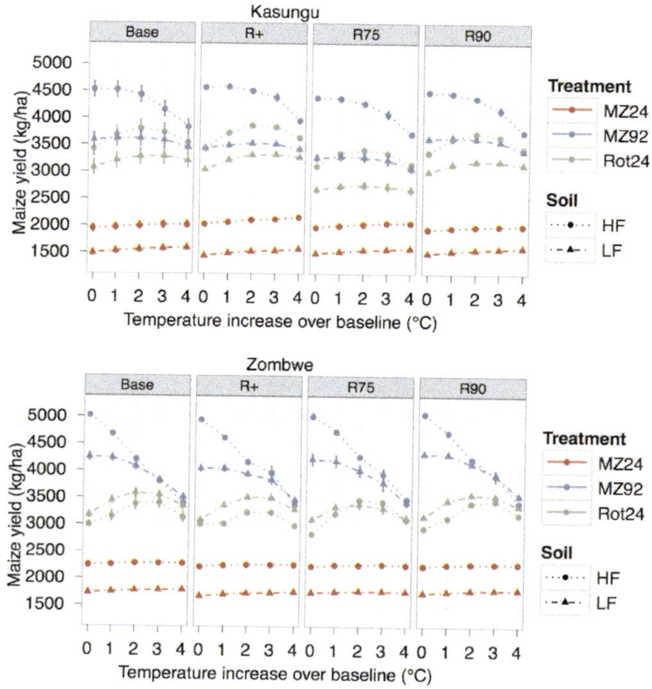

Fig. 14–1. Mean modeled maize yields in empirically generated climate scenarios based on changes in temperature and rainfall amount. Points are mean simulated yields averaged over 83 yr in Kasungu and 66 yr in Zombwe, northern Malawi. MZ24 is continuous maize with 24 kg ha^{-1} N, MZ92 is continuous maize with 92 kg ha^{-1} N, and Rot24 is a maize–pigeonpea rotation with 24 kg ha^{-1} N applied in maize years. Properties of HF (high fertility) and LF (low fertility) soils are described in Materials and Methods. Error bars in base rainfall are standard errors of the mean, describing year-to-year variability. Error bars in other rainfall groups are standard errors among scenario means within the group, describing group variability. Descriptions of rainfall scenarios are found in Table 14–1.

Yields are shown relative to the simulated yield under baseline rainfall in Fig. 14–2. This isolates the effect of rainfall from the effect of temperature and allows for comparison of this effect across treatments and soil types. With a 25% reduction in rainfall (R75 scenarios), yields in rotation at Kasungu were as much as 23% below yields at baseline rainfall, while at Zombwe yields remained within 15%. High N continuous maize had maximum yield reductions in R75 scenarios of 16% at Kasungu and 9% at Zombwe. The largest impacts in both rotation and high N maize occur for scenarios with yield reductions focused in the first 60 d of the season. The impact of reduced rainfall was greatest in the sandier soil (LF soil).

Rainfall increases did not always result in increased yields relative to baseline rainfall. For most rainfall and temperature combinations at Zombwe, impact of increased rainfall was negative, resulting in a maximum 10% reduction in

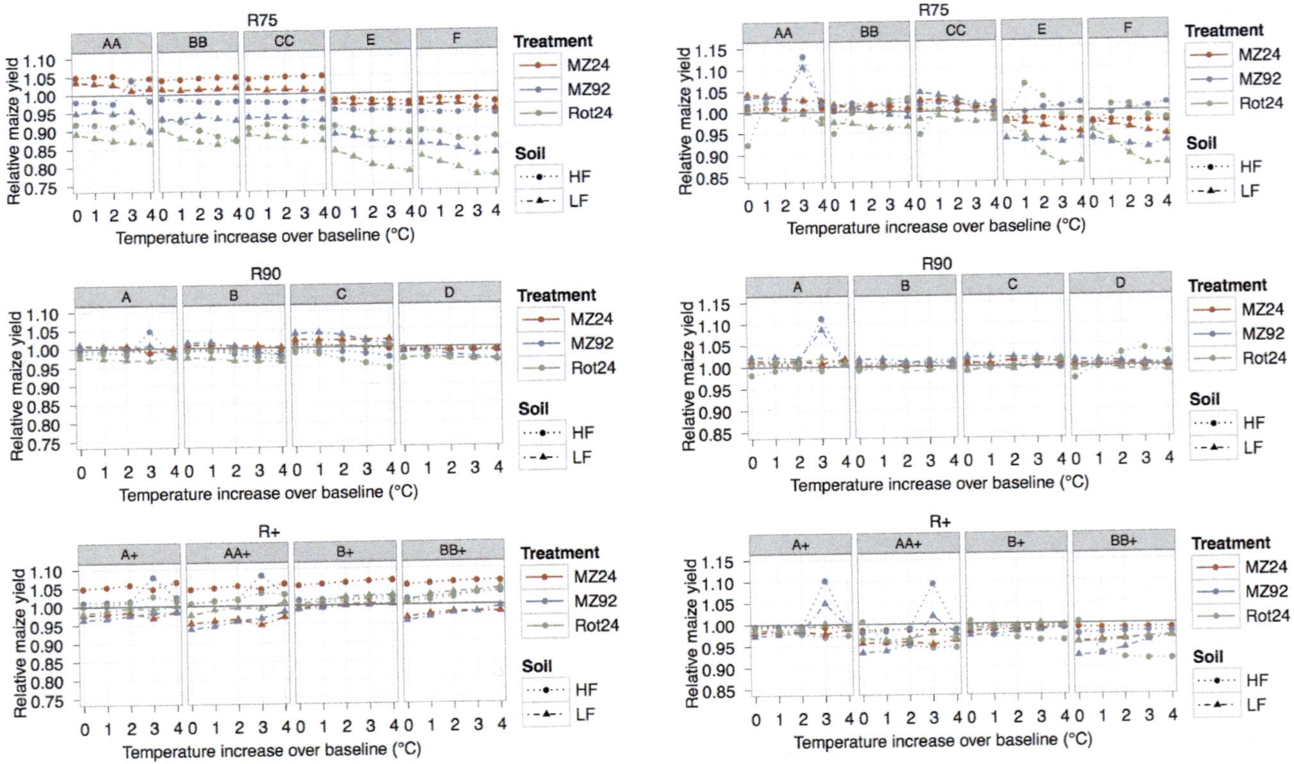

Fig. 14–2. Mean maize yields shown for 13 climate scenarios as a fraction of baseline maize yield (equals one for all treatments, soil types, and temperatures). Model results are from (a) 83 yr in Kasungu and (b) 66 yr in Zombwe. Treatments: MZ24 is continuous maize with 24 kg ha^{-1} N, MZ92 is continuous maize with 92 kg ha^{-1} N, and Rot24 is a maize–pigeonpea rotation with 24 kg ha^{-1} N applied to maize. Climate rainfall scenarios are per Table 14–1, and the two soil types are high fertility (HF) and low fertility (LF).

scenario BB+, where 25% rainfall increases are concentrated to one-fifth of rainy days (Table 14–1). At Kasungu, yields increased in increased-rain scenarios for rotation systems and for high N continuous maize in the higher fertility soil, but were reduced for the high N maize system in the sandier soil, particularly at low temperatures. In some A-type scenarios (Table 14–1), where the rainfall amount is adjusted on each rainy day, there are spikes in relative yield measurements at 3°C temperature increases; however, from Fig. 14–2 it is clear that these spikes correspond not to actual yield increases but to smaller decreases than in baseline rainfall scenarios. Overall, the combined effect of rainfall and temperature changes is to decrease yield differences due to farming practice.

Pigeonpea Yield

Pigeonpea yields were influenced substantially by location and soil type (Fig. 14–3). Maximum yields at the drier Kasungu site were just above 5 t ha^{-1}, whereas yields plateaued at 7 t ha^{-1} at the Zombwe site. The LF sandy soil was consistently

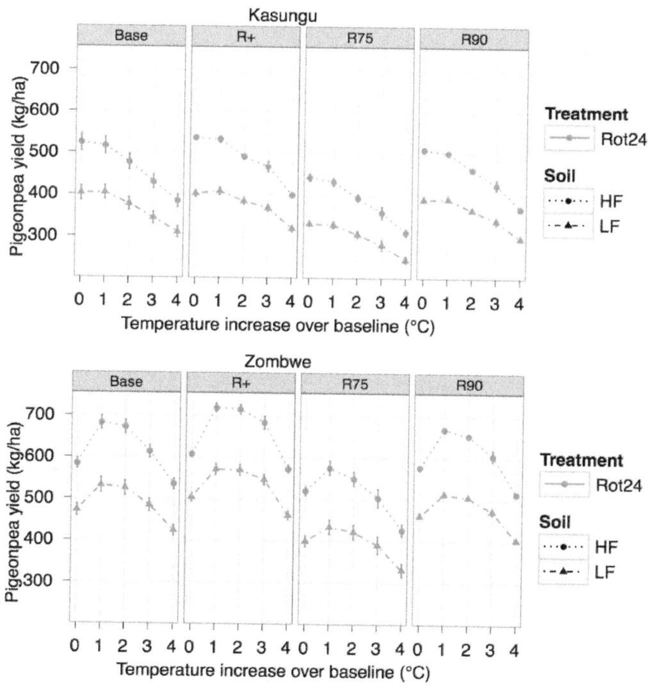

Fig. 14–3. Mean modeled pigeonpea yields in empirically generated climate scenarios based on changes in rainfall and temperature. Model results are from 83 yr in Kasungu and 66 yr in Zombwe. Pigeonpea is in rotation with maize, with 24 kg ha^{-1} N applied to maize. Properties of HF (high fertility) and LF (low fertility) soils are described in Materials and Methods. Error bars in base rainfall are standard errors of the mean, describing year-to-year variability. Error bars in other rainfall groups are standard errors among scenario means within the group, describing group variability. Descriptions of rainfall scenarios are found in Table 14–1.

associated with yields ~ 0.1 t ha^{-1} less than the HF soil (Fig. 14–3). Pigeonpea yield declined in scenarios with increased temperature at Kasungu, reaching a 25% yield decline at the maximum 4°C temperature increase. The Kasungu site has a baseline mean temperature 1°C higher than Zombwe and is drier. One cause of lower yields at higher temperatures is the decrease in length of the growing season. At Zombwe, season length was reduced by approximately 25 d for maize and 74 d for pigeonpea with a 4°C temperature increase. At Kasungu, for the same temperature increase, season length was reduced by approximately 20 d for maize, and 45 d for pigeonpea (baseline rainfall scenario).

Pigeonpea yield response to rainfall scenarios followed similar patterns, but were more marked than in maize (Fig. 14–3). Pigeonpea yields were reduced by up to 26% in low-rainfall scenarios at Kasungu and increased up to 17% above yields in base rainfall in increased rainfall scenarios. Pigeonpea yields at Zombwe were more affected by rainfall, with yield reductions of up to 35%, from a higher baseline yield (Fig. 14–3).

Water and Nitrogen Stress in Maize

Water and N stress scenarios are presented if Fig. 14–4, 14–5, 14–6, and 14–7. The APSIM model calculates a daily soil water stress value ranging from zero to one for photosynthesis, and we used this water stress value, summed over the growing season, as a measure of water stress in the crop. Figure 14–4 shows the percentage of years for the worst-case R75 scenarios, in which cumulative water stress is greater than 10 (equivalent to 10% yield reductions due to water stress). Water stress in maize is much more severe for continuous high N and rotated maize than for low N maize, as expected because of the N-limited yields of the latter (Fig. 14–6). Other factors play a role, such as the water demand associated with pigeonpea in the rotation system, which was generally associated with the highest water stress (at intermediate maize yield levels).

Water stress increases with both rising temperature and declining rainfall, contributing to the yield reductions seen in such scenarios. Yields decrease linearly so that at a cumulative water stress level of 25, yields were reduced by nearly 50%. Both rotation and high N maize show high levels of water stress in scenarios with temperature stress and rainfall deficit, and water stress levels in high N maize were highest in the sandier soil (LF soil). Interestingly, in LF soil maize in rotation has a lower percentage of stressed years than high N maize, while the pattern was reversed in HF soil (Fig. 14–6).

Cumulative N stress, as with cumulative soil water stress, was calculated as the sum of daily values of nitrogen stress for photosynthesis, which ranged from zero to one, over the growing season. Nitrogen stress reflected treatments

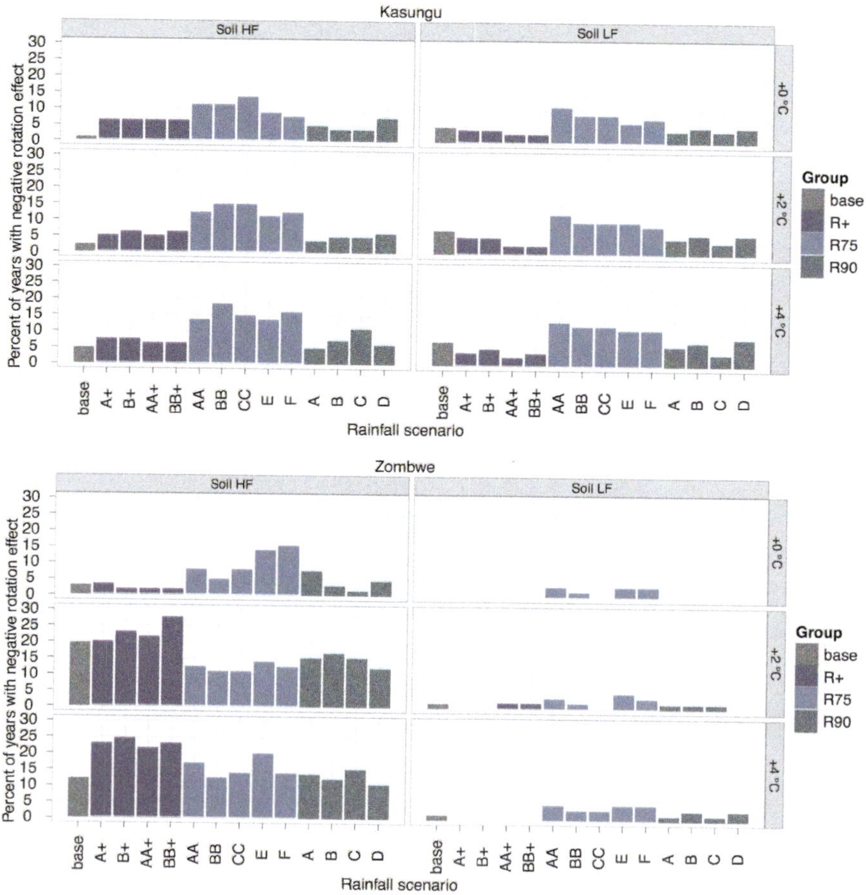

Fig. 14–4. Risk of yield reduction due to legume diversification in empirically generated climate scenarios. Risk is defined as the percentage of years where rotation yield is lower than continuous maize yield at the same N fertilization rate (24 kg ha^{-1} N). Model results are from 83 yr in Kasungu and 66 yr in Zombwe, northern Malawi. Descriptions of rainfall scenarios are found in Table 14–1. Soil types are defined in Materials and Methods. HF, high fertility; LF, low fertility.

imposed, and were markedly high in the continuous maize low N treatment (Fig. 14–5). It was possible to minimize N stress to low levels and produce twofold higher maize yields in rotated maize. Trends in N stress were similar for all rainfall scenarios at Kasungu, and only the baseline rainfall is shown. Decreases in N stress were slightly lower at Zombwe in high rainfall scenarios (data not shown).

Nitrogen stress, unlike water stress, decreased with increased temperature in climate scenarios (Fig. 14–5). The reduction in N stress with temperature gain was large in the rotation system, around 70% at Kasungu and 84% at Zombwe. In low N continuous maize, the reduction in N stress was 37% at Kasungu and 58% at Zombwe in the HF soil. Levels were reduced in high N maize as well; however,

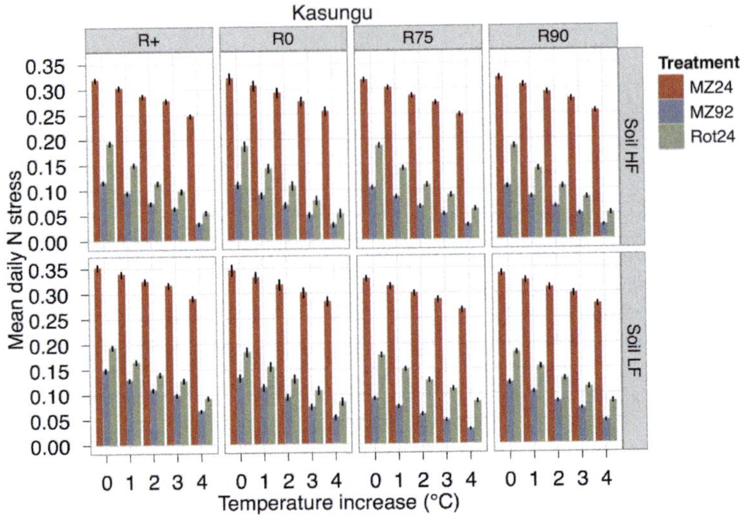

Fig. 14–5. Mean daily N stress under increases in temperature. Model results are from 83 yr in Kasungu, northern Malawi. MZ24 is continuous maize with 24 kg ha^{-1} N, MZ92 is continuous maize with 92 kg ha^{-1} N, and Rot24 is a maize–pigeonpea rotation with 24 kg ha^{-1} N applied in maize years. Soil types are defined in Materials and Methods. HF, high fertility; LF, low fertility.

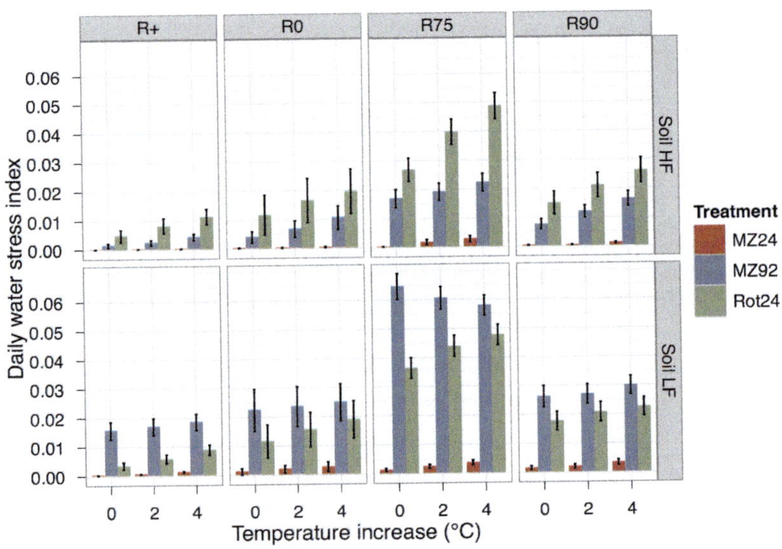

Fig. 14–6. Mean daily soil water stress under changes in rainfall and temperature. Model results are from 83 yr in Kasungu, northern Malawi. MZ24 is continuous maize with 24 kg ha^{-1} N, MZ92 is continuous maize with 92 kg ha^{-1} N, and Rot24 is a maize–pigeonpea rotation with 24 kg ha^{-1} N applied to maize. Descriptions of rainfall scenarios are found in Table 14–1. Soil types are defined in Materials and Methods. HF, high fertility; LF, low fertility.

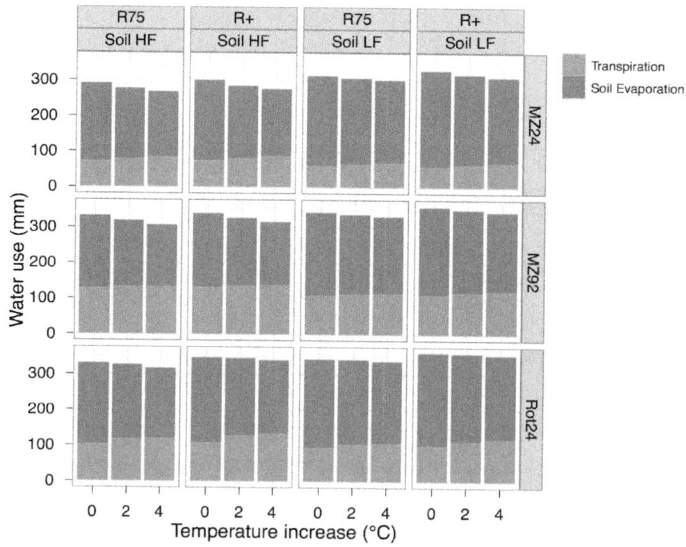

Fig. 14–7. Water use for maize transpiration and soil evaporation under changes in rainfall and temperature. Model results are from 83 yr in Kasungu, northern Malawi. MZ24 is continuous maize with 24 kg ha^{-1} N, MZ92 is continuous maize with 92 kg ha^{-1} N, and Rot24 is a maize–pigeonpea rotation with 24 kg ha^{-1} N applied to maize. Descriptions of rainfall scenarios are found in Table 14–1. Soil types are defined in Materials and Methods. HF, high fertility; LF, low fertility.

this was from a lower baseline than the other two treatments and thus had a small impact on yields.

Rotation Effects under Climate Change

Maize yields in rotation are between 150 and 200% of yields in continuous maize systems with the same fertilization rate at base rainfall and temperature, with greater increases seen in the sandier soil type (Fig. 14–1). This simulated rotation response is reduced under lower rainfall and higher temperature scenarios at the more arid site, Kasungu, but almost no effect is seen at Zombwe. This is presumably due to differences in water availability and thus maize water stress at these two sites.

Because of the risk-averse nature of smallholder farmers, it is important to examine risk of decreased yield (Fig. 14–4). This risk represents the chance that a farmer will see reduced yields following pigeonpea compared with the yield following maize. Risk is low at baseline temperature and rainfall, particularly for the Zombwe LF soil (<5% chance of yield reductions in any scenario). At Kasungu, risk generally increased with rising temperature and diminished rainfall, and the mean effect of rotation was similar for both soil types. At Zombwe a quite different response was observed (Fig. 14–4). A rotation on HF soil was risky:

increased rainfall—particularly rainfall concentrated to a few extreme events—and decreased rainfall increased risk relative to baseline rainfall.

Discussion

It is clear that the dominant constraint to crop production in typical smallholder systems is soil fertility, and that will remain the case for low fertility systems in southern Africa under even relatively extreme climate change predictions. As discussed elsewhere (Dimes et al., 2008), the challenge of intensification of smallholder crop production remains, even as climate change adaptation receives increased attention. We see here that for continuous maize with low fertilizer use, productivity remains very low, with minimal impact from climate change. This concurs with findings of a recent review that nitrogen fertility is critical for improving agricultural productivity and conservation on African smallholder farms (Vanlauwe et al., 2014).

Climate change scenarios that imposed the greatest stress (+4 temperature and 25% reduction in rainfall) did have a negative impact on crop yields; this was most noticeable for pigeonpea and for high yield maize systems (rotation and continuous high N). The simulations run here were able to extend and deepen our understanding of the water and nitrogen stress tradeoffs associated with the maize grown under continuous versus legume-diversified management and the interaction with climate change scenarios. Interestingly, reduction in maize yield associated with temperature gains hardly occurred in rotated maize (in contrast to continuous maize), which was related to an inverse relationship between N stress and temperature stress in this treatment. That is, as temperature increased, N stress was reduced in rotated maize.

Maize rotated with pigeonpea was evaluated in the climate change scenarios at 24 kg ha^{-1} N fertilizer. This should be considered in the context of smallholder farming: this treatment uses 12 kg ha^{-1} N fertilizer on an annual basis, which could well be an affordable level for many resource-poor farmers, and is not too far off the ~10 kg ha^{-1} fertilizer reported to be used in Africa (Morris et al., 2007). The importance of moderate fertilizer doses for profitable use of this input has been highlighted for African smallholders and is of particular importance for reducing yield variability in a climate change context (Snapp et al., 2003). However, questions are frequently raised about the effectiveness of moderate doses of fertilizer, as well as the challenges associated with incorporating legumes as supplemental fertility sources, related to competition for water and other resources (Mwangi, 1997). Indeed, high doses of fertilizer applied to sole-cropped maize are widely recommended, for example, 92 kg ha^{-1} N for hybrid maize in Malawi (Mhango et al., 2013; Snapp et al., 2003).

Fertilizer recommendations may need to be modified to take into account emerging evidence from studies such as this. Our modeling results provide evidence that moderate doses of fertilizer can provide substantial climate-moderating and productivity-enhancing services, at least when applied in combination with a legume rotation. An important result was no detectable increase in crop failure risk associated with a crop rotation involving a long-lived legume. A notable exception was the combination of high fertility soil and a dry site, where crop failure risk was exacerbated by pigeonpea. In such cases, a shorter duration legume or water conserving practices may need to be implemented (Jensen et al., 2003).

It is also important to note that climate change scenarios with decreases in rainfall had a larger negative impact on maize yields grown in rotation than on continuous maize, particularly for rainfall decreases focused near the beginning of the growing season (Fig. 14–8). This vulnerability to early-season rainfall reductions is particularly of interest since downscaled climate models generally project rainfall decreases for Malawi in November and December. Therefore, early-season rainfall reductions similar to those described by scenarios E and F can be expected (Table 14–1). This illustrates a potential challenge to the adoption of legume rotations in a changing climate.

Overall, the model findings reported here are consistent with results from field experimentation in East and southern Africa (Ncube et al., 2007; Nyende

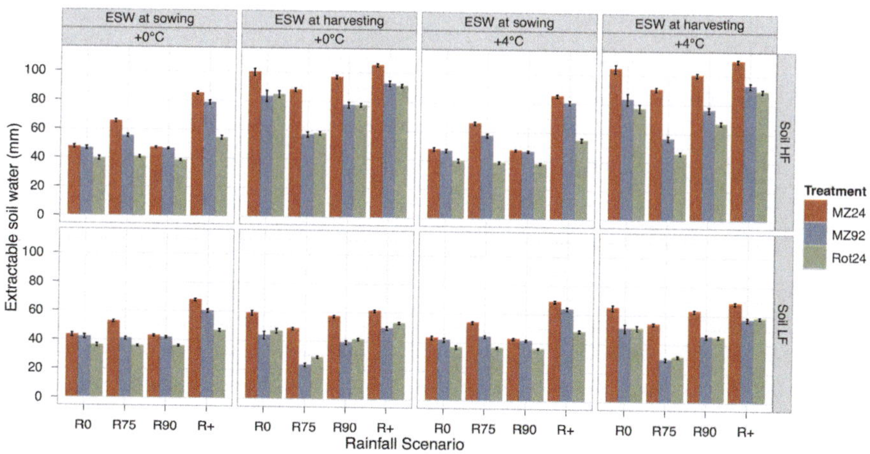

Fig. 14–8. Initial and residual soil moisture in maize growth under changes in rainfall and temperature. Model results are from 83 yr in Kasungu, northern Malawi. MZ24 is continuous maize with 24 kg ha^{-1} N, MZ92 is continuous maize with 92 kg ha^{-1} N, and Rot24 is a maize–pigeonpea rotation with 24 kg ha^{-1} N applied to maize. Descriptions of rainfall scenarios are found in Table 14–1. Soil types are defined in Materials and Methods. HF, high fertility; LF, low fertility.

and Delve, 2004; Snapp et al., 2010; Waddington and Karigwindi, 2001). These on-farm action research studies have documented production benefits and fertilizer efficiency associated with incorporating a multipurpose legume (e.g., pigeonpea or mucuna) rotation with cereal production. Of particular interest to adaptation to climate change were the results from a countrywide study in Malawi that indicated yield stability of maize response to fertilizer could be improved substantially through a rotation with a long-lived legume (Snapp et al., 2010). However, another study in southern Malawi indicated that crop failure risk was unacceptably high in some maize–legume diversified systems (Sirrine et al., 2010), and surveys have shown that low yield potential legumes can be unattractive to farmers (Snapp et al., 2002). The APSIM model can be used to extend these types of results to further explore benefits and risk associated with maize–legume rotations over space and time.

Our finding of a strong performance of maize yield in a rotation occurred despite increases in soil water stress with increased temperature and decreased rainfall. Water stress provides an explanation for the relatively better performance of rotation systems in the sandier soil, as stress levels in this soil tend to be lower for rotation systems than for continuous maize with a high N fertilization rate. This may seem counterintuitive given that the long-duration pigeonpea is extracting water for a larger fraction of the year, and soil water levels near planting are indeed lowest in these systems. However, from examination of daily water content in some simulations, it appears that while the initial soil water content is low in rotation, recharge occurs rapidly, particularly in well-drained soils. This effect was also seen in field experiments in Zimbabwe by Ncube et al. (2007). Because of this rapid recharge, the initially low water content does not lead to water stress conditions throughout the growing season. The rotation systems also include a higher amount of cover from the long-duration crop, and pigeonpea residues remain in the field until pre-sowing tillage. Increased soil cover, such as that provided by a leafy long-duration crop like pigeonpea, has been shown previously to mitigate some of the water stress effects seen under climate change scenarios (Cooper et al., 2008).

In our simulations soil type not only had a marked effect on crop response but also influenced the level of yield reduction imposed by climate change scenarios. In the sandier LF soil, the magnitude of yield reductions due to temperature increases was lower. This effect may have been mediated by a positive influence on soil N mineralization in sandy soils balancing a negative direct effect of temperature on plant growth. At the same time, in sandy soils rainfall effects were magnified. This is presumably due to more rapid drainage in sandy soils exacerbating soil moisture stress with lower rainfall. A clear effect of soil fertility was shown in Zombwe for HF soils, as rotated maize yield was lower than continuous maize yield 20% of the time. In contrast, for the LF soil at the same site, almost no risk of yield reduction occurred in rotated maize.

This result illustrates the complex interactions of plant species and climate scenario responses, as conditioned by soil fertility and site rainfall patterns. In sum, the rotation effect was influenced by tradeoffs between N and water limitations, and it was not straightforward to predict for different soil types. This provides a clear illustration of the insights that can be provided by crop and soil simulation modeling. Overall, our results indicate that regardless of changes in climate, yields in maize-based farming can be substantially increased by changing management practices. Modest use of N fertilizer can provide substantial benefits when applied to a maize crop rotated with pigeonpea, including moderation of the negative effects on yield that would otherwise accrue with a rising temperature.

Acknowledgements

We sincerely thank the McKnight Foundation's Collaborative Crops Research Program for support on multiple levels. The authors would also like to acknowledge Dr. Wezi Mhango for providing field data used in model calibration. Thank you to David and Roger Stern, who helped define the rainfall scenarios, to Mawora Thomas Mwakudisa who wrote the GenStat code that M. Ollenburger developed into what are now called the B and C scenarios, and to Danielle Zoellner for valuable assistance with editing. The Malawi farmer research groups are a source of ongoing inspiration, and we could not have done this research without technical assistance in Malawi from Mr. Ladson Chirwa, Ms. Esnart Kanyenda, and Ms. Lizzie Shumba.

References

Barron, J., J. Rockstrom, F. Gichuki, and N. Hatibu. 2003. Dry spell analysis and maize yields for two semi-arid locations in east Africa. Agric. For. Meteorol. 117:23–37. doi:10.1016/S0168-1923(03)00037-6

Beedy, T.L., S.S. Snapp, F.K. Akinnifesi, and G.W. Sileshi. 2010. Impact of *Gliricidia sepium* intercropping on soil organic matter fractions in a maize-based cropping system. Agric. Ecosyst. Environ. 138:139–146. doi:10.1016/j.agee.2010.04.008

Chikowo, R., M. Corbeels, P. Tittonell, B. Vanlauwe, A. Whitbread, and K.E. Giller. 2008. Aggregating field-scale knowledge into farm-scale models of African smallholder systems: Summary functions to simulate crop production using APSIM. Agric. Syst. 97:151–166. doi:10.1016/j.agsy.2008.02.008

Christensen, J.H., B. Hewitson, A. Busuioc, A. Chen, X. Gao, I. Held, R. Jones, R.K. Kolli, W.-T. Kwon, R. Laprise, V. Magaña Rueda, L. Mearns, C.G. Menéndez, J. Räisänen, A. Rinke, A. Sarr, and P. Whetton. 2007. Regional climate projections. In: S. Solomon et al., editors, Climate change 2007: The physical science basis. Contribution of Working Group I to the Fourth Assessment Report of the Intergovernmental Panel on Climate Change. Cambridge Univ. Press, Cambridge, UK.

Cooper, P.J.M., J. Dimes, K.P.C. Rao, B. Shapiro, B. Shiferaw, and S. Twomlow. 2008. Coping better with current climatic variability in the rain-fed farming systems of sub-Saharan Africa: An essential first step in adapting to future climate change? Agric. Ecosyst. Environ. 126:24–35. doi:10.1016/j.agee.2008.01.007

Defoer, T. 2002. Learning about methodology development for integrated soil fertility management. Agric. Syst. 73:57–81. doi:10.1016/S0308-521X(01)00100-7

Dimes, J., P. Cooper, and K. Rao. 2008. Climate change impact on crop productivity in the semi-arid tropics of Zimbabwe in the 21st century. In: Proceedings of the Workshop on Increasing the Productivity and Sustainability of Rainfed Cropping Systems of Poor, Smallholder Farmers, Tamale, Ghana. 22–25 Sept. 2008. CGIAR Challenge Program on Water and Food, Colombo, Sri Lanka. p. 189–198.

Funk, C., M.D. Dettinger, J.C. Michaelsen, J.P. Verdin, M.E. Brown, M. Barlow, and A. Hoell. 2008. Warming of the Indian Ocean threatens eastern and southern African food security but could be mitigated by agricultural development. Proc. Natl. Acad. Sci. USA 105:11081–11086. doi:10.1073/pnas.0708196105

Giller, K.E., E.C. Rowe, N. de Ridder, and H. van Keulen. 2006. Resource use dynamics and interactions in the tropics: Scaling up in space and time. Agric. Syst. 88:8–27. doi:10.1016/j.agsy.2005.06.016

Gowing, J.W., M.D.B. Young, N. Hatibu, H.F. Mahoo, F. Rwehumbiza, and O.B. Mzirai. 2003. Developing improved dryland cropping systems for maize in semi-arid Tanzania. Part 11. Use of a model to extrapolate and add value to experimental results. Exp. Agric. 39:293–306. doi:10.1017/S0014479703001297

Jensen, J.R., R.H. Bernhard, S. Hansen, J. McDonagh, J.P. Moberg, N.E. Nielsen, and E. Nordbo. 2003. Productivity in maize based cropping systems under various soil-water-nutrient management strategies in a semi-arid, alfisol environment in East Africa. Agric. Water Manage. 59:217–237. doi:10.1016/S0378-3774(02)00151-8

Kerr, R.B., S. Snapp, M. Chirwa, L. Shumba, and R. Msachi. 2007. Participatory research on legume diversification with Malawian smallholder farmers for improved human nutrition and soil fertility. Exp. Agric. 43:437–453. doi:10.1017/S0014479707005339

McCown, R.L., P.S. Carberry, Z. Hochman, N.P. Dalgliesh, and M.A. Foale. 2009. Re-inventing model-based decision support with Australian dryland farmers. 1. Changing intervention concepts during 17 years of action research. Crop Pasture Sci. 60:1017–1030. doi:10.1071/CP08455

Mhango, W.G., S.S. Snapp, and G.Y.K. Phiri. 2013. Opportunities and constraints to legume diversification for sustainable maize production on smallholder farms in Malawi. Renewable Agric. Food Syst. 38:234–244. doi:10.1017/s1742170512000178

Morris, M., V.A. Kelly, R.J. Kopicki, and D. Byerlee. 2007. Fertilizer use in African agriculture: Lessons learned and good practice guidelines. World Bank Publ., Washington, DC.

Mugwe, J., D. Mugendi, M. Mucheru-Muna, R. Merckx, J. Chianu, and B. Vanlauwe. 2009. Determinants of the decision to adopt integrated soil fertility management practices by smallholder farmers in the central highlands of Kenya. Exp. Agric. 45:61–75. doi:10.1017/S0014479708007072

Mwale, C.D., V.H. Kabambe, W.D. Sakala, K.E. Giller, A.A. Kauwa, I. Ligowe, and D. Kamalongo. 2011. Model validation through long-term promising sustainable maize–pigeon pea residue management in Malawi. In: A. Bationo et al., editors, Innovations as key to the green revolution in Africa. Springer, Dordrecht, the Netherlands. p. 325–333.

Mwangi, W.M. 1997. Low use of fertilizers and low productivity in sub-Saharan Africa. Nutr. Cycl. Agroecosyst. 47:135–147. doi:10.1007/BF01991545

Ncube, B., J.P. Dimes, M.T. van Wijk, S.J. Twomlow, and K.E. Giller. 2009. Productivity and residual benefits of grain legumes to sorghum under semi-arid conditions in south-western Zimbabwe: Unravelling the effects of water and nitrogen using a simulation model. Field Crops Res. 110:173–184. doi:10.1016/j.fcr.2008.08.001

Ncube, B., S.J. Twomlow, M.T. Wijk, J.P. Dimes, and K.E. Giller. 2007. Productivity and residual benefits of grain legumes to sorghum under semi-arid conditions in southwestern Zimbabwe. Plant Soil 299:1–15. doi:10.1007/s11104-007-9330-5

Nyende, P., and R.J. Delve. 2004. Farmer participatory evaluation of legume cover crop and biomass transfer technologies for soil fertility improvement using farmer criteria, preference ranking and logit regression analysis. Exp. Agric. 40:77–88. doi:10.1017/S0014479703001443

O'Brien, R. 1998. An overview of the methodological approach of action research. In: R. Richardson, editor, Theory and practice of action research. Univ. Federal da Paraíba, João Pessoa, Brazil.

Ollenburger, M. 2012. Modeling integrated soil fertility management options in Malawi under variable soil and climate conditions. M.S. thesis, Department of Crop and Soil Science, Michigan State Univ., Lansing.

Pickering, N.B., J.W. Hansen, J.W. Jones, C.M. Wells, V.K. Chan, and D.C. Godwin. 1994. Weatherman—A utility for managing and generating daily weather data. Agron. J. 86:332–337. doi:10.2134/agronj1994.00021962008600020023x

R Development Core Team. 2011. R: A language and environment for statistical computing. R Foundation for Statistical Computing, Vienna, Austria. http://www.R-project.org (accessed 27 Aug. 2014).

Robertson, M.J., P.S. Carberry, Y.S. Chauhan, R. Ranganathan, and G.J. O'Leary. 2001. Predicting growth and development of pigeonpea: A simulation model. Field Crops Res. 71:195–210. doi:10.1016/S0378-4290(01)00160-5

Robertson, M.J., P.S. Carberry, N.I. Huth, J.E. Turpin, M.E. Probert, P.L. Poulton, M. Bell, G.C. Wright, S.J. Yeates, and R.B. Brinsmead. 2002. Simulation of growth and development of diverse legume species in APSIM. Aust. J. Agric. Res. 53:429–446. doi:10.1071/AR01106

Robertson, M.J., W. Sakala, T. Benson, and Z. Shamudzarira. 2005. Simulating response of maize to previous velvet bean (*Mucuna pruriens*) crop and nitrogen fertiliser in Malawi. Field Crops Res. 91:91–105. doi:10.1016/j.fcr.2004.06.009

Saha, S., V.K. Sehgal, S. Nagarajan, and M. Pal. 2012. Impact of elevated atmospheric CO_2 on radiation utilization and related plant biophysical properties in pigeon pea (*Cajanus cajan* L.). Agric. For. Meteorol. 158–159:63–70. doi:10.1016/j.agrformet.2012.02.003

Sirrine, D., C. Shennan, S. Snapp, G. Kanyama-Phiri, B. Kamanga, and J.R. Sirrine. 2010. Improving recommendations resulting from on-farm research: Agroforestry, risk, profitability and vulnerability in southern Malawi. Int. J. Agric. Sustainability 8:290–304. doi:10.3763/ijas.2010.0471

Snapp, S.S., M.J. Blackie, and C. Donovan. 2003. Realigning research and extension to focus on farmers' constraints and opportunities. Food Policy 28:349–363. doi:10.1016/j.foodpol.2003.08.002

Snapp, S.S., M.J. Blackie, R.A. Gilbert, R. Bezner-Kerr, and G.Y. Kanyama-Phiri. 2010. Biodiversity can support a greener revolution in Africa. Proc. Natl. Acad. Sci. USA 107:20840–20845. doi:10.1073/pnas.1007199107

Snapp, S.S., and K.L. Heong. 2003. Scaling up: Participatory research and extension to reach more farmers. In: B. Pound et al., editors, Uniting science and participation: Managing natural resources for sustainable livelihoods. IRDC, Ontario, Canada. p. 67–87.

Snapp, S., G. Kanyama-Phiri, B. Kamanga, R. Gilbert, and K. Wellard. 2002. Farmer and researcher partnerships in Malawi: Developing soil fertility technologies for the near-term and far-term. Exp. Agric. 38:411–431. doi:10.1017/S0014479702000443

Snapp, S.S., and S.N. Silim. 2002. Farmer preferences and legume intensification for low nutrient environments. Plant Soil 245:181–192. doi:10.1023/A:1020658715648

Stern, R.D., and P.J.M. Cooper. 2011. Assessing climate risk and climate change using rainfall data—A case study from Zambia. Exp. Agric. 47:241–266. doi:10.1017/S0014479711000081

Tittonell, P., E. Scopel, N. Andrieu, H. Posthumus, P. Mapfumo, M. Corbeels, G.E. van Halsema, R. Lahmar, S. Lugandu, J. Rakotoarisoa, F. Mtambanengwe, B. Pound, R. Chikowo, K. Naudin, B. Triomphe, and S. Mkomwa. 2012. Agroecology-based aggradation-conservation agriculture (ABACO): Targeting innovations to combat soil degradation and food insecurity in semi-arid Africa. Field Crops Res. 132:168–174. doi:10.1016/j.fcr.2011.12.011

van Wijk, M.T., M.C. Ruffo, D. Enahoro, D. Parsons, S. Silvestri, R.O. Valdivia, and M. Herrero. 2012. A review on farm household modelling with a focus on climate change adaptation and mitigation. CGIAR, Copenhagen, Denmark.

Vanlauwe, B., J. Wendt, K.E. Giller, M. Corbeels B. Gerard, and C. Nolte. 2014. A fourth principle is required to define Conservation Agriculture in sub-Saharan Africa: The appropriate use of fertilizer to enhance crop productivity. Field Crops Res. 155:10–13. doi:10.1016/j.fcr.2013.10.002

Waddington, S.R., and J. Karigwindi. 2001. Productivity and profitability of maize plus groundnut rotations compared with continuous maize on smallholder farms in Zimbabwe. Exp. Agric. 37:83–98. doi:10.1017/S001447970100103X

Wickham, H. 2011. The split-supply-combine strategy for data analysis. J. Stat. Softw. 40:1–29.

Syntheses of the Current Model Applications for Managing Water and Needs for Experimental Data and Model Improvements to Enhance these Applications

L.R. Ahuja,* Liwang Ma, Robert J. Lascano, S.A. Saseendran, Q.X. Fang, David C. Nielsen, Enli Wang, and Paul D. Colaizzi

Abstract

This volume of the Advances in Agricultural Systems Modeling series presents 14 different case studies of model applications to help make the best use of limited water in agriculture. These examples show that models have tremendous potential and value in enhancing site-specific water management for different soils and climates, and evaluating cropping system sustainability over the longer term, when model results are integrated with the available field measurements in experimental studies. Here we summarize applications of 11 system models commonly reported in the literature for agricultural water management along with those presented in this volume. These 11 models vary greatly in simulating agricultural system components for water balance related processes (the differences in crop growth and N balance processes were even greater among models), which need to be kept in mind when reading about the applications of each model. A sensor-based automated irrigation scheduling system is presented. Finally, we summarize further needs for experimental data and model improvements to enhance future water management applications.

Abbreviations: ABA, abscisic acid; AgMIP, Agricultural Model Inter-Comparison and Improvement Project; APEX, Agricultural Policy Environmental eXtender model; APSIM, Agricultural Production Systems Simulator; CGMS, Crop Growth Monitoring System; CSM, cropping system model; CWPF, crop water production function; DSS, decision support system; DSSAT, Decision Support System for Agrotechnology Transfer; EPIC, Environmental Policy-Integrated Climate; ET, evapotranspiration; IRT, infrared thermometer; NCP, North China Plain; PALMS, Precision Agricultural Landscape Modeling System; PSA, plant–soil–atmosphere; RZWQM2, Root Zone Water Quality Model; STICS, Simulateur mulTIdisciplinaire pour les Cultures Standard; SWAP, Soil Water Atmosphere Plant; TSEB, two-source energy balance; WOFOST, WOrld FOod STudies; WUE, water use efficiency.

L.R. Ahuja (laj.ahuja@ars.usda.gov), Liwang Ma (Liwang.Ma@ars.usda.gov), S.A. Saseendran (Saseendran.anapalli@ars.usda.gov), USDA-ARS, Agricultural Systems Research Unit, 2150 Centre Ave., Bldg. D, Ste. 200 Fort Collins, CO 80526. *Corresponding author.

Q.X. Fang, Agronomy College, Qingdao Agricultural University, Changcheng Rd. 700, Chengyang District, Qingdao, Shandong, China, 266108 (fangqx@igsnrr.ac.cn).

Robert J. Lascano, USDA-ARS, Wind Erosion and Water Conservation Research Unit, Cropping Systems Research Laboratory, 3810 4th St., Lubbock, TX 79415 (robert.lascano@ars.usda.gov).

David C. Nielsen, USDA-ARS, Central Plains Resources Management Research Unit, 40335 County Rd. GG, Akron, CO, 80720-0400 (david.nielsen@ars.usda.gov).

Enli Wang, CSIRO Land and Water, Christian Laboratory, Clunies Ross St., Black Mountain ACT 2601, Australia (Enli.Wang@csiro.au).

Paul D. Colaizzi, USDA-ARS, Conservation and Production Research Lab., P.O. Drawer 10, 2300 Experiment Station Rd., Bushland, TX 79012-0010 (Paul.Colaizzi@ars.usda.gov).

doi:10.2134/advagricsystmodel5.c15

Practical Applications of Agricultural System Models to Optimize the Use of Limited Water
Lajpat R. Ahuja, Liwang Ma, and Robert J. Lascano, Editors
Advances in Agricultural Systems Modeling, Volume 5. Lajpat R. Ahuja, Series Editor

USDA-ARS

As we review the case studies presented in this volume, see Table 15–1 for a summary of how the 11 commonly used models differ in simulating agricultural system components.

THE AUSTRALIAN APSIM MODEL

The Agricultural Production Systems Simulator (APSIM) modeling framework (Keating et al., 2003; Wang et al., 2002) was developed to simulate the biophysical processes in farming systems as they interact with environmental changes and management intervention to determine the productivity and environmental impact of farming systems. APSIM can simulate more than 30 crops (including pasture and tree species) and provides great flexibility for the user to specify complex crop rotations and management regimes. The model has been widely applied in both Australia and China to assist in the development of management strategies for increasing water and nutrient use efficiencies.

Extensive validation of the APSIM model against experimental data was done by Chen et al. (2010a, b), Zhang et al. (2012), and J. Wang et al. (2011) for exploring the productivity and water use of the wheat (*Triticum aestivum* L.)–maize (*Zea mays* L.) double cropping system under irrigation management options in the North China Plain (NCP), where the groundwater table is rapidly declining. APSIM modeling using 40 yr of historical climate data at Luancheng, NCP showed that the current irrigation practice is often managed to maximize crop yield, and a reduction in irrigation amount would lead to significantly increased water use efficiency (WUE) (Chen et al., 2010a,b). Due to the interannual climate variation, 140 to 420 mm of irrigation water for wheat and 0 to 170 mm for maize would be needed to meet the crop water demand, which would cause the groundwater to decline at a rate about 1.5 m yr^{-1}, when other sources of groundwater recharge were not considered. This estimated rate of groundwater decline approximates the measured decline since the 1990s. If >180 mm irrigation water were available, partitioning it to wheat and maize would lead to higher total yield than applying it only to wheat. Modeling also showed an overall 4 to 6% increase in grain yield and improved WUE resulted from adoption of the "double delay" management option, that is, delayed sowing of winter wheat and harvesting of maize, due to the increase in temperature before winter, in the last three decades (X.C. Wang et al., 2011).

Table 15–1. Water balance simulated in the eleven models.

Model name	Potential evapotranspiration	Water movement in soil	Runoff	Actual water uptake	Plant water stress	Actual evaporation	Soil hydraulic properties
APSIM	Modified Priestley–Taylor method or Penman–Monteith	Tipping bucket cascade approach or Richards' equation	NRCS curve number	A function of total soil water above lower limit of plant available water	0–1 factor; actual water supply/potential plant water demand (calculated from RUE for the day).	Two-stage evaporation curve	Upper drain limit and lower limit of plant available water, K_{sat}
AquaCrop	Penman–Monteith equation	Empirical soil drainage ability function	NRCS curve number	Maximum soil water extraction multiplied by soil water stress	0–1 factor; on soil water depletion	Two-stage evaporation curve	Field capacity and wilting point, saturated hydraulic conductivity (K_{sat})
CropSyst	Priestley-Taylor method and Penman–Monteith method	Tipping bucket cascade approach or Richards' equation	NRCS curve number	Calculated from potential difference between soil water and plant xylem and soil water or root conductance	0–1 factor; actual transpiration/ potential transpiration	Two-stage evaporation curve	Field capacity and wilting point, saturated hydraulic conductivity (K_{sat}), Campbell equation
Daisy	Penman–Monteith equation	Richards' equation	When ponding exceeds surface depression storage	Steady-state radial flow to single root	0–1 factor; actual evapotranspiration/ potential evapotranspiration	Potential Exfiltration rate based on Darcy's law on the surface or energy available for evaporation	van Genuchten retention curve, Brooks–Corey, Campbell, Mualem equation for hydraulic conductivity
DSSAT	Modified Penman equation or Priestley-Taylor equation	Tipping bucket cascade approach	NRCS curve number	Empirical equation based on soil water content and root length density	0–1 factor; potential root water uptake/potential transpiration	Two-stage evaporation curve	Upper drain limit and lower limit of plant available water, K_{sat}
EPIC	Priestley–Taylor method and Penman method	Tipping bucket, cascade approach	Soil Conservation Service (SCS) curve number	Empirical equation based on potential transpiration, root depth, soil water content and an empirical water extraction distribution parameter.	0–1 factor; actual water uptake/ potential plant transpiration	Two-stage evaporation curve	Soil type, bulk density, field capacity and witling point

Model name	Potential evapotranspiration	Water movement in soil	Runoff	Actual water uptake	Plant water stress	Actual evaporation	Soil hydraulic properties
HERMES	Penman–Monteith equation, Priestley–Taylor equation	Modified capacity-based approach including capillary rise	N/A	Empirical equation depending on soil water content and root length density	0–1 factor; actual transpiration/ potential transpiration	Empirical equation based on soil water content	Capacity parameters derived from soil texture and distance from groundwater
RZWQM2	Shuttleworth–Wallace	Green-Ampt for infiltration and Richards' Equation for redistribution	Water in excess of K_{sat} and macropore flow if exists	Nimah-Hanks equation or empirical equation from DSSAT	0–1 factor; actual transpiration/potential transpiration or potential root water uptake/potential transpiration as in DSSAT	Determined by solving the Richards' equation with either constant flux or constant head upper boundary condition	Brooks–Corey soil water retention curves; Mualem equation for unsaturated hydraulic conductivity
STICS	Penman method for soil evaporation, adaptation of the Shuttleworth and Wallace method for crop potential transpiration	Tipping bucket approach	Analogy with a reservoir	Empirical equation based on potential transpiration and soil water content	0–1 factor; actual transpiration/potential transpiration	Two stage evaporation curve	Soil residual soil water, soil water content at wilting point and soil water content at field capacity)
SWAP	Penman–Monteith equation	Richards' equation	Rainfall rate exceeds the infiltration rate, or when the water table reached the soil surface.	As a function of maximum root water extraction rate	Feddes et al. (1978) function between 0 and 1.	Darcy's law calculates maximum evaporation rate for wet soil, which is further *limited for drying* soil (Black et al., 1969; Boesten and Stroosnijder, 1986).	Mualem–van Genuchten relations, with a modification near saturation.
WOFOST	Penman method	Tipping bucket approach	Exceed maximum surface storage	Potential transpiration as influenced by water shortage and oxygen reduction	0–1 factor; actual transpiration/potential transpiration	Potential evaporation rate and days since last rainfall	Soil type, initial and maximum surface storage, saturated soil hydraulic conductivity

In Australia, APSIM-based modeling has been extensively used to investigate options for improving crop management and WUE in the dryland farming systems, including crop or cultivar choice, pre-crop (crop sequence, weed control, residue management) and in-crop (sowing time, plant density, N management) management options (Kirkegaard et al., 2014), to effective use of seasonal climate forecasts combined with on-farm decision making (Wang et al., 2008a, 2009a,b). The model is also made available to growers and consultants through a simplified web interface known as Yield Prophet (Birchip Cropping Group) (Hochman et al., 2009b), enabling them to track their crop growth and soil conditions in real time.

APSIM-based modeling has also played an essential role in benchmarking WUE and defining the attainable yield targets and yield gaps of various crops across climatic zones in Australia grain regions (Hochman et al., 2009a; Wang et al., 2009b). This has enabled the separation of management impacts from climate impacts and the identification of constraints that limit crop yield and WUE (Hochman et al., 2009a, 2012). Combining observational data from 334 wheat fields and APSIM modeling, Hochman et al. (2009b) showed that further improvement in WUE (up to 21.4 kg grain ha^{-1} mm^{-1}) might be achieved by optimized plant density together with early sowing and higher N input. Kirkegaard and Hunt (2010), using modeling, further demonstrated the relative impacts and interactions of a range of pre-crop and in-crop management decisions on water productivity and showed how a novel genetic trait (long coleoptiles) that enable deeper sowing, could interact with different management options to increase the water-limited yield of wheat from 1.6 to 4.5 t ha^{-1}. Subsequently, Hunt and Kirkegaard (2011) and Hunt et al. (2013) showed that better summer fallow management (weed control and residue management) alone could lead to increased wheat yield and WUE. In a recent WUE initiative that challenged researchers and growers to increase WUE of grain-based production systems by 10% in 5 yr, APSIM-based modeling provided calculations of both a priori outcomes that were tested experimentally and extrapolation of results across sites, seasons, and up to the whole-farm scale (Kirkegaard et al., 2014). It was demonstrated through both experiments and modeling that water productivity at paddock scale can be improved by better summer weed control (37–140%), inclusion of break crops (16–83%), earlier sowing of appropriate varieties (21–33%), and matching N supply to soil type (91% on deep sands). Capturing synergies from combinations of pre- and in-crop management could increase wheat yield at the farm scale by 11 to 47% (Kirkegaard et al., 2014).

Lyon et al. (2003) applied the APSIM model to determine optimum corn plant populations and initial soil water contents at planting and recommended a plant population of 3 plants m^{-2}, with an initial soil water of 240 mm in the root zone to reduce production risk under semiarid western Nebraska conditions.

Conversely, Balwinder-Singh et al. (2011) found that the Australian APSIM wheat model failed to simulate wheat yield under deficit irrigation after calibrating for full irrigation treatments in central Punjab, India, which indicated that further improvements are needed in modeling of plant responses to water need. In this volume, Ollenburger and Snapp (2014) used APSIM to evaluate several options for sustainable intensifications of maize-based smallholder cropping at two sites in Malawi under changing climate scenarios and fertility levels. The options included intercropping, rotations of maize with pigeonpea [*Cajanus cajan* (L.) Huth], and continuous maize crop. The study showed that the tradeoffs among N and water limitations influenced yield and crop failure risk among sole crops, rotations, and intercrop systems.

The FAO AquaCrop Water Management Model

The Food and Agriculture Organization (FAO) of United Nations developed the AquaCrop model to simulate attainable yields of herbaceous crops in response to water (Steduto et al., 2009). The model is based on the concept that for a given crop species, biomass production per unit of transpiration is constant when normalized for evaporative demand and CO_2 concentration of the atmosphere. AquaCrop calculates daily first the green canopy cover from planting density and a canopy growth coefficient, bypassing leaf area index. Transpiration and soil water evaporation are simulated separately based on the fractional canopy cover and reference evapotranspiration (ET). Yield as fruit, grain, or root product is calculated from biomass and the harvest index. In the model water stress can inhibit canopy expansion and canopy transpiration, accelerate canopy senescence, and alter harvest index. Each of the effects has its own stress function with its own sensitivity threshold based on the fractional depletion of available soil water in the root zone. Water balance in the root profile is determined by a drainage function, root absorption, and soil water evaporation. AquaCrop has fewer parameters than most crop specific models, and many of its parameters are conservative. Once calibrated with extensive experimental data, the conservative parameters are applicable to diverse soils, seasons, and climates without local calibration. The model is particularly suited to assess crop production where water is the key limiting factor and is especially useful to devise water management strategies, such as timing of deficit irrigation and optimal planting time for rainfed cropping. The simplicity of the model does not permit simulation of mineral nutrient cycles, but soil fertility levels can be accommodated by calibration with local biomass data of a well-fertilized and a nutrient-deficient treatment. Low and high air temperature damage on pollination and harvestable yield are considered, as is cold inhibition of biomass production. Since its release in 2009, AquaCrop has received worldwide attention, being evaluated and applied to a variety of crops, especially in developing countries. The resulting many publications generally show that the model simulated biomass production and yield with reasonable to good fidelity, although under- or overestimations did occur, especially when water was seriously limiting. Because the simulated results were compared with measured data that are commonly not that definitive or are limited in scope, a more systematic and critical in-depth evaluation, based on extensive experimental data, is needed. The model is frequently lauded for its limited number of parameters and low input requirements, simplicity, and ease of use and has been incorporated into water management economics models (Garcia-Vila and Fereres, 2012; Donati et al., 2013) and scaled up for regional and watershed applications (Lorite et al., 2013).

The CropSyst Model

CropSyst is a process-based cropping systems model designed to be simple and user friendly for a wide range of user communities across agriculture and related scientific disciplines (Stöckle et al., 2003). The model simulates the impacts of cropping system management (e.g., cultivar selection, crop rotation, irrigation, N fertilization, salinity, and tillage operations) on productivity and the soil–water–air environment. The generic crop growth routine available within CropSyst allows simulation of various crops of interest to the user. It has already been parameterized to simulate several crops, trees, and cropping systems (e.g., Stöckle et al., 1994, 2003; Stöckle and Debaeke, 1997; Giardini et al., 1998; Pannkuk et al., 1998; Stöckle and Nelson, 1999; Confalonieri and Bechini, 2004; Sommer et al., 2012; Fumagalli et al., 2013).

CropSyst has been used for simulations of crop rotations and nutrient management across the world. Pannkuk et al. (1998) established the ability of CropSyst in modeling spring and winter wheat yields and evapotranspiration in a wheat–fallow rotation in eastern Washington region for different tillage and residue management practices. Bellochi et al. (2002) used CropSyst to simulate corn biomass, N uptake, and two soil variables (water and NO_3–N) in response to different types of tillage, N fertilization, and ground cover at Pisa, central Italy. They reported reasonable simulations of biomass, water content, and canopy area index; simulations of N dynamics needed improvement. Confalonieri and Bechini (2004) parameterized CropSyst to simulate semiperennial alfalfa (*Medicago sativa* L.) and found that the simulations were within 3 to 6% of the observed values across 3 yr. Confalonieri and Bocchi (2005) simulated flooded rice (*Oryza sativa* L.) in northern Italy within 20 to 22% of measured values. Diaz-Ambrona et al. (2005) simulated impacts of crop rotations, including fallow–wheat–pea (*Pisum *sativum* L.) and mustard (*Brassica juncea* L.)–wheat–pea, and management on the water balance of farming systems in the semiarid region of southeastern Australia, where deep drainage beyond the root zone and rising water tables contribute to salinization of soils and water streams. In a 5-yr experiment, the simulations of yield were close to measured yield for wheat, mustard, and field pea, with reasonable estimates of crop biomass, phenological development, soil water content, and water use. More recently, Sommer et al. (2008) applied CropSyst for simulating cotton (*Gossypium hirsutum* L.) production in the Khorezm region of Uzbekistan, with acceptable results for biomass.

CropSyst has also been applied extensively for developing decision support system (DSS) tools for soil–water–nutrient management in agriculture. Badini and Dioni (2001) combined CropSyst with spatial weather data in a GIS software and developed a decision support tool to determine soil suitability for crops, screen existing technologies, and build recommendation packages for a farming scenario. Dalla Marta et al. (2011) applied CropSyst for analyzing energy and water use related to the cultivation of energy crops in the Tuscany region of Italy. Long-term (50 yr) climate data from 19 locations were used in the simulations of crop production, water requirement, and other management practices to compute costs of energy crop cultivation.

Belhouchette et al. (2012) linked CropSyst to an economic optimization model and built a DSS tool for assessing the sustainability of irrigated farming systems in a Tunisian region. In the DSS, CropSyst was used to build a database to determine the relationships between agricultural management practices, crop yields, and environmental effects (salt accumulation in soil and leaching of nitrates) in a context of high climatic variability. The database was then fed into a recursive stochastic model set for a 10-yr plan that allowed analyzing the effects of cropping patterns on farm income, salt accumulation, and nitrate leaching. Lehmann and Finger (2014) developed a bio-economic model linking the CropSyst to an economic decision model at field scale. They then used this model with a genetic algorithm to optimize irrigation management decisions. Marsal and Stöckle (2012) developed a DSS for scheduling irrigation in a pear (*Pyrus communis* L.) orchard by parameterizing the generic crop growth model in CropSyst to forecast plant water potential in trees for irrigation management.

Fumagalli et al. (2013) evaluated the N leaching associated with the application of sewage sludge in rice and corn crops in the Lombardy region of northern Italy using the CropSyst model. They detected a significant effect of the sludge type and application timing on annual N leached in rice (22–154 kg N ha^{-1}). For corn, the application of sludge in the late fall period resulted in greater N leaching (61 kg N ha^{-1}) and lower yields compared to late winter fertilization and associated risks in potato (*Solanum tuberosum* L.) production in the Broye catchment in Switzerland.

The Danish Daisy Model

The Daisy model has been widely used in Europe for water and N management for groundwater quality. It was developed in Denmark by Soren Hansen (Hansen et al., 1991). In an early application of the model, Svendsen et al. (1995) found that spatial variability in irrigation amount across the field contributed to the lack of model responses to irrigation. In a study in NCP, Krobel et al. (2010) found out that simulated soil water contents did not correctly respond to irrigation events unless on-site measured soil properties were used in the Daisy model. They also showed that crop parameters had at least the same sensitivity in simulating soil water dynamics as soil parameters. A unique approach in the Daisy model is its simulation of abscisic acid (ABA) in roots for better simulation of plant water stress (Plauborg et al., 2010), although the inclusion of ABA in the model did not affect simulated yield and water use significantly in their experiments.

Kloss et al. (2012) developed a crop water production function (CWPF) for corn in Montpellier, France based on stochastically generated weather data (17 yr) using the Daisy model and found that simulated yield was affected mainly by temperature and radiation under full irrigation, but by interval of irrigation events at deficit irrigation. With the higher probability of dry seasons under the projected climate change for 2080, the Daisy model simulated reduced yield by 15%. Styczen et al. (2010) developed a plant–soil–atmosphere (PSA) management model based on Daisy to schedule irrigation from simulated soil water content, crop type, crop growth stage, and irrigation method and to schedule fertigation based on simulated plant and soil N concentrations. The threshold values to initiate and end an irrigation event were determined by experimental observations and expert opinions. An interesting feature of this PSA system was its linkage to water supply and its utilization for risk assessment when wastewater was applied (i.e., *E. coli*). In another study, Walser et al. (2011) found that the Daisy model was not adequate in simulating crop yield under severe water stress conditions. They further used an irrigation optimization algorithm in conjunction with the Daisy model to obtain optimal irrigation schedules for given irrigation amounts available for a crop season. The optimized irrigation schedule could increase production by as much as 41%.

The DSSAT Crop Systems Model

The Decision Support System for Agrotechnology Transfer (DSSAT) is a suite of process oriented cropping system models (CSM) integrated into a single software package to enable the seamless application of crop simulation in research and decision-making (Jones et al., 2003). The computer source codes of all CSMs are modular (for easy replacement or addition of modules) and incorporate more than 28 crops as modules using common soil and weather modules. The CSM models of different crops were derived mainly from their earlier versions of CROPGRO and CERES family of crop growth models (Jones et al., 2003).

The DSSAT crop models, especially CROPGRO and CERES, have been widely tested and applied in many research areas and production environments across the world (Jones et al., 2003). The CROPGRO is a process-oriented model for grain legumes, based on the SOYGRO, PNUTGRO, and BEANGRO models that consider detailed crop carbon, soil water, and soil N processes. CROPGRO evaluations and applications include work by Heinemann et al. (1999) for optimizing irrigation; Asare et al. (1996), Garrison et al. (1999), and Kizaso and Ritchie (1997) for N balance and the effects of soil water excess effects; Pang et al. (1998) for determining chemical leaching potential; Piper et al. (1998) for cultivar performance; Batchelor et al. (1994), Sau et al. (1999), and Alagarswamy et al. (2000) for calculating crop development and seed yield; and Soler and Hoogenboom (2007) for determining irrigation scheduling in peanut (*Arachis hypogaea* L.) and cotton. More recently, de Oliveira et al. (2012) tested CROPGRO-dry bean model for simulations of various dry bean cultivars in Brazil and obtained reasonable match with the experiments. Similarly, Lomeling et al. (2014) simulated rainfed cowpea [*Vigna unguiculata* (L.) Walp.] phenology in Sudan using the CROPGRO-cowpea model with reasonable accuracies for applications.

Ruiz-Nogueira et al. (2001) used the CROPGRO-soybean [*Glycine max* (L.) Merr.] model to establish best sowing windows for rainfed soybean cultivars at three locations in northwestern Spain. Jagtap and Jones (2002) combined GIS software with the CROPGRO-soybean model and developed a DSS for regional yield assessment. Using the cassava (*Manihot esculenta* Crantz) simulation model available within the DSSAT, Sarawat et al. (2004) developed a DSS for helping farmers and producers make better decisions by integrating information into a more useable form, altering production systems, enhancing management skills, and reducing costs of production. Development of DSS tool for soybean management for optimizing production in Thailand was reported by Banterng et al. (2010). Cammarano et al. (2012) developed optimum irrigation practices for water management in cotton cropping systems for maximizing economic returns in Australia.

Using the CERES-Wheat and CERES-Maize models, Yang et al. (2006) investigated agricultural water use and its impact on groundwater depletion in the piedmont region of the NCP and established a strong correlation between the two. Based on the correlation, the authors concluded that sustainable groundwater depletion is possible if better water-saving technologies are applied. In Northwest China, S. Wang et al. (2012) reported that less than 50% of flood irrigation water was used in plant transpiration. Using the CERES-Maize model, they showed that reducing or skipping irrigation events at certain growth stages increased irrigation WUE by as much as 66% in the system.

In this volume, Mercau and Otegui (2014) used the CERES-Maize model with 41 yr of historical weather data to evaluate the combined effects of late sowing, variable soil water at planting, and variable rates of N on the interannual variation in yield of a corn hybrid in a single- and wheat–maize double-crop systems at four locations in Argentina. The study defined the conditions under which late sowing was successful and increased yield stability. Soil water at sowing time was the major factor that defined the success.

Soler et al. (2007) applied CSM-CERES-Maize model for planting window determination and yield forecasting for corn grown off-season in a subtropical environment in Piracicaba, SP, Brazil. Using the developed DSS, yield forecasting was possible 45 d before the harvest date for all four maize hybrids investigated. He et al. (2012) used the CERES-Maize model and developed irrigation and N best management practices for sweet corn production on sandy soils in Florida. Iyanda et al. (2014) developed a DSS tool for identifying potential areas for corn production in Nigeria.

Thorp et al. (2008) developed a DSS tool called "Apollo" to help researchers in using DSSAT crop models to analyze precision farming and applications. Apollo has the capability to use various crop models in DSSAT and manage outputs for spatially variable land and management.

The EPIC Model and Its Modifications

The EPIC (Environmental Policy-Integrated Climate) model was first developed to simulate soil erosion effect on soil productivity (formerly known as the Erosion Productivity Impact Calculator, Williams et al., 1989) and then was extended to simulate agricultural management effects on crop production and soil and water resources (Williams et al., 1996). After that, the model was further extended to the whole farm and small watershed levels, and called APEX (Agricultural Policy Environmental eXtender model) (Williams and Izaurralde, 2006), to evaluate various land management strategies considering sustainability, erosion (wind, sheet, and channel), economics, water supply and quality, soil quality, plant competition, weather and pests.

Management capabilities of APEX include irrigation, drainage, furrow diking, buffer strips, terraces, waterways, fertilization, manure management, lagoons, reservoirs, crop rotation and selection, pesticide application, grazing, and tillage. Besides these farm management functions, APEX can be used in evaluating the effects of global CO_2 and climate change, designing environmentally safe and economical landfill sites, designing biomass production systems for energy, and other spin-off applications. The model operates on a daily time step (some processes are simulated with hourly or less time steps) and is capable of simulating hundreds of years if necessary. Farms may be subdivided into fields, soil types, landscape positions, or any other desirable configuration.

The model has been widely calibrated and tested under different climate and soil conditions (X.C. Wang et al., 2011, 2012) and used to manage irrigation water for better crop yield and WUE at different spatial and temporal levels in different countries, such as the United States, China, France, and Italy.

At the field level, an early study by Cabelguenne et al. (1995) used EPIC model to maximize maize yield with limited irrigation level of 1000 m^3 ha^{-1} based on a 20-yr simulation in the Toulouse Auzeville region of France. Later, the EPIC model was modified as EPIC-PHASE to incorporate the varying effect of water stress at different growth stages on the harvest index (Cabelguenne and Debaeke, 1996). Cabelguenne et al. (1997) used the EPIC-PHASE model to evaluate the real-time irrigation based on model calculations every 5 d. They found that the difference between actual and forecasted weather data led to different irrigation management, but using the measured weather data calculated the measured yield and reduced the risk of overirrigation and N leaching. Santos et al. (2000) used EPIC-PHASE for exploring other irrigation strategies.

In Georgia, USA, irrigation applications for the major crops were evaluated using EPIC model (Guerra et al., 2004, 2005). They found that the simulated irrigation requirement agreed well with farmer-applied irrigation amount for cotton and peanut, but more water was applied by farmers for maize than the simulated irrigation requirements in the region. The EPIC model was also demonstrated as a decision support tool to evaluate the full and deficit irrigation management for cotton and maize in southern Texas (Ko et al., 2009) and to explore irrigation schedules for optimal sunflower (*Helianthus annuus* L.) yield in Italy (Rinaldi, 2001). Recently, the model was applied for simulating crop yield and soil water dynamics for better crop irrigation management in the semiarid region of China (Wang and Li, 2010; X.C. Wang et al., 2011, 2014; Zhao et al., 2013).

The EPIC model has been used for estimating regional crop water requirements (Wriedt et al., 2009) and regional crop yield and water productivity (Liu et al., 2007a,b; Jiang et al., 2015; Van der Velde et al., 2010). Liu (2009) developed a GIS-based EPIC model (GEPIC) to study crop–water relations on large scales with high spatial resolution and applied it to estimate wheat yield and crop water productivity at a regional level (NCP) (Liu et al., 2007a) and global scale (Liu et al., 2007b). They found higher crop water productivity for irrigation wheat than rainfed wheat in both regional (North China Plain) and global scales, and the high global variability in crop water productivity suggested that global water use could be reduced through food trade among countries. In the NCP, two irrigation scenarios of reducing current irrigation amount by 20% and replacing irrigated wheat by rainfed wheat were compared. The simulation results showed similar effect on wheat yield between the two scenarios, suggesting that it is better to change irrigated wheat to rainfed wheat with less reduction in wheat yield under the same amount of water saving.

The EPIC model has also been used to explore irrigation strategies for climate variability and climate change (Dono et al., 2013a,b; Izaurralde et al., 2003; Chavas et al., 2009; Rinaldi and De Luca, 2012). For example, Dono et al. (2013b) used EPIC to evaluate the availability of irrigation water and the irrigation requirement of maize as influenced by short-term climate variability in the central Mediterranean basin, Italy. They found that the primary factor that influenced economic return was the reduced stability in the future irrigation water supply, which will likely result in higher groundwater extraction and lower demand of labor. In another study, using the GIS-based EPIC model, Dominguez-Faus et al. (2013) estimated the climate change effect on water requirement for irrigated corn ethanol production in the United States and found that irrigation rates would increase by 9% while corn yields decreased by 7% even when the projected increased irrigation requirements were met.

The German HERMES Model

HERMES is a model developed in Germany by Kersebaum (1995). Although it was developed mainly for N management, it has been used for irrigation management as well. After calibrating the HERMES model for four N treatments at three farmers' sites for winter wheat and summer maize rotation in NCP from 2009 to 2011, Michalczyk et al. (2014) applied the model to develop best water and N management scenarios for the region. They found that N rate could be reduced to 17% and irrigation water to 72% of farmer's practice without significantly affecting crop yields. As a result, N leaching was reduced to 1.8% of current level beyond the 0.9-m soil profile and 0.9% of current level beyond the

2.0-m soil profile. Using 25 yr (1987–1991) of weather data from the semiarid Saskatchewan, Canada, Kersebaum et al. (2008) found the model responded well to weather variability in simulating wheat yield and biomass, although soil water was not well simulated during winter because freezing and thawing were not considered in HERMES. Long-term crop rotations with various arable crops and permanent grassland under rainfed and irrigated conditions were simulated with HERMES for 13 sites in Czech Republic (Hlavinka et al., 2014), showing a good performance in comparison with observed crop yields and biomass production. Response of crop models on heat and water stress after flowering of winter wheat and maize was compared for seven models (DSSAT, EPIC, WOFOST, AQUACROP, FASSET, HERMES, and CROPSYST) by Eitzinger et al. (2013). While HERMES and AQUACROP showed relative moderate responses to drought conditions, WOFOST exhibited the highest sensitivity. Within the model intercomparison of Palosuo et al. (2011) the performance of eight crop models were investigated among other sites for a rainfed and irrigated plot in Germany. The highest response to the different water supplies was simulated by DAISY and HERMES. While HERMES underestimated crop biomass, which was consistent with a slight underestimation of soil water content during summer drought, the DAISY model overestimated crop growth, due to an overestimation of soil moisture. DSSAT distinctly underestimated crop biomass, although the soil moisture was significantly overestimated. Using a modified version of HERMES (MONICA), Nendel et al. (2014) found that irrigation would reduce the risk of drought in part of Germany, especially for maize, under the projected climate conditions of 2070. However, irrigation amount depended on crop type, other management practices (i.e., crop rotation), and location. Gandorfer and Kersebaum (2009) used the HERMES model to assess the economic effect of irrigation on wheat production under climate change scenarios for three sites in Bavaria. Although wheat yield risk was reduced in nearly all realizations, the financial investment to establish new irrigation equipment was only beneficial for the driest scenario.

The RZWQM2 Model

The Root Zone Water Quality Model (RZWQM2) project was initiated in the early 1990s, and the model has been used widely worldwide for water management (Ahuja et al., 2000; Ma et al., 2007). In the U.S. Great Plains, Saseendran et al. (2010) demonstrated that the calibrated RZWQM2 model could be used to identify new crop rotations to better cope with the highly variable precipitation conditions of the semiarid Great Plains without conducting time-consuming field experiments. Saseendran et al. (2013), also using RZWQM2, evaluated long-term yields and net returns of three grain crops—corn, canola [*Brassica juncea* (L.) Czern. subsp.

juncea], and proso millet (*Panicum miliaceum* L. subsp. *miliaceum*)—and two forage crops—foxtail millet [*Setaria italica* (L.) P. Beauv. subsp. *italica*] and spring triticale (× *Triticosecale rimpaui* Wittm.) at Akron, CO and Sidney, NE in the United States at various soil water contents at planting, and assessed these crops' potential use in increasing dryland cropping intensity within a wheat–summer crop–fallow rotation. Based on initial soil water content at planting, they proposed to select a summer crop in this rotation that has the probability of giving the highest net return. These results were used to develop an Excel (Microsoft) spreadsheet tool for the two locations (Fig. 15–1) to assess production risk. Later Nielsen et al. (2012) simulated canola yields at nine central Great Plains locations based on initial soil water content at planting, with 16 yr of weather records, and developed an Excel spreadsheet decision support tool (Fig. 15–2) that provides farmers with estimates of the probability of attaining or exceeding specified canola yields. In a more recent study, Saseendran et al. (2014) developed long-term CWPFs for corn yield at different levels of irrigation in Colorado for planning limited or deficit irrigation. They also found that the CWPFs were independent of soil type and location when expressed as relative yield versus relative ET or relative plant water supply.

In this volume, Fang et al. (2014a) used the RZWQM2 model to successfully reproduce the measured data of soil water content, ET, and corn yield for 4 yr at

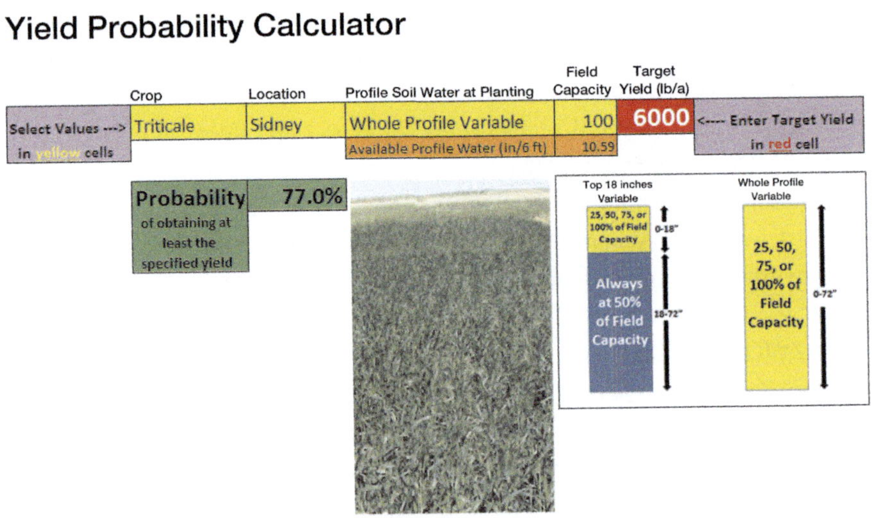

Fig. 15–1. Excel spreadsheet decision support tool for summer crop selection (canola, triticale [shown], millet, corn) based on simulated yields generated with weather data from 1948 to 2008 and four initial soil water contents at planting at Sidney, NE and Akron, CO. The simulated yields were generated using RZWQM2 and were published in Saseendran et al. (2013). Unpublished spreadsheet tool prepared by the coauthor David Nielsen, USDA-ARS, Akron, CO.

several irrigation levels in Colorado. The calibrated and tested model was then used to explore various combinations of water stress levels (created by different levels of ET-based irrigation, accounting for rainfall) applied at each of the vegetative and reproductive growth stages of corn over long-term local weather conditions, with additional constraints applying to the total seasonal water supply available. At high water supply and irrigation levels, the best grain yield and WUE were obtained with irrigations to meet 80 to 100% of the potential crop ET. At low water supply levels, best yield results were attained with irrigations to meet 60% of the potential ET demand during the vegetative stage and 100% during the reproductive stage. The model can be used to develop such location-specific recommendations for other locations in the world if some local data are available to determine the specific soil and cultivar parameters.

Saseendran et al. (2014, this volume) extended the case study of Fang et al. (2014a) to evaluate the long-term effects of irrigation interval, soil water content at

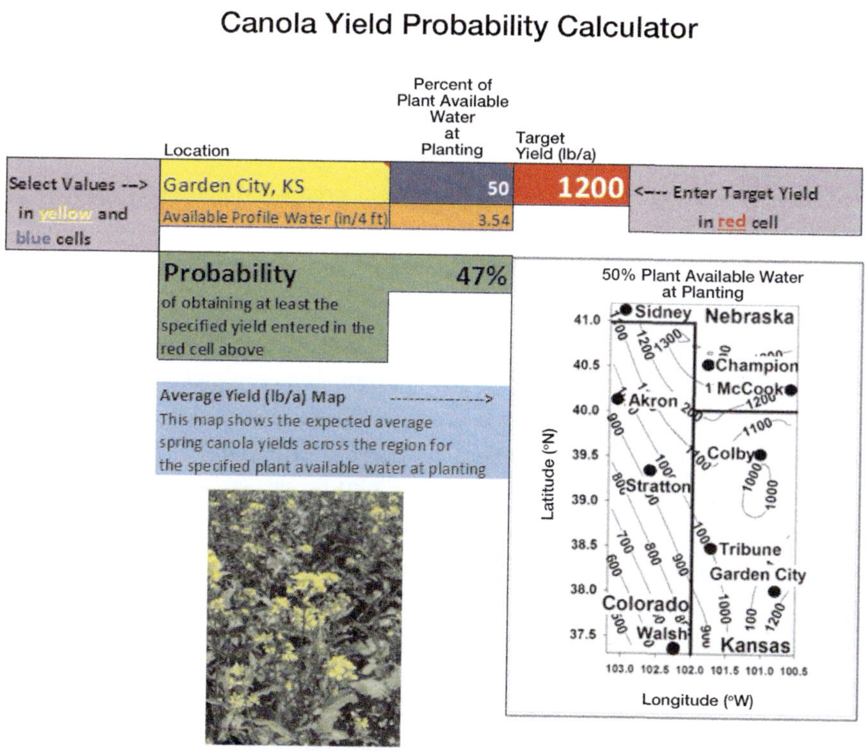

Fig. 15–2. Excel spreadsheet decision support tool for canola production at nine locations in the Great Plains based simulated yields generated with weather data from 1993 to 2008 and four initial soil water soil water contents using RZWQM2 simulation results as reported by Nielsen et al. (2012). Unpublished spreadsheet tool prepared by David Nielsen, USDA-ARS, Akron, CO.

planting, and N applications rates on corn yield at various levels of ET-based irrigation kept uniform during the entire growth period. More frequent irrigation intervals and higher initial soil water increased grain yield and WUE, especially at limited irrigation levels meeting less than 100% of crop potential ET. The N rate had a significant effect on production and water use and N use efficiencies. They developed long-term, site-specific recommendations for N rates for different levels of irrigation.

In China, Fang et al. (2010) applied the RZWQM2 model with long-term weather data (1961–1999) to investigate various irrigation strategies for the wheat–maize double cropping system in the NCP. The study showed that the preseason irrigation for wheat commonly practiced by local farmers should be postponed to the most sensitive growth stage (stem extension) for higher yield and WUE in the area. Preseason irrigation for maize was needed in 40% of the years. With different amounts of limited irrigation available (100, 150, 200, or 250 mm yr^{-1}), 80% of the water allocated to the critical wheat growth stages and 20% applied at maize planting achieved the highest WUE and the least water drainage over the long term for the two crops. In 2013, Fang et al. (2013) extended RZWQM2 applications to 15 locations in the NCP once the model was calibrated for local county level wheat and corn yield. They found that when N and/or irrigation inputs were reduced to 40 to 80% of their current levels, N leaching generally was reduced considerably without compromising crop yield greatly. Matching N input with crop requirements under limited water conditions helped achieve lower N leaching and lower soil N accumulation. Based on the long-term simulation results and water resource availability in the region, it is recommended to irrigate at 60 to 80% of the current water levels and fertilize only at 40 to 60% of the current N rate to minimizing N leaching without compromising crop yield.

In southern Portugal, Cameira et al. (1998) applied RZWQM to irrigated corn under different N management practices (fertigation vs. broadcast) and found that the model simulated yield was within 5% of experimental values. They also obtained good simulations of soil water and soil N in the growing seasons with <5% deviation. In a later application, Cameira et al. (2005) found that using an effective LAI for the stomatal resistance calculation improved ET simulation. This improvement was incorporated into the current RZWQM2 version. Simulated errors for LAI, plant height, biomass, and yield were lower than experimental errors among replicates. The calibrated model also simulated soil N and plant N uptake well (Cameira et al., 2007). Recently, Cameira et al. (2015) applied RZWQM2 to an olive grove to study water balance under full and deficit irrigation. They found that deficit irrigation could reduce N leaching by 47 to 90% without significantly affecting crop ET (or olive growth). Their simulation results also indicated that, to further minimize N leaching, N should be applied according to plant N demand.

The French STICS Model

The STICS (Simulateur mulTIdisciplinaire pour les Cultures Standard) model is a daily time-step crop model that can simulate the main processes of crop growth and development, as well as the water and N balances. The model is built from a number of components from other crop models, and can include various crops (e.g., wheat, maize, soybean, sorghum [*Sorghum bicolor* (L.) Moench], grassland) and the possibility of adding new modules (Brisson et al., 1998a, 2003). It has been recently adopted to evaluate the climate change impacts on crop yield (Butterworth et al., 2010) and soil carbon storage (Sierra et al., 2010) and other environmental issues, such as N_2O emissions (Bessou et al., 2010; Wattenbach et al., 2010) and nitrate pollution (Constantin et al., 2012). The model has also been applied to develop irrigation strategies for high WUE and water savings.

In an early study, Brisson et al. (1998b) applied the STICS model to simulate the effects of tillage and irrigation on banana (*Musa acuminata* Colla) yield, and found that the difference in irrigated banana yield between shallow and deep tillage was mainly caused by N deficit with a poor root system under shallow tillage. Under rainfed conditions, the banana yield increased with the increase in soil water exploration under deep tillage. The STICS model was also successfully evaluated for simulating grapevine phenology, yield, and soil water content under different water management strategies, and the model adequately simulated the time and duration of water stress and identified the critical periods for grapevine (*Vitis vinifera* L.) (Valdes-Gomez et al., 2009). In France, the STICS model was further applied to simulate the complex interactions between soil, plant, weather, and management on wheat yield under water-limited conditions, and the results showed that a drought-escaping strategy (early maturity) with low plant density was better for rainfed condition, while high plant density with longer growth seasons (late maturity) was better for irrigated conditions (Debaeke, 2004). Hadria et al. (2007) used STICS to develop irrigation scheduling (amount and timing) for winter wheat under optimal fertilization conditions in Marrakech and found that 70 and 40 mm of water could be saved in the cases of early and late sowing, respectively.

Recently, on the basis of a 25-yr simulation with the STICS model at three sites located in the southern Italy, Katerji et al. (2010) analyzed the "corn deficit irrigation–WUE" relations as influenced by deficit irrigation and soil properties. They found that a high total available water in the root zone led to an improvement in water and irrigation use efficiencies on a loamy clay soil, but on a clay soil there was no change or a decrease in water and irrigation use efficiencies. In another study (Khila et al., 2014), the STICS model was used to simulate effects of farmers' irrigation practices on wheat yield in the pedoclimatic conditions of the irrigated area in Tunisia over a 20-yr period. The model calculated crop yield and biomass adequately, and the farmers' irrigation practice produced higher grain yield than the rainfed system, especially on the soil with higher available soil water in the root zone.

The Dutch SWAP Model

The Soil Water Atmosphere Plant (SWAP) is a daily time-step process-oriented model that simulates soil salinity, water flow, and heat transport, and crop growth in agricultural systems in response to weather and management. SWAP was developed out of its predecessor SWATR, an agrohydrological model developed by Feddes et al. (1978). The generic crop growth module from the WOFOST model (Kroes et al., 2008) is used for simulating different crops in SWAP. SWAP has produced convincing simulations of soil salinity effects on crop productivity across many regions in the world. Ahmad et al. (2002) applied SWAP to compute soil water content and vertical soil water fluxes in the unsaturated zone for a cotton–wheat and rice–wheat cropping system in Punjab, Pakistan. Wang et al. (2004) assessed SWAP for simulating impacts of saline water irrigation on crop yield and soil salinity balance in the Hetao Irrigation District in China. Crescimanno and Garofalo (2005) simulated solute transport in a cracking soil at Sicily, Italy. Jiang et al. (2011) simulated deficit irrigation effects on soil salinity in the Shiyang River basin in the arid northwest China. Qureshi et al. (2013) simulated the effects of rising groundwater tables on soil salinity in the irrigated areas of central Iraq. Huang et al. (2013) applied SWAP for simulating groundwater fluctuation effects on soil salinity in wheat-based cropping systems in the upper Yellow River basin of Qingtongxia irrigation district in China. Martínez-Ferri et al. (2013) applied SWAP for simulating the water balance of an irrigated orchard in Southern Spain. Verma et al. (2014) evaluated SWAP for its usefulness to simulate crop growth and salinity effects on wheat production at Agra in India employing data from 12 combinations of saline and freshwater conjunctive use in irrigation. Kroes and Supit (2011) investigated the effect of groundwater salinity on grassland production in The Netherlands under present and future climate conditions. They concluded that for grassland in The Netherlands, drought stress is stronger than the stress caused by water excess, which in turn is stronger than the salinity stress. For future climatic conditions they predicted an increase of water demand for irrigation by 11 to 19%.

The SWAP model also has been used widely as a DSS for managing irrigation for salinity control and optimizing crop yields in various cropping systems across the world. Utset and Borroto (2001) linked SWAP to GIS and developed a DSS tool for assessing irrigation effects on soil salinization at San Antonio del Sur Valley, in southeast Cuba. The tool developed could be used by farmers and decision-makers to select the most suitable water management solution considering both economic and environmental impact criteria. Mandare et al. (2008) developed a DSS tool in the form of monographs for on-farm water management with saline groundwater in northwest India. The monographs were used for providing graphical methods for calculating the effects of different combinations of water quality and water depths on crop yield and soil salinity.

Sarwar and Bastiaanssen (2001) applied SWAP for decision support in water reductions for wheat and cotton crops under shallow water table conditions commonly observed in the Fourth Drainage Project in Punjab, Pakistan. Irmak and Kamble (2009) developed an on-demand irrigation management DSS tool by assimilating remote sensing-based ET into the SWAP model for wheat crops in the Sirsa Irrigation Circle of Haryana, India. Jiang et al. (2011) used SWAP and developed a DSS tool for calculating the long-term effects of deficit irrigations with saline water in the Shiyang River basin in arid northwest China. Using SWAP, Karimi et al. (2012) developed DSS tools for farm water management for checking groundwater exploitation for reducing the carbon footprint of irrigation enterprises in Iran. Qureshi et al. (2013) developed a DSS tool for assessing the need for drainage systems for managing groundwater tables for reduced soil salinity and improved wheat and corn yields in the irrigated areas of central Iraq. Huang et al. (2013) developed a DSS tool for assessing optimum groundwater table depth for controlling soil salinity for optimum grain yields in the wheat-based cropping systems in the Qingtongxia irrigation district in the upper Yellow River basin in China. Martínez-Ferri et al. (2013) developed management scenarios in response to climate change projection scenarios in Spain. Integrating SWAP with short-term field trials, Verma et al. (2014) showed that saline and freshwater conjunctive use in irrigating crops in the monsoon climate of India would help alleviate the long-term buildup of salt in the soil in the region.

The Dutch WOFOST Model

WOFOST (WOrld FOod STudies) is a simulation model for annual field crops and was originally developed for studying world food security and potential world food production (Van Diepen et al., 1989). WOFOST has been applied under various climatic and management conditions, such as climate change effects (Alexandrov and Eitzinger, 2005; Supit et al., 2012; Van Walsum and Supit, 2012), regional crop yield calculations (Supit, 1997; de Wit et al., 2010; Boogaard et al., 2013), and agricultural management effects (Rötter and Van Keulen, 1997; Rötter et al., 1997a; Wu et al., 2006). WOFOST was also used to evaluate alternative irrigation strategies at different scales and climate conditions.

The WOFOST model has been evaluated for calculating crop yield under water-limited conditions, such as winter wheat in NCP (Wu et al., 2003) and northern Italy (Marletto et al., 2005), and spring wheat and sugar beet in Prague, Czech Republic (Haberle and van Diepen, 1999). Eitzinger et al. (2004) compared the CERES, SWAP, and WOFOST models for simulating soil water contents and soil water balance on a crop field in Australia and found that CERES and SWAP models simulated the grain yield of barley (*Hordeum vulgare* L.) and wheat better than WOFOST model, but all the three models simulated soil water content with similar results and over-estimated soil water depletion. In another study, Todorovic et al. (2009) compared the AquaCrop, CropSyst, and WOFOST models for simulating sunflower growth under different irrigation regimes and found no substantial difference in the simulated yield and biomass among these models, and that the use of simpler models should be encouraged for irrigation management purposes.

The WOFOST model was also coupled with other models for studying crop water management. For example, WOFOST coupled with the SWAP model was used to determine the evapotranspiration, percolation, and rice yield in India (Govindarajan et al., 2008). They found that the highest value of water productivity was from the flooded treatment, and 66% inflow water went to recharge the groundwater. This study also concluded that water-saving irrigation alone could not improve the water productivity of a paddy crop, which depended on both groundwater level and evaporative demand. Zhou et al. (2012) evaluated the coupled model of HYDRUS-1D and WOFOST models for simulating ET, soil water, and winter wheat production in northwest China and used a water stress factor to guide irrigation scheduling and saved 27% of irrigation water compared with the current irrigation practice. In another similar study (Li et al., 2012), the coupled model was evaluated for simulating ET, soil water, and maize production in the northwest China and found that the optimal amount of irrigation for maize was 500 to 600 mm in this region.

Recently, the WOFOST model was coupled with GIS and other models to explore irrigation management at regional levels. The WOFOST model coupled with the Crop Growth Monitoring System (CGMS) was used to assess the water requirements, water consumption, and water deficit from 1976 to 2005 across Europe by Supit et al. (2010), and they found a downward trend in wheat water requirement due to a shorter growing season with higher spring temperatures, a downward trend in sugarbeet (*Beta vulgaris* L. subsp. *vulgaris*) in France and

Spain due to a lower evaporative demand with reduced solar radiation in summer and autumn, and an upward trend in the water requirement across the rest of Europe associated with high air temperature in summer and autumn. In another study, the WOFOST model coupled with the SWAP model was used to evaluate the performance of irrigation system in Sirsa district, India (Singh et al., 2006a,b). The simulation results showed that the reduction in seepage losses to 25 to 30% of the total canal inflow and reallocation of 15% canal water inflow from the northern to the central canal commands could improve the water productivity and slow declining groundwater levels. Using a similar coupled model to that described above, Mishra et al. (2013) developed irrigation strategies based on short-term rainfall forecasts in India and found 27% reductions in irrigation water application when using 5-d perfect rainfall forecasts and 32% reduction in irrigation water when using 5-d perfect rainfall forecasts under future climate conditions.

Other System Models for Water Management in this Volume

In addition to the above major agricultural system models in the literature, this volume also covered other models for water management. Yang et al. (2014, this volume) used the HybridMaize model to study the temporal dynamics of corn water use, WUE, and final corn yield over 20 yr of historical weather conditions, under both irrigated and rainfed conditions in Nebraska. The results provided quantitative relationships to maximize yield and WUE to guide management.

Lecler (2014, this volume) developed and applied ZIMSched 2.0 and its derivatives (ACRUCane) for planning and decisions for managing irrigation of sugarcane in South Africa, at field and larger scales, in irrigation systems of varying efficiencies. He showed how the models helped evaluate the design and performance of different irrigation systems with respect to soil types and climates to develop irrigation production functions for sugarcane and provided adaptation strategies under short water supplies, financial analysis of various irrigation options under changing conditions, and estimates of return flow for downstream users.

Islam et al. (2014, this volume) described how InfoCrop and other models (e.g., SWAP, APSIM, WEPP, SWAT, MIKE SHE) are being used in India for water management and food security under different climate change scenarios. They showed the application of models with respect to canal operations, water allocations, irrigation scheduling, improving water productivity, watershed management and prioritization, climate change impact on crop water demand, ET, and yields of major crops (rice, wheat, and maize), and identification of vulnerable regions and benefits of adaptations.

In southwestern France, the WaLIS model is being used to advise farmers on irrigation decisions for vineyards based on estimates of soil water in production systems with or without cover crops. Roux et al. (2014, this volume) calculated uncertainty in irrigation decisions with the vineyard model, caused by the uncertainty in estimating the input data, as a component of the overall uncertainty. The uncertainty information helps the farmers to be conscious of this production variability and account for it in some way in managing irrigations.

Lascano and Nelson (2014, this volume) used a Precision Agricultural Landscape Modeling System (PALMS) with some measurements of soil water to demonstrate that circular planting of cotton and grain sorghum under a variable topography can enhance rainfall infiltration and decrease runoff under dryland conditions. This increased infiltration was attributed to increased soil surface roughness and thus greater surface water storage under circular planting compared with linear planting.

Gomez et al. (2014, this volume) used the WABOL model to evaluate the water balance components and their impact on yield in five olive orchards under two soil management strategies (tilled bare soil and a cover crop) under current and future climates in Spain. The cover crop system was shown to decrease runoff, erosion, and deep percolation of water and nutrients. The cover crop did not influence the yield of olives if mowing or herbicides limited its transpiration in early spring; the olive transpiration was even slightly reduced.

Kiniry et al. (2014, this volume) used the ALMANAC model to evaluate environmental sustainability of bioenergy crops in the United States, by studying WUE and biomass yield stability of maize and switchgrass production systems for several cultivars and locations. The results provided the information for life-cycle analysis of bioenergy crops.

Tools for Real-Time Water Management Decisions by Producers

In this volume, Andales et al. (2014) presented a web-based decision tool for guiding farmers in making near real-time irrigation applications on different fields. The tool uses the GIS-based soil database and the Colorado Agricultural Meteorological Network for daily rainfall and reference ET. The tool gives irrigators estimates of daily soil water deficits of individual fields to determine the time and amount of irrigations. Apps for smartphones have been developed for easier use by farmers.

Fujimaki et al. (2014, this volume) developed a tool for real-time scheduling of irrigation timing and application amounts using a numerical model and forecasted weather on a day-to-day basis. The results of this irrigation scheduling method were compared with a sensor-based automatic irrigation scheme for maize in Japan and for barley in Tunisia. Calculated values of soil water content and salinity agreed well with measurements, and the net returns were similar to the automatic irrigation system. Further work is needed to improve model parameterization.

Fang et al. (2014b, this volume) used a range–livestock model and forecasted rainfall to estimate forage and livestock production for making stocking rate decisions for upcoming grazing seasons. This work has been converted into a spreadsheet decision tool, which will soon be developed as an app for smartphones.

Automatic Scheduling of Irrigation Using Sensors and Models

Some private companies market automated sprinkler irrigation systems, especially for golf courses and other managed landscapes. Some researchers have also explored the automatic irrigation systems for crops. Recent USDA-ARS research in the Texas High Plains is an example. Peters and Evett (2008) showed that infrared thermometers (IRTs) aboard center pivots could be used to schedule soybean irrigations using a simple time–temperature threshold method. O'Shaughnessy et al. (2013) developed a time-integrated version of the crop water stress index to automatically schedule irrigations by center pivot for grain sorghum. These studies showed that automatic irrigation scheduling using IRTs (wired and wireless) aboard center pivots can usually achieve the same crop yield and WUE as irrigation scheduling using a field-calibrated neutron probe, where the latter method is not feasible for commercial agricultural production due to time, labor, and regulatory requirements. However, algorithms based on a crop water stress index are generally limited to full canopy cover because they only consider the energy balance of the bulk soil–plant–atmosphere continuum (i.e., single source, or big leaf approach). When canopy cover is partial or sparse, the sunlit soil background, which is usually warmer than the canopy, can bias water stress calculations upward and result in overirrigating. Therefore, the use of a two-source energy balance (TSEB) approach (Norman et al., 1995) that will provide separate estimation of the soil evaporation and transpiration components in a partial canopy is needed. Colaizzi et al. (2012a,b, 2014) modified the TSEB approach of Norman et al. (1995) for nonrandom spatial distribution of row crops with increased accuracy for calculating evaporation and transpiration. This improved approach is in the process of being added to the center pivot controller and is being tested for various crops.

Summary of Model Applications Thus Far

All the applications described above may be summarized in the following categories:

1. Evaluate model responses to different irrigation scheduling (Fang et al., 2014a; Mishra et al., 2013).
2. Use models to propose alternative cropping sequences based on water availability and initial soil water (Saseendran et al., 2010, 2013; Fang et al., 2014b; Diaz-Ambrona et al., 2005; Hunt and Kirkegaard, 2011; Hunt et al., 2013).
3. Develop crop production function based on model results (Saseendran et al., 2014).
4. Project water demands under current and future weather conditions (Andales et al., 2014; Nendel et al., 2014; Dominguez-Faus et al., 2013).
5. Optimize current water resources (Fang et al., 2010; Walser et al., 2011).
6. Extend piece-meal experimental results to other soils and weather conditions for irrigation decision making (Lecler, 2014; Kirkegaard et al., 2014).
7. Determine crop responses to saline water irrigation (Verma et al., 2014; Jiang et al., 2011; Wang et al., 2004).
8. Assess environmental impacts of waste water irrigation (Styczen et al., 2010).
9. Quantify the interactions between water and N management (Saseendran et al., 20154; Michalczyk et al., 2014; Fang et al., 2013)
10. Investigate management effects (i.e., tillage, residue cover, plant population, row spacing or direction, planting date, intercropping, new crop cultivar, initial soil water content) on WUE (Gomez et al., 2014; Lascano and Nelson, 2014; Debaeke, 2004; Brisson et al., 1998b; Mercau and Otegui, 2014; Ollenburger and Snapp, 2014; Kirkegaard and Hunt 2010).
11. Analyze long-term stability of limited irrigation (Kiniry et al., 2014; Belhouchette et al., 2012).
12. Schedule irrigations based on simulated water stress factor and canopy temperature (Colaizzi et al., 2014; Zhou et al., 2012; Fujimaki et al., 2014; O'Shaughnessy et al., 2013; Guerra et al., 2004, 2005).
13. Address spatial distribution of soil water distribution and its effects on crop production and N leaching (Svendsen et al., 1995).
14. Evaluate cover crop effects on soil water balance (Roux et al., 2014; Gomez et al., 2014).

Needs for Experimental Data and Model Improvement

The current agricultural system models have been greatly advanced in the last few decades. Our review and synthesis showed very useful applications of models in evaluating current production systems and proposing alternative cropping systems and management practices under limited water conditions and under climate change conditions. Although the models reviewed here are at field scale, some of them have been linked to regional-scale hydrological models or GIS, such as Daisy-MIKE SHE (Boegh et al., 2004) and GEPIC (Liu, 2009).

However, there are still knowledge gaps in understanding and quantification of some processes in the models that need further work. We need to establish: (i) better understanding of the effects of water stress on different plant growth and development processes, (ii) better quantification of the interactions between water and N stresses, (iii) better understanding of root growth and water and N uptake, (iv) better mapping of crop parameters to genetic traits of a cultivar, (v) better quantification of spatial variability of soil parameters, (vi) better description of uncertainty of weather conditions on crop selection and management decisions, and (vii) accounting of extreme heat effects on pollination and harvest index. These knowledge gaps require mechanistic studies of the processes and their interactions with good experimental data from several locations. As we apply the models to data for new water and N management practices, we hope to fill some of these gaps as well.

Nonetheless, the greatest hurdle in the site-specific management applications of the models at this juncture is lack of adequate local experimental data to calibrate the soil- and cultivar-specific parameters. The models need to be integrated with field research throughout the world, and field researchers need to be made aware of the soil and crop data that should be collected for determining the model parameters and for testing the models for calculations. Application of a model at a number of locations will also fill some of the knowledge gaps in the model (Boote et al., 2012; Nendel et al., 2011).

Recent results from Agricultural Model Inter-Comparison and Improvement Project (AgMIP) indicated that great uncertainties exist in simulated ET, plant transpiration, and soil water evaporation among different models, implying potential issues in modeling water balance and plant response to water stress (Cammarano et al., 2013; Bassu et al., 2014). Experimental data sets that include measured ET, soil water balance, and crop performance under conditions from fully irrigated to severely water-limited would help greatly to improve process modeling (Ma et al., 2012). Results from model comparisons are in general not conclusive, and a model can provide good or bad simulations for many reasons. A model could have simulated crop yield adequately while poorly simulating phenology or vice versa (Rotter et al., 2012). Thus, comparison of individual processes (or modules) may reveal more knowledge gaps than comparing whole models.

As reviewed here, many of the model applications involved some kind of model calibration, including both soil and plant parameters. A model application that contributes to science and reveals knowledge gaps should require the least calibration with as many measured parameters as possible, such as soil hydraulic properties (Stulina et al., 2005). There is a need to decouple the interaction between soil and plant parameters during model calibration, so that calibrated parameters are transferable to other sites and locations. As a result, projections of modeling results to regional scales or to future climate change conditions will be more realistic.

The model applications summarized here show that the process models can be very useful to identify site-specific management practices for optimum use of limited water under different conditions. However, due to the nature of model complexity, process-based models have not been used directly as a tool for water management that can be adopted by producers. Farmers may be more likely to use the simple ET tools as shown by Andales et al. (2014) along with personal experience to make irrigation decisions. Combining sensors and simple models is more practical for seasonal water management by farmers than using a complex model. However, with improvements in the process-based models and easy-to-obtain site-independent parameters, complex system models should become more useful in the future for both strategic and tactical planning of limited irrigation water under current and climate change conditions.

References

Ahmad, M.U.D., W.G.M. Bastiaanssen, and R.A. Feddes. 2002. Sustainable use of groundwater for irrigation: A numerical analysis of the subsoil water fluxes. Irrig. Drain. 51:227–241. doi:10.1002/ird.59

Ahuja, L.R., K.W. Rojas, J.D. Hanson, M.J. Shaffer, and L. Ma, editors. 2000. Root Zone Water Quality Model. Water Resources Publications LLC, Highlands Ranch, CO.

Alagarswamy, G., P. Singh, G. Hoogenboom, S.P. Wani, P. Pathak, and S.M. Virmani. 2000. Evaluation and application of the CROPGRO-Soybean simulation model in a Vertic Inceptisol. Agric. Syst. 63:19–32. doi:10.1016/S0308-521X(99)00070-0

Alexandrov, V.A., and J. Eitzinger. 2005. The potential effect of climate change and elevated air carbon dioxide on agricultural crop production in central and southeastern Eur. J. Crop Improv. 13(1-2):291–331. doi:10.1300/J411v13n01_14

Andales, A.A., T.A. Bauder, and M. Arabi. 2014. A mobile irrigation water management system using a collaborative GIS and weather station networks. In: L.R. Ahuja, L. Ma, and R.J. Lascano, editors, Practical applications of agricultural system models to optimize the use of limited water. Adv. Agric. Systems Model. 5. ASA, SSSA, CSSA, Madison, WI. p. 53–84 (this volume). doi:10.2134/advagricsystmodel5.c3

Asare, D.K., T.W. Sammis, and T.L. Jones. 1996. Sensitivity of simulated field water balance to different soil hydraulic data. Trans. ASAE 39:2085–2092. doi:10.13031/2013.27711

Badini, O., and L. Dioni. 2001. Application of rainfall analysis, biophysical modeling and GIS to agroclimatic decision support in Madiama Commune, Mali (West Africa). Paper presented to the SANREM CRSP Research Scientific Synthesis Conference, Athens, GA. 28–30 Nov. 2001.

Balwinder-Singh, D., S. Gaydon, E. Humphreys, and P.L. Eberbach. 2011. The effects of mulch and irrigation management on wheat in Punjab, India– Evaluation of the APSIM model. Field Crops Res. 124:1–13. doi:10.1016/j.fcr.2011.04.016

Banterng, P., G. Hoogenboom, A. Patanothai, P. Singh, S.P. Wani, P. Pathak, S. Tongpoonpol, S. Atichart, P. Srihaban, S. Buranaviriyakul, A. Jintrawet, and T.C. Nguyen. 2010. Application of the Cropping System Model (CSM)-CROPGRO-Soybean for determining optimum management strategies for soybean in tropical environments. J. Agron. Crop Sci. 196:231–242. doi:10.1111/j.1439-037X.2009.00408.x

Bassu, S., N. Brisson, J.-L. Durand, K. Boote, J. Lizaso, J.W. Jones, et al. 2014. How do various maize models vary in their responses to climate change factors? Glob. Change Biol. 20:2301–2320. doi:10.1111/gcb.12520

Batchelor, W.D., J.W. Jones, K.J. Boote, and G. Hoogenboom. 1994. Carbon-based model to predict peanut pod detachment. Trans. ASAE 37:1639–1646. doi:10.13031/2013.28251

Belhouchette, H., M. Blanco, J. Wery, and G. Flichman. 2012. Sustainability of irrigated farming systems in a Tunisian region: A recursive stochastic programming analysis. Comput. Electron. Agric. 86:100–110. doi:10.1016/j.compag.2012.02.016

Bellochi, G., N. Silvestri, M. Mazzoncini, and S. Menini. 2002. Using the CropSyst model in continuous rainfed maize (*Zea mays* L.) under alternative management options. Ital. J. Agron. 6:43–56.

Bessou, C., B. Mary, J. Leonard, M. Roussel, E. Grehan, and B. Gabrielle. 2010. Modelling soil compaction impacts on nitrous oxide emissions in arable fields. Eur. J. Soil Sci. 61(3):348–363. doi:10.1111/j.1365-2389.2010.01243.x

Black, T.A., W.R. Gardner, and G.W. Thurtell. 1969. The prediction of evaporation, drainage, and soil water storage for a bare soil. Soil Sci. Soc. Am. J. 33:655–660. doi:10.2136/sssaj1969.03615995003300050013x

Boegh, E., M. Thorsen, M.B. Butts, S. Hansen, J.S. Christiansen, P. Abrahamsen, C.B. Hasager, N.O. Jensen, P. van der Keur, J.C. Refsgaard, K. Schelde, H. Soegaard, and A. Thomsen. 2004. Incorporating remote sensing data in physically based distributed ago-hydrological modelling. J. Hydrol. 287:279–299. doi:10.1016/j.jhydrol.2003.10.018

Boesten, J.J.T.I., and L. Stroosnijder. 1986. Simple model for daily evaporation from fallow tilled soil under spring conditions in a temperate climate. Neth. J. Agric. Sci. 34:425–435.

Boogaard, H., J. Wolf, I. Supit, S. Niemeyer, and M. van Ittersum. 2013. A regional implementation of WOFOST for calculating yield gaps of autumn-sown wheat across the European Union. Field Crops Res. 143:130–142. doi:10.1016/j.fcr.2012.11.005

Boote, K.J., M.R. Rybak, J.M.S. Scholberg, and J.W. Jones. 2012. Improving the CRORPGRO-Tomato model for predicting growth and yield response to temperature. HortScience 47:1038–1049.

Brisson, N., C. Gary, E. Justes, R. Roche, B. Mary, D. Ripoche, D. Zimmer, J. Sierra, P. Bertuzzi, P. Burger, F. Bussière, Y.M. Cabidoche, P. Cellier, P. Debaeke, J.P. Gaudillère, F. Maraux, B. Seguin, and H. Sinoquet. 2003. An overview of the crop model STICS. Eur. J. Agron. 18(3–4):309–332. doi:10.1016/S1161-0301(02)00110-7

Brisson, N., B. Mary, D. Ripoche, M.H. Jeuffroy, F. Ruget, P. Gate, F. Devienne-Barret, R. Antonioletti, C. Durr, B. Nicoullaud, G. Richard, N. Beaudoin, S. Recous, X. Tayot, D. Plenet, P. Cellier, J.M. Machet, J.M. Meynard, and R. Delécolle. 1998a. STICS: A generic model for the simulation of crops and their water and nitrogen balances. I. Theory and parameterization applied to wheat and corn. Agronomie 18(5–6):311–346. doi:10.1051/agro:19980501

Brisson, N., H. Ozier-Lafontaine, and M. Dorel. 1998b. Effects of soil management and water regime on banana growth between planting and flowering. Simulation using the STICS model. Acta Hort. 490:229–238.

Butterworth, M.H., M.A. Semenov, A. Barnes, D. Moran, J.S. West, and B.D.L. Fitt. 2010. North-South divide: Contrasting impacts of climate change on crop yields in Scotland and England. J. R. Soc. Interface 7(42):123–130. doi:10.1098/rsif.2009.0111

Cabelguenne, M., and P. Debaeke. 1996. EPICPHASE real time: A model for tactical irrigation management. (In French.) C. R. Acad. Agric. France 82(4):39–51.

Cabelguenne, M., P. Debaeke, J. Puech, and N. Bosc. 1997. Real time irrigation management using the EPIC-PHASE model and weather forecasts. Agric. Water Manage. 32:227–238. doi:10.1016/S0378-3774(96)01275-9

Cabelguenne, M., C.A. Jones, and J.R. Williams. 1995. Strategies for limited irrigations of maize in southwestern France– a modeling approach. Trans. ASAE 38(2):507–511. doi:10.13031/2013.27859

Cameira, M.R., R.M. Fernando, L.R. Ahuja, and L. Ma. 2007. Using RZWQM to simulate the fate of nitrogen in field soil-crop environment in the Mediterranean region. Agric. Water Manage. 90:121–136. doi:10.1016/j.agwat.2007.03.002

Cameira, M.R., R.M. Fernando, L.R. Ahuja, and L. Pereira. 2005. Simulating the fate of water in field–crop environment. J. Hydrol. 315:1–24. doi:10.1016/j.jhydrol.2005.03.020

Cameira, M.R., A. Pereira, L.R. Ahuja, and L. Ma. 2014. Sustainability and environmental assessment of fertigation in an intensive olive grove under Mediterranean conditions. Agric. Water Manage. 146:346–360.

Cameira, M.R., P.L. Sousa, H.J. Farahani, L.R. Ahuja, and L.S. Pereira. 1998. Evaluation of the RZWQM for the simulation of water and nitrate movement in level-basin, fertigated maize. J. Agric. Eng. Res. 69:331–341. doi:10.1006/jaer.1997.0253

Cammarano, D., P. Jose, B. Basso, P. Wilkens, and G. Peter. 2012. Agronomic and economic evaluation of irrigation strategies on cotton lint yield in Australia. Crop Pasture Sci. 63:647–655. doi:10.1071/CP12024

Cammarano, D., R.P. Rotter, S. Asseng, F. Ewert, C. Rosenzweig, J.W. Jones, et al. 2013. Quantifying uncertainties in modeling crop water use under climate change. Impacts World 2013, International Conference on Climate Change Effects. Potsdam, Germany. May 27–30. http://www.climate-impacts-2013.org/files/hcaw_cammarano.pdf (accessed 10 Oct. 2014).

Chavas, D.R., R.C. Izaurralde, A.M. Thompson, and X. Gao. 2009. Long-term climate change impacts on agricultural productivity in eastern China. Agric. For. Meteorol. 149(6–7):1118–1128. doi:10.1016/j.agrformet.2009.02.001

Chen, C., E. Wang, and Q. Yu. 2010a. Modelling wheat and, maize productivity as affected by climate variation and irrigation supply in north China lain. Agron. J. 102:1037–1049. doi:10.2134/agronj2009.0505

Chen, C., E. Wang, and Q. Yu. 2010b. Modelling the effects of climate variability and water management on crop water productivity and water balance in the North China Plain. Agric. Water Manage. 97:1175–1184. doi:10.1016/j.agwat.2008.11.012

Colaizzi, P.D., N. Agam, J.A. Tolk, S.R. Evett, T.A. Howell, P.H. Gowda, S.A. O'Shaughnessy, W.P. Kustas, and M.C. Anderson. 2014. Two source energy balance model to calculate E, T, and ET: Comparison of Priestley-Taylor and Penman-Monteith formulations and two time scaling methods. Trans. ASABE 57:479–498.

Colaizzi, P.D., W.P. Kustas, M.C. Anderson, N. Agam, J.A. Tolk, S.R. Evett, T.A. Howell, P.H. Gowda, and S.A. O'Shaughnessy. 2012a. Two-source energy balance model estimates of evapotranspiration using component and composite surface temperatures. Adv. Water Resour. 50:134–151. doi:10.1016/j.advwatres.2012.06.004

Colaizzi, P.D., R.C. Schwartz, S.R. Evett, T.A. Howell, P.H. Gowda, and J.A. Tolk. 2012b. Radiation model for row crops: II. Model evaluation. Agron. J. 104:241–255. doi:10.2134/agronj2011.0083

Confalonieri, R., and L. Bechini. 2004. A preliminary evaluation of the simulation model CropSyst for alfalfa. Eur. J. Agron. 21:223–237. doi:10.1016/j.eja.2003.08.003

Confalonieri, R., and S. Bocchi. 2005. Evaluation of CropSyst for simulating the yield of flooded rice in northern Italy. Eur. J. Agron. 23:315–326. doi:10.1016/j.eja.2004.12.002

Constantin, J., N. Beaudoin, M. Launay, J. Duval, and B. Mary. 2012. Long-term nitrogen dynamic in various catch crop scenarios: Test and simulations with STICS model in temperate country. Agric. Ecosyst. Environ. 147:36–46. doi:10.1016/j.agee.2011.06.006

Crescimanno, G., and P. Garofalo. 2005. Application and evaluation of the SWAP model for simulating water and solute transport in a cracking Clay Soil. Soil Sci. Soc. Am. J. 69:1943–1954. doi:10.2136/sssaj2005.0051

Dalla Marta, D.A., F. Natali, M. Mancini, R. Ferrise, M. Bindi, and S. Orlandini. 2011. Energy and water use related to the cultivation of energy crops: A case study in the Tuscany Region. 2011. Ecol. Society 16(2):2. http://www.ecologyandsociety.org/vol16/iss2/art2/ (accessed 15 Oct. 2014).

de Oliveira, E.C., J.M.N. da Costa, T.J. de Paula, Jr., W.P.M. Ferreira, F.B. Justino, and L.O. Neves. 2012. The performance of the CROPGRO model for bean (*Phaseolus vulgaris* L.) yield simulation. Maringá 34:239–246.

de Wit, A., B. Baruth, H. Boogaard, K. van Diepen, D. van Kraalingen, F. Micale, J. te Roller, I. Supit, and R. van den Wijngaart. 2010. Using ERA-INTERIM for regional crop yield forecasting in Europe. Clim. Res. 44:41–53. doi:10.3354/cr00872

Debaeke, P. 2004. Scenario analysis for cereal management in water-limited conditions by the means of a crop simulation model (STICS). Agronomie 24(6-7):315–326. doi:10.1051/agro:2004035

Diaz-Ambrona, C.G.H., G.J. O'Leary, V.O. Sadras, M.G. O'Connell, and D.J. Connor. 2005. Environmental risk analysis of farming systems in a semi-arid environment: Effect of rotations and management practices on deep drainage. Field Crops Res. 94:257–271. doi:10.1016/j.fcr.2005.01.008

Dominguez-Faus, R., C. Folberth, J. Liu, A.M. Jaffe, and P.J. Alvarez. 2013. Climate change would increase the water intensity of irrigated Corn ethanol. Environ. Sci. Technol. 47:6030–6037. doi:10.1021/es400435n

Donati, M., D. Bodini, F. Arfini, and A. Zezza. 2013. An integrated PMP model to assess the development of agro-energy crops and the effect on water requirements. Bio-based. Appl. Econ. 2:301–321.

Dono, G., R. Cortignani, L. Doro, L. Giraldo, L. Ledda, M. Pasqui, and P.P. Roggero. 2013a. An integrated assessment of the impacts of changing climate variability on agricultural productivity and profitability in an irrigated Mediterranean catchment. Water Resour. Manage. 27(10):3607–3622. doi:10.1007/s11269-013-0367-3

Dono, G., R. Cortignani, L. Giraldo, L. Doro, L. Ledda, M. Pasqui, and P.P. Roggero. 2013b. Adapting to uncertainty associated with short-term climate variability changes in irrigated Mediterranean farming systems. Agric. Syst. 117:1–12. doi:10.1016/j.agsy.2013.01.005

Eitzinger, J., S. Thaler, E. Schmid, F. Strauss, R. Ferrise, M. Moriondo, M. Bindi, T. Palosuo, R. Rötter, K.C. Kersebaum, J.E. Olesen, R.H. Patil, L. Saylan, B. Caldag, and O. Caylak. 2013. Sensitivities of crop models to extreme weather conditions during flowering period demonstrated for maize and winter wheat in Austria. J. Agric. Sci. 151:813–835. doi:10.1017/S0021859612000779

Eitzinger, J., M. Trnka, J. Hösch, Z. Žalud, and M. Dubrovský. 2004. Comparison of CERES, WOFOST and SWAP models in simulating soil water content during growing season under different soil conditions. Ecol. Modell. 171(3):223–246. doi:10.1016/j.ecolmodel.2003.08.012

Fang, Q.X., L. Ma, D.C. Nielsen, T.J. Trout, and L.R. Ahuja. 2014a. Quantifying corn yield and water use efficiency under growth stage–based deficit irrigation conditions. In: L.R. Ahuja, L. Ma, and R.J. Lascano, editors, Practical applications of agricultural system models to optimize the use of limited water. Adv. Agric. Systems Model. 5. ASA, SSSA, CSSA, Madison, WI. p. 1–24 (this volume). doi:10.2134/advagricsystmodel5.c1

Fang, Q.X., L.R. Ahuja, A.A. Andales, and J.D. Derner. 2014b. Using a model and forecasted weather to predict forage and livestock production for making stocking decisions in the coming growing season. In: L.R. Ahuja, L. Ma, and R.J. Lascano, editors, Practical applications of agricultural system models to optimize the use of limited water. Adv. Agric. Systems Model. 5. ASA, SSSA, CSSA, Madison, WI. p. 161–182 (this volume). doi:10.2134/advagricsystmodel5.c7

Fang, Q.X., L. Ma, Q. Yu, C.S. Hu, X.X. Li, R.W. Malone, and L.R. Ahuja. 2013. Quantifying climate and management effects on regional crop yield and nitrogen leaching in the North China Plain. J. Environ. Qual. 42:1466–1479. doi:10.2134/jeq2013.03.0086

Fang, Q., L. Ma, Q. Yu, L.R. Ahuja, R.W. Malone, and G. Hoogenboom. 2010. Irrigation strategies to improve the water use efficiency of wheat–maize double cropping systems in North China Plain. Agric. Water Manage. 97:1165–1174. doi:10.1016/j.agwat.2009.02.012

Feddes, R.A., P.J. Kowalik, and H. Zaradny. 1978. Simulation of field water use and crop yield. Simulation Monographs. Pudoc, Wageningen, The Netherlands.

Fujimaki, H., I. Tokumoto, T. Saito, M. Inoue, M. Shibata, T. Okazaki, K. Nagaz, and F. El-Mokh. 2014. Determination of irrigation depths using a numerical model and quantitative weather forecast and comparison with an experiment. In: L.R. Ahuja, L. Ma, and R.J. Lascano, editors, Practical applications of agricultural system models to optimize the use of limited water. Adv. Agric. Systems Model. 5. ASA, SSSA, CSSA, Madison, WI. p. 209–236 (this volume). doi:10.2134/advagricsystmodel5.c9

Fumagalli, M., A. Perego, and M. Acutis. 2013. Modelling nitrogen leaching from sewage sludge application to arable land in the Lombardy region (northern Italy). Sci. Total Environ. 461-462:509–518. doi:10.1016/j.scitotenv.2013.05.029

Gandorfer, M., and K.C. Kersebaum. 2009. Climate change impact on winter wheat production with a focus on production risk—Considering CO_2–effects and irrigation (in German with English abstract). Jahrbuch der Österreichischen Gesellschaft für Agrarökonomie 18(3):47–56. http://oega.boku.ac.at (accessed 15 Oct. 2014).

Garcia-Vila, M., and E. Fereres. 2012. Combining the simulation crop model AquaCrop with an economic model for the optimization of irrigation management at farm level. Eur. J. Agron. 36:21–31. doi:10.1016/j.eja.2011.08.003

Garrison, M.V., W.D. Batchelor, R.S. Kanwar, and J.T. Ritchie. 1999. Evaluation of the CERES-maize water and nitrogen balances under tile-drained conditions. Agric. Syst. 62:189–200. doi:10.1016/S0308-521X(99)00064-5

Giardini, L., A. Berti, and F. Morari. 1998. Simulation of two cropping systems with EPIC and CropSyst models. Ital. J. Agron. 2:29–38.

Gómez, J.A., M.T. Rodríguez Carretero, I.J. Lorite, and E. Fereres. 2014. Modeling to evaluate and manage climate change effects on water use in Mediterranean olive orchards with respect to cover crops and tillage management. In: L.R. Ahuja, L. Ma, and R.J. Lascano, editors, Practical applications of agricultural system models to optimize the use of limited water. Adv. Agric. Systems Model. 5. ASA, SSSA, CSSA, Madison, WI. p. 53–84 (this volume). doi:10.2134/advagricsystmodel5.c10

Govindarajan, S., N.K. Ambujam, and K. Karunakaran. 2008. Estimation of paddy water productivity (WP) using hydrological model: An experimental study. Paddy Water Environ. 6:327–339. doi:10.1007/s10333-008-0131-0

Guerra, L.C., G. Hoogenboom, J.E. Hook, D.L. Thomas, V.K. Boken, and K.A. Harrison. 2005. Evaluation of on-farm irrigation applications using the simulation model EPIC. Irrig. Sci. 23:171–181. doi:10.1007/s00271-005-0105-6

Guerra, L.C., G. Hoogenboom, V.K. Boken, J.E. Hook, D.L. Thomas, and K.A. Harrison. 2004. Evaluation of the EPIC model for simulating crop yield and irrigation demand. Trans. ASAE 47(6):2091–2100. doi:10.13031/2013.17794

Haberle, J., and C.A. van Diepen. 1999. Variability of potential and water-limited sugar beet and spring wheat yields simulated with Wofost model. Rostlinna Vyroba 45(10):433–438.

Hadria, R., S. Khabba, A. Lahrouni, B. Duchemin, A. Chehbouni, J. Carriou, and L. Ouzine. 2007. Calibration and validation of the STICS crop model for managing wheat irrigation in the semi-arid Marrakech/Al Haouz Plain. Arabian J. Sci. Eng. 32:87–101.

Hansen, S., H.E. Jensen, N.E. Nielsen, and H. Svendsen. 1991. Simulation of nitrogen dynamics and biomass production in winter wheat using the Danish simulation model DAISY. Fert. Res. 27:245–259. doi:10.1007/BF01051131

He, J., M.D. Dukes, G.J. Hochmuth, J.W. Jones, and W.D. Graham. 2012. Identifying irrigation and nitrogen best management practices for sweet corn production on sandy soils using CERES-Maize model. Agric. Water Manage. 109:61–70. doi:10.1016/j.agwat.2012.02.007

Heinemann, A.B., G. Hoogenboom, G.A. Georgiev, R.T. de Faria, and J.A. Frizzone. 1999. Optimization of the capacity of center pivot irrigation systems based on the dry bean model CROPGRO. http://naldc.nal.usda.gov/naldc/download.xhtml?id=IND22077233&content=PDF (accessed 12 Sept. 2014). ASAE, St. Joseph, MI.

Hlavinka, P., M. Trnka, K.C. Kersebaum, P. Cermák, E. Pohanková, M. Orság, E. Pokorný, M. Fischer, M. Brtnický, and Z. Žalud. 2014. Modelling of yields and soil nitrogen dynamics for crop rotations by HERMES under different climate and soil conditions in the Czech Republic. J. Agric. Sci. 152:188–204. doi:10.1017/S0021859612001001

Hochman, Z., D. Gobbet, D. Holzworth, T.I. McClelland, H. van Rees, O. Marinoni, J. Navarro Garcia, and H. Horan. 2012. Quantifying yield gaps in rainfed cropping systems: A case study of wheat in Australia. Field Crops Res. 136:85–96. doi:10.1016/j.fcr.2012.07.008

Hochman, Z., D. Holzworth, and J.R. Hunt. 2009a. Potential to improve on-farm wheat yield and WUE in Australia. Crop Pasture Sci. 60:708–716. doi:10.1071/CP09064

Hochman, Z., H. Van Rees, P.S. Carberry, J.R. Hunt, R.L. McCown, A. Gartmann, D. Holzworth, S. Van Rees, N.P. Dalgliesh, W. Long, A.S. Peake, and P.L. Poulton. 2009b. Re-inventing model-based decision support with Australian dryland farmers: 4. Yield Prophet® helps farmers monitor and manage crops in a variable climate. Crop Pasture Sci. 60:1057–1070. doi:10.1071/CP09020

Huang, G., S. Chen, P.S. Luis, R.B. Tiago, H. Quanzhong, and H. Yuanyuan. 2013. Assessing the effects of water table depth on water use, soil salinity and wheat yield: Searching for a target depth for irrigated areas in the upper Yellow River basin. Agric. Water Manage. 25:46–60.

Hunt, J.R., and J.A. Kirkegaard. 2011. Re-evaluating the contribution of summer fallow rain to wheat yield in southern Australia. Crop Pasture Sci. 62:915–929. doi:10.1071/CP11268

Hunt, J.R., C. Browne, T. McBeath, K. Verburg, S. Craig, and A.M. Whitbread. 2013. Summer fallow weed control and residue management impacts on winter crop yield through soil water and N accumulation in a winter-dominant, low rainfall region of southern Australia. Crop Pasture Sci. 64:922–934. doi:10.1071/CP13237

Irmak, A., and B. Kamble. 2009. Evapotranspiration data assimilation with genetic algorithms and SWAP model for on-demand irrigation. Irrig. Sci. 28:101–112. doi:10.1007/s00271-009-0193-9

Islam, A., P.B. Shirsath, S.N. Kumar, N. Subash, A.K. Sikka, and P.K. Aggarwal. 2014. Modeling water management and food security in India under climate change scenarios. In: L.R. Ahuja, L. Ma, and R.J. Lascano, editors, Practical applications of agricultural system models to optimize the use of limited water. Adv. Agric. Systems Model. 5. ASA, SSSA, CSSA, Madison, WI. p. 267–316 (this volume). doi:10.2134/advagricsystmodel5.c11

Iyanda, R.A., G. Pranuthi, S.K. Dubey, and S.K. Tripathi. 2014. Use of DSSAT CERES-Maize model as a tool of identifying potential zones for maize production in Nigeria. Int. J. Agric. Policy Res. 2:69–75.

Izaurralde, R.C., N.J. Rosenberg, R.A. Brown, and A.M. Thomson. 2003. Integrated assessment of Hadley Center (HaDCM2) climate-change impacts on agricultural productivity and irrigation water supply in the conterminous United States: Part II. Regional agricultural production in 2030 and 2095. Agric. For. Meteorol. 117(1–2):97–122. doi:10.1016/S0168-1923(03)00024-8

Jagtap, S.S., and J.W. Jones. 2002. Adaptation and evaluation of the CROPGRO-soybean model to predict regional yield and production. Agric. Ecosyst. Environ. 93:73–85. doi:10.1016/S0167-8809(01)00358-9

Jiang, J., S. Feng, Z. Huoa, Z. Zhaoc, and B. Jia. 2011. Application of the SWAP model to simulate water–salt transport under deficit irrigation with saline water. Math. Comput. Model. 54:902–911. doi:10.1016/j.mcm.2010.11.014

Jiang, Y., X. Xu, Q. Huang, Z. Huo, and G. Huang. 2015. Assessment of irrigation performance and water productivity in irrigated areas of the middle Heihe River basin using a distributed agro-hydrological model. Agric. Water Manage. 147:67–81.

Jones, J.W., G. Hoogenboom, C.H. Porter, K.J. Boote, W.D. Batchelor, L.A. Hunt, P.W. Wilkens, U. Singh, A.J. Gijsman, and J.T. Ritchie. 2003. The DSSAT cropping system model. Eur. J. Agron. 18:235–265. doi:10.1016/S1161-0301(02)00107-7

Karimi, P., A.S. Qureshi, R. Bahramloo, and D. Molden. 2012. Reducing carbon emissions through improved irrigation and groundwater management: A case study from Iran. Agric. Water Manage. 108:52–60. doi:10.1016/j.agwat.2011.09.001

Katerji, N., M. Mastrorilli, and H.E. Cherni. 2010. Effects of corn deficit irrigation and soil properties on water use efficiency. A 25-year analysis of a Mediterranean environment using the STICS model. Eur. J. Agron. 32(2):177–185. doi:10.1016/j.eja.2009.11.001

Keating, B.A., P.S. Carberry, G.L. Hammer, M.E. Probert, M.J. Robertson, D. Holzworth, N.I. Huth, J.N.G. Hargreaves, H. Meinke, Z. Hochman, G. McLean, K. Verburg, V. Snow, J.P. Dimes, M. Silburn, E. Wang, S. Brown, K.L. Bristow, S. Asseng, S. Chapman, R.L. McCown, D.M. Freebairn, and C.J. Smith. 2003. An overview of APSIM, a model designed for farming systems simulation. Eur. J. Agron. 18:267–288. doi:10.1016/S1161-0301(02)00108-9

Kersebaum, K.C. 1995. Application of a simple management model to simulate water and nitrogen dynamics. Ecol. Modell. 81:145–156. doi:10.1016/0304-3800(94)00167-G

Kersebaum, K.C., A. Wurbs, R. De Jong, C.A. Campbell, J. Yang, and R.P. Zentner. 2008. Long-term simulation of soil-crop interactions in semiarid southwestern Saskatchewan, Canada. 2008. Eur. J. Agron. 29:1–12. doi:10.1016/j.eja.2008.01.011

Khila, S.B., B. Douh, A. Mguidiche, F. Ruget, M. Mohsen, and A. Boujelben. 2014. Application of STICS model in assessment of the effects of irrigation practices and soil properties on yield of a durum wheat (*Triticum durum* Desf.) cultivar in the irrigated area of Oued Rmel in Tunisia. Annu. Rev. Plant Biol. 4(5):747–765. doi:10.9734/ARRB/2014/5053

Kiniry, J.R., M.N. Meki, T.E. Schumacher, C.J. Zilverberg, F.B. Fritschi, and V.G. Kakani. 2014. Modeling to evaluate and manage water and environmental sustainability of bioenergy crops in the United States. In: L.R. Ahuja, L. Ma, and R.J. Lascano, editors, Practical applications of agricultural system models to optimize the use of limited water. Adv. Agric. Systems Model. 5. ASA, SSSA, CSSA, Madison, WI. p. 139–160 (this volume). doi:10.2134/advagricsystmodel5.c6

Kirkegaard, J.A., and J.R. Hunt. 2010. Increasing productivity by matching farming system management and genotype in water-limited environments. J. Exp. Bot. 61:4129–4143. doi:10.1093/jxb/erq245

Kirkegaard, J.A., J.R. Hunt, T.M. McBeath, J.M. Lilley, A. Moore, K. Verburg, M. Robertson, Y. Oliver, W.R. Ward, S. Milroy, and A.M. Whitbread. 2014. Improving water productivity in the Australian grains industry—A nationally coordinated approach. Crop Pasture Sci. 65:583–601.

Kizaso, J.I., and J.T. Ritchie. 1997. A modified version of CERES to predict the impact of soil water excess on maize crop growth and development. In: M.J. Kropff, editor, Application of systems approaches at the field level. Kluwer Academic Publ., Dordrecht, The Netherlands. p. 153–167.

Kloss, S., R. Pushpalatha, K.J. Kamoyo, and N. Schutze. 2012. Evaluation of crop models for simulating and optimizing deficit irrigation systems in arid and semi-arid countries under climate variability. Water Resour. Manage. 26:997–1014. doi:10.1007/s11269-011-9906-y

Ko, J., G. Piccinni, and E. Steglich. 2009. Using EPIC model to manage irrigated cotton and maize. Agric. Water Manage. 96(9):1323–1331. doi:10.1016/j.agwat.2009.03.021

Krobel, R., Q. Sun, J. Ingwersen, X. Chen, F. Zhang, T. Muller, and V. Romheld. 2010. Modeling water dynamics with DNDC and DAISY in a soil of the North China Plain: A comparative study. Environ. Model. Softw. 25:583–601. doi:10.1016/j.envsoft.2009.09.003

Kroes, J.G., and I. Supit. 2011. Impact analysis of drought, water excess and salinity on grass production in The Netherlands using historical and future climate data. Agric. Ecosyst. Environ. 144(1):370–381.

Kroes, J.G., J.C. Van Dam, P. Groenendijk, R.F.A. Hendriks, and C.M.J. Jacobs. 2008. SWAP version 3.2. Theory and description and user manual. Aterra Rep. 1649. Alterra, Wageningen, The Netherlands.

Lascano, R.J., and J.R. Nelson. 2014. Circular planting to enhance rainfall capture in dryland cropping systems at a landscape scale: Measurement and simulations. In: L.R. Ahuja, L. Ma, and R.J. Lascano, editors, Practical applications of agricultural system models to optimize the use of limited water. Adv. Agric. Systems Model. 5. ASA, SSSA, CSSA, Madison, WI. p. 85–112 (this volume). doi:10.2134/advagricsystmodel5.c4

Lecler, N.L. 2014. Irrigation and water management decisions involving sugarcane in southern Africa. In: L.R. Ahuja, L. Ma, and R.J. Lascano, editors, Practical applications of agricultural system models to optimize the use of limited water. Adv. Agric. Systems Model. 5. ASA, SSSA, CSSA, Madison, WI. p. 317–350 (this volume). doi:10.2134/advagricsystmodel5.c12

Lehmann, N., and R. Finger. 2014. Economic and environmental assessment of irrigation water policies: A bio economic simulation study. Environ. Model. Softw. 51:112–122. doi:10.1016/j.envsoft.2013.09.011

Li, Y., W. Kinzelbach, J. Zhou, G.D. Cheng, and X. Li. 2012. Modelling irrigated maize with a combination of coupled-model simulation and uncertainty analysis, in the northwest of China. Hydrol. Earth Syst. Sci. 16(5):1465–1480. doi:10.5194/hess-16-1465-2012

Liu, J. 2009. A GIS-based tool for modelling large-scale crop-water relations. Environ. Model. Softw. 24(3):411–422. doi:10.1016/j.envsoft.2008.08.004

Liu, J., D. Wiberg, A.J.B. Zehnder, and H. Yang. 2007a. Modelling the role of irrigation in winter wheat yield, crop water productivity, and production in China. Irrig. Sci. 26(1):21–33. doi:10.1007/s00271-007-0069-9

Liu, J., J.R. Williams, A.J.B. Zehnder, and H. Yang. 2007b. GEPIC– modelling wheat yield and crop water productivity with high resolution on a global scale. Agric. Syst. 94(2):478–493. doi:10.1016/j.agsy.2006.11.019

Lomeling, D., M.M. Kenyi, A.A. Abass, S.M. Otwari, and Y.M. Khater. 2014. Using the CROPGRO model to predict phenology of cowpea under rain-fed conditions. Int. J. Plant Soil Sci. 3:824–844. doi:10.9734/IJPSS/2014/9491

Lorite, I.J., M. Garcia-Vila, C. Santos, M. Ruiz-Ramos, and E. Fereres. 2013. Aquadata and aqua-GIS: Two computer utilities for temporal and spatial simulations of water-limited yield with AquaCrop. Comput. Electron. Agric. 96:227–237. doi:10.1016/j.compag.2013.05.010

Lyon, D.J., G.L. Hammer, G.B. McLean, and J.M. Blumenthal. 2003. Simulation supplements field studies to determine no-till dryland corn population recommendations for semiarid western Nebraska. Agron. J. 95:884–891. doi:10.2134/agronj2003.0884

Ma, L., L.R. Ahuja, and R.W. Malone. 2007. Systems modeling for soil and water research and management: Current status and further needs in the 21st century. Trans. ASABE 50:1705–1713. doi:10.13031/2013.23962

Ma, L., T.J. Trout, L.R. Ahuja, W.C. Bausch, S.A. Saseendran, R.W. Malone, and D.C. Nielsen. 2012. Calibrating RZWQM2 model for maize responses to deficit irrigation. Agric. Water Manage. 103:140–149. doi:10.1016/j.agwat.2011.11.005

Mandare, A.B., S.K. Ambast, N.K. Tyagi, and J. Singh. 2008. On-farm water management in saline groundwater area under scarce canal water supply condition in the Northwest India. Agric. Water Manage. 95:516–526. doi:10.1016/j.agwat.2007.12.010

Marletto, V., F. Zinoni, L. Criscuolo, G. Fontana, S. Marchesi, A. Morgillo, and U. Andersen. 2005. Evaluation of downscaled DEMETER multi-model ensemble seasonal hindcasts in a northern Italy location by means of a model of wheat growth and soil water balance. Tellus, Ser. A, Dyn. Meterol. Oceanogr. 57(3):488–497. doi:10.1111/j.1600-0870.2005.00109.x

Marsal, J., and C.O. Stöckle. 2012. Use of CropSyst as a decision support system for scheduling regulated deficit irrigation in a pear orchard. Irrig. Sci. 30:139–147. doi:10.1007/s00271-011-0273-5

Martínez-Ferri, E., J.L. Muriel-Fernández, and J.A. Rodríguez Díaz. 2013. Soil water balance modelling using SWAP—An application for irrigation water management and climate change adaptation in citrus. Outlook Agric. 42:93–102. doi:10.5367/oa.2013.0125

Mercau, J.L., and M.A. Otegui. 2014. A modeling approach to explore water management strategies for late-sown maize and double-cropped wheat–maize in rainfed pampas region of Argentina. In: L.R. Ahuja, L. Ma, and R.J. Lascano, editors, Practical applications of agricultural system models to optimize the use of limited water. Adv. Agric. Systems Model. 5. ASA, SSSA, CSSA, Madison, WI. p. 351–374 (this volume). doi:10.2134/advagricsystmodel5.c13

Michalczyk, A., K.C. Kersebaum, M. Roelcke, T. Hartmann, S.C. Yue, X.P. Chen, and F.S. Zhang. 2014. Model-based optimization of nitrogen and water management for wheat-maize systems in North China Plain. Nutr. Cycling Agroecosyst. 98:203–222. doi:10.1007/s10705-014-9606-0

Mishra, A., C. Siderius, K. Aberson, M. van der Ploeg, and J. Froebrich. 2013. Short-term rainfall forecasts as a soft adaptation to climate change in irrigation management in north-east India. Agric. Water Manage. 127:97–106. doi:10.1016/j.agwat.2013.06.001

Nendel, C., M. Berg, K.C. Kersebaum, W. Mirschel, X. Specka, M. Wegehenkel, K.O. Wenkel, and R. Wieland. 2011. The MONICA model: Testing predictability for crop growth, soil moisture and nitrogen dynamics. Ecol. Modell. 222:1614–1625. doi:10.1016/j.ecolmodel.2011.02.018

Nendel, C., K.C. Kersebaum, W. Mirschel, and K.O. Wenkel. 2014. Testing farm management options as climate change adaptation strategies using the MONICA model. Eur. J. Agron. 52:47–56. doi:10.1016/j.eja.2012.09.005

Nielsen, D.C., S.A. Saseendran, L. Ma, and L.R. Ahuja. 2012. Simulating the production potential of dryland spring canola in the central Great Plains. Agron. J. 104:1182–1188. doi:10.2134/agronj2012.0048

Norman, J.M., W.P. Kustas, and K.S. Humes. 1995. Source approach for estimating soil and vegetation energy fluxes in observations of directional radiometric surface temperature. Agric. For. Meteorol. 77(3–4):263–293. doi:10.1016/0168-1923(95)02265-Y

Ollenburger, M., and S. Snapp. 2014. Model applications for sustainable intensification of african maize-based smallholder cropping in a changing world. In: L.R. Ahuja, L. Ma, and R.J. Lascano, editors, Practical applications of agricultural system models to optimize the use of limited water. Adv. Agric. Systems Model. 5. ASA, SSSA, CSSA, Madison, WI. p. 375–398 (this volume). doi:10.2134/advagricsystmodel5.c14

O'Shaughnessy, S.A., S.R. Evett, P.D. Colaizzi, and T.A. Howell. 2013. Wireless sensor network effectively controls center pivot irrigation of sorghum. Appl. Eng. Agric. 29(6):853–864.

Palosuo, T., K.C. Kersebaum, C. Angulo, P. Hlavinka, M. Moriondo, J.E. Olesen, R.H. Patil, F. Ruget, C. Rumbaur, J. Takac, M. Trnka, M. Bindi, B. Caldag, F. Ewert, R. Ferrise, W. Mirschel, L. Saylan,

B. Siska, and R. Rötter. 2011. Simulation of winter wheat yield and its variability in different climates of Europe: A comparison of eight crop growth models. Eur. J. Agron. 35(3):103–114. doi:10.1016/j.eja.2011.05.001

Pang, X.P., S.C. Gupta, J.F. Moncrief, C.J. Rosen, and H.H. Cheng. 1998. Evaluation of nitrate leaching potential in Minnesota glacial outwash soils using the CERES-Maize model. J. Environ. Qual. 27:75–85. doi:10.2134/jeq1998.00472425002700010012x

Pannkuk, C.D., C.O. Stockle, and R.I. Papendick. 1998. Evaluating CropSyst simulations of wheat management in a wheat–fallow region of the U.S. Pacific Northwest. Agric. Syst. 57:121–134. doi:10.1016/S0308-521X(97)00076-0

Peters, R.T., and S.R. Evett. 2008. Automation of a center pivot using the temperature-time-threshold method of irrigation scheduling. J. Irrig. Drain. Eng. 134(3):286–291. doi:10.1061/(ASCE)0733-9437(2008)134:3(286)

Piper, E.L., K.J. Boote, and J.W. Jones. 1998. Evaluation and improvement of crop models using regional cultivar trial data. Appl. Eng. Agric. 14:435–446. doi:10.13031/2013.19391

Plauborg, F., P. Abrahamsen, B. Gjettermann, M. Mollerup, B.V. Iversen, F. Liu, M.N. Andersen, and S. Hansen. 2010. Modeling of root ABA synthesis, stomatal conductance, transpiration and potato production under water saving irrigation regimes. Agric. Water Manage. 98:425–439. doi:10.1016/j.agwat.2010.10.006

Qureshi, A.S., W. Ahmad, and A.A. Ahmad. 2013. Optimum ground water table depth and irrigation schedules for controlling soil salinity in Central Iraq. Irrig. Drain. 62:414–424.

Rinaldi, M. 2001. Application of EPIC model for irrigation scheduling of sunflower in southern Italy. Agric. Water Manage. 49:185–196. doi:10.1016/S0378-3774(00)00148-7

Rinaldi, M., and D. De Luca. 2012. Application of EPIC model to assess climate change impact on sorghum in southern Italy. Ital. J. Agron. 7(1):74–85.

Rötter, R.P., T. Palosuo, K.C. Kersebaum, C. Angulo, M. Bindi, F. Ewert, R. Ferrise, P. Hlavinka, M. Moriondo, C. Nendel, J.E. Olesen, R.H. Patil, F. Ruget, J. Takac, and M. Trnka. 2012. Simulation of spring barley yield in different climatic zones of Northern and Central Europe: A comparison of nine crop models. Field Crops Res. 133:23–36. doi:10.1016/j.fcr.2012.03.016

Rötter, R., and H. Van Keulen. 1997. Variations in yield response to fertilizer application in the tropics: II. Risks and opportunities for smallholders cultivating maize on Kenya's arable land. Agric. Syst. 53:69–95. doi:10.1016/S0308-521X(96)00037-6

Rötter, R., H. Van Keulen, and M.J.W. Jansen. 1997. Variations in yield response to fertilizer application in the tropics: I. Quantifying risks and opportunities for smallholders based on crop growth simulation. Agric. Syst. 53:41–68. doi:10.1016/S0308-521X(96)00036-4

Roux, S., X. Delpuech, G. Daudin, F. Brun, J. Wery, and D. Wallach. 2014. Providing user-oriented uncertainty information with a vineyard model used for irrigation decisions. In: L.R. Ahuja, L. Ma, and R.J. Lascano, editors, Practical applications of agricultural system models to optimize the use of limited water. Adv. Agric. Systems Model. 5. ASA, SSSA, CSSA, Madison, WI. p. 183–208 (this volume). doi:10.2134/advagricsystmodel5.c8

Ruiz-Nougueira, B., K.J. Boote, and F. Sau. 2001. Calibration and use of CROPGRO-soybean model for improving soybean management under rainfed conditions. Agric. Syst. 68:151–173.

Santos, A.M., M. Cabelguenne, F.L. Santos, M.R. Oliveira, and R.P. Serralheiro. 2000. EPICPHASE: A model to explore irrigation strategies. J. Agric. Eng. Res. 75(4):409–416. doi:10.1006/jaer.1999.0523

Sarawat, V., S. Ratanasriwong, K. Puangprakon, P. Sringam, and A. Jintrawet. 2004. MunThaiDSS: A Decision Support System for Cassava Production. 2004 AFITA/WCCA Joint Congress on IT in Agriculture. http://mis.agri.cmu.ac.th/download/publication/2368_file.pdf (verified on 10 September, 2014)

Sarwar, A., and W.G.M. Bastiaanssen. 2001. Long-term effects of irrigation water conservation on crop production and environment in semiarid areas. J. Irrig. Drain. 127:331–338. doi:10.1061/(ASCE)0733-9437(2001)127:6(331)

Saseendran, S.A., L.R. Ahuja, L. Ma, and T.J. Trout. 2014. Modeling for best management of the effects of irrigation frequencies, initial water, and nitrogen on corn. In: L.R. Ahuja, L. Ma, and R.J. Lascano, editors, Practical applications of agricultural system models to optimize the use of limited water. Adv. Agric. Systems Model. 5. ASA, SSSA, CSSA, Madison, WI. p. 25–52 (this volume). doi:10.2134/advagricsystmodel5.c2

Saseendran, S.A., D.C. Nielsen, L.R. Ahuja, L. Ma, and D.J. Lyon. 2013. Simulated yield and profitability of five potential crops for intensifying the dryland wheat–fallow production system. Agric. Water Manage. 116:175–192. doi:10.1016/j.agwat.2012.07.009

Saseendran, S.A., D.C. Nielsen, L. Ma, L.R. Ahuja, and M.F. Vigil. 2010. Simulating alternative dryland rotational cropping systems in the Central Great Plains with RZWQM2. Agron. J. 102:1521–1534. doi:10.2134/agronj2010.0141

Saseendran, S.A., T.J. Trout, L.R. Ahuja, L. Ma, G.S. McMaster, A.A. Andales, J. Chaves, and J. Ham. 2014. Development of crop water production functions for corn in Colorado using RZWQM2 model. Agric. Water Manage. (in press).

Sau, F., K.J. Boote, and B. Ruiz-Nogueira. 1999. Evaluation and improvement of CROPGRO-soybean model for a cool environment in Galicia, Northwest Spain. Field Crops Res. 61:273–291. doi:10.1016/S0378-4290(98)00168-3

Sierra, J., N. Brisson, D. Ripoche, and M. Déqué. 2010. Modelling the impact of thermal adaptation of soil microorganisms and crop system on the dynamics of organic matter in a tropical soil under a climate change scenario. Ecol. Modell. 221(23):2850–2858. doi:10.1016/j.ecolmodel.2010.08.031

Singh, R., R.K. Jhorar, J.C. Van Dam, and R.A. Feddes. 2006b. Distributed ecohydrological modelling to evaluate irrigation system performance in Sirsa district, India II: Impact of viable water management scenarios. J. Hydrol. 329(3–4):714–723. doi:10.1016/j.jhydrol.2006.03.016

Singh, R., J.G. Kroes, J.C. Van Dam, and R.A. Feddes. 2006a. Distributed ecohydrological modelling to evaluate the performance of irrigation system in Sirsa district, India: I. Current water management and its productivity. J. Hydrol. 329(3–4):692–713. doi:10.1016/j.jhydrol.2006.03.037

Soler, C.M.T., and G. Hoogenboom. 2007. Determining irrigation scheduling for cotton and peanut using cropping system models. Restoring Our Natural Habitat. Proc. of the 2007 World Environmental and Water Resources Congress 2007. p. 1–10.

Soler, C.M.T., P.C. Sntelhas, and G. Hoogenboom. 2007. Application of the CSM-CERES-Maize model for planting date evaluation and yield forecasting for maize grown off-season in a subtropical environment. Eur. J. Agron. 27:165–177. doi:10.1016/j.eja.2007.03.002

Sommer, R., K. Kienzler, C. Conrad, N.I. Ibragimov, J. Lamers, C. Martius, and P. Vlek. 2008. Evaluation of the CropSyst model for simulating the potential yield of cotton. Agron. Sustain. Dev. 28:345–354. doi:10.1051/agro:2008008

Sommer, R., C. Piggin, A. Haddad, A. Hajdibo, P. Hayek, and Y. Khalil. 2012. Simulating the effects of zero tillage and crop residue retention on water relations and yield of wheat under rainfed semiarid Mediterranean conditions. Soil Crops Res. 132:40–52. doi:10.1016/j.fcr.2012.02.024

Steduto, P., T.C. Hsiao, D. Raes, and E. Fereres. 2009. AquaCrop—The FAO crop model to simulate yield response to water: I. Concepts and underlying principles. Agron. J. 101:426–437. doi:10.2134/agronj2008.0139s

Stöckle, C.O., and P. Debaeke. 1997. Modeling crop nitrogen requirements: A critical analysis. Eur. J. Agron. 7:161–169. doi:10.1016/S1161-0301(97)00038-5

Stöckle, C.O., M. Donatelli, and R. Nelson. 2003. CropSyst, a cropping systems simulation model. Eur. J. Agron. 18:289–307. doi:10.1016/S1161-0301(02)00109-0

Stöckle, C.O., S.A. Martin, and G.S. Campbell. 1994. CropSyst, a cropping systems simulation model: Water/nitrogen budgets and crop yield. Agric. Syst. 46:335–359. doi:10.1016/0308-521X(94)90006-2

Stöckle, C.O., and R.L. Nelson. 1999. CropSyst user's manual. Biological Systems Engineering Dep., Washington State Univ., Pullman, WA, USA.

Stulina, G., M.R. Cameira, and L.S. Pereira. 2005. Using RZWQM to search improved practices for irrigated maize in Fergana, Uzbekistan. Agric. Water Manage. 77:263–281. doi:10.1016/j.agwat.2004.09.040

Styczen, M., R.N. Poulsen, A.K. Falk, and G.H. Jorgensen. 2010. Management model for decision support when applying low quality water in irrigation. Agric. Water Manage. 98:472–481. doi:10.1016/j.agwat.2010.10.017

Supit, I. 1997. Predicting national wheat yields using a crop simulation and trend models. Agric. For. Meteorol. 88(1–4):199–214. doi:10.1016/S0168-1923(97)00037-3

Supit, I., C.A. Van Diepen, H.L. Boogaard, F. Ludwig, and B. Baruth. 2010. Trend analysis of the water requirements, consumption and deficit of field crops in Europe. Agric. For. Meteorol. 150(1):77–88. doi:10.1016/j.agrformet.2009.09.002

Supit, I., C.A. van Diepen, A.J.W. de Wit, J. Wolf, P. Kabata, B. Baruth, and F. Ludwig. 2012. Assessing climate change effects on European crop yields using the Crop Growth Monitoring System and a weather generator. Agric. For. Meteorol. 164:96–111. doi:10.1016/j.agrformet.2012.05.005

Svendsen, H., S. Hansen, and H.E. Jensen. 1995. Simulation of crop production, water, and nitrogen balances in two German agro-ecosystems using the DAISY model. Ecol. Modell. 81:197–212. doi:10.1016/0304-3800(94)00171-D

Thorp, K.R., K.C. DeJonge, A.L. Kaleita, W.D. Batchelor, and J.O. Paz. 2008. Methodology for the use of DSSAT models for precision agriculture decision support. Comput. Electron. Agric. 64:276–285. doi:10.1016/j.compag.2008.05.022

Todorovic, M., R. Albrizio, L. Zivotic, M.T.A. Saab, C. Stöckle, and P. Steduto. 2009. Assessment of AquaCrop, CropSyst, and WOFOST models in the simulation of sunflower growth under different water regimes. Agron. J. 101(3):509–521. doi:10.2134/agronj2008.0166s

Utset, A., and M. Borroto. 2001. A modeling-GIS approach for assessing irrigation effects on soil salinisation under global warming conditions. Agric. Water Manage. 50:53–63. doi:10.1016/S0378-3774(01)00090-7

Valdes-Gomez, H., F. Celette, I. García de Cortázar-Atauri, F. Jara-Rojas, S. Ortega-Farias, and C. Gary. 2009. Modelling soil water content and grapevine growth and development with the STICS crop-soil model under two different water management strategies. J. Int. Sci. Vigne Vin 43(1):13–28.

Verma, A.K., S.K. Gupta, and R.K. Isaac. 2014. Calibration and validation of SWAP to simulate conjunctive use of fresh and saline irrigation waters in semi-arid regions. Environ. Model. Assess. 19:45–55. doi:10.1007/s10666-013-9379-x

Van der Velde, M., G. Wriedt, and F. Bouraoui. 2010. Estimating irrigation use and effects on maize yield during the 2003 heatwave in France. Agric. Ecosyst. Environ. 135(1–2):90–97. doi:10.1016/j.agee.2009.08.017

Van Diepen, C.A., J. Wolf, and H. Keulen. 1989. WOFOST: A simulation model of crop production. Soil Use Manage. 5:16–24. doi:10.1111/j.1475-2743.1989.tb00755.x

Van Walsum, P.E.V., and I. Supit. 2012. Influence of ecohydrologic feedbacks from simulated crop growth on integrated regional hydrologic simulations under climate scenarios. Hydrol. Earth Syst. Sci. 16:1577–1593.

Walser, S., N. Schutze, G. Marcus, L. Susanne, and U. Schmidhalter. 2011. Evaluation of the transferability of a SVAT model—Results from field and greenhouse applications. Irrig. Drain. 60:59–70. doi:10.1002/ird.669

Wang, E., H. Cresswell, J. Xu, and Q. Jiang. 2009a. Capacity of soils to buffer impact of climate variability and value of seasonal forecasts. Agric. For. Meteorol. 149:38–50. doi:10.1016/j.agrformet.2008.07.001

Wang, E., J. Xu, Q. Jiang, and J. Austin. 2009b. Assessing the spatial impact of climate on wheat productivity and the potential value of seasonal climate forecasting at regional level. Theor. Appl. Climatol. 95:311–330. doi:10.1007/s00704-008-0009-5

Wang, E., J.H. Xu, and C.J. Smith. 2008a. Value of historical climate knowledge, SOI-based seasonal climate forecasting and stored soil moisture at sowing in crop nitrogen management in south eastern Australia. Agric. For. Meteorol. 148:1743–1753. doi:10.1016/j.agrformet.2008.06.004

Wang, E., M.J. Robertson, G.L. Hammer, P.S. Carberry, D. Holzworth, H. Meinke, S.C. Chapman, J.N.G. Hargreaves, N.I. Huth, and G. McLean. 2002. Development of a generic crop model template in the cropping system model APSIM. Eur. J. Agron. 18:121–140. doi:10.1016/S1161-0301(02)00100-4

Wang, E., Q. Yu, D. Wu, and J. Xia. 2008b. Climate, agricultural production and hydrological balance in the North China Plain. Int. J. Climatol. 28:1959–1970. doi:10.1002/joc.1677

Wang, F., C.E. Fraisse, N.R. Kitchen, and K.A. Sudduth. 2003. Site-specific evaluation of the CROPGRO soybean model on Missouri claypan soils. Agric. Syst. 76:985–1005. doi:10.1016/S0308-521X(02)00029-X

Wang, J., G. Huang, H. Zhan, B.P. Mohanty, J. Zheng, Q. Huang, and X. Xu. 2014. Evaluation of soil water dynamics and crop yield under furrow irrigation with a two-dimensional flow and crop growth coupled model. Agric. Water Manage. 141:10–22. doi:10.1016/j.agwat.2014.04.007

Wang, J., E. Wang, X. Yang, F. Zhang, and H. Yin. 2011. Increased yield potential of wheat–maize cropping system in the North China Plain by climate change adaptation. Clim. Change 113:825–840. doi:10.1007/s10584-011-0385-1.

Wang, S., H. Li, Y. Yang, H. Wang, Y. Wang, and Y. Jia. 2012. Using DSSAT model to assess spring wheat and maize water use in the arid oasis of Northwest China. J. Food Agric. Environ. 10:911–918.

Wang, W.G., X.G. Wang, R.D. Zhang, and R.K. Shen. 2004. Saline water irrigation in Hetao irrigation district. Land and Water Manage: Decision Tools Pract. 1&2:1086–1093.

Wang, X.C., and J. Li. 2010. Evaluation of crop yield and soil water estimates using the EPIC model for the Loess Plateau of China. Math. Comput. Model. 51:1390–1397. doi:10.1016/j.mcm.2009.10.030

Wang, X.C., J. Li, M.N. Tahir, and M.D. Hao. 2011. Validation of the EPIC model using a long-term experimental data on the semi-arid Loess Plateau of China. Math. Comput. Model. 54:976–986. doi:10.1016/j.mcm.2010.11.025

Wang, X.C., J.R. Williams, P.W. Gassman, C. Baffaut, R.C. Izaurralde, J. Jeong, and J.R. Kiniry. 2012. EPIC and APEX: Model use, calibration, and validation. Trans. ASABE 55(4):1447–1462.

Wattenbach, M., O. Sus, N. Vuichard, S. Lehuger, P. Gottschalk, L.H. Li, A. Leip, M. Williams, E. Tomelleri, W.L. Kutsch, N. Buchmann, W. Eugster, D. Dietiker, M. Aubinet, E. Ceschia, P. Beziat, T. Grunwald, A. Hastings, B. Osborne, P. Ciais, P. Cellier, and P. Smith. 2010. The carbon balance of European croplands: A cross-site comparison of simulation models. Agric. Ecosyst. Environ. 139(3):419–453.

Williams, J.R., and R.C. Izaurralde. 2006. The APEX model. In: V.P. Singh and D.K. Frevert, editors, Watershed models. CRC Press, Boca Raton, FL. p. 437–482

Williams, J., M. Nearing, A. Nicks, E. Skidmore, C. Valentin, K. King, and R. Savabi. 1996. Using soil erosion models for global change studies. J. Soil Water Conserv. 51(5):381–385.

Williams, J.R., C.A. Jones, J.R. Kiniry, and D.A. Spanel. 1989. The EPIC crop growth model. Trans. ASAE 32:497–511. doi:10.13031/2013.31032

Wriedt, G., M. Van der Velde, A. Aloe, and F. Bouraoui. 2009. Estimating irrigation water requirements in Europe. J. Hydrol. 373:527–544. doi:10.1016/j.jhydrol.2009.05.018

Wu, D., Q. Yu, C. Lu, and H. Hengsdijk. 2006. Quantifying production potentials of winter wheat in the North China Plain. Eur. J. Agron. 24:226–235. doi:10.1016/j.eja.2005.06.001

Wu, D., Z. Ouyang, X. Zhao, and Q. Yu. 2003. The applicability research of WOFOST model in North China Plain. Acta Phytoecol. Sin. 27(5):594–602.

Yang, H.S., P. Grassini, and K. Cassman. 2014. Quantifying and managing corn water use efficiencies under irrigated and rainfed conditions in Nebraska using the hybrid-maize simulation model. In: L.R. Ahuja, L. Ma, and R.J. Lascano, editors, Practical applications of agricultural system models to optimize the use of limited water. Adv. Agric. Systems Model. 5. ASA, SSSA, CSSA, Madison, WI. p. 113–138 (this volume). doi:10.2134/advagricsystmodel5.c5

Yang, Y., M. Watanabe, X. Zhang, X. Hao, and J. Zhang. 2006. Estimation of groundwater use by crop production simulated by DSSAT-wheat and DSSAT-maize models in the piedmont region of the North China Plain. Hydrol. Processes 20:2787–2802. doi:10.1002/hyp.6071

Zhang, Y., L. Feng, E. Wang, J. Wang, and B. Li. 2012. Evaluation of the APSIM-Wheat model in terms of different cultivars, management regimes and environmental conditions. Can. J. Plant Sci. 92:937–949. doi:10.4141/cjps2011-266

Zhao, X., K. Hu, and K. Stahr. 2013. Simulation of SOC content and storage under different irrigation, fertilization and tillage conditions using EPIC model in the North China Plain. Soil Tillage Res. 130:128–135. doi:10.1016/j.still.2013.02.005

Zhou, J., G. Cheng, X. Li, B.X. Hu, and G. Wang. 2012. Numerical modeling of wheat irrigation using coupled HYDRUS and WOFOST models. Soil Sci. Soc. Am. J. 76(2):648–662. doi:10.2136/sssaj2010.0467

Advances in Agricultural Systems Modeling
Transdisciplinary Research, Synthesis, and Applications

Agriculture today is complicated by growing environmental concerns, increasingly limited water available for agriculture, market-based global competition that challenges traditional production systems, more frequent droughts and climate change, and the growing production of bioenergy crops. Whole system–based quantitative planning and decision tools are needed to guide the optimum management of resources while addressing these problems. Process-level models of agricultural systems, integrated with focused field research, are required to develop these tools and extend their application to different soils, climates, and situations.

Agricultural system modeling has made substantial progress, but there are still many critical gaps in our knowledge of various processes, and especially their interactions. Most of these gaps occur at the boundaries of disciplines, and further transdisciplinary research is needed. There is also the need for better synthesis and quantification of knowledge at the whole system level, both to improve system models and realize a more systematic and collaborative approach to building models of the future. At the same time, we need to facilitate application of these models to address and solve current real-world problems. For this purpose, we need to understand the different scales of application and the scale-dependence of parameters.

The American Society of Agronomy, Crop Science Society of America, and Soil Science Society of America are taking a leadership role in encouraging transdisciplinary and interdisciplinary research and its synthesis to solve practical problems. We believe that the future breakthroughs in science and technology lie, indeed, at the boundaries of disciplines. Therefore, the Societies have initiated the new series Advances in Agricultural Systems Modeling, A Series of Transdisciplinary Research, Synthesis, and Applications.

The purpose of the series is to:

- Encourage and advance critical transdisciplinary research, and its synthesis and quantification, through publication of anonymously peer-reviewed papers by top researchers worldwide in a given knowledge gap area. The writing of papers and publication will generally be preceded by a workshop of these researchers, where the papers will be presented and discussed.

The papers will contain both new research and new or improved concepts for synthesis.

- Encourage collaboration among the top researchers in new research, synthesis, and building and improvement of model components. As often as possible and where appropriate, the authors will be asked to provide computer code of the components to be shared by researchers and modelers.
- Encourage and advance the application of system models to solve practical problems through publishing case studies of such applications, along with illustrated instructions on how the models are used. For example, an application may be the use of particular models to optimize the water and nutrients under limited water conditions and to evaluate the effects of climate change on agriculture.
- Encourage better instruction in these models and their application. The state-of-the-science syntheses given by each author will be highly useful for both undergraduate and graduate teaching. The new research in the papers and model component codes will be valuable for graduate level teaching, research, and training of student in the use of models.

Lajpat R. Ahuja
Series Editor

Printed and bound by CPI Group (UK) Ltd, Croydon, CR0 4YY

16/09/2026

14963421-0004